Microbiology and Biochemistry of Strict Anaerobes Involved in Interspecies Hydrogen Transfer

FEDERATION OF EUROPEAN MICROBIOLOGICAL SOCIETIES SYMPOSIUM SERIES

A Continuation Order Plan is available for this series. A continuation order will bring delivery of each new volume immediately upon publication. Volumes are billed only upon actual shipment. For further information please contact the publisher.

Microbiology and Biochemistry of Strict Anaerobes Involved in Interspecies Hydrogen Transfer

Edited by

Jean-Pierre Bélaich
and Mireille Bruschi

Laboratoire de Chimie Bactérienne/CNRS
Marseille, France

and

Jean-Louis Garcia

ORSTOM/Université de Provence
Marseille, France

PLENUM PRESS • NEW YORK AND LONDON

Library of Congress Cataloging-in-Publication Data

Microbiology and biochemistry of strict anaerobes involved in
  interspecies hydrogen transfer / edited by Jean-Pierre Bélaïch and
  Mireille Bruschi and Jean-Louis Garcia.
      p.   cm. -- (FEMS symposium ; no. 54)
    Proceedings of a symposium held under the auspices of the
  Federation of European Microbiological Societies, 9/12-14/89,
  Marseille, France.
    Includes bibliographical references.
    ISBN-13: 978-1-4612-7892-4   e-ISBN-13: 978-1-4613-0613-9
    DOI: 10.1007/978-1-4613-0613-9

    1. Anaerobic bacteria--Congresses. 2. Dehydrogenases--Congresses.
  3. Biomass energy--Congresses.   I. Bélaïch, Jean-Pierre.
  II. Bruschi, Mireille. III. Garcia, Jean-Louis. IV. Federation of
  European Microbiological Societies. V. Series.
  QR89.5.M53 1990
  589.9'0128--dc20   90-7023
                                                             CIP

Proceedings of a symposium held under the auspices of the
Federation of European Microbiological Societies,
September 12–14, 1989, in Marseille, France

© 1990 Plenum Press, New York
Softcover reprint of the hardcover 1st edition 1990

A Division of Plenum Publishing Corporation
233 Spring Street, New York, N.Y. 10013

PREFACE

The belief that energy might be a limiting factor for the development
of humanity led twenty years ago to a great interest being taken in
research on anaerobic digestion. The first international symposium held
in Cardiff in 1979 was followed by the meetings in Travenmund (1981),
Boston (1983), Guangzhou (1985) and Bologna (1988). By now anaerobic
digestion has come to be recognized as an appropriate technology for waste
treatment. More recently, the increase in the carbon dioxide content of
the atmosphere and (in developed countries, especially in the EEC) the
fact that more and more land is becoming available for purposes other than
food production make biomass production economically and/or socially
feasible for industrial purposes. The possibility of using renewable
organic carbon resources in this way is of great potential interest for
developing biological techniques and could considerably increase the use
of anaerobic micro-organisms in cellulose biotransformation and energy
production from crop residues.

This FEMS Symposium is devoted to the interspecies hydrogen transfer
phenomenon involved in the mineralization of organic matter in
anaerobiosis. This process is carried out in Nature by consortia of
anaerobic micro-organisms living syntrophically. Many industrial
applications of these consortia as black boxes for biogas production and
waste treatment have been described. Although these early approaches were
fruitful, it seems likely that a better knowledge at the molecular level
of the more characteristic anaerobic bacteria which constitute these
consortia would greatly increase and improve the utilization of these
organisms.

The purpose of this Symposium was to provide an opportunity for
discussing the recent progress which has been made in the biology,
biochemistry and genetics of the anaerobic microbes which participate in
the metabolism of hydrogen. Special attention was paid to bacterial
hydrogenases, key enzymes of hydrogen metabolism, which are responsible
for hydrogen transfer between the various partners in anaerobic consortia.

The Symposium was held in Marseille, France from 12-14 September 1989,
sponsored by the Federation of European Microbiological Societies, Agence
Française pour la Maîtrise de l'Energie, Conseil Régional Provence-Alpes-
Côte d'Azur, the Municipality of Marseille, and Pharmacia of France. The
Organizing Committee of the Symposium consisted of J-P Bélaich, K H
Schleifer, Claudine Elmerich, Marie-Claire Blanchard, Pomme Lamy and V A
Jacq and they express their gratitude to the sponsors who enabled the
meeting to take place.

Jean-Pierre Bélaich

v

CONTENTS

PART 2 - CONFERENCES

2.1. - MICROBIOLOGY

2.2. - BIOCHEMISTRY

2.3. - GENETICS

PART 3 - POSTER SESSION

3.1. - MICROBIOLOGY

3.2. - BIOCHEMISTRY

3.3. - GENETICS

PLENARY LECTURES

MOLECULAR HYDROGEN AND ENERGY CONSERVATION IN METHANOGENIC AND ACETOGENIC BACTERIA

Richard Sparling and Gerhard Gottschalk

Institut für Mikrobiologie
Grisebachstr. 8
D-3400 Göttingen, FRG

Molecular hydrogen is a minor constituent of our atmosphere amounting to about 0.5 ppm (Anonymus, 1976). Nevertheless, it plays an important role in the conversion of organic matter by microorganisms. Although H_2 is produced in large amounts, it is rapidly consumed, and can be considered as a very convenient vehicle for transport electrons from one organism to the other. The efficiency of such interspecies hydrogen transfer is such that very little H_2 escapes into the immediate environment. For example, Conrad et al., 1985 could only detect 0.2 µM dissolved H_2 in sludges and 0.03 µM H_2 in sediments tested.

Molecular hydrogen is produced in a number of reactions. (i) It is a by-product of nitrogen fixation as catalyzed by the enzyme nitrogenase (Stam et al., 1987);(ii) it is produced from formate in the formate hydrogen lyase reaction which occurs, for instance, in a number of enterobacteria (Ingledew and Poole, 1984) and (iii) it is produced from reduced ferredoxin in the hydrogenase reaction (Adams et al., 1981). Reduced ferredoxin can be generated in many anaerobic bacteria by pyruvate oxidation to acetyl coenzyme A as catalyzed by the pyruvate: ferredoxin oxidoreductase (Meinecke et al., 1989). Another enzyme which gives rise to the generation of reduced ferredoxin is the NAD-ferredoxin oxidoreductase (Jungermann et al., 1971). The redox potential of NADH/NAD is $E^0{}' = -320$ mV and the one of reduced ferredoxin/oxidized ferredoxin in the order of -400 mV (Thauer et al., 1977). Therefore, reduction of ferredoxin by NADH is thermodynamically unfavourable under standard conditions. Very low partial pressures of H_2 are required to drive this reaction. In addition this enzyme system underlies very stringent regulation (Jungermann et al., 1973).

Processes depending on molecular hydrogen are very widespread among microorganisms (Table 1). Numerous aerobic bacterial species are able to oxidize H_2 with molecular oxygen as electron acceptor, they represent the so-called hydrogen-oxidizing bacteria (Schlegel, 1987). Under anaerobic conditions a number of different electron acceptors can be employed by microorganisms.

Microbiology and Biochemistry of Strict Anaerobes Involved in Interspecies Transfer
Edited by J.-P. Bélaich *et al.*
Plenum Press, New York, 1990

Table 1. Electron acceptors coupled with H_2 oxidation used for energy conservation

Electron acceptor	Microorganisms
oxygen	Knallgas bacteria *Alcaligenes eutrophus*
nitrate, nitrite	*Paracoccus denitrificans*
sulfate, sulfite	*Desulfovibrio vulgaris*
sulfur	*Thermoproteus tenax*
fumarate	*Wolinella succinogenes*
CO_2	methanogenic and aceto- genic bacteria *Methanobacterium thermo-* *autotrophicum* *Acetobacterium woodii*
methanol	*Methanosphaera stadtmanae*
trimethylamine	*Methanosarcina barkeri*

Inorganic acceptors such as nitrate, nitrite, sulfate, sulfite and elemental sulphur can be used, and in addition carbon compounds such as CO_2, methanol or fumarate. It also should be mentioned in this context that many phototrophic bacteria are able to take up molecular hydrogen and to use it for the reduction of CO_2 to the redox level of the cellular constituents (Vignais et al., 1985). The ecological importance of H_2 uptake by these various organisms is best examplified in syntrophic association whereby the thermodynamics of the reactions involving the H_2-producing member is dependent on the efficient removal of H_2 by its partner (Lee and Zinder, 1988). Indeed, low whole cell Km's for H_2 uptake have been determined for both methanogenic bacteria (2,5 - 13 µM) and sulfate reducers (0,7 - 1,9 µM) (Robinson and Tiedje, 1984).

With respect to the mechanism of ATP synthesis, organisms evolving H_2 differ from those consuming H_2. The first category of anaerobes employs substrate level phosphorylation as the principle mechanism of ATP synthesis; ATP is formed in the glycolytic breakdown of carbohydrates or in the terminal step of acetate or butyrate formation. The second category of organisms depends either partially or solely on a chemiosmotic mechanism of ATP synthesis. This is very obvious for the H_2-dependent denitrification as carried out by *Escherichia coli* (Ingledew and Poole, 1984). This redox process proceeds via the respiratory chain and is coupled with the generation of a protonmotive force across the cytoplasmic membrane as under aerobic conditions.
Regarding dissimilatory sulfate reduction, it is clear from growth yield data that the reduction of sulfite to sulfide

must be coupled with ATP synthesis (Badziong and Thauer, 1978). In this respect it is noteworthy that a hydrogenase has been detected in *Desulfovibrio vulgaris* which is oriented to the periplasmic side of the cytoplasmic membrane and which could give rise to a proton gradient across the cytoplasmic membrane (Odom and Peck, 1984). It also is noteworthy that a number of redox carriers occur in sulfate-reducing bacteria which could be part of an electron transport chain and which could be involved in the generation of a protonmotive force across the cytoplasmic membrane during sulfite reduction. A direct measurement of ATP formation as coupled to sulfite reduction in whole cells or in vesicle preparations has not been achieved.

The coupling of the reduction of fumarate by H_2 to ATP synthesis by a chemiosmotic mechanism has clearly been demonstrated (Kröger, 1980). That energy is conserved in this reaction follows from the fact that *Wolinella succinogenes* but also *Escherichia coli* are able to grow at the expense of succinate formation from fumarate + H_2. Moreover, the components involved have been characterized, and proton movement and ATP synthesis have been followed employing, for instance, vesicles (Kröger, 1980; Ingledew and Poole, 1984).

H_2-dependent fermentations involving CO_2 as electron acceptor lead either to the formation of methane or of acetate. The mechanism for net ATP synthesis in these organisms was unclear until a few years ago.

The reactions in methanogenesis from H_2 + CO_2 are summarized in Table 2.

Table 2. Thermodynamics of methane synthesis from H_2 + CO_2

Step	Reaction		$\Delta G^{0'}$[a] (kJ/reaction)
1	$H_2 + CO_2 + MF$	$\rightleftharpoons$ formyl-MF + H_2O	+16
2	Formyl-MF + H_4MPT	$\rightleftharpoons$ formyl-H_4MPT + MF	− 5
3	Formyl-H_4MPT + H^+	$\rightleftharpoons$ methenyl-H_4MPT^+ + H_2O	− 2
4	Methenyl-H_4MPT^+ + H_2	$\rightleftharpoons$ methylene-H_4MPT + H^+	− 5
5	Methylene-H_4MPT + H_2	$\rightleftharpoons$ methyl-H_4MPT	−20[b]
6	Methyl-H_4MPT + HS-CoM	$\rightleftharpoons$ methyl-SCoM + H_4MPT	−29
7	Methyl-SCoM + H_2	$\rightleftharpoons$ CH_4 + HS-CoM	−85

[a]$\Delta G^{0'}$ values taken from Keltjens and van der Drift (1986).
[b]See analogous reaction in Table 21.2.
MF, methanofuran; H_4MPT, tetrahydromethanopterin

The first two reductive steps are endergonic while the last two are exergonic. The study of the mechanisms of energy conservation and ATP synthesis for CO_2 reducing methanogens has been hampered by the requirement for energy input to drive CO_2 methanogenesis (as indicated by the inhibition of CO_2-methanogenesis by various uncouplers (Daniels et al., 1984)), the requirement of Na^+ (Perski et al., 1982) and the difficulties in studying the various steps separately in whole cells. A break-through was achieved when *Methanosarcina barkeri* was studied using the substrate combination methanol + H₂ (Blaut and Gottschalk, 1984; 1987). These substrates are converted to methane according to the following equations:

$$CH_3OH + HS\text{-}CoM \longrightarrow CH_3\text{-}S\text{-}CoM + H_2O$$
$$G^0{}' = 27{,}5 \text{ kJ/reaction}$$
$$CH_3\text{-}S\text{-}CoM + H_2 \longrightarrow CH_4 + HS\text{-}CoM$$
$$G^0{}' = -85 \text{ kJ/reaction}$$

It could be demonstrated that methanogenesis from methanol + H₂ was coupled with ATP synthesis from ADP + inorganic phosphate and that this synthesis proceeded via a chemiosmotic mechanism. In the related methanogenic strain Göl which has a protein cell wall, protoplasts (Jussofie et al., 1986) and finally inside-out vesicles have been made. In such vesicles, Peinemann et al. (this book), have shown that ATP synthesis was coupled to CH₃-S-CoM reduction through a chemiosmotic proton gradient. A model summarizing the results is depicted in Fig. 1. Since methyl-coenzyme M is an intermediate for methanogenesis from all substrates utilized by methanogenic bacteria, it can be expected that the mechanism of ATP synthesis indicated in Fig. 1 would also apply to all other organisms. Methanogens form, however, a phylogenetically very diverse group. Recently (Rouvière et al., this book), it has been proposed that *Methanosarcina* and the methyl-reducing methanogens should from a separate order. Indeed methanol reduction has only been observed among representatives of this group, with one exception: *Methanosphaera stadtmanae* (Miller and Wolin, 1985) a member of the *Methanobacteriaceae* that can reduce methanol to methane with H₂. Recent data (Sparling and Gottschalk, unpublished) showed H₂ + methanol-methanogenesis to be Na⁺ independent in this organism. It was also shown that the proton ionophore TCS depleted the intracellular ATP pool while stimulating methanogenesis and that the ATPase inhibitor DCCD inhibited both ATP synthesis and methanogenesis; the latter inhibition being relieved by the addition of TCS. These data are consistant with those discussed above for *Methanosarcina barkeri*.

It is apparent that in addition to the final reaction, the conversion of methylene-tetrahydromethanopterin into methyl-coenzyme M is associated with a negative free energy change. When the role of sodium ions in methanogenesis was studied it became clear that cells also take advantage of this reaction sequence for energy conservation. In studying *Methanosarcina barkeri* it was shown that a sodium gradient is established across the cytoplasmic membrane with low sodium inside and high sodium outside the cells (Müller et al., 1986). How is this gradient established? The side of sodium pumping could be identified when methanogenesis from formaldehyde + H₂ was studied (Müller et al., 1988). Formaldehyde reacts with tetrahydromethanopterin to form methylene-tetrahydro-methanopterin which subsequently is reduced with H₂ to methane. By comparing the formation of sodium gradients across the membrane and other energetic parameters during methanogenesis from formaldehyde + H₂ and methanol + H₂ it clearly could be shown that reactions five and six in Table 2 are coupled to a sodium pump as schematically depicted in Fig. 1.

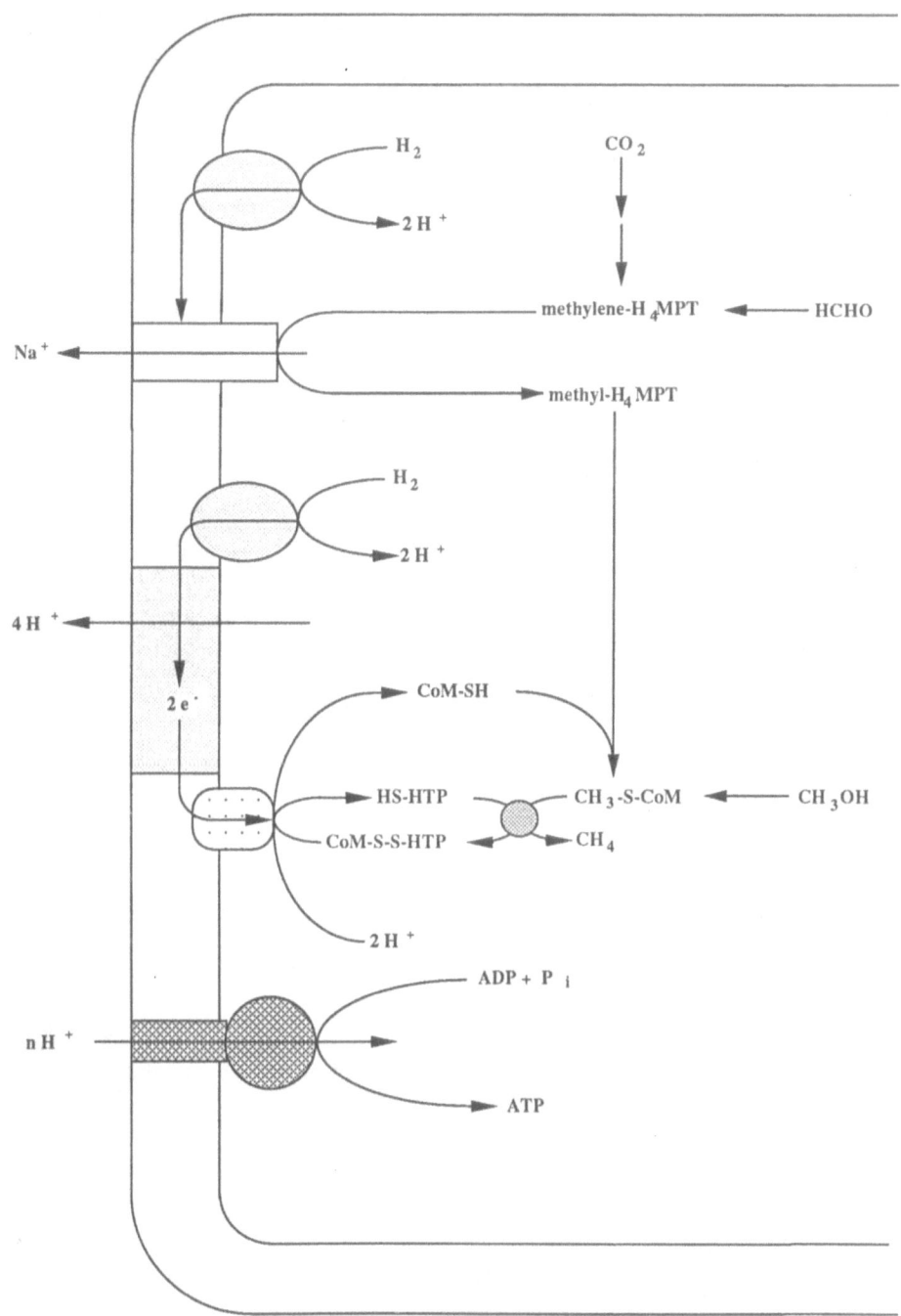

Fig. 1. Scheme for the coupling of the exergonic steps of methanogenesis to energy conservation in the form of transmembrane ion gradients. The cycling of HS-HTP here indicated was first described by Bobik et al.(1987) and Ellermann et al. (1987).

Thus, at least two ion pumps are associated with methanogenesis from CO_2 + H_2; the proton gradient established may be directly used for ATP synthesis through the F_0F_1-ATPase which has been shown to be present in *Methanosarcina barkeri* (Inatomi, 1986). The sodium gradient may be converted into a proton gradient (Müller et al., 1988) or it may be used directly to drive certain transport processes (Jarrell et al., 1984).

When the pathway of acetogenesis from CO_2 + H_2 as carried out by organisms such as *Clostridium aceticum*, *Acetobacterium woodii* and *Sporomusa ovata* is inspected (Fuchs, 1986; Ljungdahl, 1986), it is not obvious how net-ATP is gained by these organisms (Table 3).

Table 3. Thermodynamics of acetate synthesis from H_2 + CO_2

Step	Reaction		$\Delta G^{0\prime a}$ (kJ/reaction)
1	CO_2 (gas) + H_2	$\rightleftharpoons$ $HCOO^-$ + H^+	+ 3.4
2	$HCOO^-$ + H_4 folate + ATP^{4-}	$\rightleftharpoons$ N^{10}-formyl-H_4 folate + ADP^{3-} + P_i^{2-}	− 8.4
3	N^{10}-formyl-H_4 folate + H^+	$\rightleftharpoons$ N^5,N^{10}-methenyl-H_4 folate$^+$ + H_2O	− 4.0
4	N^5,N^{10}-methenyl-H_4 folate$^+$ + H_2	$\rightleftharpoons$ N^5,N^{10}-methylene-H_4 folate + H^+	−23.0
5	N^5,N^{10}-methylene-H_4 folate + H_2	$\rightleftharpoons$ N^5-methyl-H_4 folate	−57.3[b]
6	CO_2 (gas) + H_2	$\rightleftharpoons$ CO (gas) + H_2O	+20.1
7	N^5-methyl-H_4 folate + CO + CoA	$\rightleftharpoons$ acetyl CoA + H_4 folate	−21.8
8	Acetyl CoA + P_i^{2-}	$\rightleftharpoons$ acetyl phosphate$^-$ + CoA$^-$	+ 9.0
9	Acetyl phosphate$^-$ + ADP^{3-}	$\rightleftharpoons$ acetate + ATP^{4-}	−13.0

[a] $\Delta G^{0\prime}$ values taken from Thauer et al. (1977) and Fuchs (1986).
[b] A considerably lower value (−16.6 kJ/reaction) was calculated by Keltjens and van der Drift (1986).

ATP is synthesized in the acetate kinase reaction (step 9) but has to be invested in the formation of formyl-tetrahydrofolate (step 2). A clue would be if methylene-tetrahydrofolate reduction, which is analogous to the reaction involved in sodium extrusion in methanogens, would be employed by acetogens for energy conservation. A sodium dependence for CO_2 acetogenesis has indeed been observed in both *Peptostreptococcus productus* (Geerligs et al., 1989) and *Acetobacterium woodii* (Heise et al., 1989). In *Acetobacterium woodii* an association of methylene-tetrahydrofolate reduction with the generation of a sodium gradient was also indicated (Heise et al., 1989).

In conclusion it can be stated that methanogenesis as well as acetogenesis from CO_2 + H_2 are associated with ion movements across the cytoplasmic membrane. A proton pump was discovered which allows methanogens to generate a protonmotive force across the cytoplasmic membrane and a sodium pump was found in methanogens which also provides energy in form of a sodium gradient; such a pump may also play a role in acetogenesis. The involvement of these membrane-integrated pumps makes the study of H_2-dependent fermentations very difficult because such pumps can only be studied in whole cells or in subcellular vesicle preparations.

References

Adams , M.W.W., Mortenson , L.G. and Chen , J.-S., 1981 . Hydrogenase . Biochim. Biophys. Acta , 594 : 105 - 176 .

Badziong , W., and Thauer , R.K., 1978 . Growth yield and growth rates of *Desulfovibrio vulgaris* (Marburg) growing on hydrogen plus sulfate and hydrogen plus thiosulfate as the sole energy sources . Arch. Microbiol. 117 : 209 - 214 .

Blaut , M. and Gottschalk, G., 1984 . Coupling of ATP synthesis and methane formation from methanol and molecular hydrogen in *Methanosarcina barkeri* . Eur. J. Biochem. 141 : 217 - 222 .

Blaut , M., Müller , V. and Gottschalk , G., 1987 . Proton translocation coupled to methanogenesis from methanol + hydrogen in *Methanosarcina barkeri* . FEBS Lett . 215 : 53 - 57 .

Bobik , T.A., Olson , K.D., Noll , K.M. and Wolfe , R.S., 1987. Evidence that the heterodisulfide of coenzyme M and 7-mercapto-heptanoylthreonine phosphate is a product of the methylreductase reaction in *Methanobacterium* . Biochem. Biophys. Res.Commun. 140 : 455 - 460 .

Conrad , R., Phelps , T.J. and Zeikus, J.G., 1985 . Gas metabolism evidence in support of the juxtaposition of hydrogen - producing and methanogenic bacteria in sewage sludge and lake sediment . Appl. Environ. Microbiol. , 50 : 595 - 601 .

Daniels , L., Sparling , R., and Sprott , G.D., 1984 . Bioenergetics of methanogenesis . Biochim. Biophys. Acta , 768 : 113 - 163 .

Ellermann , J., Koblet , A., Pfaltz , A. and Thauer, R.K., 1987 . One the role of N - 7 - mercaptohetanoyl - O - phospho - L - threonine (component B) in the enzymatic reduction of methyl-coenzyme M to methane . FEBS Lett. 220 : 358 - 362 .

Fuchs , G., 1986 . CO2 fixation in acetogenic bacteria , variations on a theme . FEMS Microbiol. Rev. 29 : 181 - 213 .

Geerligs , G., Schönheit , P. and Diekert , G., 1989 . Sodium dependent acetate formation from CO_2 in *Peptostreptococcus productus* (Strain Marburg) . FEMS Microbiol.Lett. 57 : 253 - 258 .

Heise , R., Müller , V. and Gottschalk , G., 1989 . Sodium dependence of acetate formation by the acetogenic bacterium *Acetobacterium woodii* . J. Bacteriol. (in press) .

Inatomi , K.- I. , 1986 . Characterization and purification of the membrane-bound ATPase of the Archaebacterium *Methanosarcina barkeri* . J. Bacteriol. 167 : 837 -841.

Ingledew, W.J. and Poole , R.K., 1984. The respiratory chains of *Escherichia coli* . Microbiol. Rev. 48 : 222 - 271 .

Jarrell , K.F., Bird , S.E. and Sprott , G.D., 1984 . Sodium dependent isoleucine transport in the methanogenic archaebacterium *Methanococcus voltae* . FEBS Lett. 166 : 357 - 361 .

Jungermann , K., Leimenstoll , G., Rupprecht , E., and Thauer , R.K., 1971 . Demonstration of NADH-ferredoxin reductase in two saccharolytic *Clostridia* . Arch. Microbiol. 80 : 370 - 372 .

Jungermann , K., Thauer, R.K., Leimenstoll , G. and Decker, K., 1973. Function of pyridine nucleotide-ferredoxin oxidoreductases in saccharolytic *Clostridia* . Biochim. Biophys. Acta. 305 : 268 - 280 .

Jussofie , A., Mayer , F. and Gottschalk , G., 1986 . Methane formation from methanol and molecular hydrogen by protoplasts of new methanogenic isolates and inhibition by dicyclohexylcarbodiimide. Arch. Microbiol. 146 : 245 - 246 .

Keltjens , J.T. and van der Drift , C., 1986 . Electron transfer reactions in methanogens . FEMS Microbiol. Rev. 39 : 259 - 303 .

Kröger, A., 1980 . Bacterial electron transport to fumarate . In : " Diversity of bacterial respiratory systems " , Vol. II. pp. 1-18. C.J. Knowles (ed.) , CRC Press, Boca Raton .

Lee , M.J. and Zinder , S.H., 1988 . Hydrogen partial pressures in a thermophilic acetate-oxidizing methanogenic coculture . Appl. Environ. Microbiol., 54 : 1457 - 1461 .

Ljungdahl , L.G., 1986 . The autotrophic pathway of acetate synthesis in acetogenic bacteria . Ann. Rev. Microbiol. 40 : 415 - 450 .

Meinecke , B., Bertram , J. and Gottschalk , G., 1989 . Purification and characterization of the pyruvate - ferredoxin oxidoreductase from *Clostridium acetobutylicum* . Arch. Microbiol. 152 : 244 - 250 .

Miller , T.L. and Wolin , M.J., 1985 . *Methanosphaera stadtmaniae* gen. nov. sp. nov. : a species that forms methane by reducing methanol with H2 . Arch. Microbiol. 141 : 116 - 121 .

Müller , V., Blaut , M. and Gottschalk , G., 1986 . Generation of a transmembrane gradient of Na^+ in *Methanosarcina barkeri* . Eur. J. Biochem. 162 : 461 - 466 .

Müller , V., Winner , C. and Gottschalk , G., 1988 . Electron - transport - driven sodium extrusion during methanogenesis from formaldehyde and molecular hydrogen by *Methanosarcina barkeri* . Eur. J. Biochem. 178 : 519 - 525 .

Odom , J.M. and Peck , H.D., 1984 . Hydrogenase , electron transfer proteins , and energy coupling in the sulfate - reducing bacteria *Desulfovibrio* . Ann. Rev. Microbiol. 38 : 551 - 592 .

Perski , H.J., Schönheit , P. and Thauer, R.K., 1982 . Sodium dependence of methane formation in methanogenic bacteria . FEBS Lett. 143 : 323 - 326 .

Robinson , J.A. and Tiedje , J.M., 1984 . Competition between sulfate - reducing and methanogenic bacteria for H_2 under resting and growing conditions . Arch. Microbiol. 137 : 26 - 32 .

Schlegel , H.G., 1987 . Aerobic hydrogen - oxidizing (Knallgas) bacteria . In : " Autotrophic bacteria " , pp. 305 - 329 , H.G. Schlegel and B. Bowien (eds.) Springer Verlag , Heidelberg .

Stam , H., Stouthamer , A.H. and van Versefeld , H.W., 1987 . Hydrogen metabolism and energy costs of nitrogen fixation . FEMS. Microbiol. Rev. 46 : 73 - 92 .

Thauer , R.K., Jungermann , R. and Decker , K., 1977 . Energy conservation in chemotrophic anaerobic bacteria . Bacteriol. Rev. 41 : 100 - 180 .

Vignais , P.M., Colbeau , A., Willison , J.C. and Jouanneau , Y., 1985 . Hydrogenase , nitrogenase and hydrogen metabolism in phototrophic bacteria . Adv. Microbiol. Physiol. 26 : 156 - 234 .

APPROACHES TO GENE TRANSFER IN METHANOGENIC BACTERIA

Thomas Leisinger and Leo Meile

Mikrobiologisches Institut, ETH-Zentrum
CH-8092 Zürich, Switzerland

INTRODUCTION

Soon after methanogenic bacteria had been recognized as archaebacteria, recombinant DNA techniques were applied to this group of organisms. The ensuing molecular analysis of their genetic material has led to considerable insight into the structure, organization and expression of methanogen genes. Methanogen-specific transcription start signals have been identified, and sequence comparisons between methanogen genes and their eubacterial or eukaryotic counterparts have strengthened the view that archaebacteria are a form of life distinct from both eubacteria and eukaryotes. However, the general principles of gene organization and gene expression in methanogens appear to be similar to those of eubacteria (for a recent review see Brown et al.) [1].

Methanogenic bacteria owe their uniqueness to the ability to derive energy from the formation of methane [2]. The functional analysis of genes directly or indirectly involved in this process and the analysis of gene regulation require genetic techniques. A particular need exists for methods allowing the reintroduction into methanogenic bacteria of methanogen genes that have been altered *in vitro*. Several laboratories have been engaged in the development of genetic systems for methanogens but progress has been comparatively slow [3]. Part of the difficulties encountered in developing such systems are due to the archaebacterial biochemistry and the oxygen sensitivity of methanogens.

GENE TRANSFER SYSTEMS IN ARCHAEBACTERIA

A survey of the archaebacterial gene transfer systems presently available (Table 1) demonstrates that the halobacteria *Halobacterium volcanii* and *Halobacterium halobium* offer a number of efficient mechanisms of genetic exchange. These representatives of the extreme halophiles can

Microbiology and Biochemistry of Strict Anaerobes Involved in Interspecies Transfer
Edited by J.-P. Bélaich *et al.*
Plenum Press, New York, 1990

Table 1. Gene Transfer in Archaebacteria

Process	Organism	Reference
Mating	*Halobacterium volcanii*	[4]
Protoplast fusion	*Halobacterium volcanii*	[5]
Transformation	*Halobacterium halobium*	[7]
	Halobacterium volcanii	[6]
	Methanobacterium thermoautotrophicum (Marburg)	[11]
Transduction	*Methanococcus voltae* PS	[12]
	Methanobacterium thermoautotrophicum (Marburg)	[unpubl.]

be grown aerobically on simple media, features which - together with the presence of phages and plasmids in these strains - make them promising candidates for the development of advanced archaebacterial host-vector systems. A natural mating system providing bidirectional exchange of genetic material has been detected in *H. volcanii* [4], and in the same organism recombinants were obtained by protoplast fusion [5]. Both *H. volcanii* and *H. halobium* have been transformed at high frequency (up to 10^7 transformants per μg of DNA) using spheroplasted recipient cells and phage or plasmid DNA [6,7,8]. Most recently, shuttle vectors that can be selected and maintained in either *H. volcanii* or *Escherichia coli* have been described [9]. They are based on a resistance determinant from *H. volcanii* to mevinolin, an inhibitor of 3-hydroxy-3-methylglutaryl coenzyme A reductase, cloned on the endogenous *H. volcanii* plasmid pHV2 and on pBR322.

The development of genetic methods for methanogenic archaebacteria is less advanced. Some of the elements· for establishing cloning systems, such as auxotrophic and resistance markers as well as plasmids are available, but methods for plasmid transformation are lacking. Gene transfer has been observed so far in two methanogenic bacteria, namely in *Methanococcus voltae* and in *Methanobacterium thermoautotrophicum* (Marburg). For both organisms low efficiency transformation with genomic DNA has been reported [10,11], and a transduction-like process for *M. voltae* [12] as well as generalized transduction in *M. thermoautotrophicum* (L. Meile and T. Leisinger, unpublished) were observed (Table 1). Since these bacteria have advantageous growth properties and since they have been extensively used in studies on the biochemistry of methanogenesis (*M. thermoautotrophicum*) [2] as well as for gene-cloning studies [1], they are the favoured candidates for the development of experimental systems for molecular genetics in methanogens. The results of genetic experiments leading into this direction are reviewed in the following.

AUXOTROPHIC AND RESISTANCE MUTANTS

Methods for the induction of mutations and for the isolation of mutants play a key role in the development of genetic systems. Alkylating agents such as N-methyl-N'-nitro-N-nitrosoguanidine [13,14] and ethyl methanesulphonate [15,16] as well as irradiation with UV or gamma rays [10] have been used successfully for mutagenesis of methanogens. It has been speculated that contemporary anaerobic bacteria exhibit efficient photoreactivation and high intrinsic UV resistance as vestigal traits that appeared during early stages of evolution to protect ancient anaerobes from the damaging effects of unattenuated radiation on the primitive earth [17]. When these two properties were investigated in three methanogenic bacteria, ultraviolet resistance was not elevated relative to *Escherichia coli*, and only the *M. thermoautotrophicum* strains Marburg and ΔH but not *Methanococcus vannielii* exhibited photoreactivation. The *in vivo* action spectrum for photoreactivation suggested that in *M. thermoautotrophicum* (Marburg) a 5-deazaflavin (probably F_{420}) functions as the chromophore of the photoreactivating enzyme [18]. Oxygen sensitivity is another important parameter in the manipulation of methanogens during mutant isolation and in genetic experiments. Five different methanogenic bacteria examined with respect to this property exhibited marked differences. In *M. thermoautotrophicum* (Marburg), *Methanobrevibacter arboriphilus* and *Methanosarcina barkeri* the number of colony forming units was not affected by exposure to air up to a period of 10 to 30 hours. Longer periods of contact with oxygen led to a rapid decrease in viability.In contrast to these comparatively robust strains *M. voltae* and *M. vannielii* were highly sensitive to oxygen. They were killed without lag upon contact with air [19]. The mechanisms affording limited oxygen tolerance to some methanogens remain to be explored.

Mutants with specific requirements for an amino acid or a vitamin have been isolated in *M. voltae* and in *M. thermoautotrophicum* (Marburg) (Table 2). The average frequency of revertants per mutant bacterium amounted to $5 \cdot 10^{-8}$ or less which made these auxotrophs suitable for genetic experiments. The biochemical lesions leading to the auxotrophies listed in Table 2 are unknown. The isolation of auxotrophs in *M. thermoautotrophicum* (Marburg) is facilitated by the use of bacitracin for selectively enriching nongrowing mutants. This antibiotic is thought to interfere with pseudomurein formation and/or membrane synthesis. It preferentially kills growing cells, and its use in an enrichment procedure has led to a tenfold increase in the yield of auxotrophs. Since bacitracin also preferentially killed growing cells of *M. vannielii* its application in selective enrichments is not restricted to pseudomurein-containing methanogens [20].

The archaebacterial features of their cell envelope, their cell membrane and the protein synthesizing machinery make methanogens insensitive to many of the antibiotics used in eubacterial genetics. This has led to an intensive search for drugs which inhibit methanogen growth and could be used as selective agents in genetic experiments. Drugs of potential use in genetics should act against a specific target and spontaneous resistance should occur at a low rate. Several growth inhibitors of *M. voltae* and *M. thermoautotrophicum* (Marburg), such as the amino acid analogs azaserine, methionine sulfoximine [22], 5-methyltryptophan [41], ethionine [13] and the coenzyme M analog bromoethane sulfonate [13,41,42] did not fulfill these criteria. Low level resistance to these analogs occurred at high rates and seemed to be caused by changes in cell permeability. Table 2 lists antimetabolites with proven or potential use in methanogen genetics. Mutants resistant to these agents occur at a low rate and exhibit high resistance factors. In some cases the targets of these inhibitors are known. Pseudomonic acid has been shown to inhibit isoleucyl-tRNA synthetase of *M. voltae* [22] and *M. thermoautotrophicum* (Marburg) [14], and pseudomonic acid resistant mutants of the latter organism exhibited an insensitive enzyme [14]. A 5-fluorouracil resistant mutant of *M. thermoautotrophicum* (Marburg) was deficient in uracil-phosphoribosyltransferase activity and did not activate this compound to an inhibitory nucleotide [24]. Similarly the *M. voltae* mutants resistant to purine and pyrimidine analogs [21] (Table 2) appear to be defective in the salvage pathway for the respective bases. Both pseudomonic acid resistance [L. Meile, unpublished] and 5-fluorouracil resistance [11] have been used to select for gene transfer in *M. thermoautotrophicum* (Marburg). Since pseudomonic acid resistance is a dominant trait it may be useful in the construction of a vector plasmid.

Resistance genes for antibiotics inhibitory to methanogens may also be obtained from eubacteria, brought under methanogen expression signals and used in gene transfer experiments with *M. voltae* or *M. thermoautotrophicum* (Marburg). Possot et al. [22] have observed that both fusidic acid and puromycin are effective inhibitors of *M. voltae*. Since these compounds inhibited polypeptide synthesis in *M. voltae* cell extracts, their mode of action in the methanogen and in eubacteria was concluded to be similar. This information was recently used by the same authors to construct the first integration vector for a methanogen. It is based on the puromycin resistance gene from *Streptomyces alboniger* which was integrated with appropriate expression signals into the cloned *M. voltae* hisA gene. M. voltae was transformed with this construct to puromycin resistance via homologous recombination into the chromosomal hisA gene, and the vector was stably maintained under selective pressure [25].

METHANOCOCCUS VOLTAE

M. voltae is a mesophilic methanogen utilizing CO_2/H_2 and formate as substrates. Its genome size amounts to 1.8 ±

Table 2. Auxotrophic and Resistance Mutants in *Methanococcus voltae* and *Methanobacterium thermoautotrophicum* (Marburg)

Growth Requirement or inhibitor	Resistance factor	Reference
M. voltae		
L-Histidine		[10]
Purines		[10]
Cyanocobalamin		[10]
8-Aza-2,6-diaminopurine	1'000	[21]
8-Azaguanine	10'000	[21]
8-Azahypoxanthine	>10'000	[21]
6-Mercaptopurine	> 500	[21]
6-Azauracil	2'000	[21]
Pseudomonic acid	40	[22]
1,2,4-Triazole-3-alanine	800	[23]
M. thermoautotrophicum (Marburg)		
L-Leucine		[13]
L-Tryptophan		[20]
Adenosine		[13]
Thiamine		[13]
Pseudomonic acid	50	[14]
5-Fluorouracil	> 1'000	[24]

0.3×10^9 daltons [26]. The organism has a proteinaceous cell envelope, and DNA isolation is easy since the cells are lysed by detergents or solutions of low osmolality. *M. voltae* strain PS has been used as a source of DNA for cloning and sequencing the *nifH* [27], *hisA* [28] and the *trpBA* genes [29] as well as the genes encoding methyl coenzyme M reductase [30]. The mutants listed in Table 2 have been obtained in this strain, and strain PS is also the organism for which a transformation protocol has been developed. Up to 10^2 transformants per μg of genomic DNA were obtained by a simple procedure which did not require calcium treatment and heat shock [10]. No extrachromosomal elements have been isolated from strain PS so far. This has prevented the development of a plasmid cloning vector. Other representatives of the genus *Methanococcus*, however, carry cryptic plasmids (Table 3), and it remains to be seen whether one of these can be developed into a vector plasmid for strain PS.

Besides natural transformation another system of gene transfer has recently been discovered in *M. voltae*, strain PS [12]. Cell-free filtrates of cultures contained an agent named VTA (for Voltae Transfer Agent) capable of transferring genetic markers to mutant recipient bacteria. This activity was not affected by DNase, and the three auxotrophic mutants of *M. voltae* (Table 2) were tranduced to wild type by VTA.

Efficiencies of transfer in the range of 10^3 to 10^5 transductants per ml of donor culture were observed. VTA was more resistant to osmotic shock than cells of *M. voltae*. Further characterization of VTA, which may be a genetic transfer particle, is in progress [12].

Table 3. Plasmids of Methanogenic Bacteria

Host	Plasmid			Reference
	Designation	DNA (kb)	Copy number	
Methanobacterium thermoautotrophicum (Marburg)	pME2001	4.5	15-30	[32]
Methanobacterium thermoformicicum Z-245	pFZ1	10.5	n.d.	[33]
Methanococcus sp. AG86	pURB900	20	n.d.	[34]
Methanococcus jannaschii	pURB800	64	n.d.	[34]
	pURB801	18	n.d.	[34]
Methanococcus sp.C5	pURB500	8.7	3	[35]
Methanolobus vulcani PL12-M	pMP1	7	n.d.	[36]
Methanosarcina acetivorans C2A	pC2A	5.1	6	[37]

n.d. = not determined

With some respects the properties of VTA are reminiscent of the recently described viruslike particle (VLP) of *M. voltae* strain A3 [31]. From supernatants of late-exponential-phase cultures of this strain up to 30 VLPs per cell were recovered. The proteinaceous particles contained a 23 kb covalently closed circular DNA designated pURB600 (Table 4). Hybridization indicated the presence of a chromosomally integrated copy of pURB600 in strain A3. Although *M. voltae* PS did not produce VLPs, DNA homologous to pURB600 was detected in its chromosome. The VLP of strain A3 thus exhibits some similarities to lysogenic bacteriophages but infectivity, inducibility or VLP-mediated gene transfer could not be demonstrated so far.

Table 4. Bacteriophages and Virus-like Particles (VLPs) of
Methanogens

Host	Virus or VLP				Refer.
	Designation	Particle shape	DNA (kb)	Type	
Methanobacterium thermoautotrophicum (Marburg)	ψM1	poly-hedric	27.1	lytic	[38]
Methanobrevibacter smithii G	PG	poly-hedric	45	lytic	[39]
Methanobrevibacter smithii PS	PMS1	poly-hedric	35	lytic	[40]
Methanococcus voltae A3	pURB600	lemon	23	VLP	[31]
Methanococcus voltae PS	VTA	n.d.	n.d.	VLP	[12]

n.d. = not determined

METHANOBACTERIUM THERMOAUTOTROPHICUM

M. thermoautotrophicum (Marburg) is a thermophilic rod with
optimal growth at 65°C. Its substrate spectrum is restricted
to CO_2/H_2, and on a mineral medium without organic
supplements it reaches a doubling time of three hours. The
genome size of the organism was determined to be
approximately $1,2 \times 10^9$ daltons [43]. *Methanobacterium* sp.
are Gram-positive, and their rigid cell wall sacculus
consists of pseudomurein. This property made it difficult to
obtain high yields of undegraded chromosomal and plasmid DNA
from *M. thermoautotrophicum* until preparations of
pseudomurein endopeptidase became available [44]. This
oxygen-sensitive enzyme is found in autolysates of
Methanobacterium wolfei. It was purified approximately 500-
fold to electrophoretic homogeneity and shown to hydrolyze
the ε-Ala-Lys bond of pseudomurein. Partially purified
preparations of pseudomurein endopeptidase have proven useful
for the extraction of high molecular weight DNA from *M.
thermoautotrophicum* (Marburg).

Two strains of *M. thermoautotrophicum*, ΔH and Marburg,
are widely used in biochemical studies. They exhibit marked
differences in biochemical and physiological characteristics,
and DNA relatedness amounted to 46% binding as determined by
DNA-DNA hybridization [43]. The Marburg strain is well suited
for genetic studies since mutants (Table 2), a transformation
protocol [11], a plasmid (Table 3) and a transducing
bacteriophage (Table 4) are available in this organism. The
only polypeptide-encoding genes from strain Marburg that have
been cloned and sequenced are the genes for methyl coenzyme M
reductase [45].

Natural transformation of *M. thermoautotrophicum* (Marburg) was observed when genomic DNA from a 5-fluorouracil resistant mutant was added to wildtype cells growing on the surface of plates solidified with Gelrite gellan gum. Some of the colonies developing under these conditions contained transformants. The yield of transformants was dependent on the amount of DNA [11]. Transformation was only detected on the surface of plates solidified with Gelrite gellan gum and not on agar plates or in liquid media. The reasons for the dependence of the procedure on Gelrite (a negatively charged heteropolysaccharide) are not known.

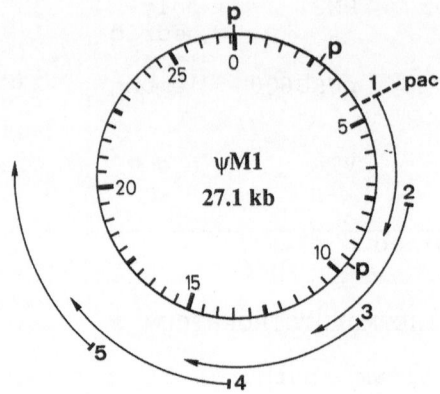

Figure 1. Simplified map of the phage ψM1 genome. The origin of the map was arbitrarily set at a *Pvu*II (=p) cleavage site. Numbered arrows indicate the average location of the left ends of phages 1 to 5 in a packaging round of concatemeric DNA as determined by restriction analysis.

The functions encoded by plasmid pME2001, the 4.5 kb multicopy plasmid of *M. thermoautotrophicum* (Marburg), are unknown [32]. In this respect pME2001 is similar to all plasmids of methanogens described so far (Table 3). Since a plasmid-free mutant of the Marburg strain grew at a slightly reduced rate but to the same extent as the wildtype strain, pME2001 seems to be dispensable for its host [L. Meile, unpublished]. When crude RNA preparations of the plasmid-carrying wildtype strain were examined for pME2001-encoded transcripts, a single prominent plasmid-encoded RNA was detected. This 611-base-pair transcript contained four possible open reading frames ranging in length from 90 to 234 bases [46]. Derivatives of pME2001 capable of replication in eubacteria and yeast have been constructed [47]. They may eventually be useful as shuttle vectors.

The other extrachromosomal element of *M. thermo-autotrophicum* (Marburg) is bacteriophage ψM1 (Table 4). This virulent, oxygen-resistant phage was isolated from an experimental anaerobic digester operated at 55°C to 60°C. The latent period of ψM1 at 62°C was 4 hours, and its burst size was 5 to 6 infective particles per cell. The phage infected none of three other thermophilic representatives of the genus *Methanobacterium* that were tested. Phage particles contain linear double-stranded DNA of 30.4 ± 1.0 kb as determined by electron microscopy. About 85 percent of these DNA molecules represent ψM1 DNA whereas the rest are multimers of plasmid pME2001 [32]. The efficient packaging of plasmid DNA suggested that the phage might also encapsidate chromosomal DNA. We have therefore tested its ability to transduce chromosomal markers and have found that *trp*, *ade*, *leu* and pseudomonic acid resistance are transduced at frequencies between 10^{-5} and 10^{-4} per plaque forming unit. To obtain high transduction frequencies it was important to use low multiplicities of infection. Evidence that the observed gene transfer is due to transduction rests on the fact that the process is insensitive to RNase and DNAse and that it is observed with phage preparations that have been purified by CsCl gradient centrifugation [Meile and Leisinger, unpublished].

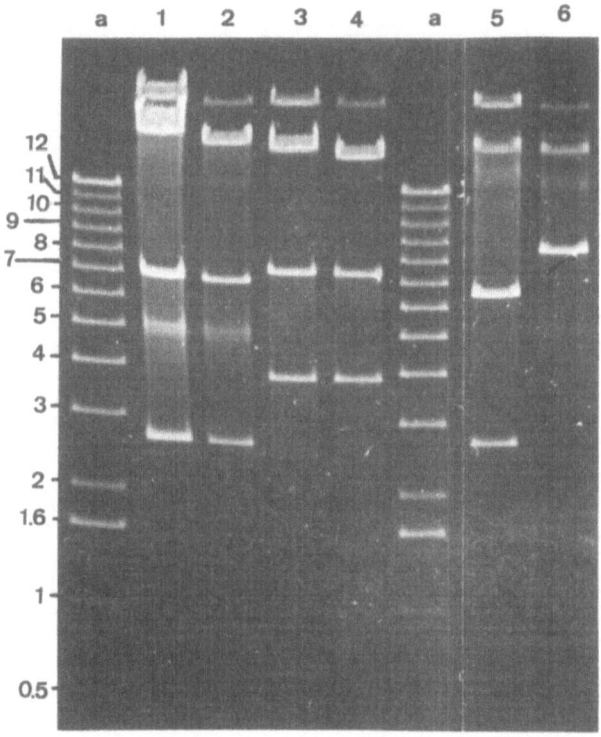

Figure 2. Restriction analysis of DNA from phage ψM1 (lanes 1,3,5) and its deletion mutant ψM2 (lanes 2,4,6). The restriction endonucleases used were *Pvu*II (lanes 1 and 2), *Hind*III (lanes 3 and 4), and *Sal*I (lanes 5 and 6). Sizes of linear DNA marker fragments (lanes a) are given in kb.

Recently we have characterized the bacteriophage ψM1 DNA using restriction endonucleases which do not cut plasmid pME2001. Endonuclease cleavage sites mapped in a circular array on the ψM1 genome, and the size of ψM1 DNA obtained by restriction analysis amounted to 27 ± 0.2 kb. Since the genome of ψM1 has been shown by electron microscopy to be linear and 30.4 kb in length, this suggests that ψM1 DNA is circularly permuted and exhibits about 10% terminal redundancy. The precursor to mature DNA thus appears to be a concatemer which is packaged by the headful mechanism.

A simplified version of the ψM1 genome map is shown in Fig. 1. The *pac* site, that is the site at which packaging of concatemeric DNA starts, was located at coordinate 4.6 kb on this map. Permutation of the phage genome extended from 6.4 kb to 21 kb on the map. This indicated that packaging proceeds from the *pac* site clockwise for up to five rounds [Jordan et al., unpublished]. Fig. 2 illustrates some features of the DNA extracted from particles of ψM1 and of ψM2, a spontaneous phage mutant with a 692-base pair deletion at position 23.25 kb on the ψM1 map. Digests obtained with three restriction endonucleases contain a substoichiometric linear DNA fragment of approx. 30 kb which represents the uncleaved multimer of plasmid pME2001. Restriction fragments ending at *pac* and therefore present in submolar concentrations are visible in the *Pvu*II and *Sal*I digests. The deletion of a *Sal*I cleavage site in the DNA of the phage variant ψM2 is evident from a comparison of the DNA fragment profiles in Fig. 2, lanes 5 and 6.

CONCLUSIONS

It appears that *M. voltae* strain PS and *M. thermoautotrophicum* (Marburg) are becoming model organisms for genetic studies in methanogens. Homologous chromosomal DNA can be reintroduced into these organisms by low-frequency transformation, and chromosomal markers are transferred within these strains by a transduction-like process (*M. voltae*) or by generalized transduction (*M. thermoautotrophicum*). These procedures are useful in mutant analysis and strain construction. The functional analysis of methanogen genes in their proper cellular environment requires plasmid cloning systems. Some of the elements for the development of such systems are available in each of the organisms. Others, such as an efficient procedure for plasmid transformation, suitable vector plasmids, and recombination-deficient recipient strains need to be developed.

Acknowledgment. Research in the author's laboratory was supported by grants 3.193.85 and 31-25177.88 from the Swiss National Foundation for Scientific Research.

REFERENCES

1. J.W. Brown, C.J. Daniels, and J.N. Reeve, Gene structure, organization, and expression in archaebacteria. CRC Crit. Rev. Microbiol. 16:287-338 (1989).

2. W.J. Jones, D.P. Nagle, Jr., and W.B. Whitman, Methanogens and the diversity of archaebacteria. Microbiol. Rev. 51:135-177 (1987).

3. D.P. Nagle, Jr., Development of genetic systems in methanogenic archaebacteria. Dev. Industrial. Microbiol. 30 (in press) (1989).

4. M. Mevarech and R. Werczberger, Genetic transfer in *Halobacterium volcanii*. J. Bacteriol. 162:461-462 (1985).

5. I. Rosenshine and M. Mevarech, Isolation and partial characterization of plasmids found in three *Halobacterium volcanii* isolates. Can. J. Microbiol. 35:92-95 (1989).

6. R.L. Charlebois, W.L. Lam, S.W. Cline, and W.F. Doolittle, Characterization of pHV2 from *Halobacterium volcanii* and its use in demonstrating transformation of an archaebacterium. Proc. Natl. Acad. Sci. USA 84:8530-8534 (1987).

7. S.W. Cline and W.F. Doolittle, Efficient transfection of the archaebacterium *Halobacterium halobium*. J. Bacteriol. 169:1341-1344 (1987).

8. N.R. Hackett and S. DasSarma, Characterization of the small endogenous plasmid of *Halobacterium* strain SB3 and its use in transformation of *H. halobium*. Can. J. Microbiol. 35:86-91 (1989).

9. W.L. Lam and W.F. Doolittle, Shuttle vectors for the archaebacterium *Halobacterium volcanii*. Proc. Natl. Acad. Sci. USA 86:5478-5482 (1989).

10. G. Bertani and L. Baresi, Genetic transformation in the methanogen *Methanococcus voltae* PS. J. Bacteriol. 169:2730-2738 (1987).

11. V.E. Worrell, D.P. Nagle, Jr., D. McCarthy, and A. Eisenbraun, Genetic transformation system in the archaebacterium *Methanobacterium thermoautotrophicum* Marburg. J. Bacteriol. 170:653-656 (1988).

12. G. Bertani, Transduction-like gene transfer in a methanogen. Abstr. Annu. Meet. Am. Soc. Microbiol., New Orleans (1989).

13. A. Kiener, C. Holliger, and T. Leisinger, Analogue-resistant and auxotrophic mutants of *Methanobacterium thermoautotrophicum*. Arch. Microbiol. 139:87-90 (1984).

14. A. Kiener, T. Rechsteiner, and T. Leisinger, Mutation to pseudomonic acid resistance of *Methanobacterium thermoautotrophicum* leads to an altered isoleucyl-tRNA synthetase. FEMS Microbiol. Lett. 33:15-18 (1986).

15. L. Bhatnagar, M.K. Jain, J.G. Zeikus, and J.-P. Aubert, Isolation of auxotrophic mutants in support of ammonia assimilation via glutamine synthetase in *Methanobacterium ivanovii*. Arch. Microbiol. 144:350-354 (1986).

16. M.K. Jain and J.G. Zeikus, Methods for isolation of auxotrophic mutants of *Methanobacterium ivanovii* and initial characterization of acetate auxotrophs. Appl. Environ. Microbiol. 53:1387-1390 (1987).

17. M.B. Rambler and L. Margulis, Bacterial resistance to ultraviolett irradiation under anaerobiosis: implications for pre-phanerozoic evolution. Science 210:638-640 (1980)

18. A. Kiener, R. Gall, T. Rechsteiner, and T. Leisinger, Photoreactivation in *Methanobacterium thermoautotrophicum*. Arch. Microbiol. 143:147-150 (1985).

19. A. Kiener and T. Leisinger, Oxygen sensitivity of methanogenic bacteria. System. Appl. Microbiol. 4:305-312 (1983).

20. T. Rechsteiner, A. Kiener, and T. Leisinger, Mutants of *Methanobacterium thermoautotrophicum*. System. Appl. Microbiol. 7:1-4 (1986).

21. T.L. Bowen and W.B. Whitman, Incorporation of exogenous purines and pyrimidines by *Methanococcus voltae* and isolation of analog-resistant mutants. Appl. Environ. Microbiol. 53:1822-1826 (1987).

22. O. Possot, P. Gernhardt, A. Klein, and L. Sibold, Analysis of drug resistance in the archaebacterium *Methanococcus voltae* with respect to potential use in genetic engineering. Appl. Environ. Microbiol. 54:734-740 (1988).

23. K.A. Sment and J. Konisky, Excretion of amino acids by 1,2,4-triazole-3-alanine-resistant mutants of *Methanococcus voltae*. Appl. Environ. Microbiol. 55:1295-1297 (1989).

24. D.P. Nagle, R. Teal, and A. Eisenbraun, 5-Fluorouracil-resistant strain of *Methanobacterium thermoautotrophicum*. J. Bacteriol. 169:4119-4129 (1987).

25. P. Gernhardt, O. Possot, M. Foglino, L. Sibold, and A. Klein, Construction of an integration vector for use in *Methanococcus voltae*. Abstr. 6th Internat. Symp. on Microbial Growth on C1 Compounds, Göttingen (1989).

26. A. Klein and M. Schnorr, Genome complexity of methanogenic bacteria. J. Bacteriol. 158:628-631 (1984).

27. N. Souillard and L. Sibold, Primary structure and expression of a gene homologous to *nifH* (nitrogenase Fe protein) from the archaebacterium *Methanococcus voltae*. Mol. Gen. Genet. 203:21-28 (1986).

28. D. Cue, G.S. Beckler, J.N. Reeve, and J. Konisky, Structure and sequence divergence of two archaebacterial genes. Proc. Natl. Acad. Sci. USA 82:4207-4211 (1985).

29. L. Sibold and M. Henriquet, Cloning of the *trp* genes from the archaebacterium *Methanococcus voltae*: Nucleotide sequence of the *trpBA* genes. Mol. Gen. Genet. 214:439-450 (1988).

30. A. Klein, R. Allmansberger, M. Bokranz, S. Knaub, B. Müller and E. Muth, Comparative analysis of genes encoding methyl coenzyme M reductase in methanogenic bacteria. Mol Gen. Genet. 213:409-420 (1988).

31. A.G. Wood, W.B. Whitman, and J. Konisky, Isolation and characterization of an archaebacterial viruslike particle from *Methanococcus voltae* A3. J. Bacteriol. 171:93-98 (1989).

32. L. Meile, A. Kiener, and T. Leisinger, A plasmid in the archaebacterium *Methanobacterium thermoautotrophicum*. Mol. Gen. Genet. 191:480-484 (1983).

33. J. Nölling, M.J. Frylink, and W.M. de Voss, Isolation and characterization of plasmid DNA from the methanogen *Methanobacterium thermoautotrophicum*. Abstr. 6th Internat. Symp. on Microbial Growth on C1 Compounds, Göttingen (1989.

34. H. Zhao, A.G. Wood, F. Widdel, and M.P. Bryant, An extremely thermophilic *Methanococcus* from a deep sea hydrothermal vent and its plasmid. Arch. Microbiol. 150:178-183 (1988).

35. A.G. Wood, W.B. Whitman, and J. Konisky, A newly-isolated marine methanogen harbors a small cryptic plasmid. Arch. Microbiol. 142:259-261 (1985).

36. M. Thomm, J.Altenbuchner, and K.O. Stetter, Evidence for a plasmid in a methanogenic bacterium. J. Bacteriol. 153:1060-1062 (1983).

37. K.R. Sowers and R.P. Gunsalus, Plasmid DNA from the acetotrophic methanogen *Methanosarcina acetivorans*. J. Bacteriol. 170:4979-4982 (1988).

38. L. Meile, U. Jenal, D. Studer, M. Jordan, and T. Leisinger, Characterization of ψM1, a virulent phage of *Methanobacterium thermoautotrophicum* Marburg. Arch. Microbiol. 152:105-110 (1989).

39. L. Baresi and G. Bertani, Isolation of a bacteriophage for a methanogenic bacterium. Abstr. Annu. Meet. Am. Soc. Microbiol., New Orleans (1989).

40. M.R. Knox and J.E. Harris, Isolation of a bacteriophage of *Methanobrevibacter smithii*. Abstr. XIV. Internat. Congr. Microbiol., Manchester (1986).

41. P. Gernhardt and A. Klein, Analogue resistant mutants of *Methanococcus voltae*. (Abstract) *in*: "Archaebacteria 85", O. Kandler and W. Zillig, eds., Gustav Fischer Verlag, Stuttgart (1986).

42. N. Santoro and J. Konisky, Charcterization of bromo-ethanesulfonate-resistant mutants of *Methanococcus voltae*: evidence of a coenzyme M transport system. J. Bacteriol. 169:600-665 (1987).

43. A. Brandis, R.K. Thauer, and K.O. Stetter, Relatedness of strains ΔH and Marburg of *Methanobacterium thermoautotrophicum*. Zbl. Bakt. Hyg. I, Abt. Orig. C2: 311-317 (1981).

44. A. Kiener, H. König, J. Winter, and T. Leisinger, Purification and use of *Methanobacterium wolfei* pseudomurein endopeptidase for lysis of *Methanobacterium thermoautotrophicum*. J. Bacteriol. 169:1010-1016 (1987).

45. M. Bokranz, G. Bäumner, R. Allmansberger, D. Ankel-Fuchs, and A. Klein, Cloning and characterization of the methyl coenzyme M reductase genes from *Methanobacterium thermoautotrophicum*. J. Bacteriol. 170:568-577 (1988).

46. L. Meile, J. Madon, and T. Leisinger, Identification of a transcript and its promoter region on the archaebacterial plasmid pME2001. J. Bacteriol. 170:478-481 (1988).

47. L. Meile and J.N. Reeve, Potential shuttle vectors based on the methanogen plasmid pME2001. Bio/Technology 3:69-72 (1985).

THE HYDROGENASE OF <u>METHANOCOCCUS VOLTAE</u>: AN APPROACH TO THE
BIOCHEMICAL AND GENETIC ANALYSIS OF AN ARCHAEBACTERIAL UPTAKE
HYDROGENASE

Erika Kothe, Sabine Halboth, Jörg Sitzmann, and
Albrecht Klein

Molecular Genetics, Dept. of Biology
Philipps University
P.O.Box 1929, D-3550 Marburg, F.R.G.

<u>Summary</u>

Investigations concerning the biochemistry and genetics
of the hydrogenase of *Methanococcus voltae* are discussed in
the context of the general knowledge of hydrogenases with
respect to their subunit structures, reactive centers, cellular localization, regulation, and evolutionary conservation.

General Roles and Characteristics of Hydrogenases

Hydrogenases are enzymes which catalyze the reversible
conversion of molecular hydrogen into protons and electrons.
The reaction can both lead to the production or consumption
of hydrogen, depending on the equilibrium of the reaction,
which is determined by the redox potentials of the reaction
partners, e.g. the concentration of hydrogen and the pH, or
the type and concentration of the acceptor molecules (see
Cammack et al.[1] and Hausinger[2] for reviews).

Hydrogenases play a key role in interspecies hydrogen
transfer. In the cases in which the enzymes are used to produce electrons and to eventually channel them into reductive
pathways, they are termed uptake hydrogenases. Uptake hydrogenases are thus involved in reductive biosynthetic or energy
generating pathways, such as anaerobic respiration in many
eubacteria or methane formation in methanogenic archaebacteria.

Heteroatoms and Cofactors Relevant to the Function of the Catalytic Centers of Hydrogenases

All known hydrogenases are iron proteins, containing
Fe/S-centers in different numbers. In many if not most hydrogenases nickel is found, which is part of the active center
and coordinated to sulfur atoms of the side chains of cysteine residues[3] and possibly nitrogen atoms of the imidazole
rings of histidine residues of the polypeptides.

Microbiology and Biochemistry of Strict Anaerobes Involved in Interspecies Transfer
Edited by J.-P. Bélaich *et al.*
Plenum Press, New York, 1990

25

In rare cases selenium has been found as constituent of hydrogenases[4,5,6], which is thought to replace one of the sulfur atoms coordinated to the nickel atom[7]. In analogy to the well studied case of formate dehydrogenase in *Escherichia coli*[8] it is suspected to be part of a selenocysteine residue replacing a cysteine in the analogous position of the poly-peptide chain. Indeed, selenocysteine has been found in the hydrogenase described from *Methanococcus vannielii*[4]. FAD is also a cofactor found in those hydrogenases which use NAD or structural analogues as electron acceptors, as will be dis-cussed in more detail below.

Subunit Structures of Different Hydrogenases

The subunit structures of different hydrogenases are of different complexities. At least two subunits have been found, but their number can be higher. This can be related to the binding of electron accepting cofactors. A general model has been put forward[1] which proposes different sites for the activation of the hydrogen and the binding of the electron accepting molecules. In this context it is noteworthy that different hydrogenases of the same organism can have differ-ent subunit structures in correlation with their different electron acceptors.

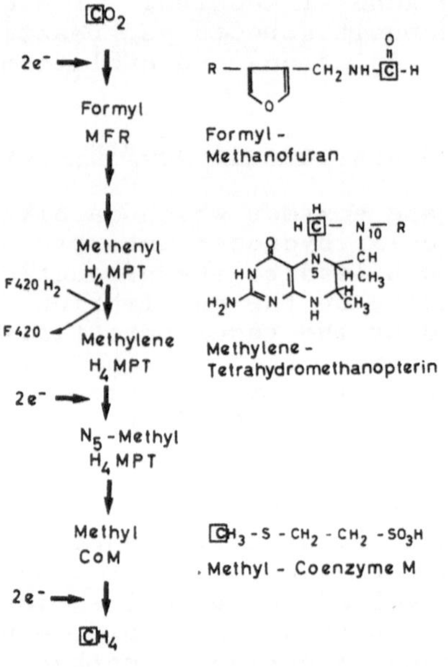

Fig. 1. Pathway of CO_2 reduction in the course of methano-genesis. The C1 moiety is bound to three coenzymes in subsequent steps of its reduction: methanofuran (MTF), tetrahydromethanopterin (H_4MPT), and thio-ethane sulfonic acid (coenzyme M). The reduction of methenyl- to methylene-tetrahydromethanopterin involves the oxidation of reduced F_{420}, a cofactor of hydrogenases found in methanogenic bacteria. (See Rouvière and Wolfe[9] for a more detailed account of the methanogenic pathway).

The Functions and Types of Hydrogenases in Methanogenic Bacteria

In methanogenic bacteria one carbon (C1) moieties can be reduced in sequential reactions from the the highest oxidation stage (CO_2) to the most reduced level, the end product methane (Fig. 1). Uptake hydrogenases are the electron generating systems, which allow these reductive steps.

Methanogenic bacteria, e.g. the most intensively studied *Methanobacterium thermoautotrophicum* strains, contain two different types of hydrogenases[10,11]. They are operationally defined according to the electron acceptors which they are able to reduce *in vitro*. One type can transfer electrons to a two electron accepting cofactor found in all methanogenic bacteria, the deazaflavin F_{420}, which is a structural analog of nicotinamide (Fig. 2).

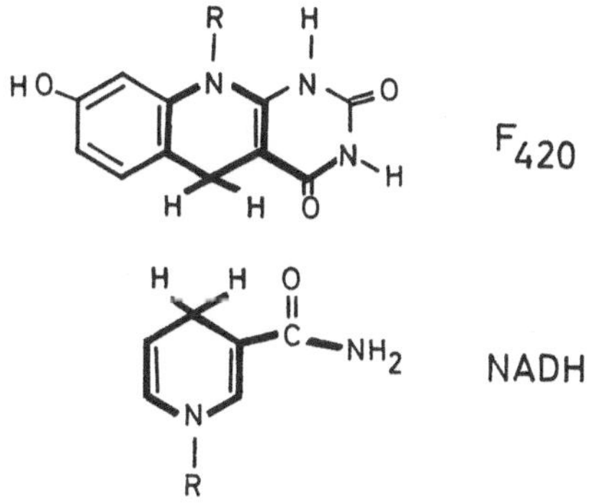

Fig. 2. Comparison of the ring systems of reduced F_{420} and NADH. The structural analogy is pointed out by heavy contours.

The other type is unable to reduce F_{420} but reduces artificial one electron acceptors such as viologens (which can also be reduced by the F420 reducing enzyme species). Recently, it has been proposed that an electron acceptor for such an enzyme may be a polyferredoxin[12]. This is a comparable situation as the one found in *Alcaligenes eutrophus*[13,14], where one hydrogenase is able to reduce NAD while the other is not.

In *Methanococci*, only one, F_{420} reducing hydrogenase species has been found to date and there are strong indications that it is the only one. It is this enzyme which we have purified and characterized from *M. voltae*[15] while the corresponding hydrogenase of *M. vannielii* has been described by Yamazaki[4].

Characterization of the F$_{420}$ Reducing Hydrogenase of
Methanococcus voltae

The hydrogenase from *M. voltae* has been purified to apparent homogeneity. Comparison of the apparent molar mass of the nondenatured active monomer with the sizes of the subunits as determined by SDS polyacrylamide gel electrophoresis resulted in the establishment of an enzyme structure comprising three subunits of estimated 43, 37, and 27 kD molecular mass. The enzyme aggregates in solution due to its hydrophobicity and forms defined large complexes. Electron microscopy shows the two size classes of the native molecules and demonstrates the regular flat ring shape of the high molecular weight form (Fig. 3).

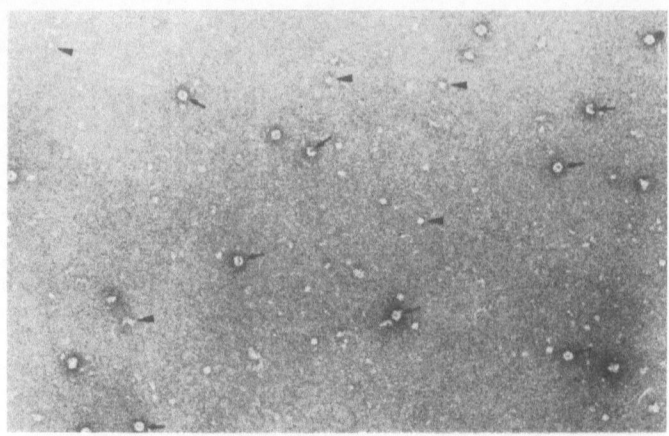

Fig. 3. Electron micrographs of *M. voltae* hydrogenase molecules. The ring shaped molecules with a central hole (arrows) are the 745 kD complexes of several minimal active native hydrogenase molecules (110 kD) indicated by arrow heads.

The hydrogenase reduces F$_{420}$. The activity towards this substrate is correlated with its chromatographic behavior on ion exchange columns, which has been interpreted in terms of an equilibrium in aqueous solution between two different conformations only one of which is able to react with the deazaflavin. The enzyme contains bound FAD, the removal of which apparently destabilizes the enzyme structure. This coenzyme is considered to mediate electron transfer from the one electron Fe/S-clusters to the two electron acceptors such as F$_{420}$. The *M. voltae* hydrogenase is most likely a FeNiSe-hydrogenase. One nickel and one selenium were found per minimal active complex employing atomic absorption spectroscopy on the purified enzyme. *M. voltae* cell extracts labelled with

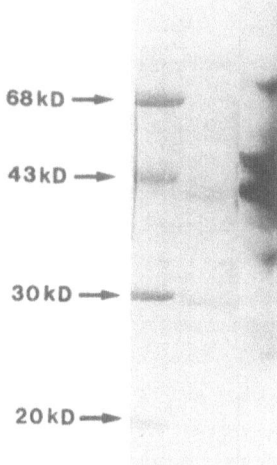

68kD →

43kD →

30kD →

20kD →

Fig. 4.   Autoradiogram of ^{75}Se labelled cell extract of *M. voltae* (right) separated on a SDS polyacrylamide gel next to purified hydrogenase (center) and size markers (left). The lower, strongly labelled band runs to the same position as the largest ("43kD") subunit band of the hydrogenase.

[75]Se show a radioactive band of the same size as the largest subunit of the hydrogenase upon SDS polyacrylamide gel electrophoresis (Fig. 4).

Localization of the *M. voltae* Hydrogenase

Immuno-gold labelling of thin sections of *M.voltae* cells using antiserum against the hydrogenase showed that the hydrogenase is associated with the cell membrane[16] (Fig. 5). This is in agreement with its hydrophobicity and its role in conjunction with the enzyme system catalyzing the last step of the C1-reduction during methanogenesis, the methyl CoM reductase system. Its central enzyme, the methyl CoM reductase (also called methyl CoM reductase component C) itself, has also been found to be membrane associated[17]. The same has been shown for the F$_{420}$ reducing hydrogenase of *Methanosarcina barkeri* (K. Fiebig, personal communication). Nothing is known about the type of anchoring of these enzymes to the membrane, but it has been pointed out that genes encoding hydrophobic polypeptide sequences characteristic of membrane integrated proteins are found closely linked to the genes encoding membrane bound hydrogenases in *A. eutrophus* (B. Friedrich, personal communication) and *Rhodobacter capsulatus*[18]. Since the hydrogenase is supposed to be involved in the generation of a proton gradient[19] across the membrane, it would be interesting to establish to which side of the membrane the enzyme is bound.

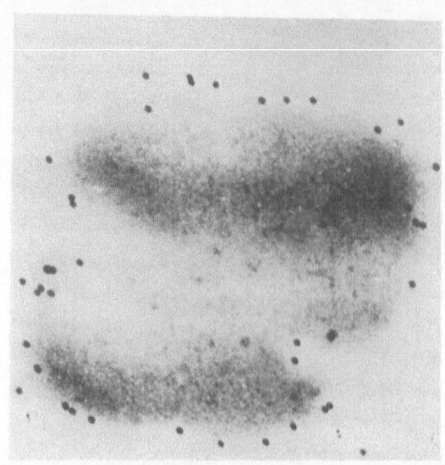

Fig. 5. Electron micrograph of a thin section of a *M. voltae* cell treated with gold-labelled anti-hydrogenase-antibodies. Controls with pre-immune-serum showed few randomly electron dense distributed grains.

Regulation of Hydrogenase Expression and the Role of Nickel

In facultative anaerobic bacteria, hydrogenases are induced only in the state of anaerobiosis (see Stewart[20] for review). Dual control by general positive regulators and by specific transcription factors is likely in these cases. In principle, methanogenic bacteria are expected to express the hydrogenases constitutively. However, it can be shown for *M. voltae* that both cell growth and hydrogenase activity per total cell protein are also correlated with the nickel concentration in the medium (table 1).

Table 1 Hydrogenase Contents in *Methanococcus voltae* Cells as a Function of the Nickel Concentration in the Growth Medium.

Nickel concentration [μM]	final cell density [cells/ml]	hydrogenase activity [units/mg protein]
0	1.2×10^6	0.2
0.8	2.5×10^6	0.4
1.5	5.0×10^6	4
2.3	2.0×10^7	5
3.0	3.8×10^7	10
3.8	8.0×10^7	46

This might indicate that nickel is involved in transcriptional or translational control of hydrogenase expression. Alternatively, the stability of the hydrogenase in the cell could be a function of its proper folding which, in

Table 2   Complementation of Pleiotropic Hydrogenase Deficient
E. coli Mutant by a Cloned *Methanococcus voltae* Genomic Fragment.

Strain	Ni added to medium [a] [mM]	hydrogenase activity [units/mg protein]
JM83	0	5.0
JM83	0.6	5.5
JM83h1 [b]	0	0
JM83h1	0.6	6.0
JM83h1(pUC8)	0	0
JM83h1(pMU1) [c]	0	4.7

[a] The cells were grown in rich medium which contains enough
nickel to allow anaerobic growth and hydrogenase formation in
wild type cells. [b] The pleiotropic mutant h1 was derived from
the *E. coli* strain JM 83, which is a wild type with respect
to hydrogenase production. [c] pMU1 is a pUC8 derived clone
containing a *M. voltae* DNA fragment.

turn, would be dependent on the presence of nickel. The
values given in table 1 indicate that a specific nickel
uptake system exists in *M. voltae* with an apparent K_M in the
micromolar range.

This reminds of the situation found in the facultative
anaerobic bacterium *E. coli*, which has two nickel uptake
systems, one of which is also specific for nickel and shows a
similarly high affinity A second one which transports nickel
and magnesium has a much higher K_M for nickel[21]. Among hydrogenase deficient strains of this bacterium pleiotropic mutants have been detected[22-25] which are devoid of any hydrogenase activity but do show such activity when the medium is
supplemented with high amounts of nickel.

We assume that this reflects the same effect which we
see in *M. voltae*. We have therefore tried whether newly isolated pleiotropic hydrogenase deficient *E. coli* mutants might
be complemented by genetic material from *M. voltae*, cloned in
an *E. coli* expression plasmid, which guarantees its transcription in the host cell. Such complementation was indeed
possible with one of the mutants. Table 2 shows that the mutant was pleiotropic, lacking hydrogenase activity in the absence of high nickel concentrations in the growth medium.

This deficiency was overcome upon introduction of a
plasmid carrying cloned *M. voltae* sequences. Fig. 6 shows a
nickel uptake experiment, which demonstrates that the hydrogenase deficiency was correlated with reduced nickel uptake
by the mutant and that this was partially overcome in the
transformed cell.

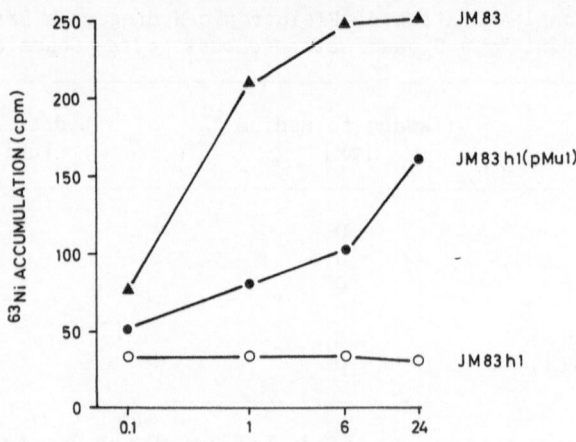

Fig. 6.   Nickel uptake by *E. coli* cells during anaerobic
growth. The cells were grown in synthetic medium
supplemented with ^{63}Ni. Equal samples taken after
the indicated times were washed and counted.

This finding should allow to follow up the nickel effect
and characterize the gene encoding the involved nickel transport or storage functions, possibly in *M. voltae* and *E. coli*.

Structural Similarity and Phylogenetic Relationship between
Hydrogenases

Hydrogenases can be assumed to have been present early
in evolution, when hydrogen was still a major component of
the atmosphere, and to have adapted to various complex pathways of different organisms. This does not necessarily mean
that all enzymes showing hydrogenase activity are of mono-
phyletic descent. It has been noted, however, that the conservation of the polypeptide sequences among subunits with
shown or supposed equivalent functions is remarkable. We have
exploited this situation and have derived probes from conserved regions of the methylviologen reducing hydrogenase of
M. thermoautotrophicum (Fig. 7) taking into account the different codon usages of this bacterium and *M. voltae*.

MT	- - IVPRICGIC - 45 - HFYHLAAPD - -
DB	- - IVQRICGVC - 42 - HFYHLAALD - -
RC	- - FTERICGVC - 42 - HFYHLHALD - -
BJ	- - FTERICGVC - 42 - HFYHLHALD - -

Fig. 7.   Conserved amino acid sequences close to the N-
termini of the subunits believed to contain the Ni-
binding sites of hydrogenases of *M. thermoautotrophicum* (MT), *D. baculatus* (DB), *R. capsulatus*
(RC), *B. japonicum* (BJ) used to derive oligonucleotides for hybridization against *M. voltae* DNA
fragments. The data were taken from Reeve et al.[12].

In hybridization experiments with genomic restriction fragments we have found that these probes do hybridize to identical restriction fragments and also to identical phages of a genomic library, i.e. that they recognize homologous sequences. This indicates a high degree of homology since mispairing due to the use of wrong codons in the design of the probes must be expected and therefore only very few amino acid exchanges in the polypeptide sequences could be tolerated in order to still allow hybridization of the probes.

Genetic Mapping

Mapping of genes in methanogens has been a problem. Conjugation is not found, transduction has only recently been established for one strain[26], transformation efficiencies are low[27,28] and, worst of all, very few mutants are available. Their number is presently far surpassed by the number of known genes or gene probes, which allow physical mapping. We have therefore undertaken to construct a physical map of the complete *M. voltae* chromosome. This is feasible on the basis of restriction analysis of undegraded chromosomal DNA.

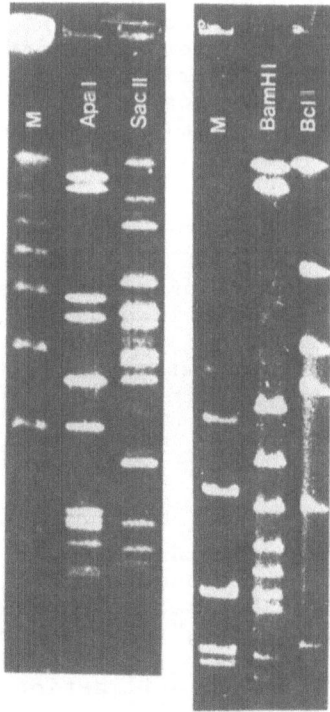

Fig. 8.   *M. voltae* chromosomal DNA digested with rarely cutting restriction enzymes, and separated by pulse field agarose gel electrophoresis. Oligomerized bacteriophage lambda DNA is shown as a marker (M). The fragments add up to 1880 kb.

Due to the very low GC contents of the *M. voltae* DNA, hexanucleotide recognition sequences consisting of G and C only are very rare, and only a few fragments are seen after treatment with such restriction endonucleases. In addition, restriction sites with the central GATC sequence also occur at a low frequency in *M. voltae* DNA (Fig. 8).

It is therefore possible to derive a circular restriction map of the chromosome, which is shown in a linearized form in Fig. 9. We have used gene probes to locate the respective genes on this map. They include a probe for the putative gene nit for nickel uptake. Since oligonucleotides cannot be used in such experiments due to the small amounts of DNA transferred in the large molecular weight range from

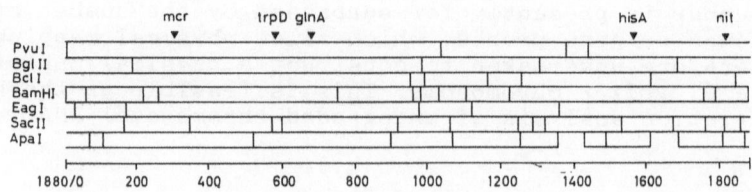

Fig. 9.   Physical and genetic map of the *M. voltae* genome.
The map was derived from experiments such as the one
exemplified in fig. 8 and hybridization of gene
probes to Southern blots of similar digests. Inter-
ruptions in the horizontal lines indicate that fragments of unknown order are found within the borders
of the adjacent restriction sites. glnA, glutamine
synthetase; hisA, phosphoribosyl(formimino)-5-amino-
1-phosphoribosyl-4-imidazolecarboxamide isomerase;
mcr, methylreductase; nit, "nickel uptake"; trpD,
anthranilate phosphoribosyl transferase.

the gels, we will have to wait for hydrogenase gene probes
before we can check on the linkage of these genes to either
the nickel uptake or the methyl reductase genes, which is
particularly interesting in view of the recent finding that
in *A. eutrophus* genetic determinants for a nickel specific
transport system are indeed linked to a hydrogenase gene
cluster[29].

References

1.   R. Cammack, D. O. Hall, and K. Rao, Hydrogenases:
     Structure and applications in hydrogen production, in:
     "Microbial gas metabolism: mechanistic, metabolic and
     biotechnological aspects", R. K. Poole and C. Dow,
     eds., Academic Press Inc., New York (1985).
2.   R. P. Hausinger, Nickel utilization by microorganisms,
     <u>Microbiol</u>. <u>Rev</u>. 51:22 (1987).
3.   S. P. J. Albracht, A. Kröger, J. W. van der Zwaan, G.
     Unden, R. Böcher, H. Mell, and R. D. Fontijn, Direct
     evidence for sulphur as a ligand to nickel in hydro-
     genase: an EPR study of the enzyme from *Wolinella
     succinogenes* enriched in ^{35}S, <u>Biochim</u>. <u>Biophys</u>. <u>Acta</u>
     874:116 (1986).

4. S. Yamazaki, A selenium-containing hydrogenase from *Methanococcus vannielii*, J. Biol. Chem. 257:7926 (1982).

5. M. Teixeira, G. Fauque, I. Moura, P. A. Lespinat, Y. Belier, B. Prickril, H. D. Peck, A. V. Xavier, J. LeGall, and J. J. G. Moura, Nickel-[iron-sulfur]-selenium-containing hydrogenases from *Desulfovibrio baculatus* (DSM 1743), Eur. J. Biochem. 167:47 (1987).

6. P. Boursier, F. J. Hanus, H. Papen, M. M. Becker, S. A. Russell, and H. J. Evans, Selenium increases hydrogenase expression in autotrophically cultured *Bradyrhizobium japonicum* and is a constituent of the purified enzyme, J. Bacteriol. 170:5594 (1988).

7. M. K. Eidsness, R. A. Scott, B. C. Prickril, D. V. DerVartanian, J. Legall, I. Moura, J. J. G. Moura, and H. D. Peck, Evidence for selenocysteine coordination to the active site nickel in the [NiFeSe]hydrogenases from *Desulfovibrio baculatus*, Proc. Natl. Acad. Sci. USA 86:147 (1989).

8. F. Zinoni, A. Birkmann, T. C. Stadtman, and A. Böck, Nucleotide sequence and expression of the selenocys-teine-containing polypeptide of formate dehydrogenase (formate-hydrogen-lyase-linked) from *Escherichia coli*, Proc. Natl. Acad. Sci. USA 83:4650 (1986).

9. P. E. Rouvière and R. S. Wolfe, Novel biochemistry of methanogenesis, J. Biol. Chem. 263:7913 (1988).

10. E.-G. Graf and R. K. Thauer, Hydrogenase from *Methano-bacterium thermoautotrophicum*, a nickel containing enzyme, FEBS Lett. 136:165 (1981)

11. F. S. Jacobson, L. Daniels, J. A. Fox, C. T. Walsh, and W. H. Orme-Johnson, Purification and properties of a 8-hydroxy-5-deazaflavin-reducing hydrogenase from *Methanobacterium thermoautotrophicum*, J. Biol. Chem. 257:3385 (1982).

12. J. N. Reeve, G. S. Beckler, D. S. Cram, P. T. Hamilton, J. W. Brown, J. A. Krzycki, A. K. Kolodziej, L. Alex, W. H. Orme-Johnson, and C. T. Walsh, A hydrogenase-linked gene in *Methanobacterium thermoautotrophicum* strain delta H encodes a polyferredoxin, Proc. Natl. Acad. Sci. USA 86:3031 (1989).

13. K. Schneider and H. G. Schlegel, Purification and properties of soluble hydrogenase from *Alcaligenes eutrophus* H16, Biochim. Biophys. Acta 452.66 (1976).

14. B. Schink and H. G. Schlegel, The membrane-bound hydro-genase of *Alcaligenes eutrophus*: I. Solubilization, purification and biochemical properties, Biochim. Biophys. Acta 567:315 (1979).

15. E. Muth, E. Mörschel, and A. Klein, Purification and characterization of an 8-hydroxy-5-deazaflavin-reducing hydrogenase from the archaebacterium *Methanococcus voltae*, Eur. J. Biochem. 169:571 (1988).

16. E. Muth, Localization of the F_{420}-reducing hydrogenase in *Methanococcus voltae* by immuno-gold technique, Arch. Microbiol. 150:205 (1988).

17. R. Ossmer, T. Mund, P. L. Hartzell, U. Konheiser, G. W. Kohring, A. Klein, R. S. Wolfe, G. Gottschalk, and F. Mayer, Immunocytochemical localization of component C of the methylreductase system in *Methanococcus voltae* and *Methanobacterium thermoautotrophicum*, Proc. Natl. Acad. Sci. USA 83:5789 (1986).

18. P. Vignais, Hydrogenase of *Rhodobacter capsulatus*, this volume.
19. M. Blaut and G. Gottschalk, Evidence for a chemiosmotic mechanism of ATP synthesis in methanogenic bacteria, Trends Biochem. Sci. 10:486 (1985).
20. V. Stewart, Nitrate respiration in relation to facultative metabolism in *Enterobacteria*, Microbiol. Rev. 52:190 (1988).
21. M. Webb, Interrelationships between the utilization of magnesium and the uptake of other bivalent cations by bacteria, Biochim. Biophys. Acta 22:428 (1977).
22. R. Waugh and D. H. Boxer, Pleiotropic hydrogenase mutants in *Escherichia coli* K12: growth in the presence of nickel can restore hydrogenase activity, Biochimie 68:157 (1986).
23. L. F. Wu and M.-A. Mandrand-Berthelot, Genetic and physiological characterization of new *Escherichia coli* mutants impaired in hydrogenase activity, Biochimie 68:167 (1986).
24. A. Chaudhuri and A. I. Krasna, Isolation of genes required for hydrogenase synthesis in *Escherichia coli*, J. Gen. Microbiol. 133:3289 (1987).
25. K. Stoker, L. F. Oltmann, and A. H. Stouthamer, Randomly induced *Escherichia coli* K-12 Tn5 insertion mutants defective in hydrogenase activity, J. Bacteriol. 171:831 (1989).
26. T. Leisinger, Approaches to gene transfer in methanogenic bacteria, this volume.
27. G. Bertani and L. Baresi, Genetic transformation in the methanogen *Methanococcus voltae* PS, J. Bacteriol. 169:2730 (1987).
28. V. E. Worrell, D. P. Nagle, D. McCarthy, and A. Eisen-braun, Genetic transformation system in the archae-bacterium *Methanobacterium thermoautotrophicum* Marburg, J. Bacteriol. 170:653 (1988).
29. G. Eberz, T. Eitinger, and B. Friedrich, Genetic deter-minants of a nickel-specific transport system are part of the plasmid-encoded hydrogenase gene cluster in *Alcaligenes eutrophus*, J. Bacteriol. 171:1340 (1989).

Supported by Deutsche Forschungsgemeinschaft, Fonds der Chemischen Industrie and Bundesministerium für Forschung und Technologie.

HYDROGENASE GENES IN DESULFOVIBRIO

Gerrit Voordouw

Division of Biochemistry
Department of Biological Sciences
The University of Calgary
Calgary, Alberta, T2N 1N4
Canada

INTRODUCTION

The objective of this mini-review is to discuss the different types of
hydrogenases that are found in sulfate-reducing bacteria from a molecular
biological perspective. Since molecular biology may mean different things to
different people, it will be defined for the purpose of this mini-review as
the field of science dealing with the (1) cloning, (2) sequence determination, and (3) expression of genes encoding hydrogenases. Progress in areas
(1) and (2) has been swift in the last four years and has led to the determination of the amino acid sequences of three different types of hydrogenases.
These were derived from the nucleotide sequences of the cloned structural
genes. Progress in area (3) has been slower: although it appears possible
to express hydrogenase structural genes from Desulfovibrio in a foreign host
such as Escherichia coli, no expression of enzyme activity has presently been
achieved. This indicates that the biosynthesis of a functional hydrogenase,
which always involves the insertion of redox prosthetic groups and usually
the export of the hydrogenase to the periplasmic space of the gram-negative
Desulfovibrio, is a complex process, which may require the presence of more
players than just the structural genes.

CLASSIFICATION OF DESULFOVIBRIO HYDROGENASES

The hydrogenases that have been isolated from species of the genus
Desulfovibrio fall in two different classes (Li et al., 1987; Lissolo et al.,
1986; Prickril et al., 1987). Members of the first class, the iron-only
hydrogenases ([Fe] hydrogenases), lack nickel and appear to contain only
iron-sulfur clusters as prosthetic groups. The structural genes for the
enzyme from D. vulgaris (Hildenborough) have been cloned and sequenced
(Voordouw et al., 1985; Voordouw and Brenner, 1985). Members of the second
class, the nickel containing hydrogenases, contain a single atom of nickel
per mole of enzyme as well as iron-sulfur clusters as prosthetic groups.
Some of these hydrogenases also contain selenium allowing a further
subdivision. This class does, therefore, contain two different types of
enzyme, the [NiFe] and [NiFeSe] hydrogenases. The structural genes of the
[NiFe] hydrogenase from D. gigas and the [NiFeSe] hydrogenase from D.
baculatus have been cloned and sequenced (Li et al., 1987; Menon et al.,

Table 1. Properties of the three types of hydrogenases from Desulfovibrio.

Property	D. vulgaris (Hildenborough) [Fe] hydrogenase	D. gigas [NiFe] hydrogenase	D. baculatus [NiFeSe] hydrogenase
M_r (pro-β, β)	13624, 9633	33983, 28435	34221, 30841
n (pro-β, β)	123, 89	314, 264	315, 283
M_r (α)	45820	61349	56797
n (α)	420	550	514
M_r ($\alpha + \beta$)	55453	89784	87638
Localization	Periplasm	Periplasm	Periplasm or cytoplasm
Nickel	0	1	1
Selenium	0	0	1
Nonheme iron	14–16	12	8
Fe_4S_4	2	2	2
Fe_xS_x	1	1	0
Specific activity[a]			
H_2 evolution	4800	440	466
H_2 utilization	50000	1500	120
$K_m(H_2)$, μM	30–300[b]	1[c]	–

[a] Values are in μmol H_2/min/mg protein, Fauque et al., 1987 and references quoted therein.

[b] For [Fe] hydrogenases I and II from Clostridium pasteurianum (Adams and Mortenson, 1984).

[c] For [NiFe] hydrogenase from Bradyrhizobium japonicum (Evans et al., 1987).

1987; Voordouw et al., 1989a). All three hydrogenases are two subunit enzymes and contain a small (β) and large (α) subunit. The molecular weights (M_r) in Daltons and number of amino acids per subunit polypeptide chain (n), derived from the nucleic acid sequences, are indicated in Table 1 for all three hydrogenases. Comparison of the translated nucleotide sequence of the small subunit gene and the NH_2-terminus determined for the mature protein has indicated that the small subunit is synthesized as a larger precursor protein (pro-β) from which the mature small subunit (β) is derived by proteolytic cleavage of an NH_2-terminal signal peptide for all three of these hydrogenases. The values of M_r and n of these precursor proteins have also been entered in Table 1. The metal ion contents in Table 1 have been rounded off to the nearest integer value (see references in Li et al., 1987 and Fauque et al., 1987). The assignment of the number and types of iron-sulfur clusters is generally derived from a combination of analytical chemistry and spectroscopic data and the conclusions of even the most exhaustive studies may not be definitive. All three hydrogenases contain two Fe_4S_4 clusters and one abnormal (x≠4) Fe_xS_x cluster. The article of Hagen et al. (1986) provides a good example of the experimental difficulties encountered in a precise determination of x, which was found to be x = 5–7 for the [Fe] hydrogenase from D. vulgaris (Hildenborough). A value x=3 is quoted for D. gigas [NiFe] hydrogenase, while a value x=0 has been indicated for [NiFeSe] hydrogenase meaning that a third non-standard cluster is thought to be lacking from this type of hydrogenase. Using the data of Table 1 with some caution, it is nevertheless clear that the biochemical properties of the [NiFe] and [NiFeSe]

hydrogenases are quite different. The assignment of these enzymes as two
members of the same class rests on the observation of a considerable sequence
homology between their respective α and β subunit sequences. The [Fe]
hydrogenase subunits are, on the contrary, not homologous with either [NiFe]
or [NiFeSe] hydrogenase justifying a separate class which is discussed in
detail below.

[Fe] HYDROGENASES

    The [Fe] hydrogenase from D. vulgaris (Hildenborough) was first charac-
terized in detail by Haschke and Campbell (1971). A systematic study of the
extractability of this hydrogenase by van der Westen et al. (1978), indicated
that it could easily be removed from D. vulgaris (Hildenborough) cells by
washing with Tris-EDTA buffers at pH 9. When properly conducted, the
procedure leads to removal of cytochrome c_3 and c_{553}, as well as [Fe]
hydrogenase from the D. vulgaris cells in the absence of cell lysis. This
strongly indicates a periplasmic location for [Fe] hydrogenase, since these
two c-type cytochromes are known to be located in the periplasm (Voordouw and
Brenner, 1986; van Rooijen et al., 1989; Pollock et al., 1989; LeGall and
Peck, 1987). Subsequent characterization of this periplasmic [Fe] hydrogenase indicated an enzyme consisting of a single polypeptide chain (M_r =
50000) and binding 3 Fe_4S_4 clusters (Mayhew and O'Connor, 1982). The enzyme
is highly active in both hydrogen uptake and hydrogen production, with the
uptake activity being approximately 10-fold higher than the production
activity (Table 1).

    The nucleic acid sequence of the structural gene for [Fe] hydrogenase
was interesting in two respects: a) It appeared that [Fe] hydrogenase is not
a single but a two subunit enzyme. The gene for the larger α subunit was
found to precede that for a smaller polypeptide, now referred to as the β
subunit. The α and β subunits are products of the hydA and hydB genes,
respectively. The designation α and β will be used throughout this mini-
review to indicate the large and small subunits of all types of hydrogenase.
In the case of the [NiFe] and [NiFeSe] hydrogenases, the nomenclature LS and
SS has been used (Menon et al., 1987; Li et al., 1987; Voordouw et al.,
1989a). Although this is more descriptive, it does not allow a simple
designation of new genes (e.g. hydC) and their products (γ). The structural
genes for the α and β subunits of [NiFe] hydrogenase could be referred to as
hynA and hynB, those of [NiFeSe] hydrogenase as hysA and hysB: different
names are necessary since some species contain several hydrogenases. The two
subunit nature of the [Fe] hydrogenase from D. vulgaris (Hildenborough) has
been confirmed by biochemical experiments (Voordouw et al., 1985; Hagen et
al., 1986) and the enzyme is now known to be a 1:1 complex of the two subunits. b) The NH_2-terminal sequence of the [Fe] hydrogenase α-subunit was
found to be homologous to bacterial 8Fe-8S ferredoxin (Voordouw and Brenner,
1985). This small electron transfer protein has been extensively characterized. Many 8Fe-8S ferredoxin sequences have been determined (see Bruschi and
Guerlesquin, 1988 for a review) and most of these have two groups of four
cysteine residues in their amino acid sequence in the characteristic pattern
C-I-X-C-X-X-C-X-X-X-C-P-X-X-A-I, where X is a variable amino acid. The
nature of the coordination of two Fe_4S_4 clusters by the two groups of
cysteines has been elucidated by the classic X-ray crystallographic work of
Adman et al. (1973). Two groups of four cysteine residues (C-35, C-38, C-41,
C-45 and C-66, C-69, C-72, C-76) were found at the NH_2-terminus of the
α-subunit of [Fe] hydrogenase in a sequence homologous to 8Fe-8S ferredoxin.
Comparison with a large number of ferredoxin sequences revealed the highest
homology with 8Fe-8S ferredoxin from D. desulfuricans Norway (Voordouw,
1987). It is, in view of this homology, likely that the NH_2-terminus is the
site of coordination of two Fe_4S_4 clusters in a structure resembling that of
8Fe-8S ferredoxin and this in turn allows a definition of the third Fe_xS_x

cluster, which is thought to be part of the H_2 binding active center. A value x=6 seems at present most probable, based on analysis of iron and acid labile sulfur contents of many preparations (Hagen et al., 1986). However, this value is not uniformly accepted and other workers represent the active site cluster as Fe_4S_4 (Li et al., 1987). Whatever the precise value of x in the active site cluster, the 10 remaining cysteine residues present in the COOH-terminal part of the α-subunit should be sufficient for its coordination. The mature β subunit of [Fe] hydrogenase does not contain cysteines and can thus not participate in covalent iron-sulfur cluster coordination.

It was not clear from the sequence how [Fe] hydrogenase is exported to the periplasm since an NH_2-terminal signal sequence was found to be absent from the α subunit (Voordouw et al., 1985; Voordouw and Brenner, 1985). Prickril et al. (1986) determined the NH_2-terminal sequence of the β subunit and obtained evidence for the presence of a complex 34 amino acid signal sequence. Thus, one of the roles of the small subunit could be in the export of hydrogenase to the periplasm of Desulfovibrio. However, this cannot be its only role since after export and processing a mature β subunit of 89 amino acid residues remains tightly associated with α. Although, as indicated above, all cysteines participating in the coordination of the three iron-sulfur clusters are present in the α subunit sequence, there must be some amino acid residues essential for activity in β, since the dissociation of the two subunits with retention of activity has not been achieved to date. The problem of cluster insertion and export of [Fe] hydrogenase has been studied by examining the fate of α and β subunits when expressed from recombinant plasmids in E. coli (Voordouw et al., 1987a; Voordouw, 1987; van Dongen et al., 1988). Both polypeptides were expressed in E. coli transformed with plasmid pHV150, which contains a 1.9 kb insert with just the hydA,B genes. Breaking the E. coli cells by French press treatment and purification resulted in the isolation of an αβ dimer. It was shown by SDS gel electrophoresis that the molecular weight of the β subunit in purified recombinant hydrogenase was identical to that in the enzyme isolated from Desulfovibrio, but it was erroneously stated (Voordouw et al., 1987a) that M_r=13.5 kDa, since this work was published prior to the finding by Prickril et al. (1986) of processing of the small subunit. The molecular weight of the mature small subunit present in fully functional [Fe] hydrogenase isolated from Desulfovibrio is now known to be M_r=9.6 kDa (Table 1). It appears that an αβ dimer in which the β subunit is properly processed (van Dongen et al., 1988) can be isolated from E. coli (pHV150) and it was shown by metal analyses and ESR spectroscopy that the ferredoxin clusters are at least partially inserted in this E. coli product. However, the active site cluster is not incorporated causing the recombinant αβ-dimer to be inactive. One possible interpretation of this result is that a specific gene product is required for the assembly of the active site Fe_xS_x cluster and that E. coli is not capable of assembling a functional periplasmic [Fe] hydrogenase because it lacks the gene for this specific 'insertase'. This interpretation is in analogy to the assembly of the nitrogenase Fe- (product of the nifH gene) and the MoFe-proteins (encoded by nifD,K) which require the products of either nifM or nifE,N for activation (Howard et al., 1986; Brigle et al., 1987). Interestingly, the nifE,N gene products were found to be homologous with those of the structural genes nifD,K. Recent work (Voordouw, 1987; van Dongen et al., 1988) indicates that the fraction of hydrogenase that can be isolated from E. coli as a soluble, inactive αβ dimer is small. The α subunit synthesized in E. coli appears to be mainly present in aggregated form in the cytoplasm. The β subunit is found predominantly in the unprocessed 13.5 kDa form and this pro-β precursor appears bound to the inner membrane presumably via the signal sequence. When the small subunit is expressed in the absence of the large subunit, it is not exported to the periplasm and vice versa (van Dongen et al. 1988). Thus E. coli, although an imperfect host for the production of a functional periplasmic [Fe] hydrogenase, is nevertheless useful in elucidating the biogenesis pathway of this enzyme. The picture that emerges from the E. coli studies is that, following synthesis of α and pro-β, pro-β becomes membrane bound but is not processed or exported. The α-subunit is acted upon by the cluster insertion machinery, folds and associates with the membrane-bound pro-β. Formation of the (α, pro-β) complex at the inner face of the cytoplasmic membrane leads to formation of a pore that allows passage and processing to a periplasmic αβ dimer. It is especially this latter aspect of the assembly mechanism that needs to be better defined. Bacterial proteins are generally thought to be excreted co-translationally and to fold while they emerge in the periplasm. Such a secretion mechanism is clearly incompatible with the present proposal. However, the observation of a lack of export and processing of pro-β when synthesized in the absence of expression of the hydA gene (van Dongen et al., 1988) seems to rule out the possibility of independent export of the α and pro-β polypeptides. The observation that an inactive αβ dimer can be isolated from E. coli indicates that an (α-proβ) hydrogenase lacking the active site Fe_xS_x cluster can be exported. The biosynthesis of a fully functional [Fe] hydrogenase in D. vulgaris (Hildenborough) may require, therefore, a perfect timing between Fe_xS_x active site cluster insertion activity and translocation to prevent the export of inactive enzyme. It could be demonstrated recently, perhaps for this reason, that even in D. vulgaris (Hildenborough) the biosynthesis of active hydrogenase is not easily increased. Extra copies of the [Fe] hydrogenase structural genes, cloned in a broad host range vector, were introduced in D. vulgaris, following the development of a system to genetically conjugate D. vulgaris (Hildenborough) and E. coli (van den Berg et al., 1989). However, the level of hydrogenase activity increased only marginally (1.5-fold), although the amount of subunits synthesized in the resulting D. vulgaris transconjugants was shown to increase up to 10-fold. This could mean that active site cluster insertion activity is rate-limiting in these transconjugants. It has not yet been shown by purification and determination of the enzyme's specific activity that the inactive hydrogenase is located in the Desulfovibrio periplasm. The development of a genetic transfer system for D. vulgaris (Hildenborough) will no doubt allow a much more detailed analysis of the biosynthesis of [Fe] hydrogenase in Desulfovibrio and the experiments by van den Berg et al. (1989) open many new avenues towards a solution of this problem.

An interesting gene (hydC) was reported recently immediately downstream from the hydA,B genes of D. vulgaris (Hildenborough) by Stokkermans et al. (1989). HydC appears to encode a protein product (γ) of 65.8 kDa. Expression of this reading frame (which has yet to be demonstrated in D. vulgaris) requires transcription in a direction opposite to that of the hydA,B genes. The putative hydC gene product, γ, is homologous with the α and β subunits of [Fe] hydrogenase and the hydC gene can be regarded as an in frame fusion of the hydA,B genes. This observation raises interesting questions on the evolution of the hydA,B genes in Desulfovibrio. Stokkermans et al. (1989) discussed and investigated the possible functions of γ which could be an alternative hydrogenase or a helper protein in the assembly of the regular periplasmic [Fe] hydrogenase (e.g. compare with nifE,N and nifD,K discussed above), by expressing either the hydC gene alone or a combination of the hydA,B and hydC genes in E. coli. Neither leads to the expression of hydrogenase activity in E. coli and no firm conclusions can therefore be drawn from these experiments. The conclusion by Voordouw et al. (1989b) that D. vulgaris subsp. oxamicus (Monticello), which has a hydA,B operon encoding [Fe] hydrogenase that shows 70-80% amino acid sequence identity to the enzyme from D. vulgaris (Hildenborough) appears to lack a hydC gene, seems to preclude an essential function of γ in the assembly of periplasmic [Fe] hydrogenase in Desulfovibrio. This conclusion was based on the observation that partial sequencing of the region downstream from the Monticello hydA,B operon, as well as hybridization of Monticello DNA with a Hildenborough hydC probe, did not reveal hydC homologous sequences either downstream from the hydA,B genes or elsewhere in the Monticello genome. Further experiments are thus required in order to delineate the possible function of hydC. The availability of a conjugation system (van den Berg et al., 1989) allows in principle the construction of deletions in the chromosome of D. vulgaris (Hildenborough). Construction of a hydC-deletion mutant appears as the most straightforward way of settling the question whether γ has an essential role in either the assembly of [Fe] hydrogenase or the hydrogen metabolism of D. vulgaris (Hildenborough).

Comparison of primary structures of proteins isolated from different sources allows one to define regions that are conserved or variable. In the case of [Fe] hydrogenase we can compare the sequence obtained for the enzyme from D. vulgaris (Hildenborough), with that from D. vulgaris subsp. oxamicus (Monticello), and the sequence for the hydC gene product. The [Fe] hydrogenases are highly homologous: the two α-subunits share 79%, while the unprocessed pro-β polypeptides share 71% sequence identity, respectively (Voordouw et al., 1989b). The two pro-β signal sequences are 34 amino acid residues long and highly homologous. This high degree of sequence homology makes it difficult to identify conserved regions. As an example, of the 18 cysteine residues present in the Hildenborough α-subunit, 17 have been conserved in the Monticello sequence. The exception is C-102, which is replaced by a valine residue in the Monticello sequence. We can thus conclude that C-102 is unlikely to coordinate to the active site cluster but cannot draw conclusions for the other 9 conserved cysteines (C-142, C-178, C-179, C-200, C-234, C-360, C-378, C-382, C-384) in the COOH-terminal region of the α-subunit. The eight cysteine residues in the NH_2-terminal region, which are thought to coordinate the ferredoxin clusters (C-35, C-38, C-41, C-45 and C-66, C-69, C-72, C-76) are, as expected, conserved in the two sequences. The value of the hydC-sequence is that the derived protein sequence for γ is far less homologous to α and β than the pairs of α and β sequences determined for the two periplasmic hydrogenases. Comparison of three domains of the Hildenborough (α, β) and γ sequences (the numbering corresponds to that in α and β) gives the following results (Stokkermans et al., 1989): Domain I, residues 1-105 of the α-subunit containing the 8Fe-8S ferredoxin homologous region, shows a sequence identity of 19% (20 of 106 residues, including the 8 'ferredoxin' cysteines). Domain II, residues 106-420 of the α-subunit containing the cysteine residues coordinating the active site cluster, has a sequence identity of 47% (147 of 315 residues). Importantly, only 5 cysteine residues (C-178, C-179, C-234, C-382 and C-384) are conserved. Domain III, residues 35-123 of the β-subunit, is 21% identical (19 of 89 residues). Comparison of the hydA,B and hydC derived amino acid sequences thus indicates that there is a total of 13 conserved cysteine residues in [Fe] hydrogenase. Eight of these are located in domain I and coordinate the two ferredoxin-clusters, while the remaining five in domain II could serve as ligands to the active site cluster. As discussed elsewhere (Voordouw, 1987), the hydA gene could have arisen by a gene fusion event. The discovery of hydC allows this hypothesis to be expanded and the discussion of the [Fe] hydrogenase genes in Desulfovibrio is concluded by considering the following path for their evolution:
a) A gene encoding a hydrogen-binding polypeptide (domains II, III) fused with that encoding an 8Fe-8S ferredoxin to form the hydC-precursor gene. The polypeptide encoded by this gene fusion may have functioned as a bidirectional, cytoplasmic [Fe] hydrogenase with 3 iron-sulfur clusters in a single polypeptide chain of 60-70 kDa.
b) Gene duplication of the hydC precursor gene into an inverted repeat sequence allowed an independent evolution of the two copies. Mutations in the electron transfer domain (I) were less critical than mutations in the hydrogen-binding domain (II, III), which remained more conserved.
c) Evolution of one of the copies into the genes for a periplasmic [Fe] hydrogenase required the insertion of a signal sequence. This sequence was inserted internally and two separate subunits (α and pro-β) evolved.
d) The second copy evolved into the present hydC sequence in D. vulgaris

(Hildenborough) but was deleted from D. vulgaris subsp. oxamicus (Monticello). The present function of the hydC gene in the Hildenborough strain is
obscure and it may be in the process of being deleted.

[NiFe] AND [NiFeSe] HYDROGENASES.

The presence and importance of nickel in this class of hydrogenases is a
relatively recent finding and was not realized when these enzymes were first
isolated from Desulfovibrio. In an extensive characterization of the hydrogenase from D. gigas, Hatchikian et al. (1978) describe it as a two subunit
enzyme (M_r = 62000 and M_r = 26000) of molecular weight 89,500 with three
Fe_4S_4 clusters. As shown in Table 1, these results provided an essentially
complete and correct picture of the enzyme except for the now recognized
essential presence of nickel. The amino acid composition of the D. gigas
enzyme was found to be similar to that of the enzyme from D. vulgaris
Miyazaki F (Hatchikian et al., 1978), which is also known to be a two subunit
enzyme (M_r = 59000 and M_r = 28000; Yagi et al. 1976). A preliminary
crystallographic study of the Miyazaki enzyme has been reported (Higuchi et
al., 1987). The enzyme from D. gigas is periplasmic while the enzyme from D.
vulgaris (Miyazaki) is thought to be membrane-bound, requiring trypsin for
solubilization. Another difference between the two enzymes is that following
its discovery in hydrogenases from Methanobacterium thermoautotrophicum (Graf
and Thauer, 1981) the presence of nickel in the D. gigas enzyme was reported
by several workers (Cammack et al., 1982, Le Gall et al., 1982; Moura et al.,
1982) but it has not been found in stoichiometric amounts in the enzyme
isolated from D. vulgaris Miyazaki F (T. Yagi, personal communication) which
is therefore referred to as an [Fe] hydrogenase. Since the properties
of the Miyazaki enzyme resemble those of the D. gigas enzyme, it seems
appropriate to include it together with the other [NiFe] hydrogenases. The
question whether this enzyme is fundamentally different will be definitively
settled when amino acid sequence and structural information allow a more
extensive comparison with [NiFe] hydrogenases isolated from D. gigas and
other sources.

It is now realized that [NiFe] hydrogenases are relatively common in
gram-negative bacteria and the enzyme has been shown to be present in
Azotobacter (Seefeldt and Arp, 1986; Yates and Robson, 1985), Bradyrhizobium
japonicum (Sayavedra-Soto et al., 1988), Escherichia coli (Ballantine and
Boxer, 1985; Sawers and Boxer, 1986) as well as gram-negative photosynthetic
bacteria such as Rhodobacter capsulatus (Leclerc et al., 1988). The above
list is by no means exhaustive. Homology between [NiFe] hydrogenases from
these various sources has now been shown by cloning and sequencing of the
structural genes as discussed below. Like the enzyme from D. vulgaris
Miyazaki F, the [NiFe] hydrogenases from the sources listed above are found
to be membrane-bound, requiring trypsin or detergent treatment for
solubilization (Sawers and Boxer, 1986; Colbeau and Vignais, 1983; Seefeldt
and Arp, 1986).

The purification of a hydrogenase containing selenium, in addition to
nickel, from Desulfovibrio desulfuricans (strain Norway 4) was reported by
Rieder et al. (1984). This [NiFeSe] hydrogenase was shown to consist of two
subunits (M_r = 56 and 29 kDa) and have properties different from those of a
membrane-bound [NiFe] hydrogenase (subunit M_r = 60 and 27 kDa) purified from
the same organism. Isolation and characterization of a [NiFeSe] hydrogenase
was also reported for D. baculatus (Texeira et al., 1987) and it is now known
that some Desulfovibrio's such as D. vulgaris (Hildenborough) contain all
three hydrogenases: the [Fe], [NiFe] and [NiFeSe] hydrogenase (Lissolo et
al., 1986; Prickril et al., 1987). The localization of the [NiFeSe]
hydrogenase is not entirely clear. It was described as a soluble cytoplasmic
enzyme in D. desulfuricans by Rieder et al. (1984), while the [NiFeSe]

hydrogenases of D. baculatus and D. salexigens have been indicated as periplasmic/cytoplasmic (Li et al., 1987) and periplasmic (Fauque et al., 1987), respectively.

The cloning and sequencing of the structural genes for the [NiFe] hydrogenase from D. gigas and the [NiFeSe] hydrogenase from D. baculatus were reported by Li et al. (1987) and Menon et al. (1987). These hydrogenases were found to be encoded by an operon in which the β-subunit gene precedes that for the α-subunit, both genes being in close proximity. The arrangement of the two genes is the reverse of that in the [Fe] hydrogenase operon (Voordouw and Brenner, 1985). The sequences as originally published by Li et al. (1987) and Menon et al. (1987) did contain a number of errors leading to reading frame shifts and to incorrect assignments of the amino acid sequence of the α and β subunits of D. gigas [NiFe] hydrogenase and the α subunit of D. baculatus [NiFeSe] hydrogenase. This prevented a proper comparison of their amino acid sequences. Alignment with the sequences of the [NiFe] hydrogenases from Bradyrhizobium japonicum and Rhodobacter capsulatus that were subsequently published (Leclerc et al., 1988; Sayavedra-Soto et al., 1988) were, for this reason, also only partially relevant. The nucleotide sequences for the two Desulfovibrio hydrogenase operons were recently corrected, following detection of reading frame shifts by the codon probability method of Staden and McLachlan (1982) and renewed nucleotide sequencing. The corrected sequences of the two α and β subunits were found to share a significant degree of sequence homology (Voordouw et al., 1989a). Following alignment, the pairs of small and large subunits were found to share 38% and 34% sequence identity, respectively. The mature β-subunits were found to be preceded by a complex NH_2-terminal signal sequence (32 amino acid residues for the [NiFeSe] and 50 amino acid residues for the [NiFe] hydrogenase), just as for [Fe] hydrogenase while the α-subunits were found to lack an NH_2-terminal signal sequence. Although, as indicated above, there is at present no biochemical evidence to support the export of [NiFeSe] hydrogenase in all cases, the [NiFe] enzyme from D. gigas is considered to be periplasmic and one must assume that the 50 amino acid pro-β signal sequence functions in hydrogenase export, possibly by a mechanism that is similar to that discussed for [Fe] hydrogenase. It is encouraging in this respect that, despite a lack of sequence homology between the [Fe] and the [NiFe]/[NiFeSe] hydrogenases, alignment of the signal sequences indicates the presence of a consensus box (R-R-X-F-X-K, where X is a variable amino acid residue). This box is also present in the pro-β sequence of the [NiFe] hydrogenases from Bradyrhizobium japonicum (Sayavedra et al., 1988) and Rhodobacter capsulatus (Leclerc et al., 1988) and is likely to represent a conserved feature in the mechanism of hydrogenase assembly and export. It is unusual for signal sequences of bacterial exported proteins to show a conserved element and such a feature has not been found in the signal sequences of other periplasmic proteins (Benson et al., 1985). A possible mechanism for the export of [NiFe] hydrogenase will be discussed below, following a focus on the sequence homology of small and large subunits in the class of nickel containing hydrogenases.

Alignment of the corrected sequences shows that of the 12 cysteine residues present in the mature β-subunits of D. gigas [NiFe] and D. baculatus [NiFeSe] hydrogenase, 10 have been conserved. These are (numbering of the unprocessed D. gigas β subunit; Voordouw et al., 1989a) C-67, C-70, C-162, C-198, C-238, C-263, C-269, C-278, C-296, C-299. Inspection of the [NiFe] hydrogenase sequences from Bradyrhizobium japonicum and Rhodobacter capsulatus also indicates these cysteine residues to be conserved. Surprisingly, the large subunit contains fewer conserved cysteines which are found (numbering for the D. gigas α-subunit) in two groups of two residues at the NH_2-terminus (C-65, C-68) and the COOH-terminus (C-530 and C-533) of the α-subunit. Comparison of the nucleic acid sequences encoding the COOH-terminus of the α-subunits for the [NiFe] hydrogenase from D. gigas and the

[NiFeSe] hydrogenase from D. baculatus indicated that the amino acid residue corresponding to C-530 is encoded by a TGA (stop) codon in [NiFeSe] hydrogenase, rather than the TGC codon for C-530 (Voordouw et al., 1989a). There is well documented evidence in several other selenium-containing proteins that the selenium is present as selenocysteine and that the position of the selenocysteine in the polypeptide chain corresponds to the presence of a TGA codon in the gene. Examples include Escherichia coli formate dehydrogenase (Zinoni et al., 1986; Shuber et al., 1986) and mammalian glutathione peroxidase (Chambers et al., 1986; Sukenaga et al., 1987). The mechanism for selenocysteine incorporation will not be reviewed here.

A critical base, determining whether a nickel containing hydrogenase is of the [NiFe] or [NiFeSe] type, has thus been identified. Although the biochemical differences between [NiFe] and [NiFeSe] hydrogenase may not be caused by this single base change alone, it would nevertheless be interesting to change the TGA codon of a [NiFeSe] hydrogenase gene to TGC or the TGC codon of a [NiFe] hydrogenase gene to TGA and investigate the enzymatic properties of the resulting mutant hydrogenases. These experiments are feasible now that conjugational gene transfer from E. coli to Desulfovibrio has been achieved (van den Berg et al., 1989). It has been demonstrated by spectroscopic studies (Eidsness et al., 1989 and He et al., 1989) that the selenocysteine serves as a ligand to the nickel in the [NiFeSe] hydrogenase of D. baculatus and it is reasonable to expect that the homologous C-530 serves as a nickel ligand in the [NiFe] hydrogenase of D. gigas. The four conserved cysteines (C-65, C-68, C-530 and C-533) of the α-subunit are again also conserved in the [NiFe] hydrogenases from Bradyrhizobium japonicum and Rhodobacter capsulatus. In summary, there appear to be 14 conserved cysteine residues in the α and β subunits of [NiFe] hydrogenases. One of these (C-530 of α) is a ligand to the nickel leaving 13 cysteine residues for the coordination of iron-sulfur clusters, which is interestingly the same number as in [Fe] hydrogenase. In [Fe] hydrogenases, all of these conserved cysteines are present in the larger α-subunit. However, in the nickel containing hydrogenases 10 conserved cysteines are found in the small, while only 3 are present in the large subunit. The β-subunit of [NiFe]/[NiFeSe] hydrogenases must therefore contribute to the coordination of the two Fe_4S_4 and the single active site Fe_xS_x cluster (Table 1). It is not possible to assign the cysteine residues, which coordinate the two electron transferring Fe_4S_4 clusters, since an 8Fe-8S ferredoxin homologous pattern (C-X-X-C-X-X-C-X-X-X-C as in [Fe] hydrogenase) is not found in the nickel containing hydrogenases. It is the author's guess that the β subunit coordinates the two Fe_4S_4 clusters, for which 8 cysteine residues are required. This leaves 2 β together with 3 α subunit cysteine residues for the coordination of the Fe_xS_x cluster. This active site cluster, which is probably in close proximity to the nickel, would then provide a crosslink between the two subunits (Seefeldt and Arp, 1987).

The assembly and export of [NiFe] hydrogenase could, in analogy to the path discussed for [Fe] hydrogenase, be achieved by: (a) synthesis of the small subunit and its association to the inner face of the cytoplasmic membrane via the pro-β signal sequence, (b) insertion of two Fe_4S_4 clusters in the pro-β polypeptide, (c) synthesis of the large subunit and insertion of nickel, (d) insertion of the active site cluster at the interface of the two subunits, and (e) translocation of the [NiFe] hydrogenase across the cytoplasmic membrane and processing of the pro-β signal sequence by the signal peptidase. Support for a cytoplasmic location of step (c) is provided by the observation that some of the Escherichia coli mutants that are defective in hydrogenase synthesis appear to be impaired in nickel import (Boxer, 1988). Expression of the genes for [NiFe] or [NiFeSe] hydrogenase from D. gigas and D. baculatus has been demonstrated by maxicell experiments (Li et al., 1987) and Western blotting (Menon et al., 1987). As in the case of [Fe] hydrogenase expression in E. coli, this has not yet led to the expression of a functional enzyme. The pro-β subunit of [NiFeSe] hydrogenase was not processed upon synthesis in E. coli (Menon et al., 1987). The inability of E. coli to synthesize a functional [NiFe] hydrogenase is more puzzling than its failure to synthesize an active [Fe] hydrogenase, since E. coli synthesizes its own [NiFe] hydrogenase which is homologous to the D. gigas enzyme, as shown by nucleic acid hybridization (Li et al., 1987). Periplasmic [Fe] hydrogenase is, on the other hand, restricted to Desulfovibrio and its synthesis in functional form may require Desulfovibrio specific factors. The observation by van den Berg et al. (1989) that increasing the dosage of [Fe] hydrogenase genes in Desulfovibrio leads to increased subunit synthesis but not increased hydrogenase activity indicates that it is perhaps naive to expect that introduction of foreign [NiFe] hydrogenase genes into E. coli will lead to their functional expression. Successful expression of D. gigas [NiFe] hydrogenase genes in E. coli may require more elaborate genetic constructions in which the E. coli structural genes (but not their upstream and/or downstream regulatory regions) are swapped for those of D. gigas to ensure proper gene dosage and regulation of expression. Finally, it must be noted that the β-subunits of [NiFe] hydro-genase from Badyrizobium japonicum and Rhodobacter capsulatus have a COOH terminal extension of 40-50 amino acid residues, when compared with the β-subunit sequences of the [NiFe] and [NiFeSe] hydrogenases from Desulfo-vibrio (Sayavedro-Soto et al., 1988). This COOH-terminal sequence is hydro-phobic and it has been proposed to contribute to the binding of these hydrogenases to the membrane (Sayavedra-Soto et al., 1988). The difference may originate from the need of the D. gigas enzyme to interact with a non-membrane bound periplasmic electron carrier (cytochrome c_3), while the B. japonicum enzyme may transfer electrons to an as yet unidentified electron carrier that is present as an integral membrane component. The difference between a periplasmic and membrane bound [NiFe] hydrogenase could thus be relatively slight and be dictated by functional electron transport chain requirements. The αβ dimer is largely present in the periplasmic space with hydrophobic sequences in the β-subunit providing a firm anchor to the outer face of the cytoplasmic membrane in some but not all cases. Although this unifying view of the cellular localization of [NiFe] hydrogenases in gram-negative bacteria is attractive, it must be noted that the enzyme from Rhodobacter capsulatus is exceptional since it is thought to protrude into the cytoplasm (Colbeau et al., 1983).

STRUCTURE AND FUNCTION OF HYDROGENASES IN DESULFOVIBRIO

The application of the techniques of molecular biology to the hydro-genase genes in Desulfovibrio has considerably increased our understanding of the structure and function of these enzymes in this genus. The use of gene probes for the three hydrogenase types that have thus far been found in Desulfovibrio allows a rapid identification of their genes in known or newly isolated species. Using this approach, it was shown that [Fe] hydrogenase genes are present in 7 of 16 species of Desulfovibrio (Voordouw et al., 1987b). The genes for [NiFe] hydrogenase have been found by Southern blotting using a D. gigas probe in D. vulgaris (Hildenborough), D. desulfur-icans (Norway 4), D. desulfuricans 27774 and D. baculatus, as well as in E. coli (Li et al., 1987). It was found recently (Voordouw, unpublished) that 15 of the 16 species, that were probed with the [Fe] hydrogenase genes in the earlier study (Voordouw et al., 1987b), contain the [NiFe] hydrogenase genes. The only exception was D. thermophilus but this organism has recently been reclassified and does now no longer reside in the genus Desulfovibrio. Thus, all Desulfovibrio's appear to contain a periplasmic [NiFe] hydrogenase, while some contain in addition the periplasmic [Fe] hydrogenase. Speculation on the physiological role of these two uptake hydrogenases, which could explain this distribution, is hampered by the fact that comparative studies of the value for the K_m for H_2 have not been reported for Desulfovibrio hydrogenases. A low K_m of 1 μM has been reported for [NiFe] hydrogenase from Bradyrhizobium japonicum relative to the values reported for the [Fe] hydrogenases from Clostridium pasteurianum (Table 1). Concentrations of dissolved hydrogen in the environment can be expected to range from 0–100 μM. Assuming K_m (H$_2$) values of the order of 1 μM for Desulfovibrio [NiFe] hydrogenase and 100 μM for [Fe] hydrogenase leads to the suggestion that every Desulfovibrio has a low activity, high affinity enzyme which allows it to compete for hydrogen with other Desulfovibrio species and other genera of bacteria in environments in which hydrogen is scarce. Approximately half of the Desulfovibrio strains investigated (Voordouw et al., 1987b) have a high activity, low affinity enzyme, which allows more efficient hydrogen harvesting in environments where this substrate is abundant. One expects that deletion of the [Fe] hydrogenase genes from these strains would not lead to serious disruption of their metabolism, if the above hypothesis on the function of the two types of hydrogenase in Desulfovibrio is correct.

Further progress in our understanding of these uptake hydrogenases will depend to a large extent on a successful determination of the three-dimensional structures of [Fe] and [NiFe] hydrogenase. The lack of sequence homology between the two types of enzyme and the different subunit molecular weights point to two quite different structures. However, the observation, that the same number of 13 residues has been conserved in both enzymes, indicates the possibility of conservation of structural elements. These could include the spatial relationships between the two electron transferring Fe_4S_4 and the active site Fe_xS_x cluster. The structural design of [NiFe] hydrogenase could well have the same basic architecture as that of the simpler [Fe] hydrogenase. The larger size of the two subunits and the introduction of nickel may have served to perfectionate the structure in terms of hydrogen affinity at the expense of hydrogen turnover. These remarks should not be interpreted as a suggestion that the [Fe] hydrogenase has been around longer than the [NiFe] enzyme and that the latter evolved from the former, since there are no homologies between the two sequences to support such an evolutionary path. It is more likely that the two enzymes evolved independently and that their distribution in species of gram-negative bacteria may have changed with a change in the average concentration of hydrogen in the atmosphere. This concentration is likely to be smaller now than in times past and as a result the distribution of periplasmic [Fe] hydrogenase may have become more restricted.

The [NiFeSe] hydrogenase has not been included in the above discussion, since several basic facts with respect to its structure and function in Desulfovibrio must first be clarified. These include the function of the pro-β signal sequence in its localization and the suggested absence of an Fe_xS_x cluster (Table 1), which seems hard to reconcile with the observation of structural homology with [NiFe] hydrogenase.

Acknowledgements. The author's research is supported by the Natural Sciences and Engineering Research Council of Canada.

REFERENCES

Adams, M.W.W., and Mortenson, L.E., 1984, The physical and catalytic properties of hydrogenase II of Clostridium pasteurianum, J. Biol. Chem., 259: 7045-7055.
Adman, E.T., Sieker, L.C., and Jensen, L.H., 1973, The structure of a bacterial ferredoxin, J. Biol. Chem., 248: 3987-3996.
Ballantine, S.P., and Boxer, D.H., 1985, Nickel-containing hydrogenase isoenzymes from anaerobically grown Escherichia coli K-12, J. Bacteriol., 163: 454-459.

Benson, S.A., Hall, M.N., and Silhavy, T.J., 1985, Genetic analysis of protein export in Escherichia coli K12, Annu. Rev. Biochem., 54: 101–134.

Brigle, K.E., Weis, M.C., Newton, W.E., and Dean, D.R., 1987, Products of the iron-molybdenum cofactor-specific biosynthetic genes, nifE and nifN, are structurally homologous to the products of the nitrogenase molybdenum-iron protein genes, nifD and nifK, J. Bacteriol., 169: 1547–1553.

Bruschi, M., and Guerlesquin, F., 1988, Structure, function and evolution of bacterial ferredoxins, FEMS Microbiol. Rev., 54: 155–176.

Boxer, D.H., 1988, Nickel and hydrogen metabolism in Escherichia coli, Abstract 2nd International Symposium on the Molecular Biology of Hydrogenase, Unico State Park, Georgia.

Cammack, R., Patil, D., Aguirre, R., and Hatchikian, E.C., 1982, Redox properties of the ESR-detectable nickel in hydrogenase from Desulfovibrio gigas, FEBS Lett., 142: 289–292.

Chambers, I., Frampton, J., Goldfarb, P., Affara, N., McBain, W., and Harrison, P.R., 1986, The structure of the mouse glutathione peroxidase gene: the selenocysteine in the active site is encoded by the "termination" codon, TGA, EMBO J., 5: 1221–1227.

Colbeau, A., Chabert, J., and Vignais, P.M., 1983, Purification, molecular properties and localization in the membrane of the hydrogenase of Rhodopseudomonas capsulata, Biochim. Biophys. Acta, 748: 116–127.

Colbeau, A., and Vignais, P.M., 1983, The membrane-bound hydrogenase of Rhodopseudomonas capsulata is inducible and contains nickel, Biochim. Biophys. Acta, 748: 128–138.

Eidsness, M.K., Scott, R.A., Prickril, B., DerVartanian, D.V., LeGall, J., Moura, I., Moura, J.J.G., and Peck Jr., H.D., 1989, Evidence for selenocysteine coordination to the active site nickel in the [NiFeSe] hydrogenase from Desulfovibrio baculatus, Proc. Natl. Acad. Sci. USA, 86: 147–151.

Fauque, G.D., Berlier, Y.M., Czechowski, M.H., Dimon, B., Lespinat, P.A., and LeGall, J., 1987, A proton-deuterium exchange study of three types of Desulfovibrio hydrogenases, J. Industrial Microbiol., 2: 15–23.

Evans, H.J., Harker, A.R., Papen, H., Russell, S.A., Hanus, F.J., and Zuber, M., 1987, Physiology, biochemistry and genetics of the uptake hydrogenase in Rhizobium, Annu. Rev. Microbiol., 41: 335–361.

Graf, E.G., and Thauer, R.K., 1981. Hydrogenase from Methanobacterium thermoautotrophicum, a nickel-containing enzyme, FEBS Lett., 136: 165–169.

Hagen, W.R., van Berkel-Arts, A., Kruse-Wolters, K.M., Voordouw, G., and Veeger, C., 1986, The iron-sulfur composition of the active site of hydrogenase from Desulfovibrio vulgaris (Hildenborough) deduced from its subunit structure and total iron-sulfur content, FEBS Lett., 203: 59–63.

Haschke, R.H., and Campbell, L.L., 1971, Purification and properties of a hydrogenase from Desulfovibrio vulgaris, J. Bacteriol., 105: 249–258.

Hatchikian, E.C., Bruschi, M., and LeGall, J., 1978, Characterization of the periplasmic hydrogenase from Desulfovibrio gigas, Biochim. Biophys. Res. Commun., 82: 451–461.

He, S.-H., Texeira, M., LeGall, J., Patil, D.S., DerVartanian, D.V., Huyn, B.H., and Peck Jr., H.D., 1989, EPR studies with [77]Se enriched [NiFeSe] hydrogenase of Desulfovibrio baculatus. Evidence for a selenium ligand to the active-site nickel, J. Biol. Chem., 264: 2678–2682.

Higuchi, Y., Yasuoka, N., Kakudo, M., Katsube, Y., Yagi, T., and Inokuchi, H., 1987, Single crystals of hydrogenase from Desulfovibrio vulgaris Miyazakif, J. Biol. Chem., 262: 2823–2825.

Howard, K.S., McLean, P.A., Hansen, F.B., Lemley, P.V., Koblan, K.S., and Orme-Johnson, W.H., 1986, Klebsiella pneumoniae nifM gene product is required for stabilization and activation of nitrogenase iron protein in Escherichia coli, J. Biol. Chem., 261: 772–778.

Leclerc, M., Colbeau, A., Cauvin, B., and Vignais, P.M., 1988, Cloning and
    sequencing of the genes encoding the large and the small subunits of the
    H$_2$ uptake hydrogenase (hup) of Rhodobacter capsulatus, Mol. Gen. Genet.,
    214: 97-108.
LeGall, J., Ljungdahl, P.O., Moura, I., Peck Jr., H.D., Xavier, A.V., Moura,
    J.J.G., Texeira, M., Huyn, B.H., and Dervartanian, D.V., 1982, The
    presence of redox-sensitive nickel in the periplasmic hydrogenase from
    Desulfovibrio gigas, Biochem. Biophys. Res. Commun., 106: 610-616.
LeGall, J., and Peck Jr., H.D., 1987, Amino-terminal amino acid sequences of
    electron transfer proteins from gram-negative bacteria as indicators of
    their cellular localization: the sulfate-reducing bacteria, FEMS
    Microbiol. Rev., 46: 35-40.
Li, C., Peck Jr., H.D., LeGall, J., and Przybyla, A.E., 1987, Cloning,
    characterization and sequencing of the genes encoding the large and
    small subunits of the periplasmic [NiFe] hydrogenase of Desulfovibrio
    gigas, DNA, 6: 539-551.
Lissolo, T., Choi, E.S., LeGall, J., and Peck Jr., H.D., 1986, The presence
    of multiple intrinsic membrane nickel containing hydrogenase in
    Desulfovibrio vulgaris (Hildenborough), Biochem. Biophys. Res. Commun.,
    139: 701-708.
Mayhew, S.G., and O'Connor, M.E., 1982, Structure and mechanism of bacterial
    hydrogenase, Trends in Biochemical Science, 7: 18-21.
Menon, N.K., Peck Jr., H.D., LeGall, J., and Przybyla, A.E., 1987, Cloning
    and sequencing of the genes encoding the large and small subunits of the
    periplasmic (NiFeSe) hydrogenase of Desulfovibrio baculatus, J.
    Bacteriol., 169: 5401-5407.
Moura, J.J.G., Moura, I., Huynh, B.H., Kruger, H.J., Texeira, M., DuVarney,
    R.C., Dervartanian, D.V., Xavier, A.V., Peck Jr., H.D., and LeGall, J.,
    1982, Unambiguous identification of the nickel epr signal, Biochem.
    Biophys. Res. Commun., 108: 1388-1393.
Pollock, W.B.R., Chemerika, P.J., Forrest, M.E., Beatty, J.T., and Voordouw,
    G., 1989, Expression of the gene encoding cytochrome c$_3$ from
    Desulfovibrio vulgaris (Hildenborough) in Escherichia coli: export and
    processing of the apoprotein, J. Gen. Microbiol., 135: in press.
Prickril, B.C., Czechowski, M.H., Przybyla, A.E., Peck Jr., H.D., and LeGall,
    J., 1986, Putative signal peptide on the small subunit of the peri-
    plasmic hydrogenase from Desulfovibrio vulgaris, J. Bacteriol., 167:
    722-725.
Prickril, B.C., He, S.-H., Li, C., Menon, N., Choi, E.S., Przybyla, A.E.,
    DerVartanian, D.V., Peck Jr., H.D., Fauque, G., LeGall, J., Texeira, M.,
    Moura, I., Moura, J.J.G., Patil, D., and Huyn, B.J., 1987,
    Identification of three distinct classes of hydrogenase in the genus
    Desulfovibrio, Biochem. Biophys. Res. Commun., 149: 369-377.
Rieder, R., Cammack, R., and Hall, D.O., 1984, Purification and properties of
    the soluble hydrogenase from Desulfovibrio desulfuricans (strain Norway
    4), Eur. J. Biochem., 145: 637-643.
Sawers, R.G., and Boxer, D.H., 1986, Purification and properties of membrane
    bound hydrogenase isoenzyme 1 from anaerobically grown Escherichia coli
    K12, Eur. J. Biochem., 156: 1324-1331.
Sayavedra-Soto, L.A., Powell, G.K., Evans, H.J., and Morris, R.O., 1988,
    Nucleotide sequence of the genetic loci encoding subunits of
    Bradyrhizobium japonicum uptake hydrogenase, Proc. Natl. Acad. Sci. USA,
    85: 8395-8399.
Seefeldt, L.C., and Arp, D.J., 1986, Purification to homogeneity of
    Azotobacter vinelandii hydrogenase: a nickel and iron containing αβ
    dimer, Biohimie, 68: 25-34.
Seefeldt, L.C., and Arp, D.J., 1987, Redox dependent subunit dissociation of
    Azotobacter vinelandii hydrogenase in the presence of SDS, J. Biol.
    Chem., 262: 16816-16822.

49

Shuber, A.P., Orr, E.C., Recny, M.A., Schendel, P.F., May, H.D., Schauer, N.L., and Ferry, J.G., 1986, Cloning, expression and nucleotide sequence of the formate dehydrogenase genes from Methanobacterium formicum, J. Biol. Chem., 261: 12942–12947.

Staden, R., and McLachlan, A.D., 1982, Codon preference and its use in identifying protein coding regions in long DNA sequences, Nucleic Acids Res., 10: 141–156.

Stokkermans, J., van Dongen, W., Kaan, A., van den Berg, W., and Veeger, C., 1989, hydγ, a gene from Desulfovibrio vulgaris (Hildenborough) encodes a polypeptide homologous to the periplasmic hydrogenase, FEMS Microbiol. Lett., 58: 212–217.

Sukenaga, Y., Ishida, K., Takeda, T., and Takagi, K., 1987, cDNA sequence coding for human glutathione peroxidase, Nucleic Acids Res., 15: 7178.

Texeira, M., Fauque, G., Moura, I., Lespinat, P.A., Berlier, Y., Prickril, B., Peck Jr., H.D., Xavier, A.V., LeGall, J., and Moura, J.J.G., 1987, Nickel-[iron-sulfur]-selenium containing hydrogenases from Desulfovibrio baculatus (DSM 1743). Redox centers and catalytic properties, Eur. J. Biochem., 167: 47–58.

van den Berg, W.A.M., Stokkermans, J.P.W.G., and van Dongen, W.M.A.M., 1989, Development of a plasmid transfer system for the anaerobic sulphate reducer Desulfovibrio vulgaris, J. Biotechnol., in press.

van der Westen, H.M., Mayhew, S.G., and Veeger, G., 1978, Separation of hydrogenase from intact cells of Desulfovibrio vulgaris, FEBS Lett., 86: 122–126.

van Dongen, W., Hagen, W., van den Berg, W., and Veeger, C., 1988, Evidence for an unusual mechanism of membrane translocation of the periplasmic hydrogenase of Desulfovibrio vulgaris (Hildenborough) as derived from expression in Escherichia coli, FEMS Microbiol. Lett., 50: 5–9.

van Rooijen, G.J.H., Bruschi, M., and Voordouw, G., 1989, Cloning and sequencing of the gene encoding cytochrome c_{553} from Desulfovibrio vulgaris Hildenborough, J. Bacteriol., 171: 3575–3578.

Voordouw, G., 1987, Molecular biology of redox proteins in sulphate reduction, in: "The Nitrogen and Sulphur Cycles", J.A. Cole and S. Ferguson (eds.), Society for General Microbiology Symposium, 42: 147–160.

Voordouw, G., and Brenner, S., 1985, Nucleotide sequence of the gene encoding the hydrogenase from Desulfovibrio vulgaris (Hildenborough), Eur. J. Biochem., 148: 515–520.

Voordouw, G., and Brenner, S., 1986, Cloning and sequencing of the gene encoding cytochrome c_3 from Desulfovibrio vulgaris (Hildenborough), Eur. J. Biochem., 159: 347–351.

Voordouw, G., Hagen, W.R., Kruse-Wolters, M., van Berkel-Arts, A., and Veeger, C., 1987a, Purification and characterization of Desulfovibrio vulgaris (Hildenborough) hydrogenase expressed in Escherichia coli, Eur. J. Biochem., 162: 31–36.

Voordouw, G., Kent, H.M., and Postgate, J.R., 1987b, Identification of the genes for hydrogenase and cytochrome c_3 in Desulfovibrio, Can. J. Microbiol., 33: 1006–1010.

Voordouw, G., Menon, N.K., LeGall, J., Choi, E.-S., Peck Jr., H.D., and Przybyla, A.E., 1989a, Analysis and comparison of nucleotide sequences encoding the genes for [NiFe] and [NiFeSe] hydrogenase from Desulfovibrio gigas and Desulfovibrio baculatus, J. Bacteriol., 171: 2894–2899.

Voordouw, G., Strang, J.D., and Wilson, F.R., 1989b, Organization of the genes encoding [Fe] hydrogenase in Desulfovibrio vulgaris subsp. oxamicus Monticello, J. Bacteriol., 171: 3881–3889.

Voordouw, G., Walker, J.E., and Brenner, S., 1985, Cloning of the gene encoding the hydrogenase from Desulfovibrio vulgaris (Hildenborough) and determination of the NH_2-terminal sequence, Eur. J. Biochem., 148: 509–514.

Yagi, T., Kimura, K., Daidoji, H., Sakai, F., Tamura, S., and Inokuchi, H., 1976, Properties of purified hydrogenase from the particulate fraction of Desulfovibrio vulgaris (Miyazaki), J. Biochem., 79: 661-671.

Yates, M.G., and Robson, R.L., 1985. Mutants of Azotobacter chroococcum defective in hydrogenase activity, J. Gen. Microbiol., 131: 1459-1466.

Zinoni, F., Birkman, A., Stadtman, T.C., and Bock, A., 1986, Nucleotide sequence and expression of the selenocysteine containing polypeptide of formate dehydrogenase (formate-hydrogen-lyase linked) from Escherichia coli, Proc. Natl. Acad. Sci. USA, 83: 4650-4654.

THE HYDROGENASES OF SULFATE-REDUCING BACTERIA : PHYSIOLOGICAL, BIOCHEMICAL

AND CATALYTIC ASPECTS

E.C. HATCHIKIAN[1] , V.M. FERNANDEZ[2] and R. CAMMACK[3]

1.LCB/CNRS BP71, 13277 Marseille Cedex 9, F. 2.CSIC, Serrano

119, 28006 Madrid, Spain.3. King's College, London W87AH,UK

INTRODUCTION. Three metabolic groups of bacteria are involved in the anaerobic degradation of complex organic materials into methane (1). Fermentative bacteria hydrolyze lipids, protein and polysaccharides and ferment most products with excretion of acetate, saturated fatty acids, hydroxyacids, alcohols, CO_2 and H_2 as major endproducts. A second group, termed H_2-producing acetogenic bacteria,transform these endproducts into H_2, acetate and CO_2. Finally, methanogenic bacteria generate methane and CO_2 . The acetogenic and methanogenic bacteria grow in syntrophic associations through the process of interspecies H_2 transfer, and some acetogenic bacteria can only be cultured in the presence of hydrogen-utilizing microorganisms (1-4) (Figure 1). The H_2-producing acetogenic bacteria include some of the sulfate-reducing bacteria (2,3) and the species of obligate syntrophes isolated from cocultures (5-11).

Interspecies H_2 transfer can be simply described as the transfer of molecular hydrogen from H_2-producing to H_2-utilizing bacteria in mixed culture with the maintenance of a low partial pressure of hydrogen. This can result in altered fermentation products or growth by fermentations which are formally unfavorable on thermodynamic grounds (1-4). Hydrogen is thus an extremely important intermediate in these anaerobic fermentations. The physiology and biochemistry of its production and utilization have been intensively studied in species of the sulfate-reducing bacteria of the genus Desulfovibrio which can

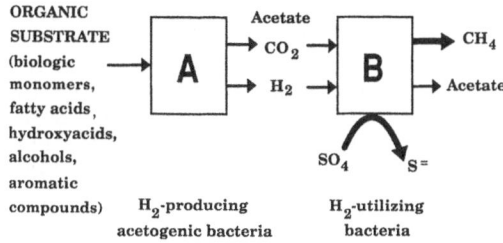

Figure 1. Interspecies H_2 transfer between H_2-producing acetogenic bacteria (A) and H_2-utilizing bacteria (B). Sulfate-reducers function both as (A) or (B) depending on sulfate concentration in the environment.

Microbiology and Biochemistry of Strict Anaerobes Involved in Interspecies Transfer
Edited by J.-P. Bélaich *et al.*
Plenum Press, New York, 1990

uniquely function both as hydrogen producing and hydrogen utilizing
bacteria in these microbial consortia, depending on environmental
conditions (Fig.1). Hydrogenase, a key enzyme in interspecies hydrogen
transfer processes has been intensively investigated in the past decade
(12-14). The present work will focus on the physiological, biochemical
and catalytic aspects of hydrogenases from Desulfovibrio.

I. PHYSIOLOGY OF HYDROGENASES IN DESULFOVIBRIO

Sulfate-reducing bacteria of the genus Desulfovibrio are strict
anaerobes that obtain energy for growth from the oxidation of a limited
number of organic substrates and molecular H_2 (15,16). The reducing
equivalents are used in the reduction of sulfur compounds (SO_4^{2-},
$S_2O_3^{2-}$, SO_3^{2-}) to H_2S with coupling of electron transfer to oxidative
phosphorylations. They exhibit an extremely active hydrogen metabolism
which plays a central role in energy generating mechanisms of these
microorganisms. Several types of hydrogen metabolism of Desulfovibrio
species in different environmental conditions have to be considered.

**Table 1 . Anaerobic oxidation of pyruvate and lactate by pure cultures
of Desulfovibrio or mixtures of Desulfovibrio with
methanogens.**

Equation	$\Delta G_0'$ (kcal/reaction[a])
1. $CH_3 CO COO^- + 2H_2 \rightleftharpoons CH_3COO^- + HCO_3^- + H^+ + H_2$	– 45.1
2. $2CH_3 CHOH COO^- + 4H_2O \rightleftharpoons 2CH_3COO^- + 2HCO_3^- + 2H^+ + 4H_2$	– 1.9
3. $4H_2 + HCO_3^- + H^+ \rightleftharpoons CH_4 + 3 H_2O$	– 32.4
2+3 $2CH_3 CHOH COO^- + H_2O \rightleftharpoons 2CH_3 COO^- + HCO_3^- + H^+ + CH_4$	– 34.3

[a] data from (2)

A. H_2 utilization as sole energy source in the presence of sulfate

Most of the Desulfovibrio species grow on hydrogen as electron
donor and sulfate or thiosulfate as terminal acceptors (17,18). The
cellular localization of oxireductases and electron carriers indicates
that H_2 oxidation and sulfate reduction take place on opposite sides of
the cytoplasmic membrane (19) : hydrogenase and cytochrome c_3 were
found in the periplasmic space whereas the reductases were localized in
the cytoplasm. Dissimilatory sulfate reduction is a transmembrane redox
process with external oxidation of H_2 and vectorial electron transport.
This metabolism requires the presence of only one hydrogenase located
in the periplasmic space (19). External H_2 oxidation coupled to
vectorial electron transfer is believed to be responsible for the
generation of a proton gradient which supports ATP synthesis. This
mechanism does not require a direct coupling of proton translocation to
electron transfer for dissimilatory sulfate reduction. A similar
bioenergetic mechanism probably occurs when Desulfovibrio species are
involved as H_2-utilizing sink in syntrophic associations with obligate
H_2-producing acetogenic bacteria (7,8,10).

B. H_2 production by pyruvate and lactate fermentation

Some Desulfovibrio species can produce hydrogen when growing fermentatively on pyruvate in the absence of sulfate (20-22). The oxidation of pyruvate to acetate, CO_2 and H_2 is thermodynamically favorable (Table 1, equation 1) (2). Energy is derived from ATP produced through substrate level phosphorylation.

The degradation of lactate in the absence of sulfate is not favorable on thermodynamic grounds (Table 1, eq. 2) (2). The fermentation can proceed only when it is coupled with another reaction such as that occurring when Desulfovibrio species are associated with H_2-utilizing methane bacteria (eq. 3). The free energy change for equation 2 becomes progressively more negative and the reaction proceeds more effectively when the partial pressure of H_2 is maintained at a much lower level than the standard conditions (eq. 2 + 3). When growing on lactate by interspecies hydrogen transfer with methanogenic bacteria, Desulfovibrio species can be described as purely fermentative organisms deriving their energy from substrate level phosphorylation. H_2 production from pyruvate and lactate fermentation has been correlated with the presence of an internal hydrogenase activity (23,24), not yet characterized.

C. H_2 metabolism during growth on lactate-sulfate medium

Production and consumption of H_2 during growth on lactate-sulfate medium has been described with a number of Desulfovibrio species (Table 2). D. vulgaris Hildenborough produces high amounts of hydrogen which accumulates during the growth (25,26) : H_2 production accounts for 25% of the electrons derived from lactate oxidation. D. vulgaris Miyazaki (27) and D. vulgaris Madison (28) produce low amount of H_2 in the early stage of the growth which is subsequently consumed. D. desulfuricans ATCC 7757 and D. fructosovorans evolve trace amounts of hydrogen during growth on lactate-sulfate (Hatchikian, E.C., unpublished) whereas no H_2 production was observed with D. gigas and D. africanus (29). It is to be noted that the net production of H_2 during growth does not exclude H_2 cycling.

Table 2 . H_2 metabolism on lactate-sulfate medium by Desulfovibrio species.

Organism	H_2 metabolism[a]	References
D. vulgaris Hildenborough	production	25, 26
D. vulgaris Miyazaki	production and consumption	27
D. vulgaris Madison	production and consumption	28
D. desulfuricans ATCC 7757	production	(b)
D. fructosovorans	production	(b)
D. gigas	no production	29
D. africanus	no production	29

(a) H_2 metabolism represents hydrogen produced or consumed during growth as measured by gas chromatography. [b]) Hatchikian, E.C., unpublished results.

II. BIOENERGETICS

Two bioenergetic mechanisms have been described for the growth of Desulfovibrio on organic substrates plus sulfate. On the basis of H_2 production during growth on lactate plus sulfate (22-24), vectorial electron transfer (19) and experiments with spheroplasts, a chemiosmotic hydrogen cycle was proposed by Odom and Peck (4,13,23) as a mechanism by which Desulfovibrio produces the ATP required for growth. This mechanism involves the following steps : i) cytoplasmic formation of molecular hydrogen from oxidation of lactate; ii) diffusion of H_2 across the cytoplasmic membrane; iii) oxidation of H_2 by a periplasmic hydrogenase and iv) vectorial electron transfer across the membrane for cytoplasmic reduction of the electron acceptor. This process results in proton translocation and ATP synthesis without direct coupling of proton translocation with electron transfer. This scheme explains the unique ability of Desulfovibrio to grow in consortia with either H_2-producing or H_2-consuming bacteria (4,23).

An alternative bioenergetic mechanism termed "trace hydrogen transformation redox model" has been proposed more recently (28). It is proposed that the metabolism of hydrogen during the growth on lactate or pyruvate as electron donor and sulfate as electron acceptor is to control the redox level of internal electron carriers. In this model, hydrogen production is a side reaction consequence of the oxidation of reduced electron carriers which link to electron transfer mediated phosphorylation via sulfate reduction. The functions of hydrogenase in this model differ significantly from those in the obligate H_2 cycling hypothesis. The cytoplasmic hydrogenase plays a role in trace H_2 production and the periplasmic hydrogenase can consume this hydrogen preventing energy loss.The periplasmic hydrogenase may play a major catabolic role when Desulfovibrio grows on H_2 plus sulfate. In the obligate H_2 cycling model the periplasmic hydrogenase itself generates the proton motive force whereas in the alternative model, proton translocation is coupled to electron transfer through a typical Mitchell loop.

The "trace H_2" model is supported by the following lines of evidence : i) the finding that H_2 did not competitively inhibit organic substrate oxidation in the presence of sulfate implies that H_2 is not an obligate intermediate, ii) higher levels of H_2 were formed from lactate-sulfate than from pyruvate-sulfate, suggesting that the more positive lactate-pyruvate half reaction ($E^{o'}=-190mV$) requires more H_2 production than the pyruvate-acetyl CoA half reaction ($E^{o'}=-540mV$) to prevent overreduction of the electron carriers involved in substrate oxidation. Neither model accounts for the accumulation of H_2 by D. vulgaris Hildenborough and the failure of molecular hydrogen to inhibit growth on lactate-sulfate medium (30). However, a direct evidence of hydrogen cycling has been recently presented. Utilizing membrane-inlet mass spectrometry, simultaneous production and consumption of hydrogen was demonstrated during the metabolism of pyruvate plus sulfate by washed intact cells of D. vulgaris Hildenborough (31). Hydrogen cycling is also strongly suggested by the metabolism of carbon monoxide in the presence of sulfate by D. vulgaris Madison (32). Hydrogen which is presumably produced during the oxidation of CO, accumulated even in the presence of low concentration of CO (4%) and was not consumed before CO had been exhausted. Concentrations of 4% CO were very inhibitory to the growth of Desulfovibrio due to uncoupling of energy linked H_2-dependent reduction of sulfate. H_2 production and consumption during metabolism of small amounts of CO in the presence of sulfate can be explained assuming the presence of two hydrogenases : an internal enzyme linked to CO dydrogenase and insensitive to CO and a periplasmic uptake hydrogenase sensitive to CO that initiates vectorial electron transfer and chemiosmotic synthesis of ATP during sulfate reduction (32).

One of the more controversial aspects of hydrogen cycling is related to the oxidation of lactate to pyruvate. Hydrogen can be formed from lactate under the conditions of interspecies hydrogen transfer. However, the E_0' of the lactate/pyruvate couple is high (-190 mV) compared to hydrogen electrode (-420 mV) which should make growth on lactate plus sulfate susceptible to inhibition by hydrogen, assuming hydrogen cycling (33). Growth on lactate plus sulfate is not inhibited by H_2 suggesting that hydrogen cycling is not involved in the oxidation of lactate to pyruvate (32, 34, 35).

It is noteworthy that energy metabolism and biochemical pathways of the microorganisms belonging to the genus Desulfovibrio show an extreme diversity. Generalizations concerning the mechanisms of energy coupling and the function of hydrogenases and electron transfer components may not be valid for every species of Desulfovibrio (30). This is illustrated by a detailed study of the metabolic balance of different species of Desulfovibrio grown on lactate-sulfate media under conditions in which the energy source was the sole growth limiting factor (26, 29) (Table 3). D. gigas and D. africanus utilize all the reducing equivalents from lactate to reduce sulfate according to the usual reaction proposed so far for all dissimilatory sulfate-reducers (36) D. vulgaris Hildenborough produces 1 mol of hydrogen from 2 mol of lactate whereas D. desulfuricans Norway 4 derives a signifiant amount of the electrons for the formation of butanol (29).

Table 3 . Metabolism balance of D. gigas and D. africanus (reaction 1), D. vulgaris Hildenborough (reaction 2) and D. desulfuricans Norway 4 (reaction 3) (from ref. 26, 29).

$$CH_3CHOHCOO^- + 0.5\ SO_4^{2-} \longrightarrow CH_3COO^- + CO_2 + 0.5\ S^{2-} + H_2O \qquad \text{(reaction 1)}$$

$$CH_3CHOHCOO^- + 0.37\ SO_4^{2-} + 0.56\ H^+ \longrightarrow CO_2 + 0.98\ CH_3COO^-$$
$$+ 0.02\ CH_3CH_2OH + 0.16\ H_2S + 0.215\ HS^- + 0.5\ H_2O + 0.48\ H_2 \quad \text{(reaction 2)}$$

$$CH_3CHOHCOO^- + 0.375\ SO_4^{2-} \longrightarrow CO_2 + 0.8\ CH_3\ COO^-$$
$$+ 0.375\ S^{2-} + 0.1\ CH_3\ (CH_2)_2\ CH_2OH + 0.8\ H_2O \qquad \text{(reaction 3)}$$

In contrast to D. vulgaris Miyazaki strain (27) which produces trace amounts of hydrogen only at the early stage of the growth, high amounts of H_2 are formed by D. vulgaris Hildenborough throughout growth (26). The fermentation balance of D. vulgaris Hildenborough and D. desulfuricans Norway 4 strongly suggest that H_2 and butanol production are side reactions of the mechanism of regulation of the levels of reduced electron carriers.

Two different regulatory redox processes function in these Desulfovibrio species. On one hand, D. vulgaris Hildenborough eliminates the excess of reducing power as hydrogen via cytochrome c_3 and hydrogenase. On the other hand, D. desulfuricans Norway 4 couples the oxidation of reduced ferredoxin to the reduction of pyridine nucleotides involved in the reduction of acetyl-CoA to butanol (37).

The two bioenergetic mechanisms (23, 28) are dependent upon the existence of at least two functionally distinct hydrogenases : one located in the cytoplasm or the inner aspect of the cytoplasmic membrane involved in H_2 production, and another, periplasmic, involved in H_2 consumption. The three classes of hydrogenases so far identified in Desulfovibrio species, (Fe), (NiFe) and (NiFeSe) hydrogenases (ref

14, and section II) have been found to be randomly distributed among the species of Desulfovibrio.Table 4 shows the cellular localizations of these distinct hydrogenases within various Desulfovibrio species. The (Fe) type appears to be present only in the periplasmic space (38-40) whereas the other hydrogenases may be found in any cellular compartment. A second hydrogenase which appears to be distinct from the major periplasmic (Fe) and (NiFe) hydrogenases of D. desulfuricans ATCC 7757 and D. fructosovorans respectively, has been immunologically detected in each microorganism (Table 4). In this context, D. vulgaris Hildenborough is particularly interesting since it contains the three types of hydrogenase (33).

Table 4 . Cellular localization of the three classes of hydrogenases in various Desulfovibrio species.

Organism	Localization			Ref.
	Periplasm	Membrane	Cytoplasm	
D. vulgaris	Fe	NiFeSe NiFe	n.d.	(14)
D. gigas	NiFe	n.d.	NiFe	(14,a)
D. baculatus	NiFeSe	NiFe NiFeSe	NiFeSe	(14)
D. desulfuricans (Norway 4)	(+)[1]	NiFe	NiFeSe	(41,42)
D. desulfuricans (ATCC 7757)	Fe	(+)[1]	n.d.	(39,a)
D. fructosovorans	NiFe	n.d.	NiFe (+)[1]	(43,a)

(a) Hatchikian, E.C., Forget, N., Nivière, V. and André, D. unpublished
(1) uncharacterized hydrogenase.

III. BIOCHEMICAL AND CATALYTIC PROPERTIES OF THE HYDROGENASES FROM DESULFOVIBRIO

A. General model of hydrogenase. The H_2 uptake reaction catalyzed by the enzyme hydrogenase represents the resultant of at least two consecutive reactions : the activation of H_2 that takes place in the active center H, followed by electron transfer to the acceptor-binding site (site A of Fig 8-1 in ref 44). H_2 evolution is the result of the reverse flux of electrons from the site A to H. The exchange reaction of hydrogen with deuterium or tritium does not involve additional electron donors or acceptors (45). Therefore this activity reflects the functional state of the H site. In those hydrogenases which contain only iron-sulfur clusters, it is proposed that the site H is a specialized iron-sulfur cluster (14,46). In the nickel-containing hydrogenases, there is increasing evidence that this metal is involved in the activation of H_2 (44). The site of interaction with external electron donors and acceptors (A site) appears to be distinct from the H site on the basis of differential inhibition (47). Both sites communicate by intra-molecular electron transfer; this may be through electron carrier groups such as (Fe-S) clusters. Most hydrogenases and in particular those containing nickel, are inactive in the exchange assay in the oxidized state. Activation through reductive treatment and oxidative deactivation of hydrogenases appear to be linked to the redox properties of the H site (44).

B. Types of hydrogenases found in Desulfovibrio. Up to date three soluble molecular forms of hydrogenases have been isolated from various Desulfovibrio species (14). Well characterized hydrogenases representative of each of these forms are : the periplasmic hydrogenase from D. vulgaris (Hildenborough) (38) which contains exclusively non-heme iron ((Fe) hydrogenase); the periplasmic nickel-iron hydrogenase ((NiFe) hydrogenase) from D. gigas (48-50); and the nickel-iron-selenium hydrogenase ((NiFeSe) hydrogenase) from D. desulfuricans (Norway)(42,51) and D. baculatus (DSM 1743) (52). In addition to differences in structural (metal content, type of clusters, amino acid sequences) and immunological properties, the three classes of hydrogenases so far characterized in Desulfovibrio can be differenciated by their catalytic properties, as well as their different sensitivity to inhibitors (14). The ratio of activities of H_2 evolution to H_2 uptake were found to be about 10, 1 and 0.25, respectively, for the (Fe),(NiFe) and (NiFeSe) hydrogenases.These ratio may be rather variable,depending on the different species and conditions in which the assays are performed (see also Table 4 in ref 14). The (NiFe) hydrogenase has been found to be insensitive to the inhibitor nitrite and moderately resistant to CO and NO (53,54). By contrast, (Fe) and (NiFeSe) hydrogenase is extremely sensitive to the inhibition by CO (53).

B.1. "Only-Iron" hydrogenases. (Fe) hydrogenases have been isolated from the periplasm of D. vulgaris Hildenborough (38,46,55,56), D. desulfuricans ATCC 7757 (39) and D. vulgaris Miyazaki K (40). A membrane-bound enzyme has been extracted and purified from cells of the strain F of D. vulgaris Miyazaki (57). D. vulgaris (Hildenborough) (Fe) hydrogenase shows immunological relationships with Clostridium pasteurianum hydrogenase (58). Table 5 summarizes the properties of these enzymes. The number of iron atoms found in different preparations of D. vulgaris (Hildenborough) (Fe) hydrogenase range from 10 to 16 atoms per molecule (46,55,59). The protein has been shown to contain two (4Fe-4S) clusters of the ferredoxin type, and another iron-sulfur

Table 5. Molecular properties of some representatives (Fe) hydrogenases

	D. vulgaris Hildenborough	D. desulfuricans ATCC 7757		D. vulgaris Miyazaki F
Localization	periplasm	periplasm	periplasm	membrane
$M_r \times 10^{-3}$	57	56	55.5	89
Subunits $M_r \times 10^{-3}$	46 11	47	43 12.5	60 29
NH_2-terminal	Ser Ala	Pro	Ser Ala	n.r.
Iron (atoms/mol)	10-16	12	10	8
E 400 $mM^{-1}cm^{-1}$	45	n.r.	44	47
Activity[a]				
H_2 evolution	4.6-3.8	9	6	0.6
H_2 uptake	50	n.r.	20	n.r.
References	(38,46,55,56)	(39)	(b)	(57)

a) mMol $H_2 \times mg^{-1}$ protein $\times min^{-1}$ with methyl viologen as electron carrier.
b) Hatchikian, E.C., Forget, N. and André, C. (unpublished).

cluster involved in the activation of hydrogen (H site) (55,56,59). It has been proposed that this H center could be a (6Fe-6S) cluster (46,60) or a particular (4Fe-4S) cluster (55). The uncertainty in the content of iron makes it difficult to give an answer to the question of the nature of H site.

The periplasmic hydrogenase from D. desulfuricans (ATCC 7757) has been described as a monomeric protein of 47 kDa of molecular mass estimated by SDS-PAGE and of 56 kDa estimated by equilibrium sedimentation analysis (39). The properties of this enzyme have been reexamined and the enzyme has been found to be composed of two subunits of 43 and 12.5 kDa (Table 5). The N-terminal amino acid sequences of both subunits determined up to 25 residues are the same as those of the subunits of D. vulgaris (Hildenborough) hydrogenase (55); both enzymes are also similar in that their small subunits lack cysteine residues (Hatchikian, E.C., Forget, N., André, C. and Cammack, R. unpublished results). The content of non-heme iron of D. desulfuricans hydrogenase (10 atoms per mol of enzyme) equals the lowest value found in D. vulgaris enzyme (Table 5). However, EPR spectra of D. desulfuricans hydrogenase resemble those of the D. vulgaris enzyme since it exhibits a complex signal in its fully reduced state due to spin-spin interaction between two $(4Fe-4S)^{+1}$ clusters, and an axial signal with g values of 2.06 and 2.00 after treatment with CO. This signal has been postulated to represent the hydrogen binding site with CO as a ligand in D. vulgaris enzyme (14).

B.2. Nickel-containing hydrogenases. Two immunologically and biochemically distinct Ni-hydrogenases, the (NiFe) and (NiFeSe) hydrogenases, have been described in species of the genus Desulfovibrio (14). In the following we will focus on the (NiFe) enzyme isolated from D. gigas .

Reversible activation. This extensively studied protein, purified from the periplasmic space, contains 1 atom of Ni, 11 atoms of Fe and 12 atoms of acid-labile sulfide in a molecule of 89.5 kDa consisting of two subunits of 62 and 26 kDa (48-50,61). The ESR, Mössbauer and MCD spectra of the protein as isolated have been interpreted as arising from an Ni(III) ion, a paramagnetic $(3Fe-4S)^{+1}$ cluster and two diamagnetic $(4Fe-4S)^{+2}$ clusters (49,50,62,63). Two ESR signals have been identified in the native enzyme as due to nickel by isotopic substitutions : Ni-A and Ni-B (Table 7) (50). The relative amounts of these signals can be varied by reduction and reoxidation (64,65) and varies among Ni-hydrogenases from different species. All the paramagnetic centers of native D. gigas hydrogenase detectable by ESR are reversibly reduced at redox potentials less negative than the hydrogen potential (Fig. 2). The enzyme, when isolated under aerobic conditions, is inactive towards hydrogen, although some activity is detected in the assay of H_2-evolution with reduced methyl viologen (66). Prolonged incubation with strong reductants (H_2, dithionite) increases notably the evolution (66) and uptake activity (67,68) and renders the enzyme active in the exchange assay (69,70). The slow process of reactivation ($E_a = 88$ KJ.mol^{-1}) can be considered as a general conversion from a form which is essentially inert towards hydrogen, Unready state, to an Active state. The active enzyme can be inactivated, producing mostly the unready state. By contrast, anaerobic reoxidation by DCIP was able to convert the enzyme to the Ready state (68). Unlike the unready form, the ready enzyme is rapidly converted to the active form by strong reductants. These three forms can be identified by their different reactivity in the assays of hydrogenase (Table 6) and by their Ni-ESR spectra (Table 7).

Table 6. Catalytic forms of D. gigas hydrogenase

Hydrogenase Activity	Unready	Ready	Active
Isotopic Exchange	–	–	+
Evolution of H_2	–	+[a]	+
Uptake of H_2			
Low Potential Acceptor (MV)	–	+[a]	+
High Potential Acceptor (DCIP)	–	–	+

[a]) An induction period is observed before full activity is achieved

Table 7. Properties of nickel ESR signals of D. gigas hydrogenase forms

	Unready	Ready	Active
Signal	Ni–A	Ni–B	Ni–C
g Values	2.32, 2.23, 2.01	2.34, 2.16, 2.01	2.19, 2.16, 2.01
E_m (pH 7.0)[a]	–150	N.D.	–270/–390[b]
E_m/pH (mV)	60	N.D.	120/60
Light Sensitivity	–	–	+
CO Sensitivity	–	–	+
Splitting (T < 20K)	–	–	+[c]

a) SHE; b) appearance and disappearance of the Ni-C signal. c) the splitting of the Ni-C signal is dependent on the presence of reduced (4Fe-4S) clusters.

    The activation-deactivation process observed with the isolated enzyme could also be functional in vivo : freshly grown cells of D. gigas treated with DCIP or flushed with argon were converted to the ready state (Nivière, V., Hatchikian, E.C. and Fernandez, V.M., unpublished).

    However, the molecular basis for the interconversion among the different forms is not well understood. ESEEM spectra of D. gigas hydrogenase (72) indicate that the nickel environment of the active protein is different from that in the unready state. Comparison of the spectra collected for enzyme samples in H_2O and in D_2O established that the metal environment is not accessible to the solvent in the unready state, while it becomes accessible to deuterium during the activation (72).

    Three iron-sulfur-cluster. Huynh et al. (62) have evidenced that the 3Fe cluster remained intact and was not converted into a (4Fe-4S) cluster under the reducing conditions which resulted in enzyme activation. Moreover, our results on Active hydrogenase purified under an atmosphere of 10% H_2/90% Ar, showed that this preparation as put under argon produced the usual signal of the three-iron cluster at g= 2.01. This indicates that the presence of this cluster in the protein is not an artefact due to oxidative damage of 4Fe clusters during aerobic purification. However, its function is unknown. Its redox potential (Fig. 2) is too positive as to be involved in the

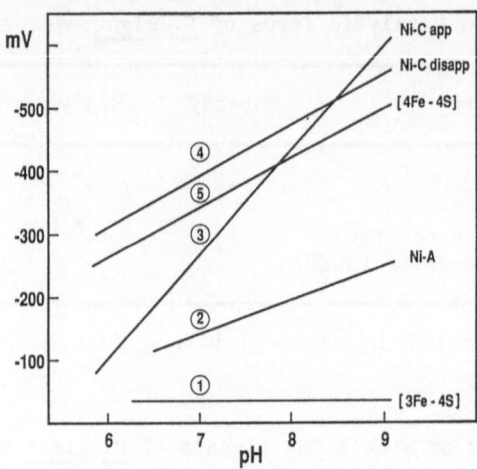

Figure 2. Plots of the midpoint potentials of the redox centers of D. gigas hydrogenase as a function of pH.

reaction of H_2 evolution. Also, our results of treatment of D. gigas hydrogenase with Cu (II)-ascorbate or Hg (II) suggest that the integrity of this cluster is not essential for the reduction of methyl viologen with hydrogen (73). One interesting possibility is the coexistence of two electron pathways in this enzyme : One between H_2 and low redox potential substrates, bidirectional, involving nickel and the two 4Fe clusters; another, unidirectional, between H_2 and high potential substrates involving nickel and 3Fe cluster eventually connected through one of the 4Fe clusters.

Coordination of the Nickel site. The lineshapes of the ESR spectra of Ni in those hydrogenases in which it is detectable are remarkably similar. This is a strong indication that the nickel environment is highly conserved. It therefore seems reasonable to correlate spectroscopic information on nickel in hydrogenases from different species in order to obtain a composite picture.

EXAFS studies of the F420-reducing hydrogenase from M. thermoautotrophicum (74) indicated that sulfur was the principal scattering nucleus. Best fits to the data were obtained with 3 sulfur atoms. Low temperature MCD spectra of this enzyme support at least one S-Ni bond on the basis of charge transfer bands (75). In D. gigas hydrogenase, EXAFS spectra of the nickel in the oxidized state were best fitted with four sulfur atoms at a distance of 2.2 A (76). However, lighter atoms coordination could be masked by the predominance of sulfur scattering in the data (76).

Coordination of sulfur atoms has been also ascertained by investigating the hyperfine structure of ESR spectra of enzyme from Wolinella succinogenes enriched in ^{33}S (77). The data indicated hyperfine interaction with one sulfur nucleus. However, the presence of another sulfur cannot be discounted if the coordination geometry and electron distribution are such that the hyperfine interaction is very weak. In addition, ESR spectroscopy of ^{77}Se enriched hydrogenase from D. baculatus (78) and EXAFS of this (NiFeSe) hydrogenase (79) have established that the selenium atom of a selenocysteine residue located on the large subunit of the protein (80) is coupled to nickel.

Further insight into nickel coordination of D. gigas hydrogenase has been obtained with ESEEM spectroscopy (72). The electron spin echo envelope modulation technique of pulsed EPR spectroscopy is particularly sensitive for detecting weak hyperfine interactions between paramagnetic metal centres and nuclei such as ^{14}N or ^{2}H (81, 82). Samples of D. gigas hydrogenase in the unready (Ni-A) and active (Ni-C) states both in H_2O and $2H_2O$ were examined with this technique (72). Fourier transforms of the 3-pulse ESEEM taken at 8.7 GHz, for Ni-A and Ni-C in H_2O provided good evidence for the presence of ^{14}N, possibly from imidazole, linked to nickel both in the unready and active enzyme (72).

To synthesize the converging information gathered from different studies on various Nickel-hydrogenases (72,74-82), one can propose a minimal structural basis in which high symmetry is hardly expected.

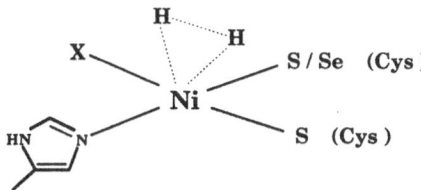

Figure 3. Tentative structure of nickel environment in hydrogenases.

Taking into account that the coordination number of stable Ni complexes will vary from four to six either in d7 Ni (III) or d8 Ni (II) configurations, three permanent ligand positions can be postulated: S (cysteine), S/Se (cysteine), N (imidazole). One more ligand X could occupy the fourth permanent site in the Ni-A state. This would allows the approach of hydrogen molecules in the course of activation to the Ni-C state. The coordination number would then raise to 5 or 6 (Figure 3).

The role of Nickel in hydrogenase catalysis. Several results from nickel-hydrogenases strongly support the involvement of nickel in the H site. These are the association of Ni-C ESR signal with the active enzyme (64,65), the deuterium kinetic isotope effect in photolysis of this signal and the effect of carbon monoxide, an inhibitor competitive with hydrogen, on the spectrum of Ni-C. These early findings of Albracht and coworkers (83,84) with the enzyme from C. vinosum were also found with the enzyme from D. gigas (65). Also the apparent accessibility of the Ni-C site to solvent protons is consistent with catalytic function towards hydrogen. This fact has been taken into account in the model for activation and reaction cycle presented in Fig. 4 based on the following postulates :
i) hydrogen production and consumption occur at the nickel site; ii) the (4Fe-4S) clusters are involved as secondary electron carriers; iii) nickel exists in the oxidation states 3^+ and 2^+ (other possible mechanisms have been drawn up which differ mainly in the oxidation levels of nickel (83,85)); iv) the Ni(III) site is closed to the exterior, Ni(II) is open; v) the Ni-C signal associated with the Active state is a Ni(III)-H_2 complex; vi) each reduction step involves one electron and one proton.

In this scheme it is proposed that the activation of the two deactivated forms of the enzyme (unready and ready) involves a change in the oxidation state of nickel. In the case of the unready state, the activation occurs through a conformational change after reduction of the enzyme. The reaction cycle includes at least one hydride intermediate in order to explain the hydrogen-deuterium exchange reaction catalyzed by the enzyme (Fig. 4). The formal oxidation level of nickel center varies from +3 (Ni(III)) to +1 ($\overline{Ni(III)}$-H_2) during the activation process and from +2 (Ni(II)) to zero (Ni(II)-H_2^-) during the reaction cycle.

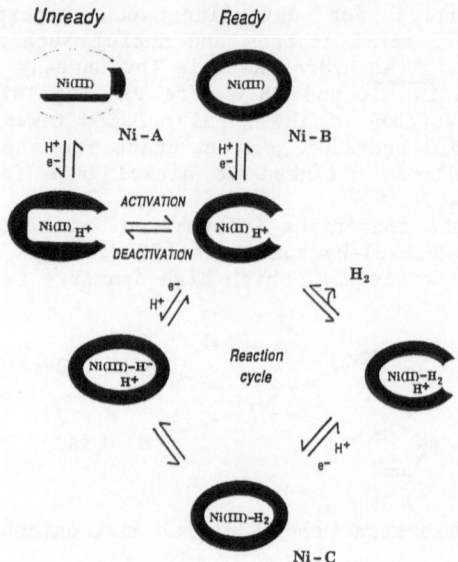

Figure 4. D. gigas hydrogenase activation and reaction cycle.

Primary structure of (NiFe) hydrogenases. The nucleotide sequence encoding the (NiFe) hydrogenase from D. gigas (80,86) and from D. fructosovorans (87) have been determined. The amino acid sequence of the D. gigas enzyme shows nearly 70% homology with that of D. fructosovorans (Fig. 5) and only 35% with the sequence of the (NiFeSe) hydrogenase from D. baculatus to which it has been found to be related (80,88).

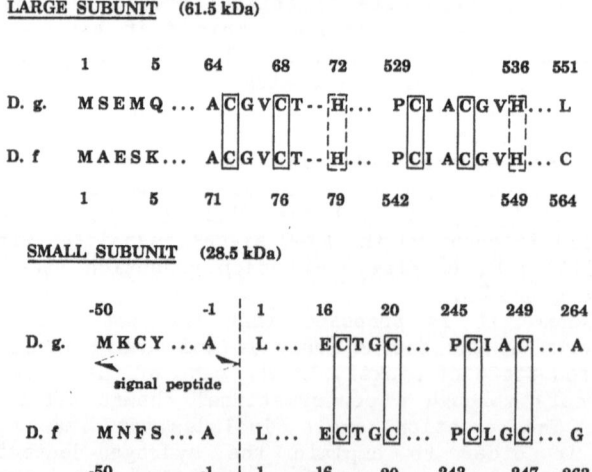

Figure 5. Outline structure of the periplasmic (NiFe) hydrogenases from D. gigas (80) and D. fructosovorans (87). Identical cysteine and histidine residues are boxed.

Figure 6. Comparison of C-terminal amino acid sequences of large subunit of hydrogenases from D. gigas (D.g.) (80), D. fructosovorans (D.f.) (87), D. baculatus (D.b.) (80,88), Rhodobacter capsulata (R.c.) (91), Bradyrhizobium japonicum (B.j.) (92), Methanobacterium thermoautotrophicum (M.t.) (93). Identical residues are boxed. Dashed boxes enclose the locations at which the same residue occurs in five of the sequences.

```
                520              530              540              550
D. g.   P V E I L R T V H S Y D P C I A C G V H V I D P E S N Q V H K F R I L
D. f.   P V E I L R T V H A F D P C I A C G V H V I E P E T N E I L K F K V C
D. b.   P V N V G R L V R S Y D P U L G C A V H V L H A E T G E E H V V N I D
R. c.   P V E I L R T L H S F D P C L A C S T H V M S A E G P P D H R Q G P V ...
B. J.   P L E I L R T I H S F D P C L A C S T H V M S P D G Q E L A K V K V R
M. t.   F N L M E M V I R A Y D P C L S C A T H T I D S Q M R L A T L E V Y D ...
```

In D. gigas hydrogenase as well as in other (NiFe) enzymes, 11 cysteine residues are required to chelate two (4Fe-4S) and one (3Fe-4S) clusters and at least two more are involved in the coordination of nickel (76-80). There are 10 conserved cysteines on the small subunit and 6 more on the large subunit of (NiFe) hydrogenases from Desulfovibrio (80,87).It is evident that none of the subunits contains enough conserved cysteines to link all the metal centers. The binding of nickel to the large subunit of both (NiFe) and (NiFeSe) hydrogenases can be presumed from comparative data obtained with Alcaligenes eutrophus hydrogenase whose large subunit, homologous to that of Desulfovibrio hydrogenases (58) has been reported to contain one nickel ion and one (4Fe-4S) cluster (89).The presence of selenocysteine in the C-terminal region of the large subunit of D . baculatus enzyme has allowed the identification of an homologous cysteine of (NiFe) hydrogenases (80) with an analogous function of binding of nickel (78-80). The amino acid sequence of the C-terminal region of the large subunit has been found to be highly conserved in all the Nickel-hydrogenases so far sequenced (80,87,90-92)(Fig. 6).

A comparative analysis of these sequences suggests that the motif Cys530-X-X-Cys533-X-X-His536 (D. gigas hydrogenase numbering) present in all the sequences of Fig.6 is involved in the coordination of nickel. Another similar motif present in the N-terminal region of the large subunit (Cys65-X-X-Cys68-X-X-X-His72) is also highly conserved and therefore could be involved in the binding of nickel. However, this hypothesis is less plausible since it would imply that six possible coordination positions of nickel are occupied by permanent ligands, which is very unlikely for the catalytic activity of the enzyme.

The remaining conserved cysteines in the small and large subunits could accomodate the iron-sulfur clusters. The 10 conserved residues found in the small subunit are sufficient to chelate two (4Fe-4S) clusters (80,87) whereas the (3Fe-4S) cluster could be accommodated by the cysteine residues of the large subunit not involved in nickel coordination. Such a distribution of iron-sulfur clusters between both subunits is substantiated by the presence of two Cys-X-X-Cys motives in the small subunit and one more in the N-terminal region of the large subunit. These motives appear to be the minimal structural basis specially required for (4Fe-4S) clusters accommodation since the spacing between second and third as well as between third and fourth cysteines involved in the binding of clusters is highly variable among iron-sulfur proteins (93-96).

Acknowledgements. The authors would like to thank V. Nivière, N. Forget, C. André and A. Chapman for experimental contributions. We are also grateful to Prof. D. Benlian and Dr. M. Frey for helpful discussions on structural aspects of hydrogenase and to Mrs. C. Allard for processing the manuscript. This research was supported by PIRSEM under grant N° AIP 3064 (E.C.H.), by the UK Science and Engineering Research Council (R.C.), by CICYT under grant PB 0204 (V.M.F.) and Action Intégrée Franco-Espagnole N° 078 (E.C.H. and V.M.F.).

References

1. **Bryant, M.P.** (1979). Microbial methane production-theoretical aspects . S. Anim. Sci. 48 : 193-201.
2. **Bryant, M.P. Campbell, L.L., Reddy, C.A., Crabill, M.R.** (1977). Growth of Desulfovibrio in lactate or ethanol media low in sulfate in association with H_2-utilizing methanogenic bacteria. Appl. Environ. Microbiol. 33 : 1162-1169.

3. **Mc Inerney, M.J. and Bryant, M.P.** (1981). Anaerobic degradation of lactate by syntrophic associations of <u>Methanosarcina barkeri</u> and <u>Desulfovibrio</u> species and effect of H_2 on acetate degradation. Appl. Environ. Microbiol. 41 : 346-354.

4. **Peck, H.D. Jr. and Odom, J.M.** (1984). Hydrogen cycling in <u>Desulfovibrio</u> : a new mechanism for energy coupling in anaerobic microorganisms. In H.O. Halvorsen and Y. Cohen (eds), Microbiol mats : stromatolites. (pp. 215-243). Alan R. Liss, New York.

5. **Bryant, M.P., Wolin, E.A., Wolin, M.J. and Wolfe, R.S.** (1967). <u>Methanobacillus omelianski</u>, a symbiotic association of two species of bacteria. Arch. Microbiol. 59 : 20-31.

6. **Schink, B. and Plennig, N.** (1982). Fermentation of trihydroxybenzenes by Pelobacter acidigallia gen. nov. sp. nov., a new strictly anaerobic, non spore-forming bacterium. Arch. Microbiol. 133 : 195-201.

7. **Boone, D.R. and Bryant, M.P.** (1980). Propionate-degrading bacterium, Syntrophobacter wolinii sp. nov. gen. nov., from Methanogenic Ecosystems. Appl. Environ. Microbiol. 40: 626-632.

8. **Mc Inerney, M.J., Bryant, M.P., Hespell, R.B. and Costerton, J.W.** (1981). <u>Syntrophomonas wolfei</u> gen. nov. sp. nov., an anaerobic syntrophic fatty acid-oxidizing bacterium. Appl. Environ. Microbiol. 41 : 1029-1039.

9. **Krumholtz, L.R. and Bryant, M.P.** (1986). <u>Syntrophococcus sucromutans</u> sp. nov. gen. nov. uses carbohydrates as electrons donor and formate, Methoxymonobenzenoids or <u>Methanobrevibacter</u> as electrons acceptor systems. Arch. Microbiol. 144 : 8-14.

10. **Mounfort, D.O. and Bryant, M.P.** (1982). Isolation and characterization of an anaerobic syntrophic benzoate-degrading bacterium from sewage sludge. Arch. Microbiol. 133 : 249-256.

11. **Stieb, M. and Schink, B.** (1985). Anaerobic oxidation of fatty acids by <u>Clostridium bryantii</u> sp. nov. a spore-forming, obligately syntrophic bacterium. Arch. Microbiol. 140 : 387-390.

12. **Adam, M.W.W., Mortenson, L.E. and Chen, J.S.** (1980). Hydrogenase. Biochim. Biophys. Acta. 594 : 105-176.

13. **Odom, J.M. and Peck, H.D.Jr.** (1984). Hydrogenase, electron transfer proteins, and energy coupling in the sulfate-reducing bacteria <u>Desulfovibrio</u>. Ann.Rev.Microbiol. 38 : 551-592.

14. **Fauque, G., Peck, H.D. Jr. Moura, J.J.G., Huynh, B.H., Berlin, Y. Der Vartanian, D.V., Teixeira, M., Przybyla, A.E., Lespinat P.A., Moura, I. and Le Gall, J.** (1988). The three classes of hydrogenases from sulfate-reducing bacteria of the genus Desulfovibrio. FEMS Microbiol. Rev. 54 : 299-344.

15. **Plennig, N., Widdel, F. and Truper, H.G.** (1981). The dissimilatory sulfate-reducing bacteria. In "The Prokariotes ; starr M.P., Solp H., Truper H.G., Balows A., Schlegel, G. (eds) : Berlin, Springer-Verlag, Vol. 1, pp. 926-940.

16. **Widdel, F.** (1988).Microbiology and ecology of sulfate- and sulfur-reducing bacteria. In Biology of anaerobic microorganisms. pp. 469-585. A.J.B. Zehnder, ed. T. Wilez and sous, Inc. New York.

17. **Badziong, W., Thauer, R.K., and Zeikus, J.G.** (1978).Isolation and characterization of Desulfovibrio growing on hydrogen plus sulfate as the sole energy source. Arch. Microbiol. 116 : 41-49.

18. **Brandis, A. and Thauer, R.K.** (1981). Growth of <u>Desulfovibrio</u> species on hydrogen and sulfate as sole energy source. J. Gen. Microbiol. 126 : 249-252.

19. **Badziong, W. and Thauer, R.K.** (1980).Vectorial electron transport in Desulfovibrio vulgaris (Marburg) growing on hydrogen plus sulfate as sole energy source. Arch. Microbiol. 125 : 167-174.

20. **Postgate, J.R.** (1952). Growth of sulfate-reducing bacteria in sulfate-free media. Research (London) 5 : 189-190.

21. **Vosjan, J.H.** (1975). Respiration and fermentation of the sulfate-reducing bacterium Desulfovibrio desulfuricans in continuous culture. Plant soil. 43 : 141-152.

22. **Traore, S.A.** (1981).Energètique de la croissance des bactéries sulfato-réductrices lorsque l'accepteur final d'électrons est le sulfate ou une bactérie méthanigène, transfert interspécifique d'hydrogène, énergie de maintenance. Thèse de Doctorat ès Sciences, Université d'Aix-Marseille, 215 p.

23. **Odom, J.M. and Peck, H.D. Jr.** (1981). Hydrogen cycling as a general mechanism for energy coupling in the sulfate-reducing bacteria Desulfovibrio sp. FEMS Microbiol. Lett. 12 : 47-50.

24. **Odom, J.M. and Peck, H.D. Jr.** (1981).Localization of dehydrgenases, reductases and electron transfer components in the sulfate-reducing bacterium, Desulfovibrio gigas. J. Bacteriol. 147 :161-169.

25. **Hatchikian, E.C., Chaigneau, M. and Le Gall, J.** (1976).Analysis of gas production by growing cultures of three species of sulfate-reducing bacteria. In Microbiol. production and utilization of gases. pp. 109-118. H.G. Schlegel, G. Gottschalk and N. Plennig (ed.) Microbiol. production and utilization of gases. Erich Goltze, Göttingen.

26. **Traore, A.S., Hatchikian, E.C., Belaich, J.P. and Le Gall, J.** (1981). Microcalorimetric studies of the growth of sulfate-reducing bacteria : Energetics of Desulfovibrio vulgaris growth. J. Bacteriol. 145 : 191-199.

27. **Tsuji, K. and Yagi, T.** (1980). Significance of hydrogen burst from growing cultures of Desulfovibrio vulgaris Miyazaki and the role of hydrogenase and cytochrome c_3 in energy producing system. Arch. Microbiol, 125 : 35-42.

28. **Lupton, F.S., Conrad, R. and Zeikus, J.G.** (1984). Physiological function of hydrogen metabolism during growth of sulfidogenic bacteria on organic substrates. J. Bacteriol. 159 : 843-849.

29. **Traore, S.A., Hatchikian, E.C., Le Gall, J. and Belaich, J.P.** (1982). Microcalorimetric studies of the growth of sulfate-reducing bacteria : comparison of the growth parameters of some Desulfovibrio species. J. Bacteriol. 149 : 606-611.

30. **Le Gall, J. and Fauque, G.** (1988).Dissimilatory reduction of sulfur compounds. In Biology of Anaerobic Microorganisms (Zehnder, A.J.B., ed.) pp. 587-639. John Wiley and Sons, Inc., New York.

31. **Peck, H.D., Jr. Le Gall, J., Lespinat, P.A., Berlier, Y. and Fauque, G.** (1987). A direct demonstration of hydrogen cycling by Desulfovibrio vulgaris Hildenborough employing membrane-inlet mass spectrometry. FEMS Microbiol. Lett. 40 : 295-299.

32. **Lupton, F.S., Conrad, R. and Zeikus, J.G.** (1984). CO metabolism of Desulfovibrio vulgaris strain Madison : Physiological function in the absence or presence of exogenous substrates. FEMS Microbiol. Lett. 23, 263-268.

33. **Peck, H.D. Jr. and Lissolo, T.** (1988).Assimilatory and dissimilatory sulphate reduction : enzymology and bioenergetics. In the Nitrogen and Sulphur Cycles (Cole, J.A. and Ferguson, S.J. eds.) pp. 99-132, 42nd Symposium of

the Society for General Microbiology, Cambridge University Press, Cambridge.

34. **Odom, J.M. and Wall, J.D.** (1986). Properties of a hydrogen-inhibited mutant of Desulfovibrio desulfuricans (ATCC 27774) J. Bacteriol. 169 : 1335-1337.

35. **Panklania, I.P., Gow, L.A. and Hamilton, W.A.** (1986).The effect of hydrogen on the growth of Desulfovibrio vulgaris (Hildenborough) on lactate. J. Gen. Microbiol. 132 : 3549-3556.

36. **Peck, H.D. Jr.** (1962). Symposium on metabolism of inorganic compounds. V. Comparative metabolism of inorganic sulfur compounds in microorganisms. Bacteriol. Rev. 26 : 67-94.

37. **Gottschalk, G.** (1978). Bacterial fermentations. In Bacterial Metabolism (Starr, M.P.ed), Chap. 8, pp. 186-188 ; Springer verlag, New York.

38. **Van der Westen, H., Mayhew, S.G. and Veeger, C.** (1978). Separation of hydrogenase from intact cells of Desulfovibrio vulgaris. Purification and properties. FEBS Lett. 86 : 122-126.

39. **Glick, B.R., Martin, W.G. and Martin, S.M.** (1980). Purification and properties of the periplasmic hydrogenase from Desulfovibrio desulfuricans. Can. J. Microbiol. 26 : 1214-1223.

40. **Aketagawa, J., Kobayashi and Ishimoto, M.** (1983). Characterization of periplasmic hydrogenase from Desulfovibrio vulgaris Miyazaki K. J. Biochem. 93 : 755-762.

41. **Lalla-Maharajh, W.V., Hall, D.O., Cammack, R., Rao, K.K. and Le Gall, J.** (1983). Purification and properties of the membrane bound hydrogenase from Desulfovibrio desulfuricans. Biochem. J., 209, 445-454.

42. **Rieder, R., Cammack, R. and Hall, D.O.** (1984). Purification and properties of the soluble hydrogenase from Desulfovibrio desulfuricans Eur. J. Biochem. 145 : 637-643.

43. **Hatchikian, E.C., Traore, A.S., Fernandez, V.M. and Cammack, R.** (1989). Characterization of the nickel-iron periplasmic hydrogenase from Desulfovibrio desulfuricans. Submitted to Eur. J. Biochem.

44. **Cammack, R., Fernandez, V.M. and Schneider, K.** (1988). Nickel in hydrogenases from sulfate-reducing, photosynthetic, and hydrogen-oxidizing bacteria. In "The bioinorganic Chemistry of nickel" (Lancaster, J.R. Jr., ed.) pp. 167-190. VCH. Publishers, Deerfield Beach Florida.

45. **Mortenson, L.E. and Chen, J.-S.** (1974). Hydrogenase. In "Microbial iron metabolism", (Neilands, J.B., ed.). pp. 231-282. Acedemic Press, New York.

46. **Hagen, W, Van Berkel-Arts, A., Kruse-Wolters, K.M., Voordouw, G. and Veeger, C.** (1986). The iron-sulfur composition of the active site of hydrogenase from Desulfovibrio vulgaris deduced from its subunit structure and total iron-sulfur content. FEBS Lett. 203, 59-63.

47. **Hallahan, D.L., Fernandez, V.M., Hatchikian, E.C. and Hall, D.O.** (1986). Differential inhibition of catalytic sites in Desulfovibrio gigas hydrogenase. Biochimie 68 : 49-54.

48. **Hatchikian, E.C., Bruschi, M. and Le Gall, J.** (1978). Characterization of the periplasmic hydrogenase from Desulfovibrio gigas. Biochem. Biophys. Res. Commun. 82 : 451-461.

49. **Cammack, R., Patil, D., Aguirre, R. and Hatchikian, E.C.** (1982). Redox properties of the ESR-detectable nickel in hydrogenase from Desulfovibrio gigas. FEBS Lett. 142 : 289-292.

50. **Moura, J.J.G., Moura, I., Huynh, B.H., Kruger, H.-J., Teixeira, M., Du Varney, R.C., Der Vartanian, D.V., Xavier, A.V., Peck, H.D. Jr and Le Gall, J.** (1982). Unambigous identification of the nickel EPR signal in ^{61}Ni-enriched Desulfovibrio gigas

hydrogenase. Biochem. Biophys. Res. Commun. 108 :
1388-1393.

51. **Bell, S.H., Dickson, D.P.E., Johnson, C.E., Rieder, R., Cammack, R., Hall, D.O. and Rao, K.K.** (1984). Spectroscopic studies of the nature of the iron clusters in the soluble hydrogenase from Desulfovibrio desulfuricans (strain Norway 4). Eur. J. Biochem. 145 : 645-651.

52. **Teixeira, M., Fauque, G., Moura, I., Lespinat, P.A., Berlin, Y., Prickril, B., Peck, H.D. Jr., Xavier A.V., Le Gall, F. and Moura, J.J.G.** (1987). Nickel-(iron-sulfur)-selenium-containing hydrogenases from Desulfovibrio baculatus (DSM 1743). Eur. J. Biochem. 167 : 47-58.

53. **Berlier, Y., Fauque, G.D., Le Gall, J. Choi, E.S., Peck, Jr. H.D. and Lespinat, P.A.** (1987). Inhibition studies of three classes of Desulfovibrio hydrogenase : Application to the further characterization of the multiple hydrogenases found in Desulfovibrio vulgaris Hildenborough. Biochem. Biophys. Res. Commun. 146 : 147-153.

54. **Fauque, G., Berlier, Y., Choi, E.S., Peck, Jr. H.D., Le Gall, J. and Lespinat, P.A.** (1987). The carbon monoxide inhibition of the proton-deuterium exchange activity of iron, nickel-iron and nickel-iron-selenium hydrogenases from Desulfovibrio vulgaris Hildenborough. Biochem. Soc. Trans. 15 : 1050-1051.

55. **Huynh, B.H., Czechowski, M.H., Kruger, H.-J., Der Vartanian, D.V., Peck, H.D. Jr. and Le Gall, J.** (1984). Desulfovibrio vulgaris hydrogenase : A non heme iron enzyme lacking nickel that exhibits anomalous EPR and Mössbauer spectra. Proc. Nat. Acad. Sci. USA 81 : 3728-3732.

56. **Voordouw, G. and Brenner, S.** (1985). Nucleotide sequence of the gene encoding the hydrogenase from Desulfovibrio vulgaris (Hildenborough). Eur. J. Biochem. 148 : 515-520.

57. **Yagi, T. Kimura, K., Daidoji, H. Sakai, F., Tamura, S. and Inokushi, H.** (1976). Properties of purified hydrogenase from the particulate fraction of Desulfovibrio vulgaris Miyazaki. J. Biochem. 79 : 661-671.

58. **Kovacs, K.L., Seefeldt, L.C., Tigyi, G., Doyle, C.M., Mortenson, L.E. and Arp, D.J.** (1989). Immunological relationship among hydrogenases. J. Bacteriol. 171 : 430-435.

59. **Grande, H.J., Dunham, W.R., Averill, B., Van Dijk, C. and Sands, R.H.** (1983). Electron paramagnetic resonance and other properties of hydrogenases isolated from Desulfovibrio vulgaris (strain Hildenborough) and Megasphaera elsdenii. Eur. J. Biochem. 136 : 201-207.

60. **Kanatzidis, M.G., Hagen, W.R., Dunham, W.R., Lester, R.K. and Coucouvanis, D.** (1985). Metastable Fe/S clusters. The synthesis, electronic structure, and transformations of the $(Fe_6S_6(L)_6)^{3-}$ clusters (L = Cl^-, Br^-, I^-, RS^-, RO^-) and the structure of $((C_2H_5)_4N)_3(Fe_6S_6Cl_6)$. J. Am. Chem. Soc. 107, 953-961.

61. **Le Gall, J., Ljungdahl, P.O., Moura, I., Peck, H.D. Jr., Xavier, A.V., Moura, J.J.G., Teixeira, M., Huynh, B.H. and Der Vartanian, D.V.** (1982). The presence of redox-sensitive Nickel in the periplasmic hydrogenase from Desulfovibrio gigas. Biochem. Biophys. Res. Commun. 106 : 610-616.

62. **Huynh, B.H., Patil, D.S., Moura, I., Teixeira, M., Moura, J.J.G., Der Vartanian, D.V., Czechowski, M.H., Prickril, B.C., Peck, H.D. Jr. and Le Gall, J.** (1987). On the Active sites of the (NiFe) hydrogenase from Desulfovibrio gigas ; Mössbauer and redox-titration studies. J. Biol. Chem. 262 : 795-800.

63. **Johnson, M.K., Zambrano, I.C., Czechowski, M.H., Peck, H.D. Jr., Dervartanian, D.V., Le Gall, J.** (1986). Magnetic circular dichroism and electron paramagnetic resonance studies of

nickel-containing hydrogenases. In "Frontiers in Bioinorganic chemistry", pp. 36-44. Xavier, A.V. (Ed.), VCH Publishers, Weinheim.

64. Fernandez, V.M., Hatchikian, E.C., Patil, D.S. and Cammack, R. (1986). ESR-detectable nickel and iron-sulphur centres in relation to the reversible activation of Desulfovibrio gigas hydrogenase. Biochim. Biophys. Acta. 883 : 145-154.

65. Cammack, R., Patil, D.S., Hatchikian, E.C. and Fernandez, V.M. (1987). Nickel and iron-sulphur centres in Desulfovibrio gigas hydrogenase : ESR spectra, redox properties and interactions. Biochim. Biophys. Acta. 912 : 98-109.

66. Fernandez, V.M., Aguirre, R. and Hatchikian E.C. (1984). Reductive activation and redox properties of hydrogenase from Desulfovibrio gigas . Biochim. Biophys. Acta. 790 : 1-7.

67. Lissolo, T., Pulvin, S. and Thomas, D. (1984). Reactivation of the hydrogenase from Desulfovibrio gigas by hydrogen. Influence of redox potential. J. Biol. Chem. 259 : 11725-11730.

68. Fernandez, V.M., Hatchikian, E.C., and Cammack, R. (1985). Properties and reactivation of two different deactivated forms of Desulfovibrio gigas hydrogenase. Biochim. Biophys. Acta. 832 : 69-79.

69. Berlier, Y.M., Fauque, G., Lespinat, P.A. and Le Gall, J. (1982) Activation, reduction and proton-deuterium exchange reaction of the periplasmic hydrogénase from Desulfovibrio gigas in relation with the role of cytochrome c_3. FEBS Lett. 140 : 185-188.

70. Hallahan, D., Fernandez, V.M., Hatchikian, E.C., Cammack, R. (1986). Proton-tritium exchange activity of activated and deactivated forms of Desulfovibrio gigas hydrogenase. Biochim. Biophys. Acta. 874 : 72-75.

71. Ref. deleted.

72. Chapman,A. Cammack,R., Hatchikian,E.C., McCracken,J. and Peisach J. (1988). A pulsed EPR study of redox-dependent hyperfine interactions for the nickel centre of Desulfovibrio gigas hydrogenase. FEBS Lett. 242 : 134-138.

73. Fernandez, V.M., Rua, M.L., Reyes, P., Cammack, R. and Hatchikian, E.C. (1989). Inhibition of Desulfovibrio gigas hydrogenase with copper salts. Eur. J. Biochem., in press.

74. Lindahl, P.A., Kojima, N., Hausinger, R.P., Fox, J.A., Teo, B.K., Walsh, C.T. and Orme-Johnson, W.H. (1984). Nickel and Iron EXAFS of F_{420}-reducing hydrogenase from Methanobacterium thermoautotrophicum. J. Am. Chem. Soc. 106 : 3062-3064.

75. Johnson, M.K., Zambrano, I.C., Czechowski, M.H., Peck, H.D.Jr., Dervartanian, D.V. and Le Gall, J. (1985). Low temperature magnetic circular dichroism spectroscopy as a probe for the optical transitions of paramagnetic nickel in hydrogenase. Biochem. Biophys. Res. Commun. 128, 220-225.

76. Scott, R.A., Wallin, S.A., Czechowski, M., Der Vartanian, D.V., Le Gall, J., Peck, H.D. Jr. and Moura, I. (1984). X-ray absorption spectroscopy of nickel in the hydrogenase from Desulfovibrio gigas. J. Am. Chem. Soc. 106 : 6864-6865.

77. Albracht, S.P.J., Kroger, A., Van der Zwaan, J.W., Unden, G., Bocher, R., Mell, H. and Fontijn, R.D. (1986). Direct evidence for sulphur as a ligand to nickel in hydrogenase : an EPR study of the enzyme from Wolinella succinogenes enriched in ^{33}. Biochim. Biophys. Acta. 874 : 116-127.

78. He, S.H., Teixeira, M., Le Gall, J., Patil, D.S., Moura, I., Moura, J.J.G., Dervartanian, D.V., Huynh, B.H. and Peck, H.D. Jr. (1989). EPR studies with ^{77}Se enriched (NiFeSe) hydrogenase of Desulfovibrio baculatus. Evidence for a selenium ligand to the active-site nickel. J. Biol. Chem. 264, 2678-2682.

79. Eidsness, M.K., Scott, R.A., Prickril, B.C., Dervartanian, D.V., Le Gall, J., Moura, I., Moura, J.J.G. and Peck, H.D.Jr. (1989). Evidence for selenocysteine coordination to the active site nickel in the (NiFeSe) hydrogenase from Desulfovibrio baculatus. Proc. Nat. Acad. Sci. USA 86, 147-151.

80. Voordouw, G., Menon, N.K., Le Gall, J., Choi, E.-S., Peck, H.D. Jr. and Przybyla, A.E. (1989). Analysis and comparison of nucleotide sequences encoding the genes for (NiFe) and (NiFeSe) hydrogenases from Desulfovibrio gigas and Desulfovibrio baculatus. J. Bacteriol. 171 : 2894-2899.

81. Mims, W.B. and Peisach, J. (1980). Pulsed EPR studies of metalloproteins. In Biological Applications of Magnetic Resonance (Shulman, R.G. ed.) pp.221-269, Academic Press, New York.

82. Tan, S.L., Fox, J.A., Kojima, N., Walsh, C.T. and Orme-Johnson, C.T. (1984). Nickel coordination in deazaflavin and viologen-reducing hydrogenases from Methanobacterium thermoautotrophicum : investigation by electron spin echo spectroscopy. J. Am. Chem. Soc. 106, 3064-3066.

83. Van der Zwaan, J.W., Albracht, S.P.J., Fontijn, R.D. and Slater, E.C. (1985). Monovalent nickel in hydrogenase from Chromatium vinosum; light sensitivity and evidence for direct interaction with hydrogen. FEBS Lett. 179 : 271-277.

84. Van der Zwaan, J.W., Albracht, S.P.J., Fontijn, R.D. and Roelofs, Y.B.M. (1986). EPR evidence for direct interaction of carbon monoxide with nickel in hydrogenase from Chromatium vinosum. Biochim. Biophys. Acta. 872 : 208-215.

85. Teixeira, M., Moura, I., Xavier, A.V., Huynh, B.H., DerVartanian, D.V., Peck, H.D.Jr., Le Gall, J. and Moura, J.J.G. (1985). Electron paramagnetic resonance studies on the mechanism of activation and the catalytic cycle of the nickel-containing hydrogenase from Desulfovibrio gigas. J. Biol. Chem. 260 : 8942-8950.

86. Li, C., Peck, H.D.Jr., Le Gall, J. and Przybyla, A.E. (1987). Cloning, characterization and sequencing of the genes encoding the large and small subunits of the periplasmic (NiFe) hydrogenase of Desulfovibrio gigas. DNA 6 : 539-551.

87. Rousset, M., Dermoun, Z., Hatchikian, E.C., and Belaich, J.P. (1989). Cloning and sequencing of the locus encoding for the large and small subunits genes of the periplasmic (NiFe) hydrogenase of Desulfovibrio fructosovorans. J. Bacteriol., submitted.

88. Menon, N.K., Peck, H.D.Jr., Le Gall, J. and Przybyla, A.E. (1987). Cloning and sequencing of the genes encoding the large and small subunits of the periplasmic (NiFeSe) hydrogenase of Desulfovibrio baculatus. J. Bacteriol. 169 : 5401-5407.

89. Hornhardt, S., Schneider, K. and Schlegel, H.G. (1986). Characterization of a native subunit of the NAD-linked hydrogenase isolated from a mutant of Alcaligenes eutrophus H16. Biochimie, 68, 15-24.

90. Leclerc, M., Colbeau, A., Cauvin, B. and Vignais, P.M. (1988). Cloning and sequencing of the genes encoding the large and the small subunits of the H_2 uptake hydrogenase (hup) of Rhodobacter capsulatus. Mol. Gen. Genet. 214 : 97-107.

91. Sayavedra-Soto, L.A., Powell, G.K., Evans, H.J. and Morris, R.O. (1988). Nucleotide sequence of the genetic loci encoding subunits of Bradyrhizobium japonicum uptake hydrogenase. Proc. Nat. Acad. Sc. USA, 85 : 8395-8399.

92. Reeve, J.N., Beckler, G.S., Gram, D.S., Hamilton, P.T., Brown, J.W., Krzycki, J.A., Kolodziej, A.F., Alex, L., Orme-Johnson, W.H. and Walsh, C.T. (1989). A hydrogenase-linked gene in Methanobacterium thermoautotrophicum strain ΔH encodes a polyferredoxin. Proc. Nat. Acad. Sci. USA 86, 3031-3035.

93. Adman, E.T., Sieker, L.C. and Jensen, L.H. (1972). The structure of a bacterial ferredoxin. J. Biol. Chem. 248 : 3987-3996.

94. Carter, C.W.Jr., Kraut, J., Freer, S.T., Xuong, N.H., Alden, R.A. and Bartsch, R. (1974). Two-angstrom crystal structure of oxidized Chromatium high potential iron protein. J. Biol. Chem. 249, 4212-4225.

95. Stout, G.H., Turley, S., Sieker, L.C. and Jensen, L.H. (1988). Structure of ferredoxin I from Azotobacter vinelandii. Proc. Nat. Acad. Sci. USA 85, 1020-1022.

96. Bruschi, M. and Guerlesquin, F. (1988). Structure, function and evolution of bacterial ferredoxins. FEMS Microbiol. Rev. 54, 155-176.

THE F_{420}-REDUCING HYDROGENASE OF *Methanospirillum hungatei* STRAIN GP1

G. Dennis Sprott

Division of Biological Sciences, National Research Council
of Canada, Ottawa, Ontario, Canada K1A 0R6

The F_{420}-reducing hydrogenase (H_2ase) of <u>Methanospirillum</u> <u>hungatei</u>
was isolated from spheroplast lysates by sedimentation, followed by
either sucrose gradients or nickel-affinity chromatography, and gel
filtration. Most of the enzyme was free of the cytoplasmic membrane,
although isolated membranes retained ca. 15% of the F_{420}-reducing H_2ase
activity. The brown H_2ase protein had an absorption spectrum
characteristic of a nonheme iron protein. In electron micrographs it was
a coin-shaped, multisubunit protein of 15.9 nm diameter with a central
depression on one surface. During chromatography on phenyl Sepharose the
H_2ase exhibited hydrophobic properties. Labeling with the isoprenoid
precursor, [^{14}C]-mevalonate, and lipid analysis of the H_2ase $CHCl_3$/MeOH
extracts, established that lipid was associated with the enzyme. This
association appears to result from membrane contamination of the
hydrophobic enzyme. The holoenzyme was about 720 kDa and contained 6-7
Ni^{2+} atoms. H_2-dependent reduction of F_{420} activity was readily, but
transiently, reactivated by anaerobic conditions following exposure of
the enzyme to air. Mg^{2+} or Ca^{2+} were stimulatory. The holoenzyme was
composed of α-subunits of 51 kDa, and 30-31 kDa β and γ subunits of
nearly identical mass. The N-terminal amino acid sequences of the first
20-25 residues were very similar in β and γ subunits. Comparisons made
to sequences known for other H_2ases, established that the <u>M. hungatei</u>
H_2ase was quite different. Antibody raised against the purified
hydrogenase of strain GP1 gave negative reactions with extracts of nine
other methanogens, and a reaction of identity with <u>M. hungatei</u> strain JF1
and <u>Methanosarcina</u> <u>barkeri</u> strain MS.

INTRODUCTION

Hydrogenase is a key enzyme required by methanogens during growth on
H_2 and an oxidized form of carbon, usually CO_2. In these methanogens the
energy yielding reaction is $4H_2 + CO_2 \rightarrow CH_4 + 2H_2O$. The physiological
electron acceptor for the hydrogenase reaction ($H_2 \rightarrow 2H^+ + 2e^-$) is an
8-hydroxy-5-deazaflavin cofactor called F_{420} (Tzeng et al. 1975), which
in reduced form is implicated in activating at least one step in the CO_2
reduction pathway (Hartzell et al. 1985) and in the reduction of NADP
(Tzeng et al. 1975; Baron and Ferry 1989a). Further, the protons in CH_4
are derived largely from H_2O, rather than from H_2 (Spencer et al. 1980),
in keeping with a possible role of hydrogenase in the conservation of
energy by establishing a proton gradient across, or within, the
cytoplasmic membrane.

Microbiology and Biochemistry of Strict Anaerobes Involved in Interspecies Transfer
Edited by J.-P. Bélaich *et al.*
Plenum Press, New York, 1990

75

F_{420}-reducing H_2ases have been purified from several methanogens; namely, Methanococcus vannielii (Yamazaki 1982); Methanosarcina barkeri (Fauque et al. 1984); Methanobacterium thermoautotrophicum (Fox et al. 1987), Methanospirillum hungatei (Sprott et al. 1987), Methanococcus voltae (Muth et al. 1987), and Methanobacterium formicicum (Baron and Ferry 1989b). In addition, a particulate H_2ase reducing benzyl viologen was purified from membranes of Methanobacterium strain G2R (McKellar and Sprott 1979). Important properties of these enzymes which are exploited to advantage in the various purification schemes are the large size of the enzyme, its hydrophobicity, and metal binding. The purification is carried out in aerobic buffers to maintain the enzyme in a relatively stable, inactive form, which can be reactivated under appropriate reducing conditions. Activity may be best preserved by removing H_2 prior to exposing the cells to air (Fox et al. 1987). Since localization will be discussed later, it is adequate to note here that upon cell breakage most of the F_{420}-reducing H_2ase activity is readily separated from the cytoplasmic membrane (Scheme 1). The enzyme can be released gently from the cells of M. hungatei by osmotic lysis of spheroplasts. In most other methanogen genera one, or more, pass of the cells through a French pressure cell is used with similar results. Following breakage, 93-96% of the activity was sedimented upon high speed centrifugation, and was readily separated from the cytoplasmic membrane and methylcoenzyme M reductase in sucrose gradients (Scheme 1).

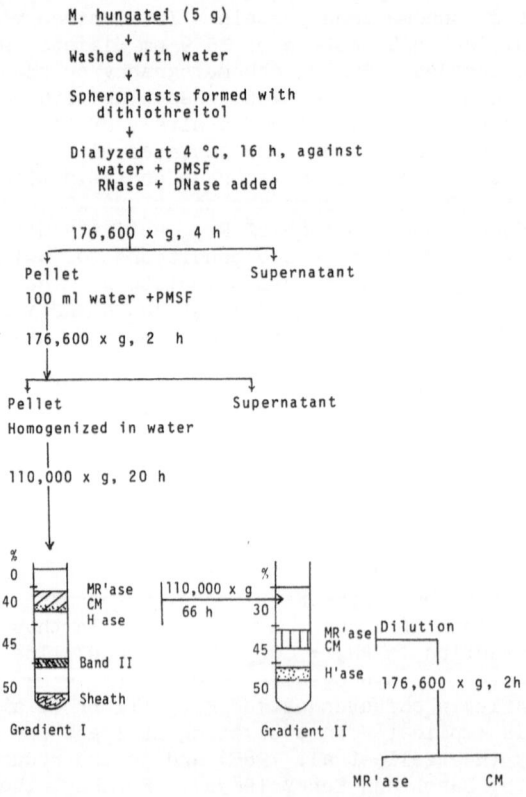

Scheme 1. Purification of F_{420}-reducing H_2ase from M. hungatei using sucrose gradients (Sprott et al. 1987). MRase, methylcoenzyme M reductase; CM, cytoplasmic membrane; PMSF, phenylmethylsulfonyl fluoride.

Crude methanogen extracts, upon electrophoresis under non-dissociating conditions and activity staining, often contain more than one H_2ase activity band (McKellar and Sprott 1979). Multiple bands may occur for several reasons. First, this phenomenon may reflect the aggregation of an α, β, γ, low molecular weight species (ca. 109,000) to the larger form of the F_{420}-reducing hydrogenase as found in M. thermo autotrophicum and M. formicicum (Fox et al. 1987; Baron and Ferry 1989b). Second, two H_2ases may be present, both capable of reducing viologen dyes, but only one active for F_{420}. This distinction in two H_2ases has been made in M. formicicum (Jin et al. 1983), and M. thermoautotrophicum (Jacobson et al. 1982), where the two H_2ases are likely to be distinct enzymes. Presently, there is little evidence to support the contention that a loss of one of the 3 subunits of the F_{420}-reducing H_2ase represents the F_{420}-inactive species (see Kojima et al. 1983).

In the case of M. hungatei, F_{420} and benzyl viologen-reducing activities co-purified indicating the presence of only one predominant H_2ase (Table 1). Similar conclusions were made for Methanococcus vannielii (Yamazaki 1982) and Methanosarcina barkeri (Fauque et al. 1984).

A new method designed for the rapid purification of the F_{420}-reducing H_2ase from M. hungatei is shown in Scheme 2. The procedure involves lysis of spheroplasts by low pressure (5,000-8,000 p.s.i.), differential centrifugations, and Ni-affinity column chromatography. The F_{420}-reducing H_2ase bound to the metal chelating sepharose 6B (Pharmacia) by affinity to the Ni-matrix since no binding occurred in a column lacking Ni. The brown H_2ase was eluted by a decreasing pH gradient at pH 5.8-6.0 (Fig. 1). The purification is illustrated in Table 2. Unfortunately, electrophoretic homogeneity was not achieved without a further step (i.e. gel filtration), but Ni-affinity chromatography was useful in replacing time-consuming sucrose gradient centrifugations. An abundant particulate component emerged from the Ni-column without binding. On SDS-PAGE gels the polypeptide pattern compared closely to that of cytoplasmic membrane, purified according to scheme 1. Autoradiograms were compared of lipids extracted from this fraction and cytoplasmic membrane, following labeling by growth with the phytanyl precursor [^{14}C] mevalonic acid. The complex pattern of ether lipids in both cases were similar, confirming structural similarity to the

Table 1. Purification of hydrogenase from M. hungatei spheroplast lysates[a]

Purification stage	Protein mg	Activity μmol.min^{-1}		Specific activity μmol.min^{-1}.mg^{-1}	
		F_{420}	BV	F_{420}	BV
Spheroplast lysate	4715	315	1387	0.0668	0.294
Washed HSP	2520	323	1262	0.128	0.479
Gradient I	103	16.8	600	0.163	5.83
Gradient II	41	14.6	260	0.356	6.34
Biogel A-15 M Agarose	36	30.5	398	0.847	11.1

Note: BV, benzyl viologen; HSP, high-speed pellet.
[a]Reproduced from Sprott et al. (1987).

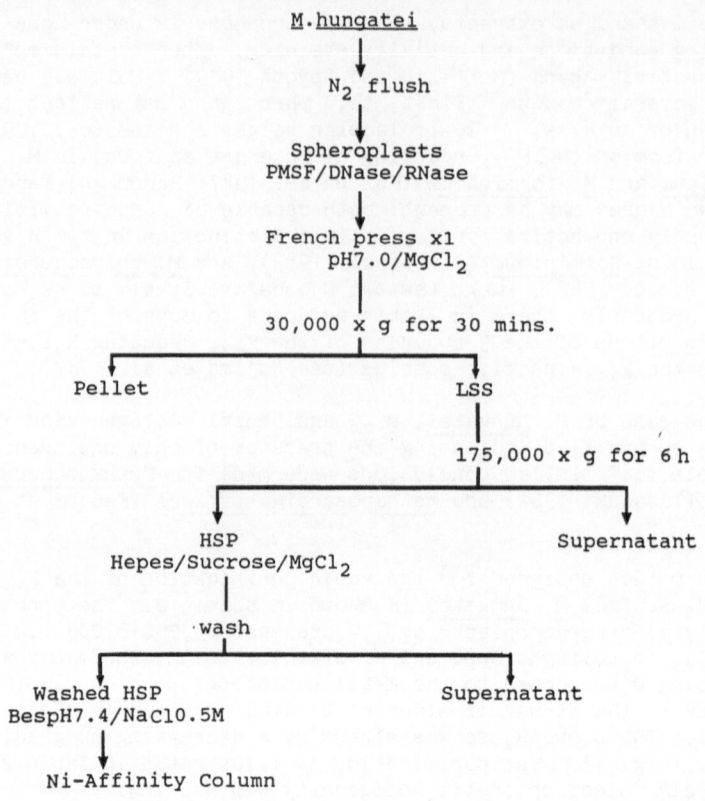

M.hungatei

↓

N₂ flush

↓

Spheroplasts
PMSF/DNase/RNase

↓

French press x1
pH7.0/MgCl₂

↓

30,000 x g for 30 mins.

Pellet   LSS

↓

175,000 x g for 6 h

HSP   Supernatant
Hepes/Sucrose/MgCl₂

↓

wash

Washed HSP   Supernatant
BespH7.4/NaCl0.5M

↓

Ni-Affinity Column

Scheme 2. Purification of the F_{420}-reducing H_2ase of <u>M. hungatei</u> using Ni-affinity chromatography.

cytoplasmic membrane. A yellow component bound weakly at pH 7.4 , and was identified as methylcoenzyme M reductase by reaction in double immunodiffusion with anti <u>M. hungatei</u> reductase, absorbance spectrum, and banding pattern of α, β, γ subunits on SDS-PAGE (Sprott et al. 1987).

ELECTRON MICROSCOPY

F_{420}-reducing H_2ase is a flattened sphere of 15.9 nm diameter with a central pocket on one surface (Sprott et al. 1987). The enzyme from <u>M. thermoautotrophicum</u> was of similar diameter, but ring shaped with a central channel of 4 nm (Wackett et al. 1987). Ring-shaped molecules of 18 nm diameter have been described for <u>M. voltae</u> as well (Muth et al. 1987).

Freeze-thawing (-20°C) of the purified H_2ase from <u>M. hungatei</u> without cryoprotectant caused a loss of F_{420}-reducing activity and different morphological forms could be separated by anion exchange chromatography (Sprott et al. 1987).

MOLECULAR WEIGHT

The molecular weight of the 15.9 nm particle was 720 kDa by agarose 15 M chromatography. All methanogens examined have similar sized

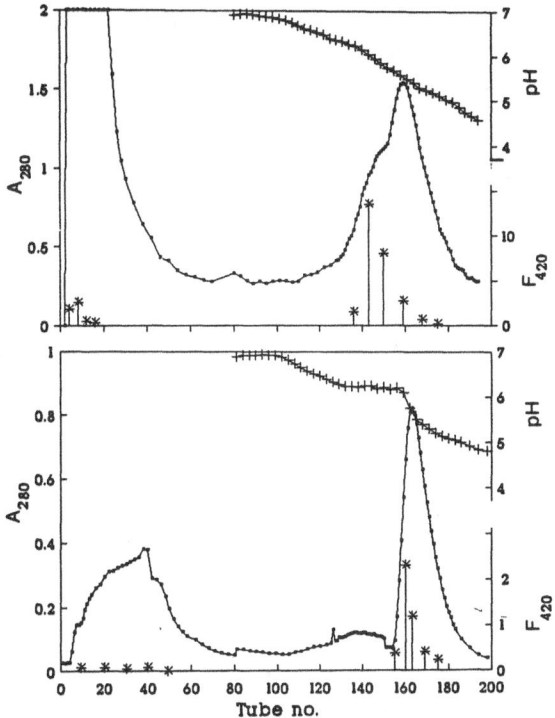

Fig. 1.   Ni affinity chromatography step in purification of F_{420}-
          reducing H_2ase from M. hungatei. Fractions 136-174 (upper
          panel) were loaded on a second similar column (lower
          panel). A_{280} (●), pH gradient (+), μmoles F_{420} reduced/
          min. per 2.9 ml fraction (*).

F_{420}-reducing H_2ases (Table 3). The high molecular weight form is
considered not to be an artifact of the aerobic purification, since it is
retained upon electrophoresis under reducing conditions (Fox et al.
1987). Most F_{420}-reducing H_2ases are composed of three polypeptide
subunits, most likely of 1:1:1 stoichiometry in M. thermoautotrophicum
(Fox et al. 1987). In M. hungatei three subunits are detected as well,
although the two smaller subunits separate only slightly on SDS-PAGE gels
(Fig. 2).

Table 2.   Rapid purification of F_{420}-reducing H_2ase of M. hungatei

Purification Stage	Protein mg	Activity, F_{420} μmol.min^{-1}	Specific activity μmol.min^{-1}mg^{-1}
Spheroplast lysate	3942	631	0.160
LSS	2621	665	0.254
Washed HSP	442	411	0.930
Ni-affinity, first	125	125	1.00
Ni-affinity, second	13.2	36	2.73

Note: LSS, low speed supernatant; HSP, high speed pellet

Table 3. Native molecular weight and subunit composition of F_{420}-reducing H_2ases

Methanogen	Holoenzyme	Subunits				Reference
		α	β	γ	δ	
		M_R (KDaltons)				Sprott et al.
Methanospirillum hungatei GP1	720	50.7	31	30.2	–	(1987), Fig. 2
Methanobacterium thermoautotrophicum	800[+]	47	31	26	–	Fox et al. (1987)
Methanobacterium formicicum MF	600	42.6	34	23.5	–	Jin et al. (1983)
Methanobacterium formicicum JF-1	1020[+]	43.6	36.7	28.8	–	Baron & Ferry (1989)
Methanococcus vannielii	1300⇄340	56	42	35	–	Yamazaki (1982)
Methanococcus voltae	745[+]	55	45	37	27	Muth et al. (1987)
Methanosarcina barkeri	800	60	–	–	–	Fauque et al. (1984)

[+]A lower molecular weight form (ca. 105-115 kDa) was observed in lesser amount and interpreted as the minimal sized F_{420}-reactive species.

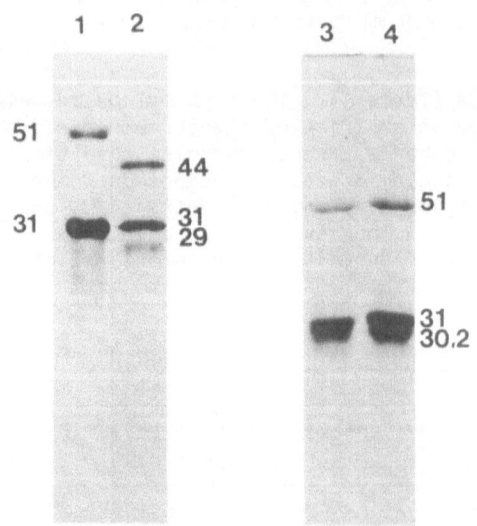

Fig. 2. SDS-PAGE (sodium dodecylsulfate-polyacrylamide gel electrophoresis) of F_{420} reducing H_2ases purified by the sucrose gradient method, scheme 1. Lane 1, M. hungatei H_2ase, 9 µg; lane 2, M. thermoautotrophicum H_2ase, 7.5 µg; lanes 3 and 4, M. hungatei H_2ase, 7.5 and 15 µg, respectively. In lanes 3 and 4 the electrophoresis run was extended to achieve a separation in the 30.2 and 31 kDa polypeptides. Bands were visualized by silver staining.

HYDROPHOBICITY

Aggregation of the polypeptide subunits to form the high molecular weight species may be aided by hydrophobic interaction. Indeed, those F_{420}-reducing H_2ases subjected to phenyl sepharose column chromatography have bound strongly. Elution has been accomplished with ethyleneglycol (Muth et al. 1987), dimethyl sulfoxide (Fox et al. 1987), Triton X-100 (Baron and Ferry 1989b), or, in the case of M. hungatei, with ethanol (Sprott et al. 1987). Hydrophobic interactions between subunits may explain the resistance of the H_2ase complex to separation in denaturants (Fox et al. 1987). Purification of the subunits of the M. hungatei H_2ase by column chromatography in the presence of detergents at 23°C have so far been unsuccessful. The only effective means of accomplishing dissociation and separation has been with hot sodium dodecylsulfate plus β-mercaptoethanol followed by SDS-PAGE.

ABSORBANCE SPECTRUM

The oxidized H_2ase of M. hungatei is brown having an absorbance spectrum typical of a nonheme iron protein, and identical to the M. barkeri H_2ase (Fauque et al. 1984).

REACTIVATION

Methanogen H_2ases quickly lose activity upon exposure to air, but reactivation is possible. This was first demonstrated in methanogens for Methanobacterium strain G2R, where reactivation occurred during incubation of the H_2ase under an H_2 atmosphere in the presence of either dithionite or glucose/glucose oxidase (McKellar and Sprott 1979). The enzyme was not stable after reactivation although storage at 4°C prolonged activity. An instability in reactivated F_{420}-reducing H_2ase occurs in the enzyme purified from M. hungatei within several hours of reactivation (Table 4). Stability was improved in the presence of FAD^+, F_{420} and 40 mM KCl.

Table 4. Reactivation of F_{420}-reducing H_2ase from M. hungatei

Addition to reactivation mix	Activation time (h)		
	2	4	24
	F_{420} reduced (%)		
H_2	71	100	0
H_2, FAD, F_{420}, KCl (40 mM)	83	96	38
H_2, KCl (40 mM)	74	-	10
H_2, KCl (200 mM)	66	-	5
N_2	46	71	6

Note: H_2ase from the Ni-affinity purification stage was reactivated at 22°C for the times indicated. The reactivation mix consisted of H_2ase 0.25 mg/ml, 25 mM Hepes, pH 7.0, $MgCl_2$ 5 mM, NaCl 250 mM, and β-mercaptoethanol 10 mM. Additions were FAD 30 μM, F_{420} 32 μM, and KCl 40 or 200 mM.

The conditions for reactivation have been refined by Muth et al. (1987), also by Baron and Ferry (1989b), and especially by Fox et al. (1987), and can be summarized as follows. Enzyme of at least 0.12 mg/ml concentration is mixed with substrate (F_{420} 50 μM or methylviologen 0.1 mM), and degassed with argon or nitrogen in a serum bottle. KCl (0.8 - 1M) and FAD (30 μM-0.1 mM) may be beneficial depending on the enzyme source and stage of purity. A thiol reducing agent is added, often β-mercaptoethanol at 10 mM, and the inert gas in then replaced by H_2. Incubation for 30-60 min is followed by cooling on ice and replacement of the H_2 gas phase. For H_2ase from M. thermoautotrophicum this basic procedure resulted in activated enzyme which maintained >80% specific activity after 24 h at room temperature (Fox et al. 1987).

In the case of M. hungatei, reactivation of the H_2ase in buffer containing 10 mM β-mercaptoethanol did not require FAD, F_{420} or KCl (Table 4). An H_2 atmosphere enhanced the activity recovered, but considerable reactivation occurred in the absence of H_2.

It is important to stress that reactivated enzyme in the presence of H_2 is irreversibly inactivated once reexposed to air. Fox et al. (1987) noted, however, that reexposure to air is harmless if H_2 is first removed to cause the reoxidation of Fe/S, nickel and FAD centers. These reduced centers are presumably the source of damaging oxygen radicals. A second possibility is that reactivation involves the irreversible reductive modification of a nickel center and incorporation of metal ions (Adams et al. 1986).

REDOX COFACTORS

Present knowledge of redox cofactors and metal ions in methanogen H_2ases is summarized in Table 5. Absorbance spectra indicate that these enzymes have Fe/S centers, although quantitative data on sulfur is often lacking. Similarly Ni has been found in cases where it has been sought. The presence of a flavin, usually FAD, is common and bound FAD may serve as a $1e^-/2e^-$ redox intermediate between the Fe, S clusters and F_{420} (Walsh 1980). Se, Zn and Cu have been detected in H_2ases isolated from specific methanogens. In Desulfovibrio baculatus evidence has been presented for selenocysteine coordination to active site nickel (Eidsness et al. 1989).

LOCALIZATION

Fractionation of cell extracts suggests that part, at least, of the F_{420}-reducing H_2ase activity is membrane associated (McKellar and Sprott 1979; Baron et al. 1987; Sprott et al. 1987). The property of hydrophobicity, and tendency towards self-aggregation, implies a hydrophobic interaction with membrane components. Indeed, it is difficult to remove all traces of membrane lipid from the F_{420}-reducing H_2ase of M. hungatei, and re-addition of lipid results in an increase in rates of F_{420} reduction. Immunogold techniques indicate a cytoplasmic membrane localization for the F_{420}-reducing H_2ases of Methanococcus voltae (Muth 1988) and Methanobacterium formicicum (Baron et al. 1989).

N-TERMINAL SEQUENCING OF SUBUNITS

Polyclonal antibody was prepared against the purified H_2ase of M. hungatei and tested by double immunodiffusion for cross reactivity with other methanogen crude extracts (Sprott et al. 1987). A reaction of

Table 5. Redox cofactors of methanogen hydrogenases

Methanogen	H$_2$ase-type	Flavin	Fe/S	Ni	Se	Other	per KD$_a$ size	Reference
Methanococcus vannielii	F$_{420}$				3.8		340	Yamazaki (1982)
Methanosarcina barkeri	F$_{420}$	1 FMN or riboflavin	8-10 Fe	0.6-0.8			60	Fauque et al. (1984)
Methanobacterium formicicum	Viologen		10 Fe, 8S	1		1 Zn, 2Cu	70	Adams et al. (1986)
Methanobacterium formicicum	F$_{420}$	1 FAD	12-14 Fe, 11S	1	0	0.55 Zn	109	Baron & Ferry (1989b)
Methanococcus voltae	F$_{420}$	1 FAD	4.5 Fe	0.6-0.7	0.6-0.7		105	Muth et al. (1987)
Methanobacterium thermoautotrophicum	F$_{420}$	0.8-0.9 FAD	13-14 Fe, +S	0.6-0.7	0.6-0.7		115	Fox et al. (1987)
Methanospirillum hungatei	F$_{420}$			6-7			720	Sprott et al. (1987)

identity was found for the closely related JF-1 strain of $\underline{M}$. $\underline{hungatei}$. Cross reaction occurred also for $\underline{Methanosarcina}$ $\underline{barkeri}$ extract, and a weak reaction of non-identity was seen with extract of $\underline{Methanobacterium}$ strain G2R. Nine other methanogen extracts were negative, including $\underline{M}$. $\underline{formicicum}$ and $\underline{M}$. $\underline{thermoautotrophicum}$, suggesting considerable dissimilarity to most other F_{420} reducing H_2ases.

Subunits of the F_{420}-reducing H_2ase of $\underline{M}$. $\underline{hungatei}$ were separated by SDS-PAGE followed by electroblotting and sequencing of the N-terminal amino acids. Subunits β and γ separated only slightly and in the first 22 amino acids differed in only 2-4 residues from each other. In contrast to β and γ subunits, the N-terminal sequence of the α-subunit was ca. 50% homologous to that of $\underline{M}$. $\underline{formicicum}$ (Baron and Ferry 1989b) and $\underline{M}$. $\underline{thermoautotrophicum}$ (Fox et al. 1987).

CONCLUSION

In this report I have attempted to summarize much of the data on F_{420}-reducing H_2ases, with emphasis on, and comparison to, our own work with $\underline{M}$. $\underline{hungatei}$. Much data exists on H_2ases from non-methanogens, from which we can obtain inspiration and guidance. The reader is directed to other sources, such as Grahame (1988), for a treatise on these H_2ases.

ACKNOWLEDGEMENTS

Excellent technical support was provided by Kathleen Shaw. N-terminal amino acid sequencing was performed by David Watson.

REFERENCES

Adams, M.W.W., Jin, S.-L.C., Chen, J.-S., and Mortenson, L.E., 1986, The redox properties and activation of the F_{420}-non-reactive hydrogenase of $\underline{Methanobacterium}$ $\underline{formicicum}$, Biochem. Biophys. Acta, 869:37-47.
Baron, S.F., Brown, D.P., and Ferry, J.G., 1987, Locations of the hydrogenases of $\underline{Methanobacterium}$ $\underline{formicicum}$ after subcellular fractionation of cell extract, J. Bacteriol. 169:3823-3825.
Baron, S.F., and Ferry, J.G., 1989a, Reconstitution and properties of a coenzyme F_{420}-mediated formate hydrogen-lyase system in $\underline{Methanobacterium}$ $\underline{formicicum}$, J. Bacteriol. 171:3854-3859.
Baron, S.F., and Ferry, J.G., 1989b, Purification and properties of the membrane-associated coenzyme F_{420}-reducing hydrogenase from $\underline{Methanobacterium}$ $\underline{formicicum}$, J. Bacteriol, 171:3846-3853.
Baron, S.F., Williams, D.S., May, H.D., Patel, P.S., Aldrich, H.C., and Ferry, J.G., 1989, Immunogold localization of coenzyme F_{420}-reducing formate dehydrogenase and coenzyme F_{420}-reducing hydrogenase in $\underline{Methanobacterium}$ $\underline{formicicum}$, Arch. Microbiol. 151:307-313.
Eidsness, M.K., Scott, R.A., Prickril, B.C., DerVartanian, D.V., Legall, J., Moura, I., Moura, J.J.G., and Peck, H.D., 1989, Evidence for selenocysteine coordination to the active site nickel in the [NiFeSe] hydrogenases from $\underline{Desulfovibrio}$ $\underline{baculatus}$, Proc. Natl. Acad. Sci. U.S.A. 86:147-151.
Fauque, G., Teixeira, M., Moura, I., Lespinat, P.A., Xavier, A.V., Der Vartanian, D.V., Peck, H.D., LeGall, J., and Moura, J.G., 1984, Purification, characterization and redox properties of hydrogenase from $\underline{Methanosarcina}$ $\underline{barkeri}$ (DSM800), Eur. J. Biochem. 142:21-28.

Fox, J.A., Livingston, D.J., Orme-Johnson, W.H., and Walsh, C.T., 1987, 8-Hydroxy-5-deazaflavin-reducing hydrogenase from Methanobacterium thermoautotrophicum; 1. Purification and characterization, Biochemistry, 26:4219-4227.

Grahame, D.A., 1988, A summary of new findings in research on hydrogenase, BioFactors, 1:279-283.

Hartzell, P.L., Zvilius, G., Escalante-Semerena, J.C., and Donnelly, M.I., 1985, Coenzyme F_{420} dependence of the methylene-tetramethanopterin dehydrogenase of Methanobacterium thermoautotrophicum, Biochem. Biophys. Res. Commun. 133:884-890.

Jacobson, F.S., Daniels, L.D., Fox, J.A., Walsh, C.T., and Orme-Johnson, W.H., 1982, Purification and properties of an 8-hydroxy-5-deazaflavin-reducing hydrogenase from Methanobacterium thermoautotrophicum. J. Biol. Chem. 257: 3385-3388.

Jin, S.-L.C., Blanchard, D.K., and Chen, J.-S., 1983, Two hydrogenases with distinct electron-carrier specificity and subunit composition in Methanobacterium formicicum, Biochim. Biophys. Acta, 748:8-20.

Kojima, N., Fox, J.A., Hausinger, R.P., Daniels, L., Orme-Johnson, W.H., and Walsh, C., 1983, Paramagnetic centers in the nickel-containing, deazaflavin-reducing hydrogenase from Methanobacterium thermoautotrophicum, Proc. Natl. Acad. Sci. USA, 80:378-382.

McKellar, R.C., and Sprott, G.D., 1979, Solubilization and properties of a particulate hydrogenase from Methanobacterium strain G2R, J. Bacteriol. 139:231-238.

Muth, E., 1988, Localization of the F_{420}-reducing hydrogenase in Methanococcus voltae cells by immuno-gold technique. Arch. Microbiol. 150:205-207.

Muth, E., Morschel, E., and Klein, A., 1987, Purification and characterization of an 8-hydroxy-5-deazaflavin-reducing hydrogenase from the archaebacterium Methanococcus voltae, Eur. J. Biochem. 169:571-577.

Spencer, R., Daniels, L., Fulton, G., and Orme-Johnson, W.H., 1980, Product isotope effects on in vivo methanogenesis by Methano-bacterium thermoautotrophicum, Biochemistry, 19:3678-3683.

Sprott, G.D., Shaw, K.M., and Beveridge, T.J., 1987, Properties of the particulate enzyme F_{420}-reducing hydrogenase isolated from Methanospirillum hungatei, Can. J. Microbiol. 33:896-904.

Tzeng, S.F., Wolfe, R.S., and Bryant, M.P., 1975, Factor 420-dependent pyridine nucleotide-linked hydrogenase system of Methanobacterium ruminantium, J. Bacteriol. 121:184-191.

Wackett, L.P., Hartwieg, E.A., King, J.A., Orme-Johnson, W.H., and Walsh, C.T., 1987, Electron-microscopy of nickel-containing methanogenic enzymes: Methyl reductase and F_{420}-reducing hydrogenase, J. Bacteriol. 169:718-727.

Walsh, C., 1980, Flavin coenzymes: At the crossroads of biological redox chemistry, Acc. Chem. Res. 13:148-155.

Yamazaki, S., 1982, A selenium containing hydrogenase from Methanococcus vannielii, J. Biol. Chem. 257:7926-7929.

ECOLOGICAL IMPACT OF SYNTROPHIC ALCOHOL AND FATTY ACID OXIDATION

Alfons J.M. Stams and Alexander J.B. Zehnder

Department of Microbiology
Agricultural University
Hesselink van Suchtelenweg 4
7603 CT Wageningen; The Netherlands

INTRODUCTION

Interspecies electron transfer is irrefutable associated with methanogenic environments. Methanogens can only to use a very limited range of substrates and therefore acetogens are required for the conversion of reduced organic fermentation products (ethanol, lactate, propionate, butyrate etc.) to methanogenic substrates. The activity of methanogens affects the metabolism of both fermentative and acetogenic bacteria. Fermentative organisms which can dispose part of the reducing equivalents as molecular hydrogen, form more oxidized and less reduced organic products in the presence of hydrogen-consuming methanogens. On the other hand the degradation of such reduced organic compounds is obligately linked with methanogenesis.
The following will focus on factors which affect the formation and degradation of reduced organic compounds in methanogenic ecosystems. Information of the literature is combined with some unpublished data obtained from our laboratory.

FORMATION OF REDUCED ORGANIC PRODUCTS

Source of reduced organic fermentation products

Ethanol and lactate are typical products formed in the fermentation of polysaccharides. No bacterial species have been described which ferment e.g. amino acids to these compounds (Barker, 1981; McInerney, 1988). It is remarkable that under moderate thermophilic conditions almost exclusively sugar-fermenting bacteria have been isolated which form lactate and ethanol as the major reduced organic products, indicating that other fermentation products are of less importance (Table 1). Clostridium thermosaccharolyticum is the only one which is able to form high amounts of butyrate, but only under sporulating conditions (Hsu and Ordal, 1970). All the other species form acetate, hydrogen and different amounts of lactate and ethanol as fermentation products.

Microbiology and Biochemistry of Strict Anaerobes Involved in Interspecies Transfer
Edited by J.-P. Bélaich et al.
Plenum Press, New York, 1990

87

Table 1. Products formed by moderate thermophilic sugar-fermenting anaerobes.

Organism	Temperature	Products
Acetogenium kivui[a]	60	acetate
Clostridium stercorarium[b]	60	acetate, ethanol, lactate, H2
Clostridium thermoaceticum[c]	60	acetate
Clostridium thermocellum[d]	60	acetate, ethanol, H2 (lactate, butyrate)
Clostridium thermohydrosulfuricum[e,f]	60	acetate, ethanol, lactate, H2
Clostridium thermosaccharolyticum[g]	56	acetate, lactate, H2 ethanol, butyrate
Fervidobacterium nodosum[h]	70	acetate, lactate, H2, (ethanol)
Thermobacteroides acetoethylicus[i]	65	ethanol, acetate, H2, (butyrate, isobuturate)
Thermoanaerobacter ethanolicus[j]	72	ethanol, (acetate, ethanol, H2)
Thermoanaerobium brockii[k]	65	acetate, ethanol, lactate, H2
Thermoanaerobium lactoethylicum[l]	65	acetate, ethanol, lactate, H2 (propionate, butyrate, isovalerate)

[a]Leigh et al., 1981; [b]Madden, 1983; [c]Fontaine et al., 1942; [d]Weimer and Zeikus, 1977; [e]Ng et al., 1981; [f]Wiegel et al., 1979; [g]Hsu and Ordal, 1970; [h]Patel et al., 1985; [i]Ben-Bassat and Zeikus, 1981; [j]Wiegel and Ljungdahl, 1981; [k]Zeikus et al., 1979; [l]Kondratieva et al., 1989.

The observations made with isolated species is reflected in incubations of complex anaerobic consortia at higher temperatures. Zoetemeyer et al. (1982) showed that ethanol and lactate became more important fermentation product in a glucose fermenting stirred tank reactor. Figure 1 shows the products formed by granular sludge from a sugar refinery incubated with sucrose at 55 °C in the presence of bromoethanesulfonic acid to inhibit methanogenesis. Acetate, ethanol, lactate, formate and hydrogen were important products, whereas propionate was only formed in minor amounts. In a similar mesophilic experiment propionate accumulated to a concentration of about 5 mM. Under thermophilic conditions butyrate started to accumulate after sucrose had been degraded; most likely it is formed from lactate and acetate in a similar way as in the ethanol fermentation by Clostridium kluyveri (Bornstein and Barker, 1948).

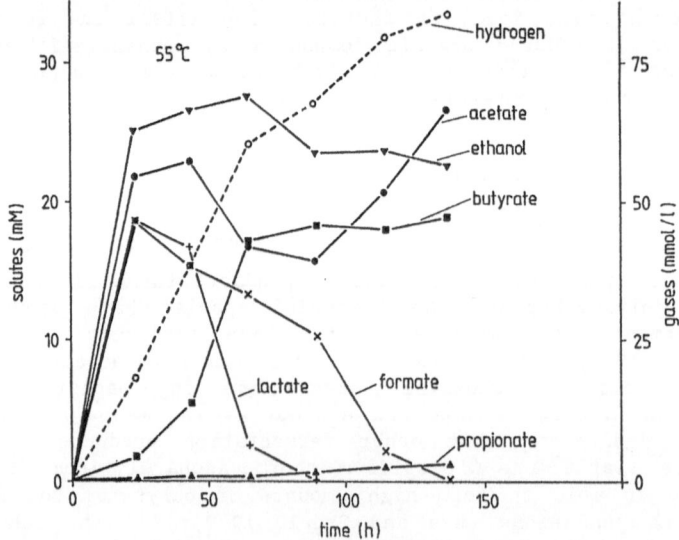

Figure 1. Product formation by granular sludge (5 % wet weight) incubated with 20 mM sucrose at 55 ° in the presence of 10 mM Bres.

Propionate can be formed from several sources. The conversion of sugars, lactate, glycerol and ethanol via a propionic acid type fermentation as carried by Propionibacterium is the best-known way of forming propionate (Schink, 1984; 1986; Laanbroek et al., 1982). Reducing equivalents formed in the oxidation of the substrate are disposed by the reductive conversion of pyruvate to propionate.

Both a reductive and oxidative way of propionate formation is possible in the conversion of amino acids. The fermentation of aspartate, serine and alanine (Hansen and Stams, 1989; Schweiger and Buckel, 1984) resembles the above mentioned reductive formation. Acidaminobacter hydrogenoformans is able to oxidize glutamate and histidine to propionate (Stams and Hansen, 1984). The conversion of glutamate occurs via an oxidative deamination to α-ketoglutarate followed by an oxidative decarboxylation to succinyl-CoA. Recently a thermophilic succinate-degrading bacterium was isolated from granular sludge at our laboratory which was able to degrade a wide variety of amino acids (Cheng Guangsheng, unpublished). This oxidative propionate formation is interesting especially because it may be a rather common way of propionate formation under thermophilic conditions (Skrabanja and Stams, this book).

Butyrate is a compound which can be formed both from sugars (Gottschalk, 1986), some amino acids like e.g. glutamate or lysine (Barker, 1981; McInerney, 1988) and from C1-compounds (Zeikus, 1983). Some bacteria use acetate as sink for reducing equivalents and form butyrate (Bornstein and Barker, 1948; Nanninga and Gottschal, 1985). In all cases butyrate is formed via a reduction of crotonyl-CoA and is therefore a reduced product. Recently it was found that in a mixed anaerobic consortium butyrate can be formed from propionate and vice verse (Tholozan et al., 1988). Bacteria and pathway involved in this process are still unknown, but based on labelling studies it occurs via a direct (de)carboxylation of the C1-atom.

Role of the hydrogen partial pressure on the formation of reduced organic compounds

Fermentative bacteria can be devided according to their ability to form molecular hydrogen. Bacteria which are not able to form hydrogen have a fixed mass balance. Typical examples are some bacteria with a propionic or lactic acid fermentation (Gottschalk, 1986). Many fermentative bacteria do form besides reduced organic compounds also hydrogen as a sink for reducing equivalents. Such organisms are per definition affected by the hydrogen partial pressure i.e. the presence of hydrogen-consuming methanogens regulates which products are formed and to which extent they are formed. The effect of the addition of methanogens is often not recognized, either because of ignorance or because of the faster growth of fermentative organisms in comparison with methanogens. Table 2 summarizes some observed shifts in product formation by fermenting bacteria in the presence of hydrogenotrophic bacteria.

The classical example is the fermentation of glucose by Ruminococcus albus as described by Ianotti et al. (1973). In pure culture this organism forms acetate, ethanol and hydrogen as fermentation products, whereas in the presence of Wolinella (Vibrio) succinogenes acetate and hydrogen are the only fermentation products. This shift in product formation has been explained by the unfavourable energetics for NADH oxidation linked to hydrogen formation. Under standard conditions the $\Delta G^{\circ}{}'$ value is 18.8 kJ per mol, but at a pH_2 of 10^{-3} atm about zero. From the above it should be clear that in methanogenic ecosystems

Table 2. Shifts in product formation by fermentative bacteria in the presence of hydrogenotrophic bacteria. Products in parentheses are not formed or formed in lower amounts in the presence of a hydrogenotroph.

Organism	substrate	Products	Reference
Ruminococcus albus	glucose	(ethanol), acetate, H2	Ianotti et al., 1973
Clostridium thermocellum	cellulose	(ethanol), acetate, H2	Weimer and Zeikus, 1977
Bacteroides xylanolyticus	xylose	(ethanol), acetate, H2	Scholten-Koerselman et al., 1986
Selenomonas ruminantium	glucose	(lactate), acetate, (propionate), H2	Scheifinger et al., 1975 Chen and Wolin, 1977
Acetobacterium woodii	fructose	(acetate), H2	Winter and Wolfe, 1980

where the potential methanogenic activity is higher than the input of fermentable substrates, reduced organic compounds like ethanol, lactate and butyrate are of less importance and in case of propionate it depends from which substrate it is formed. The most abundant glucose-fermenting bacterium which can be isolated from methanogenic granular sludge of a UASB reactor fed glucose at a low input rate, forms ethanol, acetate, and H_2 as products, but in coculture with methanogens acetate is the only organic product (Plugge, unpublished results). Recently anaerobic bacteria were described which, in the absence of a potential electron acceptor, degrade glucose to acetate and H_2 only in the presence of hydrogenotrophic bacteria (Brulla and Bryant, 1989; Krumholz and Bryant, 1986). The quantitative importance of such obligate syntrophic sugar-fermenting bacteria remains to be investigated.

DEGRADATION OF REDUCED ORGANIC PRODUCTS

Role of interspecies electron transfer

Interspecies hydrogen transfer. Under methanogenic conditions propionate and butyrate can only be degraded syntrophically. Ethanol, however, is a versatile substrate which in the absence of an inorganic electron acceptor can be degraded via different types of metabolism (Table 3). The direct oxidation to acetate is most important in anaerobic environments with high methanogenic activity; only in the case of shock loadings intermediairy products are formed (Grotenhuis et al., 1986; Smith and McCarty, 1989). Reactions involved in the conversion of ethanol, propionate and butyrate coupled to H_2 transfer and the respective $\Delta G°'$ values [in kJ] are given in equation 1 to 5.

$$\text{Ethanol} + H_2O \longrightarrow \text{Acetate}^- + H^+ + 2H_2 \qquad + 9.6 \quad (1)$$
$$\text{Propionate}^- + H_2O \longrightarrow \text{Acetate}^- + HCO_3^- + H^+ + 3H_2 \qquad + 76.1 \quad (2)$$
$$\text{Butyrate}^- + H_2O \longrightarrow 2\text{Acetate}^- + H^+ + 2H_2 \qquad + 48.1 \quad (3)$$
$$4H_2 + HCO_3^- \longrightarrow CH_4 + H_2O \qquad -135.6 \quad (4)$$
$$\text{Acetate}^- + H_2O \longrightarrow CH_4 + HCO_3^- \qquad - 31.0 \quad (5)$$

Under standard conditions the conversion of ethanol, butyrate and propionate have a positive $\Delta G'$ and are energetically not possible. Several reports have already discussed extensively how the hydrogen partial pressure affects these conversions and how the actual $\Delta G'$ values can reach values which allow growth for both syntrophic partners (Dolfing, 1988; McCarty, 1981; Zinder, 1984; Stams et al., 1989). This aspect is not reiterated here and the reader is referred to the excellent overview of Dolfing (1988).

Table 3. Ethanol degradation by anaerobic bacteria.

Organism	Reference
Pelobacter carbinolicus ethanol $\longrightarrow$ acetate + 2 H2	Schink, 1984
Acetobacterium carbinolicum ethanol + 1 CO2 $\longrightarrow$ 1.5 acetate	Eichler and Schink, 1984
Desulfobulbus propionicus Pelobacter propionicus 3 ethanol + 2 CO2 $\longrightarrow$ acetate + 2 propionate	Laanbroek et al., 1982 Schink, 1984
Clostridium kluyverii ethanol + acetate $\longrightarrow$ butyrate	Bornstein and Barker, 1948
Methanogenium organophilum ethanol + 0.5 CO2 $\longrightarrow$ acetate + 0.5 CH4	Widdel et al., 1988

A variety of proton-reducing acetogenic bacteria have been described (Table 4). Ethanol-degrading species can also use other substrates like e.g. acetoine, butanediol or pyruvate (Eichler and Schink, 1984; Schink 1984; Plügge and Stams, 1989), whereas butyrate-oxidizers can grow on crotonate (Beaty and McInerney, 1987). Syntrophobacter wolini is the up to now only described species which was obtained in a defined biculture (Boone and Bryant, 1980). Thermophilic syntrophic butyrate-degrading cultures have been described, but these have not yet been classified taxonomically (Ahring and Westermann, 1987a,b). No stable thermophilic propionate-oxidizing cultures have been described so far. This may be due to the sensitivity of these cocultures for changes in environmental conditions or to a high starvation rate of thermophilic hydrogenotrophic methanogens in the absence of substrate.

Formate transfer. Recently, the hypothesis was put forward that besides hydrogen transfer also formate transfer must play an important role (Thiele and Zeikus, 1987; 1988; Ozturk et al., 1988). Several arguments were given to support interspecies formate transfer in syntrophic ethanol, lactate and butyrate oxidation. i) At the same molar concentration of formate and H_2 syntrophic electron transfer coupled to formate formation is energetically in favour compared with the coupling to hydrogen formation. This because the $\Delta G°'$ values are calculated at 1 M concentration for solutes and at 1 atm for gases. At a pH_2 of 1 atm the concentration of H_2 in solution is about 0.85 mM. ii) The diffusion coefficient for formate in water was mentioned to be higher than that of hydrogen (Boone et al., 1989). iii) Most hydrogenotrophic methanogens are also able to use formate. iv) Some proton-reducing bacteria are able to form formate. v) The apparent hydrogen oxidation rates in syntrophic cultures are reported to be higher than can calculated from the diffusion coefficient for hydrogen and the distance between hydrogen-producing and hydrogen-consuming bacteria (Thiele and Zeikus, 1987; Ozturk et al., 1988; Boone et al. 1989), vi) The hydrogen oxidation by the most abundant hydrogen-consuming methanogen at a certain low pH_2 value is not high enough to account for the observed rate of methanogenesis. vii) ^{14}C-formate is formed from [14]CO$_2$ in a syntrophic ethanol-degrading culture (Thiele and Zeikus, 1988), viii) formate dehydrogenase activity was demonstrated in the syntrophic butyrate-oxidizing Syntrophobacter wolfei (Boone et al., 1989).

Table 4. Syntrophic ethanol-, butyrate- and propionate-degrading bacteria

Compound	Organism	Source	Reference
Ethanol:	S-organism	methanogenic culture	Bryant et al, 1967
	Pelobacter carbinolicus	sediments	Schink, 1984
	Acetobacterium carbinolicum	sediments	Eichler and Schink, 1984
	Isolate EE121	methanogenic sludge	Plugge et al., 1989
Propionate	Syntrophobacter wolinii	digestors	Boone and Bryant, 1980
Butyrate	Syntrophomonas wolfei	digestors, rumen, sediments	McInerney et al. 1977 McInerney et al. 1981
	Syntrophomonas sapovorans	digestors	Roy et al. 1986
	Clostridium bryantii	sediments, digestors	Stieb and Schink, 1985

In a complex community it is difficult to distinguish between inter-species hydrogen- and formate transfer because formate and H_2/CO_2 are in close equilibrium and both are easily oxidized to hardly detectable concentrations. Untill the role of hydrogen and formate in syntrophic degradation in not fully understood, the term electron transfer is probably most appropriate. However, to elucidate the importance of H_2 or formate transfer in detail it should be considered that: i) In complex anaerobic communities presumably many different bacteria are present which have either a formate lyase or both a hydrogenase and formate dehydrogenase. Such enzyme systems will result in a rapid interconversion of H_2/CO_2 and formate, especially if the system is perturbated by e.g. high nutrient and/or hydrogen concentrations. Methanothrix soehngenii is able to form H_2 from formate (Huser et al., 1982), whereas Desulfovibrio in the absence of sulfate interconverts formate and H_2/CO_2 (Stams, unpublished results). ii) A comparison of kinetic data obtained for pure cultures of methanogens and mixed methanogenic consortia should be interpreted with caution. E.g. Methanobrevibacter arboriphilus will not be isolated and its relative importance will be underestimated if media are used without the essential nutrient cysteine (Zehnder and Wuhrmann, 1977). A clear argument against the necessity of formate transfer is that syntrophic ethanol and butyrate-degrading cultures exist which grow in the presence of methanogens that are unable to oxidize formate (Bryant et al., 1967; McInerney et al., 1981; Ahring and Westermann, 1987a,b). There is at least clear evidence that in propionate adapted methano-genic granular sludge formate transfer is not important because in thin sections propionate oxidizers were found to be surrounded by Methano-brevibacter arboriphilus, a methanogen which can not use formate (Zehnder and Wuhrmann, 1977), and the distance between the two types of bacteria is shorter, than the required distance which can be calculated from propionate turover rates and the hydrogen diffusion coefficient (Stams et al, 1989).

Acetate conversion. Up to now little attention has been paid to the role of acatate cleavage in syntrophic degradation. In methanogenic ecosystems acetate is cleaved by the methanogenic bacteria Methano-sarcina and Methanothrix (Zinder and Mah, 1979, Huser et al, 1982; Mah et al., 1978; Zinder et al., 1984; Hang Min and Zinder; 1989; Scherer and Sahm, 1981; Westermann et al., 1989). Under thermophilic conditions acetate can also be oxidized syntrophically to 2 molecules of CO_2 and 4 H_2 (Zinder and Koch, 1984). A remarkable physiological difference between the acetoclastic Methanosarcina and Methanotrhix is their affinity for acetate. Table 5 summarizes reported treshold values for acetate of the two types of methanogens. The K_m for acetate of Methanothrix is far lower than that of Methanosarcina, a property which was attributed to the different mechanisms for acetate activation (Jetten et al. 1989; Jetten, unpublished). The former activates acetate via an acetyl-CoA synthetase, an enzyme which requires the investment of 2 molecules of ATP per acetyl-CoA formed, whereas the latter organism requires only one ATP for activation via an acetate kinase-phosphotransacetylase system. Methanothrix can create acetate concentrations as low as 7 μM, a value which is very close to treshold values for acetate (3 μM) found in methanogenic granular sludge (Jetten, unpublished). Low acetate concentrations may have a strong effect on syntrophic alcohol and fatty acid oxidation. The last step in syntrophic degradation is the conversion of acetyl-CoA to acetate via the phosphoroclastic split, a reaction which is stoichiometrically and obligately linked to ATP formation. The $\Delta G^{\circ\prime}$ value for the conversion of acetyl-CoA + ADP to acetate + ATP is about -4 kJ per mol (Thauer et al., 1977). Assuming that for living cells the concentration of ATP and ADP are about equal and the acetyl-CoA concentration is 1 mM, an intracellular concentration of 10 mM acetate would already inhibit grows of the acetogen.

If granular methanogenic sludge is incubated with ethanol, propionate or butyrate, these compounds are degraded to methane and CO_2 without the intermediate formation of high amounts of hydrogen and acetate, indicating that both are extremely low ($pH_2 < 10^{-4}$ atm; acetate < 0.1 mM). Figure 2 shows the computed development of the $\Delta G'$ values for ethanol and fatty acid oxidation in the absence of hydrogen consumption but in the presence or absence of Methanotrix, maintaining an acetate concentration of 10 μM. In the absence of the acetoclastic methanogen the $\Delta G'$ values for propionate, butyrate and ethanol oxidation would become positive after about 0.2, 2.5 and 10 mM of substrate is degraded, respectively. Maintaining an acetate concentration of 10 μM has significant effects on all conversions. The $\Delta G'$ value for butyrate oxidation would still be negative after about 15 mM of substrate had been degraded, despite the fact that the hydrogen partial pressure has increased to 0.24 atm. Up to now clear data of the effect of acetate removal on defined syntrophic cultures are scarce. Ahring and Westermann (1987a) showed that in the presence of an acetoclastic methanogen butyrate was faster degraded by a thermophilic butyrate-degrading coculture, although still some acetate accumulated during degradation of butyrate. Boone and Xun (1987) showed that the addition of 20 mM acetate slowed down syntrophic propionate oxidation, while no effect was found by the addition of a similar concentration of NaCl.

Table 5. Affinity of acetoclastic methanogens for acetate.

Organism	Strain	Treshold [μM]	Reference
Methanothrix	opfikon	7	Jetten, unpublished
	spec.	69	Westermann et al., 1989
	CALS-1	12	Hang Min and Zinder, 1989
Methanosarcina	Fusaro	200	Scherer and Sahm, 1981
	CALS-1	190	Hang Min and Zinder, 1989
	227	1180	Westermann et al., 1989
	mazei	397	Westermann et al., 1989

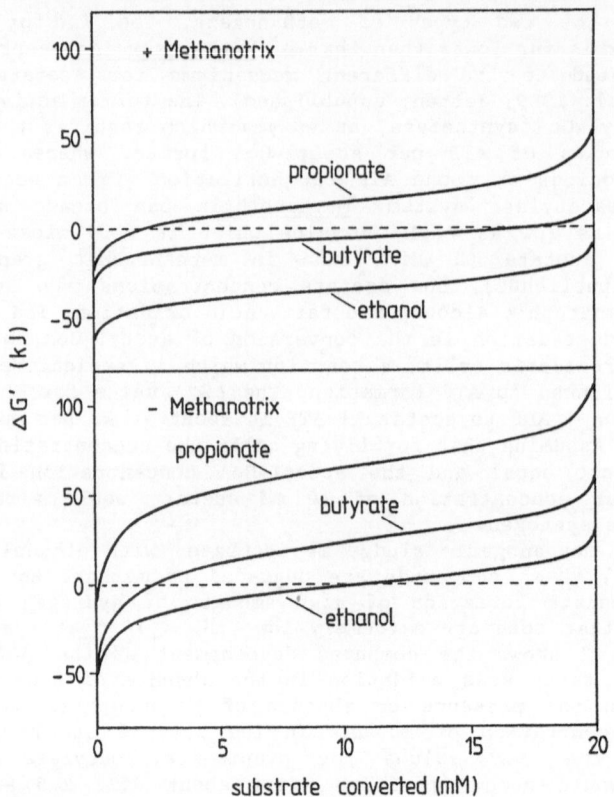

Figure 2. Computation of the ΔG' of propionate, butyrate and ethanol oxidation in the absence of hydrogenotrophs but in the presence of <u>Methanothrix</u> maintaining an acetate concentration of 10 μM.
Conditions: closed bottles containing 50 ml medium with 20 mM substrate and 50 mM HCO_3^- (pH 7) and 70 ml (1.5 atm) N_2/CO_2 (80/20).

CONLUDING REMARKS

Both the formation and degradation of reduced organic products in methanogenic environments is affected by the activity methanogens. Reduced organic compounds are generally less produced in the presence of hydrogen-consuming methanogens. This is clearly the case with ethanol and butyrate. Propionate takes in an exceptional position because it is either an oxidized or a reduced product, whose formation is either stimulated or inhibited by methanogens.

Syntrophic degradation of alcohols and fatty acids is an intriguing phenomenon, which needs to be studied in more detail both on ecological and physiological level. Besides hydrogen transfer also formate and acetate transfer may have a strong impact on syntrophic degradation. Defined syntrophic cultures and sophisticated analytical methods have become available in the recent past which may allow to study the effects of substrate concentration and possible intermediates in more detail.

ACKNOWLEDGMENT

This research was made possible by a grant of the Royal Netherlands Academy of Arts and Sciences.

REFERENCES

Ahring, B.K., and Westermann, P., 1987. Thermophilic anaerobic degradation of butyrate by a butyrate-utilizing bacterium in coculture and triculture with methanogenic bacteria, Appl. Environ. Microbiol., 53:429.

Ahring, B.K., and Westermann, P., 1987. Kinetics of butyrate, acetate, and hydrogen metabolism in a thermophilic, anaerobic, butyrate-degrading triculture, Appl. Environ. Microbiol., 53:434.

Barker. H.A., 1981, Amino acid degradation by anaerobic bacteria, Ann. Rev. Biochem., 50:23.

Beaty, P.S., and McInerney, M.J., 1987, Growth of Syntrophomonas wolfei in pure culture on crotonate, Arch. Microbiol., 147:389.

Ben-Bassat, A., and Zeikus, J.G., 1981, Thermobacteroides acetoethylicus gen.nov. and spec.nov., a new chemoorganotrophic anaerobic bacterium. Arch. Microbiol., 128:365.

Boone, D.R., and Bryant, M.P., 1980, Propionate-degrading bacterium, Syntrophobacter wolinii sp.nov. gen.nov. from methanogenic ecosystems, Appl. Environ. Microbiol., 40:626.

Boone, D.R., Johnson, R.L., and Yitai Liu, 1989, Diffusion of the interspecies electron carriers H_2 and formate in methanogenic ecosystems and its implications in the measurement of K_m for H_2 or formate uptake, Appl. Environ. Microbiol., 55:1735.

Boone, D.R., and Xun, L., 1987, Effects of pH, temperature and nutrients on propionate degradation by a methanogenic enrichment culture, Appl. Environ. Microbiol., 53:1589.

Bornstein, B.T., and Barker, H.A., 1948, The enrgy metabolism of Clostridium kluyveri and the synthesis of fatty acids, J. Biol. Chem., 172:659.

Brulla, W.J., and Bryant, M.J., 1989, Growth of syntrophic anaerobic acetogen strain PA-1 with glucose or succinate as energy source, Appl. Environ. Microbiol. 55:1289.

Bryant, M.P., Wolin, E.A., Wolin, M.J., and Wolfe, R.S., 1967, Methanobacillus omelianskii, a symbiotic association of two species of bacteria, Arch. Microbiol., 59:20.

Chen, M., and Wolin, M.J., 1977, Influence of CH_4 production by Methanobacterium ruminantium on the fermentation of glucose and lactate by Selenomonas ruminantium, Appl. Environ. Microbiol. 34:756.

Dolfing, J., 1988, Acetogenesis, in: "Biology of anaerobic Microorganisms" Zehnder, A.J.B., ed, John Wiley and Sons, New York.

Eichler, B., and Schink, B., 1984, Oxidation of primary aliphatic alcohols by Acetobacterium carbinolicum sp.nov. a homoacetogenic anaerobe. Arch. Microbiol., 147:152.

Fontaine, F.E., Peterson, W.H., McCoy, E., Johnson, M.J., and Ritter, G.J., 1942, A new type of glucose fermentation by Clostridium thermoaceticum N. Sp., J. Bacteriol., 43:701.

Gottschalk, G., 1986, "Bacterial metabolism", Springer Verlag, New York.

Grotenhuis, J.T.C., Houwen, F.P., Plugge, C.M., and Zehnder, A.J.B., 1986, Microbial interactions in granular sludge, Proc. IV Int. Symp. Microb. Ecol., Ljubljana, pp. 163-168.

Hang Min, and Zinder, S.H., 1989, Kinetics of acetate utilization by two thermophilic acetotrophic methanogens: Methanosarcina sp. strain CALS-1 and Methanothrix sp. strain CALS-1, Appl. Environ. Microbiol., 55:488.

Hansen, T.A., and Stams, A.J.M., 1989, A rod-shaped, Gram negative, propionigenic bacterium with a wide substrate range and the ability to fix nitrogen. Arch. Microbiol., Submitted.

Hsu, E.J., and Ordal, Z.J., 1970, Coparative metabolism of vegetative and sporulating cultures of Clostrdium thermosaccharolyticum, J. Bacteriol., 102:369.

Huser, B.A., Wuhrmann, K, and Zehnder, A.J.B., 1982, Methanothrix soehngenii gen.nov. spec.nov., a new acetotrophic non-hydrogen-oxidizing methane bacterium, Arch. Microbiol. 132:1.

Ianotti, E.L., Kafkewitz, D., Wolin, M.J., and Bryant, M.P.,1973, Glucose fermentation products by Ruminococcus albus grown in continuous culture with Vivrio succinogenes: changes caused by interspecies transfer of H_2, J. Bacteriol., 114:1231.

Jetten, M.S.M., Stams, A.J.M., and Zehnder, A.J.B., 1989, Isolation and characterization of acetyl-CoA synthetase from Methanothrix soehngenii, J. Bacteriol., In press.

Kondratieva, E.N., Zacharova, E.V., Duda, V.I., and Krivenko, V.V., 1989, Thermoanaerobium lactoethylicus spec.nov., a new anaerobic bacterium from a hot spring of Kamchatka, Arch. Microbiol. 151:177.

Krumholz, L.R., and Bryant, M.P., 1986, Syntrophococcus sucromutans sp.nov. gen.nov. uses carbohydrates as electron donors and formate, methoxymonobenzoids or Methanobrevibacter as electron acceptor systems, Arch. Microbiol., 143:313.

Laanbroek, H.J., Abee, T. and Voogd, I.L., 1982, Alcohol conversions by Desulfobulbus propionicus Lindhorst in the presence and absence of sulfate and hydrogen, Arch. Microbiol., 133:178.

Leigh, J.A., Mayer, F., Wolfe, R.S., 1981, Acetogenium kivui, a new thermophilic hydrogen-oxidizing, acetogenic bacterium, Arch. Microbiol., 129:275.

Madden, R.H., 1983, Isolation and characterization of Clostridium stercorarium sp.nov., a cellulolytic thermophile, Int. J. Syst. Bacteriol., 33:837.

Mah, R.A., Smith, M.R., and Baresi, L., 1978, Studies on an acetate-fermenting strain of Methanosarcina, Appl. Environ. Microbiol., 35:1174.

McCarty, P.L., 1981, One hundred years of anaerobic treatment, in: "Anaerobic digestion 1981," Hughes, D.A. et al., ed, Elsevier, Amsterdam.

McInerney, M.J., 1988, Anaerobic hydrolysis and fermentation of fats and proteins, in: "Biology of anaerobic Microorganisms" Zehnder, A.J.B., ed, John Wiley and Sons, New York.

McInerney, M.J., Bryant, M.P., and Pfennig, N., 1979, Anaerobic bacterium that degrades fatty acids in syntrophic association with methanogens. Arch. Microbiol., 122:129.

McInerney, M.J., Bryant, M.P., Hespell, R.B., and Costerton, 1981, Syntrophomonas wolfei gen.nov. sp.nov, an anaerobic syntrophic, fatty acid-oxidizing bacterium, Appl. Environ. Microbiol., 41:1029.

Nanninga, H.J., and Gottschal, J.C., 1985, Amino acid fermentation and hydrogen transfer in mixed cultures, FEMS Microbiol. Ecol., 31:261.

Ng., T.K., Ben-Bassat, A., Zeikus, J.G., 1981, Ethanol production by thermophilic bacteria: fermentation of cellulosic substrates by cocultures of Clostridium thermocellum and Clostridium thermohydrosulfuricum, Appl. Environ. Microbiol., 41:1337.

Ozturk, S.S., Palsson, B.O., and Thiele, J.H., 1988,Control of interspecies electron transfer during anaerobic digetion: Dynamic diffusion reaction models for hydrogen gas trasfer in microbial flocs. Biotechnol. Bioeng., 33:745.

Patel, B.K.C., Morgan, H.W., and Daniel, R.M., 1985, Fervidobacterium nodosum gen.nov. and spec.nov., a new chemoorganotrophic, caldoactive, anaerobic bacterium, Arch. Microbiol., 141:63.

Roy, F., Samain, E., Dubourguier, H.C., Albagnac, G., 1986, Syntrophomonas sapovorans sp.nov. a new obligate proton reducing anaerobe oxidizing saturated and unsaturated long chain fatty acids, Arch. Microbiol., 145:142.

Plugge, C.M., Grotenhuis, J.T.C., and Stams, A.J.M., 1989, Isolation and characterization of an ethanol-degrading anaerobe from methanogenic granular sludge. This book.

Scheifinger, C.C., Linehan, B., and Wolin, M.J., 1975, H_2 production by Selenomonas ruminantium in the absence and presence of methanogenic bacteria, Appl. Microbiol., 29:480.

Scherer, P., and Sahm, H., 1981, Influence of sulfur-containing compounds on growth of Methanosarcina barkeri in a defined medium. Eur. J. Appl. Microbiol., 12:29.

Schink, B., 1986, New Aspects of fatty acid metabolism in anaerobic digestion, Proc. IV Int. Symp. Microbiol. Ecol. Ljubljana. pp. 180-184.

Schink, B., 1984, Fermentation of 2,3-butanediol by Pelobacter carbinolycus sp.nov. and Pelobacter propionicus and evidence for propionate formation from C_2-compounds, Arch. Microbiol., 137:33.

Schink, B., and Pfennig, N., 1982, Propionigenium modestum gen.nov. sp.nov., a new strictly anaerobic, nonsporing bacterium growing on succinate, Arch. Microbiol., 133:209.

Scholten-Koerselman, I, Houwaard, F., Janssen, P., and Zehnder, A.J.B., 1986, Bacteroides xylanolyticus sp.nov., a xylanolytic bacterium from methane producing cattle manure, Ant. van Leeuwenhoek, 52:543..

Schweiger, G., and Buckel, W., 1984, On the degradtion of (R)-lactate in the fermentation of alanine to propionate by Clostridium propionicum, FEBS Lett., 171:79.

Skrabanja, A.T.P., and Stams, A.J.M., 1989, Oxidative propionate formation by anaerobic bacteria, This book.

Smith, D.P., and McCarty, P.L., 1989, Energetic and rate effects on methanogenesis of ethanol and propionate in perturbed CSTRs, Biotech. Bioeng., 34:39.

Stams, A.J.M., and Hansen, T.A., 1984, Fermentation of glutamate and other compounds by Acidaminobacter hydrogenoformans gen.nov. spec. nov., an obligate anaerobe isolated from black mud. Studies with pure cultures and mixed cultures with sulfate-rediucing and methanogenic bacteria, Arch. Microbiol.,137:329.

Stams, A.J.B., Grotenhuis, J.T.C., and Zehnder, A.J.B., 1989, Structure-function relationship in granular sludge. Proc. V Int. Symp. Microbiol. Ecol., Kyoto, in press.

Stieb, M., and Schink, B., 1985, Anaerobic oxidation of fatty acids by Clostridium bryantii sp.nov., a sporeforming, obligately syntrophic bacterium, Arch. Microbiol., 140:387.

Thauer, R.K., Jungermann, k., and Decker, K., 1977, Energy conservation in chemotrophic anaerobic bacteria. Bacteriol. Rev., 41:100.

Thiele, J.H., and Zeikus, J.G., 1987, Interactions between hydrogen- and formate-producing bacteria and methanogens during anaerobic digestion, in: "Handbook on anaerobic fermentations", Erickson, L.E. and Fung, D., ed, Marcel Dekker, New York, pp. 537-595.

Thiele, J.H., and Zeikus, J.G., 1988, Control of interspecies electron flow during anaerobic digestion: significance of formate transfer versus hydrogen transfer during syntrophic methanogenesis in flocs, Appl. Environ. Microbiol., 54:20.

Tholozan, J.L., Samain, E., Grivet, J.P., Moletta, R., Dubourguier, H.C., and Albagnac, G., 1988, Reductive decarboxylation of propionate into butyrate in methanogenic ecosystems. Appl. Environ. Microbiol., 54:441

Weimer, P.J., and Zeikus, J.G., 1977, Fermentation of cellulose and cellobiose by Clostrdium thermocellum in the presence and absence of Methanobacterium thermoautotrophicum, Appl. Environ. Microbiol. 33:289.

Westermann, P., Ahring, B.K., and Mah, R.A., 1989, Treshold acetate concentrations for acetate catabolism by aceticlastic methanogenic bacteria, Appl. Environ. Microbiol., 55:524.

Widdel, F., Rouviere, P.E. and Wolfe, R.S., 1988, Classification of secondary alhohol-utilizing methanogens including a new thermophilic isolate, Arch. Microbiol., 150:477.

Wiegel, J., and Ljungdahl, L.G., 1981, Thermoanaerobacter ethanolicus gen.nov. sp.nov., a new extreme thermophilic anaerobic bacterium, Arch. Microbiol., 128:343.

Wiegel, J., Ljungdahl, L.G., and Rawson, J.R., 1979, Isolation from soil and properties of the extreme thermophile Clostridium thermohydrosulfuricum, J. Bacteriol., 139:800.

Winter, J., and Wolfe, R.S., 1980, Methane formation from fructose by syntrophic associations of Acetobacterium woodii and differnt strains of methanogens, Arch. Microbiol., 124:73.

Zehnder, A.J.B., and Wuhrmann, K., 1977, Physiology of Methanobacterium strain AZ, Arch. Microbiol., 111:119.

Zeikus, J.G., 1983, Metabolism of one carbon compounds by chemoorganotrophic anaerobes. Adv. Microbiol. Physiol., 24:215.

Zeikus, J.G., Hegge, P.W., and Anderson, M.A., 1979, Thermoanaerobium brockii gen.nov. and sp.nov., a new chemoorganotrophic, caldoactive, anaerobic bacterium, Arch. Microbiol., 122:41.

Zinder, S.H., 1984, Microbiology of anaerobic conversion of organic wastes to methane: recent developments, ASM News, 50:294.

Zinder, S.H., and Koch, M., 1984, Non-acetoclastic methanogenesis from acetate: acetate oxidation by a thermophilic syntrophic coculture, Arch. Microbiol., 138:263.

Zinder, S.H., Cardwell, S.C., Anguish, T., Lee, M., and Koch, M., 1984, Methanogenesis in a thermophilic (58°C) anaerobic digestor: Methanothrix sp. as an important aceticlastic methanogen, Appl. Environ. Microbiol. 47:796.

Zinder, S.H., and Mah, R.A., 1979, Isolation and chracterization of a thermophilic strain of Methanosarcina unable to use H_2-CO_2 for methanogenesis, Appl. Environ. Microbiol., 38:996.

Zoetemeyer, R.J., Arnoldy, P., Cohen, A., and Boelhouwer, C., 1982, Influence of temperature on the anerobic acidification of glucose in a mixed culture forming part of a two-stage digestion process, Water Res., 16:312.

METHANOGENESIS FROM PROPIONATE IN SLUDGE AND ENRICHMENT SYSTEMS

R.A. Mah[1], L.-Y. Xun[1], D.R. Boone[2], B. Ahring[3]
P.H. Smith[4], and A. Wilkie[4]

[1]Department of Environmental Health Sciences
School of Public Health, UCLA
Los Angeles, CA 90024-1772

[2]Department of Environmental Science and
Engineering, Oregon Graduate Center
Beaverton, OR 97006-1999

[3]Institute of Biotechnology
The Technical University of Denmark
DK-2800 Lyngby, Denmark

[4]Department of Microbiology and Cell Science
University of Florida
Gainesville, FL 32611

INTRODUCTION

The biological formation of methane from organic matter is a
complex microbiological process involving many physiologically
dependent relationships between and among a diversity of heterotrophic fermentative and methanogenic bacteria. The methanogenic
metabolism of all organic matter leads to the formation of the
same types of intermediates namely H_2, CO_2, and formate, acetic,
propionic, and butyric acids. These compounds are, in turn,
converted directly or indirectly to methane by methanogenic
bacteria alone or acting together with non-methanogenic heterotrophs. These latter organisms may be syntrophic partners of the
methanogens or simply members of a broader food-chain. Estimates
on the sum total of the fermentative contributions by these few
metabolic intermediates account for 100% of the methane formed in
a typical digestor. The importance of acetate as a direct
methanogenic intermediate is already well established (5,6).
Evidence points to a more complicated role played by the
metabolism of butyrate and propionate in this fermentation. The

Microbiology and Biochemistry of Strict Anaerobes Involved in Interspecies Transfer
Edited by J.-P. Bélaich et al.
Plenum Press, New York, 1990

99

main focus of the present paper is to examine the role of
propionate in methanogenesis not only to re-assess its importance
as a source of methane or methanogenic precursors in digestors
but also to examine the biochemical and physiological basis for
its conversion to methane.

THE IMPORTANCE OF PROPIONATE

The contribution of the various intermediates (Table 1) to
the formation of methane was estimated by measurement of the
overall rate of methane formation and calculation of the quantity
of methane formed from each volatile acid and from H_2 oxidation/
CO_2 reduction. This estimate was accomplished by measuring
turnover rates and calculating the percentage each intermediate
contributed to methane based on the measured pool sizes (6). The
actual data (previously unpublished) from an experiment on a
digestor fermenting animal waste are shown in Table 1 below.

Table 1

Contribution of Intermediates to
Methanogenesis

Intermediate	% Methane Formed
ACETATE	
1. Via Fermentation Reactions	42
2. From Butyrate	6
3. From Propionate	20
Total from Acetate	**68**
HYDROGEN	
1. Via Fermenation Reactions	15
2. From Butyrate	2
3. From Propionate	15
Total from Hydrogen	**32**
TOTAL METHANE	**100**

Based on the data in Table 1, it is clear that the two
direct precursors of methane, acetate and H_2/CO_2, account for 68%
and 32%, respectively, or all (100%) of the methane formed. It
is also evident that 35% of the total methane is attributed to
the metabolism of propionate, with 20% of the contribution coming
through conversion of propionate to acetate and 15% through its
concomitant oxidation to form H_2. By comparison, the only other
important indirect intermediate, butyrate, accounted for only 8%
of the total methane in this experimental system. The exact
burden attributable to propionate or butyrate must, however, vary

with each type of digestor. Nevertheless, there is no question
regarding the importance of propionate (and butyrate) in the
complete methanogenic fermentation of organic compounds.

DIGESTOR STUDIES

An expected consequence of fermentation failure is an
increase in volatile fatty acid (vfa) concentration which accompanies the cessation of gas production. In fact, a sudden
increase in vfa has long been the signal of impending digestor
failure leading to accumulation of vfa's and inhibition of
methanogenesis. The question remains whether the increased
concentration of vfa's is simply due to an inability to
metabolize the products at a fast enough rate to keep pace with
production or whether some other controlling factors may be
involved. We investigated some of these issues in the present
study of propionate.

In an on-going digestor system, we examined the fate of
continuously infusing propionate into a stable fermentation being
batch-fed Napier grass on a daily basis. The infusion of
propionate on top of the on-going fermentation of Napier grass
substrate allowed us to evaluate any specific effects of the
propionate on a more or less stable basal microbial population.
Any shifts or changes in the population accompanying the infused
propionate should be exerted on a fairly substrate specific and
local scale since the metabolism of propionate is restricted
under these anaerobic conditions to only a few specialized slow-
growing organisms.

Table 2

Continuous Infusion of VFA's
into Napier Grass Digestors

Infused VFA	Infusion Rate μmol/ml/day	Acetate Equivalent	% Methane Yield
Acetic Acid	20	20	98
Propionic Acid	5	5	95
Propionic Acid	7.5	7.5	92
Propionic Acid	12.5	12.5	87

Table 2 shows the results of infusing a relatively high
concentration of acetate and different quantities of propionate
into a Napier grass-fed digestor. The infusion of vfa's was
accomplished over an extended period of time by gradually
increasing the acid concentration until the desired level was
reached. This implied that a gradual enrichment for the
specialized organisms capable of metabolizing the added vfa's

occurred during this period. The sudden infusion of high
concentrations of vfa's resulted in digestor failure. Since
propionate is metabolized methanogenically by way of acetate, we
could calculate the quantity of acetate equivalents expected from
propionate addition, and we selected the actual test concentrations of propionate at much lower acetate equivalents than the
infused acetate. The results show that the 20 μmol/ml/day of
infused acetate had very little effect on the total methane
yield, with 98% of the expected methane coming from the
fermentation of both the Napier grass substrate as well as the
infused acetate. Since the infused acetate was completely
metabolized to CH_4 and CO_2, the lower concentration (in acetate
equivalents) of infused propionate should also be metabolized.
However, at the lowest propionic acid infusion rates (5 and 7.5
μmol/ml/day) tested, 95% and 92% of the expected methane was
generated. At a propionate infusion rate of 12.5 μmol/ml/day
which yields a calculated acetate equivalent of about one-half
the rate of the infused acetate, we found a decrease in expected
methane yield to 87%, i.e., an 11-13% reduction in methanogenesis! Since we found no measureable change in the propionate pool
size and since an even higher infusion rate for acetate had no
inhibitory effect, we concluded that propionate metabolism was
not affected. However, because the methane yield was decreased
by 13%, propionate must either directly or indirectly through
action of the enriched propionate-metabolizing population,
exercise an inhibitory effect on the fermentation of the Napier
grass substrate. This was a surprising finding since we expected
the infused propionate to exert an effect on a much more
restricted microbial level. Further experiments, including the
use of radioactively labeled propionate, should clarify this
preliminary finding.

If we increased the rate of propionate infusion from 12.5 to
20 μmol/ml/day, the Napier grass fermentation remained stable, as
measured by gas production and vfa pool size, until day 35 when
gas production decreased at a rapid rate (Figure 1). At this
point, the rate of gas production continued to fall, and it was
accompanied by a corresponding increase in propionate (Figure 1).
Examination of the H_2 concentration in this failing digestor
revealed a stable concentration of H_2 up until the time of
failure at day 35, when H_2 concentration decreased rapidly
(Figure 2). Interestingly, the decrease in H_2 is accompanied by
a concomitant increase in propionate, implying the reduction of
intermediates to propio-nate and not CH_4 as the final electron
sink for the failed system. Because of the inhibitory effect of
infused propionate on the overall fermentation of Napier grass,
we conclude that the changes in H_2 concentration reflect an
effect of this inhibition rather than a cause of the inhibition.
Both the ability to form CH_4 from CO_2 reduction by H_2 oxidation
and the continued formation of H_2 were inhibited by these
changes. Thus, for practical applications in monitoring during
operational maintenance of a digestor, measurement of H_2 as an
indicator of impending digestor failure does not appear to be
justified.

THE PROPIONATE CONSORTIUM

The concentration of H_2 during the oxidation of propionate
is, nonetheless, an important factor in the thermodynamics of the
propionate oxidation reaction (1,8). H_2 must be kept at
vanishingly low concentrations in order to oxidize propionate

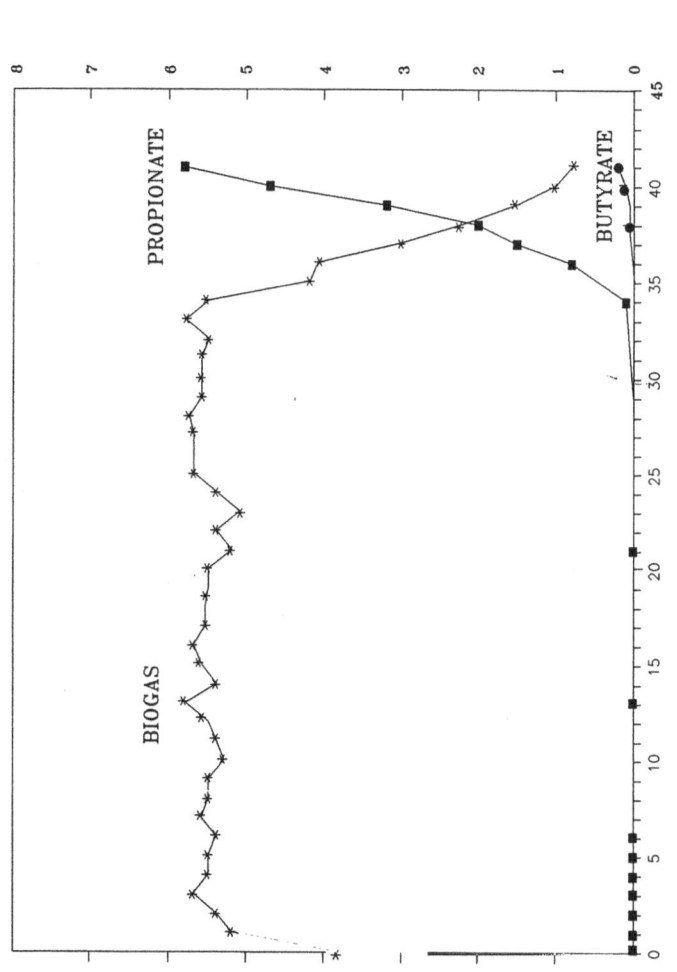

Figure 1. Propionate and Butyrate Concentrations and Gas Production Rates in a Napier Grass–fed Digestor Infused with Propionate

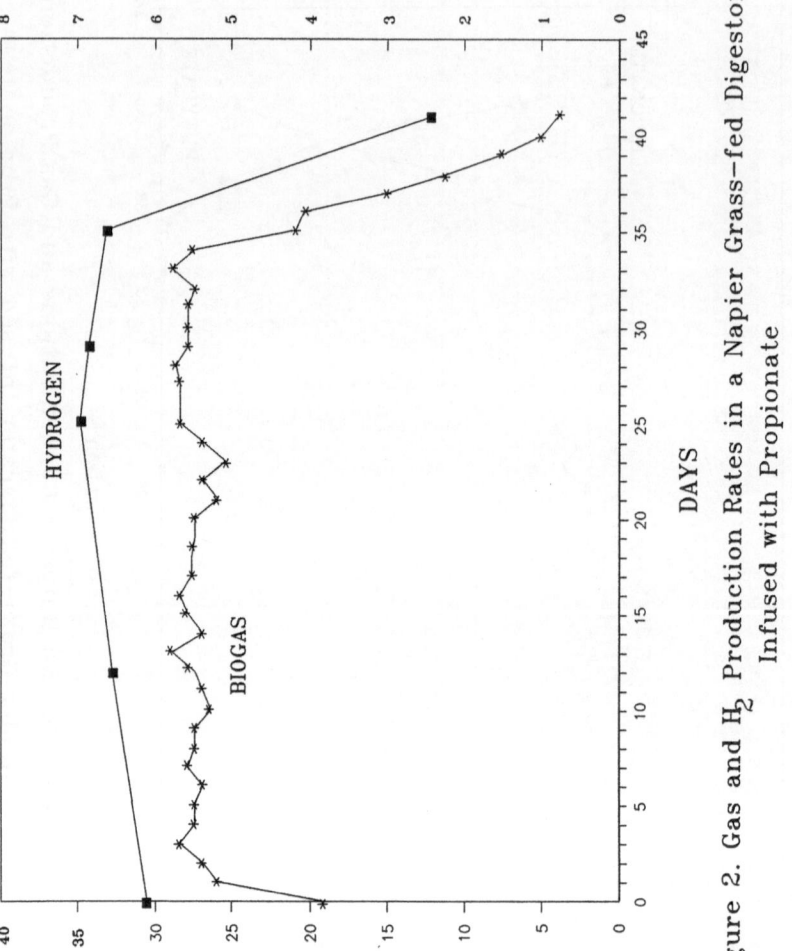

Figure 2. Gas and H$_2$ Production Rates in a Napier Grass-fed Digestor
Infused with Propionate

104

under anaerobic conditions. [Boone, et al. (2) recently calculated that the turnover of H_2 is too slow to account for its role as an oxidizible substrate in the reduction of CO_2 during methanogenesis. Formate, instead of H_2, appears to be a likelier intermediate. However, for convenience, we will continue to refer to H_2 instead of formate in interspecies transfer reactions and consider formate and H_2 to be equivalent.]

Propionate supports the slowest growth of all the known methanogenic intermediates perhaps because of the physiological interactions involved, the thermodynamics of the reaction, which is the least favorable of all the intermediates (8), and the complex pathway involved in its metabolism.

Only a handful of syntrophic co-cultures have been isolated to date. Of these, only one species, *Syntrophobacter wolinii*, was reported for propionate metabolism (1). This organism was first co-cultured by Boone and Bryant (1) using a H_2-oxidizing sulfate-reducing bacterium (SRB) and, more recently, by Xun (9), using similar techniques. There were several reasons for the difficulty in obtaining new strains of this important group of propionate-degrading syntrophs. In addition to the requirement for fastidiously anaerobic techniques, the slow growth rates of the organisms and the requirement for biological reducing systems have all contributed to the paucity of isolates. Up to and including the present, it has not been possible to grow *S. wolinii* in the absence of SRB; it is possible to reduce the numbers of SRB by eliminating sulfate from the medium and adding large concentrations of a H_2-oxidizing methanogen, but the SRB cannot be entirely eliminated from the consortium. Without the initial addition of SRB, it was not possible to obtain a co-culture of *S. wolinii*.

The data reported in the remainder of this paper were based on the study of either a highly enriched propionate culture which we carried by regular transfer for a period of over three years or a co-culture containing combinations of the following organisms: *S. wolinii* LX-2, *Desulfovibrio* G11, *Methanobacterium bourgense* LX-1, and *Methanosarcina mazei* S-6. G11 and S-6 were previously described.

SYNTROPHOBACTER WOLINII

S. wolinii LX-2 was isolated by co-inoculation of a highly enriched culture into dilution media containing *Desulfovibrio* strain G-11. Serial ten-fold dilutions of the enrichment were inoculated directly into pre-reduced roll tube media containing propionate, sulfate, and rumen fluid. Before solidifying the roll tubes, 0.2 ml of an actively growing culture of G11 was added to each tube. After 4 weeks, large (>1mm), isolated dark colonies were formed at high dilutions (0.05-5 nl). These colonies were present in tubes with propionate but absent in those without propionate. A large short rod occurring singly, in pairs, or sometimes in chains as well as the morphologically recognizable G11 were observed under phase-contrast microscopy. A well isolated colony was picked into liquid medium and roll-tube dilutions containing an active culture of G11 were again inoculated as previously described. After growth of dark colonies, the procedure was repeated until a single colony type containing only *S. wolinii* and G11 was obtained. A single colony was picked and diluted into liquid medium and the highest dilution showing growth of the co-culture was designated strain LX-2.

It was not possible to culture LX-2 by substituting a H_2-oxidizing methanogen for strain G11. Even inoculation of a H_2-oxidizing methanogen isolated from the active enrichment culture was not successful. Numerous efforts to eliminate the SRB were futile. A propionate-degrading co-culture containing large numbers of a H_2-oxidizing methanogen was transferred for three years without addition of sulfate, but *Desulfovibrio* G11 was still present in the culture. Under these conditions, the co-culture produced CH_4, CO_2, and acetate as the only products of propionate degradation. Co-culture was also attempted by using roll tubes with a methanogen lawn. Small colonies of G11 persisted in tubes with large inocula; at higher dilutions, colonies never developed even after 4 months of incubation.

Strain LX-2 was a gram-negative rod, 1.0 x 2-4 μm in size. It was similar in morphology to *S. wolinii* strain DB, the type strain. Spores were not observed. LX-2 used propionate only in co-culture with a H_2-using SRB or methanogen. Alternative substrates such as acrylic, succinic, maleic, malonic, and fumaric acid did not support growth of the co-culture. When pyruvate or lactate served as substrate, only colonies of G11 occurred, indicating that only G11 fermented pyruvate and lactate. The growth rate of the co-culture was determined by measuring acetate production, and a value of $\mu = 0.0022$ h^{-1} without added rumen fluid and $\mu = 0.0042$ h^{-1} with rumen fluid. Finally, the mol % G + C of the culture was determined by co-culturing primarily with the methanogen. By freezing and thawing cells and digesting with lysozyme, the methanogen DNA could be eliminated from the mixture. Only LX-2 DNA was detectable in the CsCl gradients, and its mol % G + C was 56.5 ± 0.5.

METHANOGENIUM BOURGENSE STRAIN LX-1

M. bourgense LX-1 was isolated from the active propionate enrichment in an attempt to obtain an appropriate H_2-oxidizing partner directly from the culture and not from a stock culture collection. Strain LX-1 was the predominant H_2-oxidizing methanogen present in the enrichment system and was isolated from high dilution using H_2/CO_2 as the only methanogenic substrate. It was a non-sporulating gram negative irregular coccus approximately 1-2.7 μm in diameter. Its pH optimum was 7.2-7.7, and it grew optimally at 37°C. It had a $\mu = 0.04$ h^{-1}, and we determined its mol % G + C = 56.2. DNA hybridization yielded a sequence similarity of 78 with *M. bourgense* MS2. According to a report of the Ad Hoc Committee of the International Committee for Systematic Bacteriology, bacterial strains wtih approximately 70% or more DNA-DNA relatedness belong to a single species (7). Thus, LX-2 and *M. bourgense* MS2 are strains of the same species (10).

STUDIES ON THE PROPIONATE ENRICHMENT

We used the propionate enrichment cultures for several studies to determine the pH and temperatuare characteristics for propionate vs acetate degradation. Propionate enrichment cultures were grown in several media containing different concentrations of Trypticase peptone and yeast extract. We found that cultures growing without any added organic nutrients grew as

rapidly as those with added nutrients, μ = 0.0048 h^{-1}. Growth
rates were also simlar in medium supplemented with rumen fluid or
sludge supernatant liquid. Although our enrichment cultures did
not require added organic growth factors, other data indicated
that individual organisms present in the enrichment mixture did.
These types of nutrients may have been supplied through cross-
feeding by prototrophic heterotrophs. If we diluted the
propionate enrichment to extinction in inorganic salts medium
methanogenesis and propionate degradation only occurred in tubes
inoculated with 5 μl of culture but not with 0.5 μl. If,
however, we supplemented the basal medium with organic compounds
(e.g., rumen fluid, sludge supernatant), propionate degradation
occurred in tubes inoculated with only 5 nl. Thus, growth
factors were probably needed to supply necessary requirements for
the most numerous organisms present at the highest dilutions of
the enrichment culture.

We adapted one series of propionate enrichment cultures to
utilize propionate without acetate degradation by transferring
the enrichment cultures as soon as the propionate was degraded
but before acetate was metabolized (3). In these cultures,
acetate accumulated because the aceticlastic methanogens were
eventually diluted out. The primary propionate enrichment
cultures were maintained as usual and were transferred only when
propionate was completely converted to CH_4 and CO_2 with no
acetate accumulating.

EFFECT OF pH

The growth rates of propionate enrichment cultures with and
without acetate conversion to CH_4 and CO_2 were determined after
adapting cultures to each of several different pH values. At pH
8.4 and higher and 6.0 and lower, propionate degradation was more
rapid than acetate degradation. When cultures were transferred
to media of different pH, growth immediately occurred at the rate
typical for that pH provided there was no carryover effect from
the inoculum on the pH of the new medium.

EFFECT OF TEMPERATURE

The growth rates at various temperatures were measured after
gradually shifting cultures at 3-5°C intervals from 37 to 55°C.
The optimum temperature occurred between 37 and 42°C; when
cultures grown at higher temperatures were shifted back to 37°C,
they resumed growth at the 37°C rate without any lag. Cultures
grown at temperatures of 45°C or higher were unable to degrade
acetate. However, propionate could still support growth at 45°C
but not apparently at 50°C although at this latter temperature,
propionate was degraded.

The adaptation of the propionate enrichments to various pH
values, temperatures and other medium conditions occurred
rapidly. The immediate growth rate was often the same as that
measured after several months of transfer and adaptation,
provided the initial conditions tested were the same as those
examined later. These data indicated that the pH and temperature

range for degradation of propionate to acetate, CO_2, and H_2 is broader than that for its complete degradation to CH_4 and CO_2. It appears that the aceticlastic methanoagens are much narrower in their adaptability to environmental change than are the propionate oxidizers.

LABELING STUDIES FOR PROPIONATE METABOLISM

When ^{14}C-carboxyl-labeled propionate was metabolized by a propionate consortium consisting of S. wolinii LX-1, M. bourgense LX-2, and numerically insignificant concentrations of Desulfo-vibrio G-11, the radioactivity ended up almost exclusively in CO_2. However, when 2-^{14}C-labeled propionate was metabolized by the same consortium, the label was now equally distributed between both carbons of acetate. We were able to determine the position of labeling because the 2-^{14}C-propionate is converted only to the level of acetate in the above consortium. The culture was then sterilized by autoclaving, cooled, and re-inoculated with an axenic culture of Methanosarcina mazei S-6 which converted the acetate to CH_4 and CO_2. We found that the label was evenly distributed between these two final end products, indicating a mechanism of conversion of propionate which conformed to the succinate-propionate pathway (4).

In the metabolism of propionate via the succinate pathway, propionate is first converted to propionyl-CoA by reaction with succinyl-CoA (see Figure 3). Propionyl-CoA then reacts with oxalacetate to form pyruvate and methylmalonyl-CoA. In a complicated intra-molecular re-arrangement reaction, the methylmalonyl-CoA is converted via a mutase reaction to form succinyl-CoA, which is converted to succinate.

If the carboxyl group of propionate is labeled, then the acyl-carbon attached to CoA is labeled in the first round of the pathway. Thus, when labeled propionyl-CoA reacts with oxal-acetate the resulting methylmalonyl-coA is also labeled in the acyl-carbon. Methylmalonyl-CoA is then re-arranged to form succinyl-CoA and then succinate. Once succinate is formed, the labeled carboxyl group is equally distributed between the two carboxyl groups because of the symmetry of succinate. On subsequent rounds of the pathway, carboxyl-labeled pyruvate gives rise exclusively to labeled CO_2 via formate hydrogen lyase.

Similarly, when 2-^{14}C-propionate is metabolized by the succinate pathway (Figure 3), methylmalonyl-CoA is labeled in the methylene carbon. After the mutase reaction, the succinate which is formed is labeled in one of the methylene carbons. Again, since succinate is symmetrical, the two methylene carbons are equally labeled. Hence, on the second and subsequent rounds of the pathway, the resulting pyruvate is labeled in the methyl and carbonyl carbons and gives rise on decarboxylation to uniformly labeled acetate. Our data support this pathway.

The metabolism of propionate by anaerobic microorganisms is a complicated process both biochemically and physiologically. When propionate is catabolized anaerobically by oxidation to acetate and CO_2, the electrons generated by this reaction are disposed of by proton reduction to form H_2, which is subsequently oxidized by CO_2-reducing methanogenic bacteria. This proton

Proposed pathway for
propionate catabolism

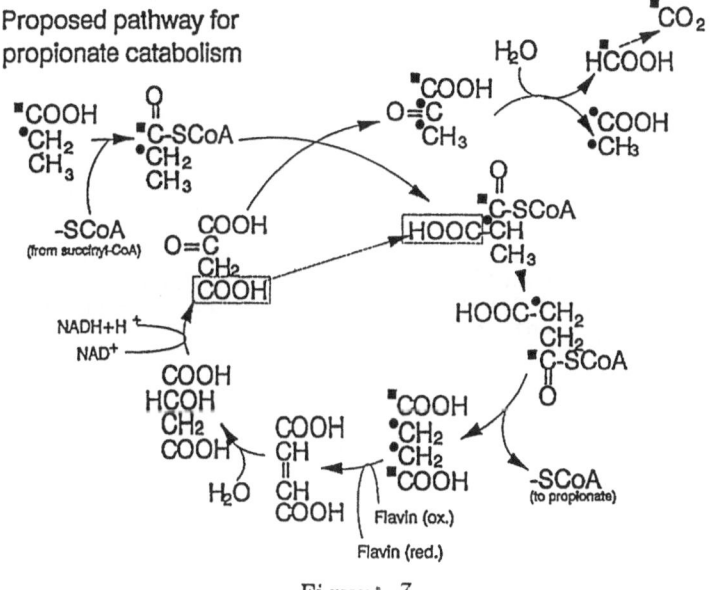

Figure 3

109

reduction is exergonic only over a narrow range of low H_2 concentrations (1.5-106 nM). Thus, the thermodynamics of this reaction make it impossible to oxidize propionate by a single organism acting alone because the H_2 product must be removed and kept at a low enough concentration to make the reaction thermodynamically feasible.

The metabolism of propionate, one of the quantitatively important methanogenic intermediates in digestor systems, remains one of the slowest reactions in the methanogenic fermentation. However, inspite of its complex nature, the ability of propionate enrichments to adapt readily to a broad range of pH, temperatures, and nutrient conditions indicates a greater flexibility than previously thought. The inhibitory effects of propionate enrichment of a subgroup within a stable basal digestor population requires further investigation into the overall regulatory reactions which govern the interplay of microorganisms in this mixed population of bacteria involved in methanogenesis.

ACKNOWLEDGMENTS

This work is part of a joint program (Methane from Biomass) sponsored by the Institute of Food and Agricultural Sciences (IFAS), University of Florida, Gainesville, FL, and the Gas Research Institute, Chicago, IL.

References

1. D. R. Boone and M. P. Bryant . Propionate degrading bacterium , *Syntropher wolinii* , sp. nov. gen. nov. , from methanogenic ecosystems . Appl. Environ. Microbiol. 40 : 626 - 632 .(1980).
2. D. R. Boone , R.L. Johnson and Y. Liu . Diffusion of the interspecies electron carriers H_2 and formate in methanogenic ecosystems and its implications in the measurement of K_m for H2 or formate uptake . Appl. Environ. Microbiol. 55 : 1735 - 1741 . (1989).
3. D. R. Boone and L. Xun . Effects of pH , temperature , and nutrients on proponiate degradation by a methanogenic enrichment culture . Appl. Environ. Microbiol. 53 : 1589 - 1592 . (1987).
4. G. Gottschalk . " Bacterial metabolism " , second edition , pp 244 - 246 . Springler - Verlag New York Inc. , 175 Fith Avenue , New York 10010 , USA .(1985).
5. J.L. Jeris and P.L. McCarty. The Biochemistry of methane fermentation using C^{14}- tracers . J. Water Pollut. Control Fed. 37 , 178 - 192 . (1965).
6. P.H. Smith and R. A. Mah . Kinetics of acetate metabolism during sludge digestion . Appl. Microbiol. 14, 368 - 371 .(1966)
7. L.G. Wayne , D.J. Brenner , R.R. Colwell , P.A.D. Grimont , O. Kandler , M.I. Grichevsky , L.H. Moore , W.E.C. Moore , R.G.E. Murray , E. Stackebrandt , M.P.Starr and H.G. Trüper . Report of the Ad Hoc Committee on reconciliation of approaches to bacterial systematics . Int. J. Syst.Bacteriol. 37, 463 - 464 . (1987).

8. M.J. Wolin . (1974). Metabolic interactions among intestinal microorganisms. Am. J. Clin. Nutr. 27 , 1320 - 1328 .

9. L.Y. Xun . Microbial physiology of an anaerobic propionate-degrading consortium . Ph.D. Dissertation , Departement of Environmental Health Sciences , UCLA . (1989).

10. L.Y. Xun , D. R. Boone and R. A. Mah . Deoxyribonucleic acid hybridation study of *Methanogenium* and *Methanocorpusculum* species , emendation of the genus *Methanocorpusculum* and transfer of *Methanogenium aggregans* to the genus *Methanocorpusculum* as *Methanocorpusculum aggregans* comb. nov. Int. J. Syst. Bacteriol. 39 , 109 -111 .(1989).

CONFERENCES - M : MICROBIOLOGY

SUBSTITUTION OF H_2-ACCEPTOR ORGANISM WITH

CATALYTIC HYDROGENATION SYSTEM IN

METHANOGEN COUPLED FERMENTATIONS

D.O. Mountfort* and H.F. Kaspar

Cawthron Institute
Private Bag, Nelson, New Zealand

INTRODUCTION

In recent years there has been considerable interest in anaerobic digestion as a waste-treatment process and for the production of biogas. However, in industrialised economies, the latter is considered mostly uneconomical. This situation could be reversed if fermentations could be modified favouring the production of more valuable products than methane. One possible alternative to biogas would be the production of volatile fatty acids and hydrogen. The organisms important to this process in digestion are the obligate proton reducing acetogens.[1-4] The energy metabolism of these organisms normally depends on hydrogen removal by methanogens or sulphate reducers.[5,6] Thus, if alternatives for the utilization of this gas are to be sought, they must provide an effective substitute for the H_2 consuming organisms.

In this presentation we describe palladium catalysed hydrogenation of unsaturated hydrocarbons in hydrogen-producing fermentation systems and demonstrate that in cocultures of the obligate proton-reducing acetogen, *Syntrophomonas wolfei* with *Methanospirillum hungatei*, the H_2-accepting methanogen can be replaced by the catalytic hydrogenation system allowing continued acetogenesis and H_2 production by the acetogen.

DEVELOPMENT OF EFFECTIVE CATALYTIC HYDROGENATION SYSTEMS

Lindlar catalyst (1 to 2% Pd on $CaCO_3$), Pd on charcoal (5%; moisture content, 45%), Pd on $BaSO_4$ (10%), Pd on Al_2O_3(5%), and Pd on polyethyleneimine beads (1 to 2%; 40 mesh) were selected for experiments on catalytic hydrogenation. When each was placed either in the gas phase of anaerobic culture tubes (18 x 150mm) as indicated in Fig.1 or directly into uninoculated anaerobic culture medium[7] (10ml) in the presence of H_2 (130 umol) and either acetylene or propylene (each at 130 umol), palladium on charcoal or Pd on $BaSO_4$ (each at 4.5 mg.ml^{-1}) were the most effective catalysts in culture media, and Lindlar catalyst (Pd on $CaCO_3$) at 50mg per tube, the most effective in the gas phase. Rates of propylene hydrogenation up to 300 umol.h^{-1} could be obtained with Lindlar catalyst which was more than 10-fold faster than the rates achievable with catalyst in liquid

Microbiology and Biochemistry of Strict Anaerobes Involved in Interspecies Transfer
Edited by J.-P. Bélaich *et al.*
Plenum Press, New York, 1990

115

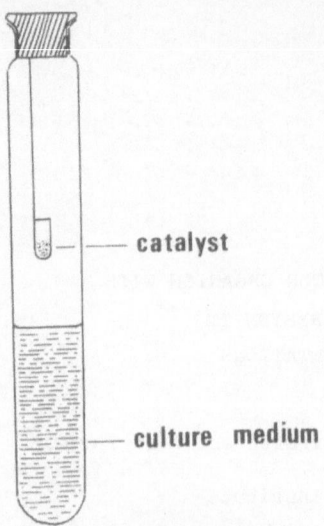

Figure 1.   Catalytic hydrogenation system with
catalyst suspended above culture medium.

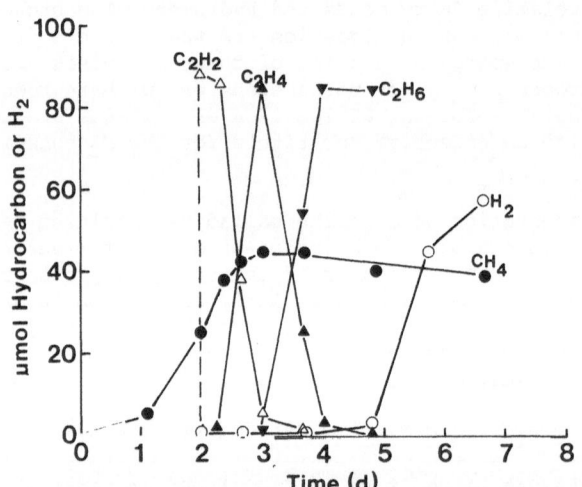

Figure 2.   Effect of Lindlar catalyst and acetylene on
gas metabolism of an *R. albus - M.
hungatei* coculture.   Medium (10 ml)
containing 100 mg of Whatman No.1 filter
paper strips was inoculated with 0.5 ml of a
4 day old *R. albus/M. hungatei* coculture
and incubated at 37°C.   Catalyst (50 mg)
was suspended above the medium in a glass
pouch.   Acetylene was added to cocultures
after 48 h.

culture. Propylene was hydrogenated at the fastest rate followed by acetylene, butene and pentene. In liquid medium, rates of acetylene and ethylene hydrogenation were similar for Pd on charcoal or Pd on $BaSO_4$ but only the former catalytic system hydrogenated the longer chained alkenes.

For further investigations Lindlar catalyst was selected for experiments with catalyst in the gas phase and Pd on charcoal or Pd on $BaSO_4$ for experiments with catalyst in culture media.

HYDROGENATION OF ALKYNES AND ALKENES DURING CELLULOSE FERMENTATION BY RUMEN CELLULOLYTIC ORGANISMS

Incubation of *Neocallimastix frontalis* with cellulose (100mg) in anaerobic growth medium (10ml) containing palladium on charcoal (40 to 200mg.tube^{-1}) and propylene resulted in no degradation of cellulose or olefin hydrogenation. In control tubes with propylene but no catalyst, cellulose degradation was normal as indicated by a maximum of 120 umol of H_2 in 6 days. With cultures of *Ruminococcus albus*, fermenting cellulose, hydrogenation of propylene did occur in the presence of Pd-charcoal but rates of propane formation were considerably slower than with catalyst (Lindlar) suspended in pouches above the culture medium.

Neither Lindlar catalyst nor Pd on charcoal significantly affected the ratio of the fermentation products acetate to ethanol from *R. albus* indicating that these catalysts are not reactive enough to cause a shift in the fermentation pattern towards more oxidized products.

Since Pd on charcoal markedly reduced cellulolytic activity of the rumen organisms and catalysed hydrogenation at a slower rate than Lindlar catalyst, the latter was chosen for further experiments.

Addition of acetylene to cultures of *N. frontalis* or *R. albus* in the presence of Lindlar catalyst, and with either organism in monoculture or in coculture with the H_2- utilizing methanogen, *Methanospirillum hungatei*, first resulted in the production of ethylene which was subsequently hydrogenated to ethane. Methanogenesis in the cocultures was inhibited. Depending on the amount of acetylene added, cultures resumed production of H_2 after the alkyne had been completely hydrogenated. Fig 2 shows the effects of acetylene on gas metabolism of an *R. albus-M. hungatei* coculture fermenting cellulose in the presence of catalyst. In the absence of catalyst, addition of acetylene to cultures resulted in inhibition of methanogenesis but not cellulose fermentation, and H_2 accumulated.

When pentene was substituted for acetylene in the catalyst experiments, similar results were obtained but with pentane being produced.

In the presence of catalyst and either propylene or butene, fermentation-methanogen cultures continued to produce methane at the same rate as controls receiving no added olefin indicating that these olefins were ineffective in inhibiting methanogenesis. However, when bromoethanesulfonic acid (BES) was added to these systems, methane production stopped and olefin reduction occurred. These results indicated that for propylene and butene H_2 availability for olefin reduction in methanogenic cocultures required the addition of BES.

When an olefin mixture (pentene, butene, and propylene) was added to cultures of *R. albus* and *M. hungatei*, inhibition of methanogenesis occurred and all three olefins were hydrogenated at rates inversely proportional to chain length.

SUBSTITUTION OF CATALYTIC OLEFIN REDUCTION FOR METHANOGEN IN SYNTROPHIC COCULTURE WITH *S. WOLFEI*

To determine whether catalytic hydrogenation of olefin could function as an alternative H_2 sink to methanogenesis in syntrophic cocultures Pd-based catalyst was placed in anaerobic culture medium together with the coculture.[8] Among six catalysts Pd on $BaSO_4$ was found to have no effect on methanogenesis in the absence of added olefin but in the presence of ethylene which inhibited methanogenesis, this catalyst was found to be the most effective in the production of ethane. Production of the alkane was possible only because of continued butyrate oxidation by the syntroph resulting in H_2 production. Acetylene, butene, and propylene were less effective than ethylene as H_2 acceptors and addition of BES was found to be necessary to inhibit methanogenesis in the case of the two longer-chained olefins.

The optimal concentrations of Pd- $BaSO_4$ and ethylene for ethane production were 7.7 $mg.ml^{-1}$ and 30 kPa respectively. Ethane production could be further improved by addition of sand to the culture medium to increase surface area, the optimal concentration of which was 0.7 $g.ml^{-1}$.

In the absence of ethylene and catalyst the syntrophic coculture produced 1.8 mol of acetate and 0.3 mol of methane per mol of butyrate oxidized. However, in the presence of catalyst and olefin, for each mol of butyrate oxidized 2.4 mol of acetate and 0.05 mol of methane were produced together with 0.8 mol of ethane. In control incubations >67% of the H_2 produced could be accounted for by methane, and with the catalytic system, more than 77% could be accounted for by methane, ethane, and H_2. catalyst.

The data demonstrate the technical feasibility of uncoupling butyrate oxidation by *S. wolfei* from H_2 utilization by *M. hungatei* by incorporation of a catalytic olefin reducing system into the culture medium.

CONCLUSION

This communication describes how incorporation of catalytic hydrogenation system may be used to modify anaerobic digestion. However, a number of difficulties must be expected if catalysts are to be used in anaerobic digestors, and the likely detrimental effect on catalyst life of organic acids, sulphide, and metal binding agents would need to be considered together with environmental and economic constraints.

Probably the best option for use of Pd-based catalysts in digestors, would be in those optimized for organic acid production such as has been described by Playne.[9,10] In digestor operations modified for this purpose, the methanogenic step is removed either by addition of BES, through high dilution rates, or by decreased pH or by a combination of these. H_2 production accompanies organic acid production and this can be used in olefin reduction by incorporation of the catalytic hydrogenation system.

SUMMARY

1. Monocultures or cocultures of rumen H_2-producing anaerobes with
   Methanospirillum hungatei, fermenting cellulose could be
   coupled to the reduction of alkynes or alkenes to alkanes in the
   presence of Pd-based hydrogenation catalysts.

2. In cocultures of *Syntrophomonas wolfei* and *M. hungatei* with
   catalytic hydrogenation system in liquid medium, butyrate oxidation
   by *S. wolfei* could be coupled to ethylene reduction with
   concomitant loss of methanogenesis.

REFERENCES

1. D.R. Boone and M.P. Bryant. Propionate-degrading bacterium,
   Syntrophobacter wolinii sp.nov.gen.nov. from methanogenic
   ecosystems. Appl. Environ. Microbiol 40:626 (1980)

2. J.M. Henson and P.H. Smith. Isolation of a butyrate-utilizing
   bacterium in coculture with *Methanobacterium
   thermoautotrophicum* from a thermophilic digestor - Appl.
   Environ. Microbiol. 49: 1461 (1985).

3. M.J. McInerney, M.P. Bryant, R.B. Hespell, and J.W. Costerton.
   Syntrophomonas wolfei gen. nov. sp. nov., an anerobic
   syntrophic, fatty acid oxidizing bacterium. Appl. Environ.
   Microbiol. 41:1029 (1981).

4. D.O. Mountfort, W.J. Brulla, L.R. Krumholtz, and, M.P. Bryant.
   Syntrophus buswellii gen.nov.sp.nov., a benzoate cataboliser
   from methanogenic ecosystems. Int.J.Syst. Bacteriol. 34:216
   (1984)

5. M.J. Wolin and T.L. Miller. Interspecies hydrogen transfer: 15
   years later. ASM News. 48:561 (1982).

6. H.F. Kaspar and K. Wuhrmann. Product inhibition in sludge
   digestion. Microbial Ecol. 4:241 (1978).

7. D.O. Mountfort and H.F. Kaspar. Palladium-mediated hydrogenation
   of unsaturated hydrocarbons with hydrogen gas released during
   anaerobic cellulose degradation. Appl. Environ. Microbiol. 52:
   744 (1986).

8. H.F. Kaspar, A.J. Holland and D.O. Mountfort. Simultaneous
   butyrate oxidation by *Syntrophomonas wolfei*, and catalytic
   olefin reduction in absence of interspecies hydrogen transfer.
   Arch. Microbiol. 147:334 (1987).

9. M.J. Playne. Microbial conversion of cereal straw and bran into
   volatile fatty acids - key intermediates in the production of
   liquid fuels. Food. Technol Aust. 32:451(1980).

10. M.J. Playne and B.R. Smith. Acidogenic fermentations of wastes to
    produce chemicals p474-509 in Proceedings of the first ASEAN
    workshop on Fermentation Technology, Kuala Lumpur 1982.

HYDROGEN TRANSFER IN MIXED CULTURES OF ANAEROBIC

BACTERIA AND FUNGI WITH METHANOBREVIBACTER SMITHII

Colin S. Stewart, Anthony J. Richardson
Roseileen M. Douglas and Corinne J. Rumney

Rowett Research Institute, Bucksburn
Aberdeen, AB2 9SB, U.K.

INTRODUCTION

The processes involved in the reduction of the oxides of sulphur, nitrogen and carbon compete for electrons. The free-energy changes involved, and the nature of the environment influence the outcome of this competition. Although methanogenesis is normally seen as the ultimate electron acceptor in anaerobic fermentations, methanogenesis may be competitively inhibited under some conditions, for example in anaerobic sediments containing high concentrations of sulphates (Widdel, 1986), or when methanogenic faecal slurries are supplemented with nitrates (Allison & Macfarlane, 1988).

The rumen fermentation is comparatively well understood, and although sulphate reducing bacteria are present (Stewart & Bryant, 1988), methanogenesis is an important route for the disposal of electrons. The approximate overall stoichiometry for the fermentation of dietary organic matter in the rumen can be described by the formula of Wolin (Miller & Wolin, 1979).

$$57.5 \ C_6H_{12}O_6 = 65 \ HAc + 20 \ HPr + 15 \ HBu + 60 \ CO_2 + 35 \ CH_4 + 25 \ H_2O.$$

A similar approximation is more difficult for the fermentation in the monogastric gut, since endogenous substrates (mucopolysaccharides, proteins, immunoglobulins etc.) provide substrates for the synthesis of 50% or more of the hind gut microbial biomass. Furthermore, in the large bowel of humans, significant amounts of methane are formed in less than 50% of the population. The fermentation in methanogenic human faeces is comparable to that in the rumen (Miller & Wolin, 1979). The distribution of methanogens and of sulphate reducers in human faeces led Gibson et al. (1988) to conclude that dissimilatory sulphate reduction and methanogenesis are not normally compatible in the human colon.

The fermentation stoichiometry does not show the flux of compounds produced by some microbial populations and utilised by others. In reviewing the turnover of intermediates in the rumen fermentation, Hungate (1975) calculated that H_2 had the most rapid flux (over 700 nmol/ml/min), and that this could account for all of the methane formed. Formate flux was estimated to be around 130 nmol/ml/min, accounting for about 20% of the H_2 formed. Succinate flux was around 40 nmol/ml/min, accounting for about one-third of the propionate formed. In the rumen, it seems that normally formate is split to H_2 and CO_2 (Carroll & Hungate, 1957). Most rumen methanogens can utilise formate,

Microbiology and Biochemistry of Strict Anaerobes Involved in Interspecies Transfer
Edited by J.-P. Bélaich et al.
Plenum Press, New York, 1990

though their affinity for formate is usually too low to account for a substantial contribution to methanogenesis. However, Lovley *et al.* (1984) described methanogenic rumen bacteria that metabolised formate at concentrations within the rumen range. Recently, Boone *et al.* (1989) calculated the potential for diffusion of H_2 and formate between dispersed microorganisms in cultures containing the butyrate oxidiser *Syntrophomonas* and *Methanobacterium formicicum*. In these cultures, formate was the major interspecies electron carrier, since formate diffuses rapidly enough to support the observed rate of methanogenesis, but H_2 does not. In the rumen, some H_2-producing ciliate protozoa exist in close physical contact with methanogens (Vogels *et al.*, 1980). Although the fate of H_2 and formate in physically associated microbial consortia may differ from that in dispersed cultures, it is clearly not necessarily correct to assume that H_2 is the main route for electron transfer to methanogens in a given co-culture.

The reoxidation of reduced nucleotides in anaerobes may be coupled to the formation of products such as ethanol, lactate, propionate, butyrate and higher fatty acids (see later Fig.1). These products are capable of being utilised by microorganisms, but a combination of thermodynamic and ecological principles ensures that in the gut certain pathways are more important than others. Fatty acids can be converted to methane in anaerobic systems with long turnover times (Wolin & Miller, 1983). However the turnover time of gut contents is so short that these acids escape microbial degradation. It is well established that active methanogenesis tends to repress the formation of other reduced products like lactate and ethanol. For bacteria like *Ruminococcus albus* and the anaerobic fungi, the presence of methanogens typically results in enhanced production of acetate by the electron donor. As the production of acetate is coupled to synthesis of ATP (Gottschalk & Andreesen, 1979), co-culture with methanogens can increase the growth yield of H_2 donating organisms (see later Fig. 1 and Table 3; reviewed by Wolin & Miller, 1988). In this paper we review the microbial species involved in H_2 transfer reactions and describe some recent experiments with co-cultures of gut microorganisms and *Methanobrevibacter smithii*, a methanogen that utilises formate and H_2/CO_2.

HYDROGEN AND FORMATE PRODUCERS IN THE GUT

Many gut microorganisms produce H_2 and formate when grown in axenic culture. Their formation varies according to the species and strain examined, the growth rate and the substrate. Some examples are shown in Table 1. The bacterium *Selenomonas ruminantium* is of special interest because most isolates produce little or no detectable H_2 in axenic culture (Table 1). However when grown in co-culture with methanogens, *S. ruminantium* supports vigorous methane production (Scheifinger *et al.*, 1975). With *Ruminococcus albus*, the production of formate varies according to the culture system used. When grown in continuous culture at a range of dilution rates, *R. albus* strain 7 did not produce formate (Table 1). In batch culture, formate production occurred in the late logarithmic and early stationary phase of growth (Miller & Wolin, 1973).

The ciliate protozoan *Isotricha prostoma* produced H_2 most rapidly from fructose sucrose and glucose; cellobiose supported little H_2 formation (Prins & Van Hoven, 1977). In contrast (Van Hoven & Prins, 1977) the rate of H_2 production by *Dasytricha ruminantium* was greatest with cellobiose as substrate. Although formate was not found among the fermentation products of *D. ruminantium* fermenting endogenous amylopectin (Table 1), *D. ruminantium* formed formate during the fermentation of glucose (Van Hoven & Prins, 1977). The maximum rate of formation of H_2 from glucose was about 10 times higher per cell in *I. prostoma* (around 250 pmol/cell/h) than in *D. ruminantium* (Prins & Van Hoven, 1977; Van Hoven & Prins, 1977). The flagellate *Trichomonas foetus* has been shown to possess a mechanism of hydrogen formation comparable to that in clostridia (Bauchop, 1971).

Table 1. Production of hydrogen and formate by gut microorganisms.

Species/strain		Substrate	H_2	For-mate	Reference
Bacteria					
Eubacterium					
cellulosolvens	261	Glucose	0.2	0.4	Prins *et al.* (1972)
"	252	"	0.4	0.5	Prins *et al.* (1972)
Selenomonas ruminantium					
var. *bryanti*	L22	Glucose	ND	0.5	Prins (1971)
var. *bryanti*	L22	Mannitol	ND	0.3	Prins (1971)
var. *bryanti*	L22	Maltose	ND	0.2	Prins (1971)
var. *bryanti*	E22	Glucose	ND	0.6	Prins (1971)
var. *bryanti*	A22	Glucose	ND	0.6	Prins (1971)
S.ruminantium	HD4	Glucose	ND	4	Chen & Wolin (1977)
S.ruminantium	HD4	Lactate	ND	2	Chen & Wolin (1977)
Clostridium butyricum		Glucose	233	ND	Cited, Wolin (1982)
Ruminococcus albus 7		Glucose	237	ND	Iannotti *et al.* (1973)
Ruminococcus albus 7		Glucose	33	59	Miller & Wolin (1973)
R. flavefaciens C94		Cellulose	37	62	Latham & Wolin (1977)
Butyrivibrio					
fibrisolvens NOR 37		Cellobiose	57	34	Latham & Legakis (1976)
" " IL 631		Cellobiose	2	27	Latham & Legakis (1976)
" " D 1		Cellobiose	32	49	Latham & Legakis (1976
Fungi					
Neocallimastix R1		Glucose	62	77	Lowe *et al.* (1987)
Neocallimastix R1		Xylose	120	61	Lowe *et al.* (1987)
Protozoa					
Isotricha prostoma		Amylopectin	138		Prins & Van Hoven (1977)
Dasytricha ruminantium		Amylopectin	139	ND	Van Hoven & Prins (1977)

ND = not detected. Concentrations are mmol/100 mmol substrate, except for Prins (1971), mmol/100 ml of medium containing 0.5% substrate.

METHANOGENS IN THE GUT

The predominant methanogenic species found in the gut are listed in Table 2, together with their main substrates. *Methanobrevibacter* species are most commonly isolated, followed by *Methanosarcina* and *Methanococcus* species. The genus *Methanogenium*, originally found in marine habitats, was detected by Miller *et al.* (1986) in the faeces of chickens and turkeys. In addition to H_2/CO_2 and formate, other substrates for methanogenesis in the gut include methanol, formed during demethylation of plant polymers (Schink & Zeikus, 1980) and utilised by *Methanosarcina* and *Methanosphaera* species. Tri-methylamine is a product of the breakdown of choline (Fiebig & Gottschalk, 1983), and can be utilised by *Methanosarcina* (Table 2).

CO-CULTURE STUDIES

The effect of the presence of methanogens on the fermentation products that accumulate in co-cultures varies according to the identity of the H_2 donor, but normally

Table 2. Some methanogens isolated from intestinal tracts and faeces

Species	Host	Substrates (other than H_2/CO_2)	Reference
Methanobrevibacter smithii	Human	Formate	Miller *et al.* (1982)
Methanobrevibacter sp	Bovine	Formate	Lovley *et al.* (1984)
Methanobrevibacter sp	Bovine	Formate	Miller *et al.* (1986)
Methanobrevibacter sp	Horse Ruminant Pig Rat Goose	Formate	Miller *et al.* (1986)
M. arboriphilus	Termite	-	Cited, Miller & Wolin (1986)
M. ruminantium	Bovine	Formate	Smith & Hungate (1958)
Methanobacterium formicicum	Bovine	Formate	Oppermann *et al.* (1967)
Methanosarcina sp.	Bovine	Methanol methyl- amines acetate	Patterson & Hespell (1979)
Methanogenium sp.	Fowl	-	Miller *et al.* (1986)
Methanosphaera stadtmaniae	Human	Methanol	Miller & Wolin (1985)
Methanomicrobium sp	Bovine	Formate	Paynter & Hungate (1986)

involves an increase in the production of H_2 at the expense of other reduced products. The basis of this effect can perhaps be most readily appreciated by considering some of the fermentation pathways in which NADH is oxidised, as shown schematically in Fig. 1. More detailed accounts of the reactions involved are given in Miller (1978), Gottschalk & Andreesen (1979), Wolin (1982), Miller & Wolin (1983) and Thauer & Kroger (1984). The synthesis of formate is not shown in Fig. 1. In the case of *Ruminococcus albus*, formate is not formed from pyruvate, but from the reduction of CO_2 (Miller & Wolin, 1979). Briefly, the production of H_2 from NADH by the action of NADH-ferredoxin oxidoreductase and ferredoxin hydrogenase is thought to be thermodynamically feasible only at low partial pressures of H_2 (Gottschalk & Andreesen, 1979; Wolin 1982). In addition, the activity of NADH-ferredoxin oxidoreductase is indirectly regulated by the H_2 concentration, being most active at the low partial pressures of H_2 that are achieved in the presence of an H_2 'sink' such as a methanogen (Thauer & Kroger, 1984). The oxidation of NADH to form H_2 and NAD^+ may divert electrons from the production of lactate, ethanol, butyrate or succinate, and favours the production of acetate from pyruvate, which does not require NADH.

The effect of co-culture of the fungus *Neocallimastix* with methanogens is summarised in Table 3. Reducing ethanol and lactate production in favour of acetate is clearly beneficial in that ATP production will be enhanced in the co-culture. Co-culture with methanogens also changes the fermentation products of other microorganisms, but it is not clear whether there is a net gain of ATP in the co-cultures compared with the axenic cultures. For example, the cellulolytic bacterium *Ruminococcus flavefaciens* shows decreased succinate and increased acetate synthesis when grown in the presence of methanogens (Table 3). Some anaerobic bacteria are thought to conserve energy by

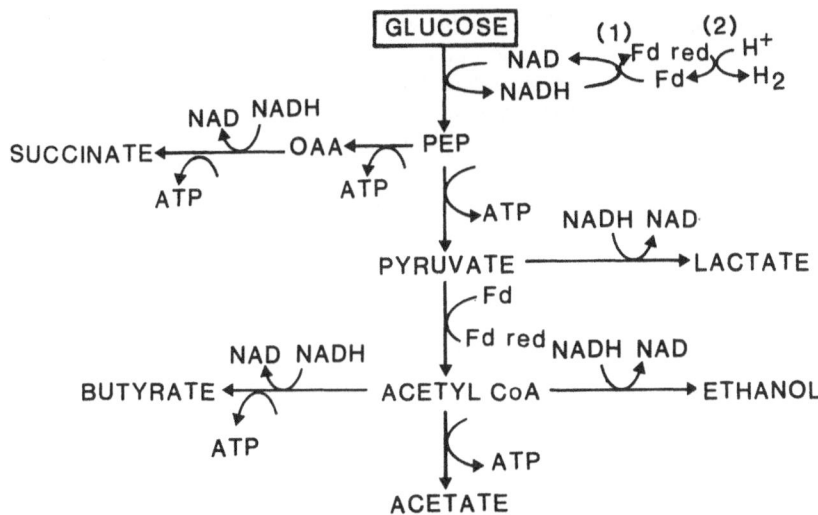

Fig. 1. Fate of reducing equivalents in some anaerobic fermentations. In this example, H_2 may be formed from NADH by NAD-ferredoxin oxidoreductase [1] linked to ferredoxin hydrogenase [2] when the partial pressure of H_2 is low. PEP = phospho-enol pyruvate, OAA = oxaloacetate, NAD = nicotinamide adenine dinucleotide.

electron-transport linked phosphorylation during the reduction of fumarate to succinate (Macy *et al.* 1975). If this happens in *R. flavefaciens*, there is no obvious benefit in replacing succinate production with that of acetate.

Co-culture with *Methanobacterium thermoautotrophicum* reduced ethanol formation by *Clostridium thermocellum* but did not reduce the formation of butyrate or lactate (Table 3). In contrast, the ciliate protozoan *Isotricha* showed reduced formation of butyrate and lactate and increased the formation of acetate in the presence of *Methanosarcina barkeri* (Hillman *et al.*, 1988).

EFFECTS OF THE PRESENCE OF M. SMITHII ON THE SUSCEPTIBILITY OF ANAEROBIC FUNGI TO IONOPHORES

Ionophores have little effect on methanogens in axenic culture (Hammes *et al.*, 1979; reviewed by Stewart & Richardson, 1989), but they inhibit the growth of many of the Gram positive rumen bacteria that provide H_2 for methanogenesis in the rumen (Chen & Wolin, 1979). Anaerobic rumen fungi growing in axenic culture are sensitive to ionophores. However, the presence of *Methanobrevibacter smithii* strain PS reduced the sensitivity of the fungi *Neocallimastix frontalis* and *Piromonas* to monensin and lasalocid (Stewart & Richardson, 1989). The effect of co-culture with *M. smithii* on glucose utilisation by *Piromonas communis* strain P and *Neocallimastix frontalis* strain RE 1 in the presence of monensin is summarised in Table 4. The presence of *M. smithii* also reduced the inhibitory effects of monensin and lasalocid on fungal growth and enhanced the production of acetate by the fungi. When *M. smithii* was present, formate did not accumulate and the production of ethanol and lactate were reduced (Stewart & Richardson, 1989).

Since the fungi gain ATP as a result of the presence of methanogens, it is assumed the effects of the ionophores are partly counteracted using processes that require ATP. It

Table 3.   Differences between the fermentation products
of axenic cultures of *Neocallimastix* sp.,
Ruminococcus flavefaciens and *Clostridium
thermocellum*, and co-cultures of these
organisms with methanogens.

Product	mmol/100 mmol hexose fermented		
	R. flave- faciens.[a]	*C. thermo- cellum.*[b]	*Neocallimastix* sp.[c]
Acetate	82 (+)	105 (+)	62 (+)
Succinate	82 (-)		
Formate	61 (-)		82 (-)
Ethanol		70 (-)	18 (-)
Lactate		1 (+)	64 (-)
H_2	37 (-)	85 (-)	35 (-)
CH_4	83 (+)	56 (+)	59 (+)
CO_2	142 (+)	2 (-)	51 (+)

References: a) Latham & Wolin (1977); b) Weimer &
Zeikus (1977); c) Bauchop & Mountfort (1981). + =
increase and - = decrease in the amount of product
detected in the presence of methanogens.

is established that monensin mediates an electroneutral exchange of certain cations and H^+.
In the bacterium *Streptococcus bovis*, the intracellular concentration of K^+ is normally
much higher than the concentration in the nutrient medium. Monensin causes an efflux of
K^+ accompanied by H^+ influx resulting in a fall in internal pH and a reduction in the
protonmotive force (Russell, 1987). Extra ATP formed in the presence of the methanogen
could be used to restore the protonmotive force so that growth can continue.

EFFECTS OF THE PRESENCE OF METHANOGENS ON CELLULOLYSIS

The presence of methanogens was found by Bauchop & Mountfort (1981) to enhance
both the rate and extent of cellulose hydrolysis by *Neocallimastix*. It seemed that this
effect was caused by enhanced growth (and enzyme production) resulting from the high
ATP yields in the mixed culture. Wood *et al.* (1986) found that co-culture of
Neocallimastix with *Methanobrevibacter smithii* increased the cellulolytic activity of cell
free enzyme preparations, and in particular increased the cotton-degrading activity detected
in supernatant fluids. Co-culture with methanogens was also found to enhance the
degradation of straw by strains of *Neocallimastix frontalis* and *Piromonas communis*,
(Joblin *et al.*, 1989) though the increase in activity was markedly less than that
found by Bauchop & Mountfort (1981) who used filter paper as the substrate for growth.
Presumably the attack of straw is limited by its chemical structure, and increased
microbial growth and enzyme production does not result in a directly proportional increase
in straw degradation.

In the human colon, various species of *Bacteroides* play a prominent role in the
digestion of polysaccharides of animal and plant origin (Salyers, 1984). Since it is known
that these bacteria produce H_2, the effect of the presence of *Methanobrevibacter smithii* on
the ability of *Bacteroides* species to degrade plant cell walls has been studied. The
results are summarised in Table 5. The strains were tested singly, both in the presence

Table 4. The effect of the presence of *M. smithii* strain PS
on the response of *Piromonas communis* strain P and
Neocallimastix frontalis strain RE 1 to monensin
(Stewart & Richardson, 1989).

Strains inoculated	Glucose utilised (mg/ml)			
	P	P+PS	RE 1	RE 1+PS
Monensin (µg/ml)				
None	1.8	1.8	1.8	1.9
0.5	NS	NS	1.1	1.9
1.0	0.4	1.8	0.4	1.5
2.0	0.3	1.7	0.1	0.2

Initial concentration of glucose was 1.8 to 1.9 mg/ml, and
incubation for 5 days at 38°C. NS = not studied.

and absence of *M. smithii*. In general, although methane was produced in the co-cultures, no enhancement of plant cell wall degradation occurred in the presence of *M. smithii*. For this reason, the data presented in Table 5 are the pooled averages of the observations with all 6 bacteria. There was only one exception to this generalisation, in that *B. thetaiotamicron* strain 2255 apparently increased the degradation of broad bean cell walls in the presence of *M. smithii* to about 150% of that which occurred in the pure culture (data not shown). The amounts of methane produced in the co-cultures were small (Table 5), being less than 5% of the quantity which could be expected if anaerobic fungi had been the H_2 donor. The acidic fermentation products of these strains were not studied, but the major products of the species are known to be acetate, succinate and propionate, with traces of higher acids; significant amounts of formate are not normally found (Holdeman *et al.*, 1986). Given the presumed role of fumarate reduction in the energy metabolism of the *Bacteroides* (Macy *et al.*, 1975), these bacteria may have evolved mechanisms to limit their H_2 production to conserve ATP production steps. These results suggest that in these *Bacteroides*, the regulation of H_2 production at low partial pressures of H_2 differs from the regulatory mechanism employed by *Ruminococcus albus* (Thauer & Kroger, 1984). In addition, succinate is reductively decarboxylated to α-oxoglutarate, which is a precursor for the synthesis of amino acids in the *Bacteroides* (Allison *et al.*, 1979), and a reduction in succinate production might impair amino acid synthesis for growth.

INHIBITION OF METHANOGENESIS BY HUMAN FAECAL BACTERIA

In experiments in batch cultures using a habitat-simulating nutrient medium, attempts were made to establish methanogenesis by incubating *M. smithii* (4×10^7 cells) with a culture inoculated with a sample of faeces from a human subject (Rumney, Henderson & Stewart, unpublished data). It had previously been established that faecal samples from this subject did not produce methane. After incubation with *M. smithii*, H_2 accumulated, but methane was not detected, although it was shown that when H_2 was added to the medium in the absence of the faecal inoculum, *M. smithii* produced methane. The 'anti-methanogenic' property of the faecal culture persisted during several subcultures, and survived on chilling the cultures to 4°C for several months. The inhibitory property was retained on filtration of the culture through 0.2 µm membrane filters, but destroyed by autoclaving (121°C, 15 min). There are a number of ways in which the growth or activity of methanogens might have been suppressed in these cultures. Some of the factors possibly involved include the following.

Table 5. Degradation of plant cell walls, H_2 and CH_4 formation in cultures of *Bacteroides* incubated in the presence and absence of *M. smithii* strain PS (Douglas, Chesson & Stewart, unpublished data).

	Wheat bran	Apple	Cabbage	Broad bean	Carrot
% loss in weight.					
Av. of 6 strains	14.5	48.1	34.0	25.4	50.9
" " + PS	15.7	49.2	31.9	27.0	52.1
H_2 prod. (μl)					
Av. of 6 strains	9.4	9.3	7.8	10.5	2.0
" " + PS	0.0	0.0	0.0	0.0	0.0
CH_4 prod. (μl)					
Av. of 6 strains	0.0	0.0	0.0	0.0	0.0
" " + PS	22.1	16.3	15.8	20.8	16.5

The strains tested were *Bacteroides ovatus* strain 1896, *B. vulgatus* strains 10583 and 1447, *B. thetaiotamicron* strains 2255 and 2079, and *B. uniformis* strain 100. Incubation was for 5 d at 38°C. The initial weight of plant cell walls was 20 mg, and the culture volume, 9 ml. Plant cell walls were sterilised by gamma irradiation.

1) Medium oxidised, nutritionally deficient, or adverse pH.
2) Competition from nitrate reduction (Gibson *et al.*, 1988).
3) Competition from sulphate reduction (Allison & Macfarlane, 1988).
4) Presence of an inhibitor such as a bacteriocin.
5) Presence of bacteriophage (Roustan *et al.*, 1986).

M. smithii grew in the nutrient medium when H_2 was added, and H_2 was produced by the faecal culture, so (1) above can be dismissed. Membrane filtration should remove sulphate- and nitrate-reducing organisms (2 and 3). One product of sulphate reducers that might be inhibitory is H_2S, but there was no evidence (i.e. smell, formation of insoluble sulphides) for excessive H_2S production in these cultures. Similarly, although ammonia was produced, the final concentration in the medium after growth (around 70 mM, double the concentration in the uninoculated medium), is within the range of concentrations tolerated by methanogens. Some methanogens possess bacteriophages, but electron microscopic examination of the cultures described here showed no evidence of their presence. The effect encountered here may be due to the presence of a heat-labile inhibitory compound of microbial origin, but of unknown identity.

CONCLUSIONS

The most obvious beneficial effect of methanogens on non-methanogens occurs with those producers of H_2 or formate such as the anaerobic fungi and *Ruminococcus albus* that benefit from the presence of active methanogenesis by improved energy conservation. Improved cellulolysis, and enhanced tolerance of ionophores are presumably consequences of this effect. The variations in the amounts of H_2 and formate produced by different species, and by different strains of the same species, point to differences in the regulation of the fermentation pathways, and demand that considerably more work be carried out to quantify the effect on ATP generation of the presence of methanogens in different mixed populations. The ability to measure physiological changes in individual species growing in co-culture with methanogens would provide the possibility of significant advances in understanding the events that occur intracellularly in co-culture, such as changes in ion permeability in the presence of ionophores. Given the mass-transfer problems involved, simply linking the gas-phases of otherwise separated cultures would provide only a partial solution, but in the absence of obvious alternatives, such experiments seem worthwhile.

In the gut, the relationships between methanogenesis, sulphate reduction and the reduction of nitrates and nitrites are clearly competitive. This competition strongly affects the microbiology of the human colon. There is a suggestion that some faecal bacteria may produce heat-labile compounds toxic to methanogens. The possible existence of such factors suggests that much remains to be learned about the interactions between methanogens and other gastrointestinal microorganisms.

REFERENCES

Allison, C., and Macfarlane, G.T., 1988, Effect of nitrate on methane production and fermentation by slurries of human faecal bacteria, J. Gen. Microbiol., 134: 1397-1405.
Allison, M. J., Robinson, I. M., and Baetz, A. L., 1979, Synthesis of α-ketoglutarate by reductive carboxylation of succinate in *Veillonella, Selenomonas* and *Bacteroides* sp., J. Bacteriol., 140: 980-986.
Bauchop, T., 1971, Mechanism of hydrogen formation in *Trichomonas foetus*, J. Gen. Microbiol., 68: 27-33.
Bauchop, T., and Mountfort, D. O., 1981, Cellulose fermentation by a rumen anaerobic fungus in both the absence and presence of rumen methanogens, Appl. Environ. Microbiol., 42, 1103-1110.
Boone, D. R., Johnson, R. L., and Liu, Y., 1989, Diffusion of the interspecies electron carriers H_2 and formate in methanogenic ecosystems and its implications in the measurement of K_m for H_2 or formate uptake, Appl. Environ. Microbiol., 55, 1735-1741.
Carroll, E. J., and Hungate, R. E., 1957, Formate dissimilation and methane production in bovine rumen contents, Arch. Biochem. Biophys., 56: 525-536.
Chen, M., and Wolin, M. J., 1977, Influence of CH_4 production by *Methanobacterium ruminantium* on the fermentation of glucose and lactate by *Selenomonas ruminantium*, Appl. Environ. Microbiol., 34: 756-759.
Chen, M., and Wolin, M.J., 1979, Effect of monensin and lasalocid-sodium on the growth of methanogenic and rumen saccharolytic bacteria, Appl. Environ. Microbiol. 38: 72-77.
Fiebig, K., and Gottschalk, G., 1983, Methanogenesis from choline by a co-culture of *Desulfovibrio* sp. and *Methanosarcina barkeri*, Appl. Environ. Microbiol., 45: 161-168.
Gibson, G. R., Macfarlane, G. T., and Cummings, J. H., 1988, Occurrence of sulphate-reducing bacteria in human faeces and the relationship of dissimilatory sulphate reduction to methanogenesis in the large gut, J. Appl. Bacteriol., 65: 103-111.
Gottschalk, G., and Andreesen, J. R., 1979, Energy metabolism in anaerobes, in "Microbial Biochemistry," J. R. Quayle ed., Int. Revs. Biochem., 21: Williams and Wilkins, Baltimore.
Hammes, W. P., Winter, J., and Kandler, O., 1979, The sensitivity of the pseudomurein-containing genus *Methanobacterium* to inhibitors of murein synthesis, Archiv. for Microbiol., 123: 275-279.
Hillman, K., Lloyd, D., and Williams, A. G., 1988, Interactions between the methanogen *Methanosarcina barkeri* and rumen holotrich ciliate protozoa, Lett. Appl. Microbiol., 7: 49-53.
Holdeman, L. V., Kelley, R. W., and Moore, W. E. C., 1986, Bacteroidaceae, in "Bergey's Manual of Systematic Bacteriology," N. R. Kreig and J. G. Holt, eds., Williams and Wilkins, Baltimore.
Hungate, R. E., 1975, The rumen microbial ecosystem, Ann. Rev. Ecol. Syst., 6: 39-66.
Iannotti, E. L., Kafkewitz, D., Wolin, M. J., and Bryant, M. P., 1973, Glucose

fermentation products of *Ruminococcus albus* grown in continuous culture with *Vibrio succinogenes*: changes caused by interspecies transfer of H₂, J. Bacteriol., 114: 1231-1240.

Joblin, K., Campbell, G. P., Richardson, A. J., and Stewart, C. S., 1989, Fermentation of barley straw by anaerobic rumen bacteria and fungi in axenic culture and in co-culture with methanogens, Lett. Appl. Microbiol., (in the press). Latham, M. J., and Legakis, N. J., 1976, Cultural factors influencing the utilisation or production of acetate by *Butyrivibrio fibrisolvens*, J. Gen. Microbiol., 94: 380-388.

Latham, M. J., and Wolin, M. J., 1977, Fermentation of cellulose by *Ruminococcus flavefaciens* in the presence and absence of *Methanobacterium ruminantium*, Appl. Environ. Microbiol., 34: 297-301.

Lovley, D. R., Greening, R. C., and Ferry, J. G., 1984, Rapidly growing rumen methanogenic organism that synthesises coenzyme M and has a high affinity for formate, Appl. Environ. Microbiol., 48: 81-87.

Lowe, S. E., Theodorou, M. K., and Trinci, A. P. J., 1987, Growth and fermentation of an anaerobic rumen fungus on various carbon sources and effect of temperature on development, Appl. Environ. Microbiol., 53: 1210-1215.

Macy, J., Probst, I., and Gottschalk, G., 1975, Evidence for cytochrome involvement in fumarate reduction and adenosine 5' triphosphate synthesis by *Bacteroides fragilis* grown in the presence of haemin, J. Bacteriol., 123: 436-442. Miller, T. L., 1978, The pathway of formation of acetate and succinate from pyruvate by *Bacteroides succinogenes*, Arch. Microbiol., 117: 145-152.

Miller, T. L., and Wolin, M. J., 1973, Formation of hydrogen and formate by *Ruminococcus albus*, J. Bacteriol., 116: 836-846.

Miller, T. L., and Wolin, M. J., 1979, Fermentations by saccharolytic intestinal bacteria, Am. J. Clin. Nutr., 164-172.

Miller, T. L., and Wolin, M. J., 1981, Fermentation by the human large intestine microbial community in an *in vitro* semicontinuous culture system, Appl. Environ. Microbiol., 42: 400-407.

Miller, T. L. and Wolin, M. J., 1985, *Methanosphaera stadtmaniae* gen. nov., sp. nov.: a species that forms methane by reducing methanol with hydrogen, Arch. Microbiol., 141: 116-122.

Miller, T. L., and Wolin, M. J., 1986, Methanogens in human and animal intestinal tracts, Syst. Appl. Microbiol., 7: 223-229.

Miller, T. L., Wolin, M. J., and Kusel, E. A., 1986, Isolation and characterisation of methanogens from animal feces, Syst. Appl. Microbiol., 8: 234-238.

Miller, T. L., Wolin, M. J., Hongxue, Z., and Bryant, M.P., 1986, Characteristics of methanogens isolated from bovine rumen, Appl. Environ. Microbiol., 51: 201-202.

Miller, T. L., Wolin, M. J., Conway de Macario, E., and Macario, A. J. L., 1982, Isolation of *Methanobrevibacter smithii* from human feces, Appl. Environ. Microbiol., 43, 227-232.

Oppermann, R. A., Nelson, W. O., and Brown, R. E., 1957, *In vitro* studies on methanogenic rumen bacteria, J. Dairy Sci., 40: 779-788.

Patterson, J. A., and Hespell, R. B., 1979, Trimethylamine and methylamine as growth substrates for rumen bacteria and *Methanosarcina barkeri*, Curr. Microbiol., 3: 79-83.

Paynter, M. J. B., and Hungate, R. E., 1968, Characterisation of *Methanobacterium mobilis* sp. nov., isolated from the bovine rumen, J. Bacteriol., 95: 1943-1951.

Prins, R. A., 1971, Isolation, culture, and fermentation characteristics of *Selenomonas ruminantium* var. *bryanti* var. n. from the rumen of sheep, J. Bacteriol., 105: 820-825.

Prins, R. A., and Van Hoven, W., 1977, Carbohydrate fermentation by the rumen ciliate *Isotricha prostoma*, Protistol., 13: 599-606.

Prins, R. A., Van Vught, F., Hungate, R. E., and Van Vorstenbosch, C. J. A. H. V., 1972, A comparison of strains of *Eubacterium cellulosolvens* from the rumen, Ant. Van Leeuwen., 38: 1-11.

Roustan, J. L., Touzel, J. P., Prensier, G., Dobourguier, H. C., and Albagnac, G., 1986, Evidence for a lytic phage for *Methanothrix* sp., in "Biology of Anaerobic Bacteria," H.

C. Dubourguier, ed., Elsevier, Amsterdam.

Russell, J. B., 1987, A proposed mechanism of monensin action in inhibiting ruminal bacterial growth: effects on flux and protonmotive force, J. Anim. Sci., 64: 1519-1525.

Salyers, A. A., 1984, *Bacteroides* of the human lower intestinal tract, Ann. Rev. Microbiol., 38: 293-313.

Scheifinger, C. C., Linehan, B., and Wolin, M. J., 1975, H_2 production by *Selenomonas ruminantium* in the absence and presence of methanogenic bacteria. Appl. Microbiol., 29, 480-483.

Smith, P. H., and Hungate, R. E., 1958, Isolation and characterisation of *Methanobacterium ruminantium* n. sp., J. Bacteriol., 75: 713-718.

Stewart, C. S., and Bryant, M. P., 1988, The Rumen Bacteria, in "The Rumen Microbial Ecosystem," P. N. Hobson, ed., Elsevier Applied Science, London.

Stewart, C. S., and Richardson, A. J.,1989, Enhanced resistance of anaerobic rumen fungi to the ionophores monensin and lasalocid in the presence of methanogenic bacteria, J. Appl. Bacteriol., 66: 85-93.

Thauer, R. K., and Kroger, A., 1984, Energy metabolism of two rumen bacteria with special reference to growth efficiency, in "Herbivore Nutrition in the Tropics and Sub-Tropics," F. M. C. Gilchrist and R.I. Mackie, eds., The Science Press, South Africa.

Van Hoven, W., and Prins, R. A.,1977, Carbohydrate fermentation by the rumen ciliate *Dasytricha ruminantium*. Protistol., 13: 599-606.

Vogels, G. D., Hoppe, W. F., and Stumm, C. K., 1980, Association of methanogenic bacteria with rumen ciliates, Appl. Environ. Microbiol., 40: 608-612.

Weimer, P.J., and Zeikus, J. G., 1977, Fermentation of cellulose and cellobiose by *Clostridium thermocellum* in the absence and presence of *Methanobacterium thermoautotrophicum*, Appl. Environ. Microbiol., 33: 289-297.

Widdel, F., 1986, Sulphate reducing bacteria and their ecological niches, in "Anaerobic Bacteria in Habitats Other Than Man," E. M. Barnes and G. C. Mead, eds., Blackwell, Oxford.

Wolin, M. J., 1982, Hydrogen transfer in microbial communities, in "Microbial Interactions and Communities,"A. T. Bull and J. H. Slater, eds., Academic Press, London.

Wolin, M. J. and Miller, T., 1983, Carbohydrate fermentation, in "Human Intestinal Microflora in Health and Disease," D. J. Hentges, ed., Academic Press, London.

Wolin, M. J., and Miller, T. L., 1988, Microbe-microbe interactions, in "The Rumen Microbial Ecosystem" P.N. Hobson, ed., Elsevier Applied Science, London.

Wood, T. M., Wilson, C. A., McCrae, S. I., and Joblin, K. N., 1986, A highly active extracellular cellulase from the anaerobic rumen fungus *Neocallimastix frontalis*, FEMS Microbiol. Lett., 34: 37-40.

OXIDATIVE PROPIONATE FORMATION BY ANAEROBIC BACTERIA

Arno T.P. Skrabanja and Alfons J.M. Stams

Department of Microbiology
Agricultural University
Hesselink van Suchtelenweg 4
6703 CT Wageningen; The Netherlands

INTRODUCTION

Hydrogen is an important intermediate in the degradation of organic material. The oxidation of hydrogen by methanogens creates a low hydrogen partial pressure, which is essential for the breakdown of e.g. propionate and butyrate and which may cause a shift in fermentation products of by hydrogen producing fermentative bacteria. When hydrogen is removed in methanogenic environments, fermentative bacteria, which dispose part of their reducing equivalents as molecular hydrogen form more oxidized and less reduced organic products. In addition, the degradation of reduced organic compounds is only possible in the presence of methanogens.
Propionate is a compound, which can be formed both via an oxidative and reductive way by the fermentative organisms. The fermentation of aspartate, serine and alanine (Hansen and Stams, 1989; Schweiger and Buckel, 1984, Naaninga and Gotschal, 1985) resembles the reductive formation of propionate as carried out by Propionibacterium (Schink, 1984; Laanbroek et al, 1982), where reducing equivalents formed in the oxidation of substrate are disposed by the reductive conversion of pyruvate to propionate. Propionate, however, can also be formed in an oxidative pathway. In this paper we present some cases in which amino acids are converted to propionate in an oxidative way, and in which propionate formation is stimulated by interspecies hydrogen transfer.

MATERIALS AND METHODS

Organisms

Acidaminobacter hydrogenoformans (DSM 2784) was kindly provided by Dr T.A. Hansen , University of Groningen, the Netherlands and Methanobrevibacter arboriphilus strain AZ (DSM 744) was purchased from the Deutsche Sammlung für Mikroorganismen, Braunschweig, Federal Republic of Germany. Methanobacterium thermoautotrophicum strain ΔH was kindly provided by Prof. G.D. Vogels, University of Nijmegen, The Netherlands. Strain Su 883 was isolated from sludge samples from an UASB reactor of the CSM sugar refinery at Breda, The Netherlands (Cheng Guansheng et al. 1989).

Microbiology and Biochemistry of Strict Anaerobes Involved in Interspecies Transfer
Edited by J.-P. Bélaich *et al.*
Plenum Press, New York, 1990

Media and cultivation

A basal bicarbonate buffered medium with vitamins as described by Huser (1981) was used, with addition of 0.02 % yeast extract. Trace elements were according to Stams et al (1983). Incubations were done in partially filled serum bottles under an atmosphere of 80 % N_2 or H_2 and 20 % CO_2. A. hydrogenoformans was cultivated at 30°C with 20 mM glutamate or α-ketoglutarate as substrate, in absence or presence of M. arboriphilus. Pure cultures of M. arboriphilus were cultivated at 37°C, with addition of 0.5 g/l cystein-HCl to the medium. Cocultures were cultivated at 30°C, in these cases cystein-HCl was omitted. Su 883 was cultivated in a similar way at 55°C (Cheng Guansheng et al, 1989).

Preparation of cell free extract

Cells of A. hydrogenoformans grown in 8 l batch cultures were harvested in the late log phase by centrifugation in a CEPA continuous centrifuge and washed twice with 100 mM Tris-HCl pH = 7.4 containing 2 mM $MgCl_2$. Extracts were prepared under anaerobic conditions. After passage through a French Pressure Cell at 100 MPa, the suspension was centrifuged for 10 min at 20,000 g and the supernatant was stored under N_2/CO_2 (80%/20%) in 100 mM Tris-HCl pH = 7.4.
Cell extracts of Su 883 were prepared by ultrasonic degradation in an anaerobic glove box and centrifuged at 20.000 g for 20 min. Supernatants were stored under N_2/CO_2 in 100 mM Tris-HCl pH = 7.4.

Enzyme assays

L-glutamate dehydrogenase was measured spectrophotometrically according to Winnacker (1970). Benzylviologen linked HSCoA dependent α-ketoglutarate dehydrogenase was measured in a similar way as pyruvate dehydrogenase described by Odom and Peck (1981). 3-Methylaspartase was assayed according to Hsiang and Bright (1969) and NAD^+ dependent 2-hydroxyglutarate dehydrogenase was determined according to Lerud and Whiteley (1971). Isocitrate dehydrogenase was determined as described by Brandis-Heep et al (1983). All enzyme activities of A. hydrogenoformans extracts were determined at 30°C under anaerobic conditions, unless it was proven that the activity was not affected by the presence of oxygen. Enzyme activities of extracts of Su 883 were determined at 50°C under anaerobic conditions.

Analytical methods

Fatty acids were determined gaschromatically on a Varian gaschromatograph with a Chromosorb 101 (80-100 mesh) column 2m x2 mm; column temperature 160°C, injection port 220°C, flame ionization detector 240°C, carrier gas (30 ml/min) nitrogen saturated with formic acid. Hydrogen and methane were determined gaschromatographically on a Packard-Becker 417 gaschromatograph with a thermal conductivity detector and a molecular sieve held at 50°C; the carrier gas was argon at a flow rate of 20 ml/min. Glutamate and α-ketoglutate were determined enzymatically with glutamate dehydrogenase according to Winnacker (1970). Protein was determined according to Bradford (1976) with bovine serum albumin as standard.

RESULTS AND DISCUSSION

Glutamate fermentation

Acidaminobacter hydrogenoformans is a fermentative bacterium capable of converting glutamate and histidine to propionate in an oxidative way (Stams and Hansen, 1984). This bacterium disposes reducing equivalents almost exclusively as hydrogen and formate and its product formation is clearly affected by the presence of hydrogen. In pure culture A. Hydrogenoformans grows slowly on glutamate and forms acetate, HCO_3^-, formate and H_2 as end products. A more rapid fermentation is obtained in mixed cultures with methanogens and under these conditions propionate is a major product.

With α-ketoglutarate as substrate, A. hydrogenoformans forms the same products as with glutamate. However, propionate is also formed in the absence of a methanogen. Conversion of α-ketoglutarate to propionate is energetically more favorable than the conversion of glutamate to the same product.

(1) $glutamate^- + 4\ H_2O \longrightarrow propionate^- + 2\ HCO_3^- + 2\ H_2 + NH_4^+ + H^+$

$$\Delta G^{o\,`} = -\ 5.8\ kJ/mol$$

(2) $\alpha\text{-ketoglutarate}^{2-} + 3\ H_2O \longrightarrow propionate^- + 2\ HCO_3^- + H_2 + H^+$

$$\Delta G^{o\,`} = -\ 65.7\ kJ/mol$$

Recently, a thermophilic succinate-degrading bacterium was isolated from granular sludge, which is able to degrade a wide variety of amino acids (Cheng Guansheng et al, 1989). This bacterium resembled A. hydrogenoformans in several aspects.

Glutamate, histidine, arginine and ornithine were degraded in a similar oxidative way to propionate as in A. hydrogenoformans; arginine and ornithine were only degraded in the presence of a hydrogen consuming organism. In the presence of a methanogen more propionate is formed, but also in its absence propionate is formed from glutamate. In this feature Su 883 differs from A. hydrogenoformans, as seen when the conversion of glutamate by the two strains is compared.

Table 1 Product formation from 20 mM glutamate (mM)

	gasphase	acetate	propionate	glut.cons.	H2	formate
Acidaminobacter hydrogenoformans	N2	17.5	0.8	9.8	4.2	6.4
	H2	8.4	0.5	4.0	n.d.	2.4
+ methanogen		24.8	6.8	18.7	0.4	0.5
Su 883	N2	8.2	7.2	16.0	8.2	
+ methanogen		5.6	14.8	15.8	16.4*	
Selenomonas acidaminophila	N2	33.8	6.7			
	H2	27.5	11.8			

* methane is expressed as hydrogen equivalents

In table 1 glutamate conversion by <u>Selenomonas acidaminophila</u>, a bacterium that forms propionate in a reductive way is included (Nanninga et al, 1987). The growth of this organism in the presence of a hydrogen containing atmosphere resulted in an increased formation of propionate.
This is completely in contrast with our findings, where propionate formation can only be enhanced by removal of hydrogen.

Table 2 Enzyme activities in cell extracts (μmol mg^{-1} min^{-1}) of glutamate grown cells

	A. hydrogenoformans	Su 883
β-methylaspartase	0.24	0
glutamate dehydrogenase	86.3	80.0
isocitrate dehydrogenase	0.12	0.16
α-ketoglutarate dehydrogenase	0.01	0.50
hydroxyglutarate dehydrogenase	0	0.06
fumarate reductase	0	0

Enzyme activities

Table 2 shows enzyme activities involved in the conversion of glutamate to acetate and propionate. Possible pathways of acetate formation from glutamate are given in figure 1. Two of these routes are described by Barker et al (1981,1974): the mesaconate route via (3-β)-methylaspartate and citramalate and the hydroxyglutarate route. A third possibility is the conversion via α-ketoglutarate and isocitrate.

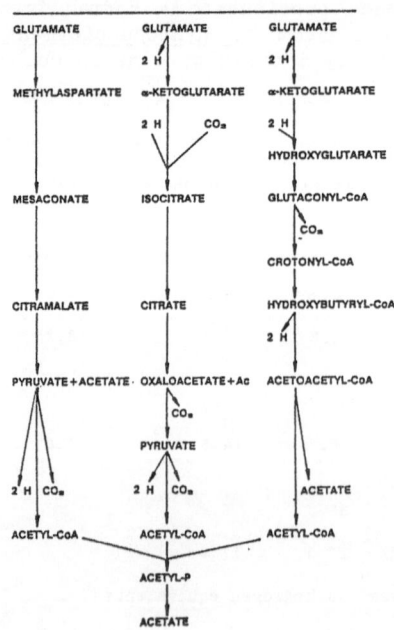

Figure 1. PATHWAYS FOR GLUTAMATE CONVERSION

Table 3. Products formed from biotrypticase (1 %) by methanogenic granular sludge (5 % ww) from a sugar refinery at 37 and 55 °C in the presence and absence of bromoethanesulfonic acid (Bres). Products were determined after 16 and 70 h of incubation.

37 °C		Ace	Prop	But	i-But	i-Val	H2	CH4
- Bres	16 h	14.0	3.2	3.4	2.2	5.7	0	61.5
	70 h	2.1	5.3	6.0	4.7	8.5	0	71.9
+ Bres	16 h	13.2	3.8	3.0	2.2	5.8	0.2	4.1
	70 h	37.6	8.2	7.3	4.4	8.4	0.3	10.7
55°C								
- Bres	16 h	21.7	3.9	3.0	2.1	5.9	0	9.3
	70 h	41.7	12.2	6.2	5.0	10.3	1.4	22.0
+ Bres	16 h	14.7	3.3	2.8	2.2	4.9	5.6	0
	70 h	29.8	6.3	5.2	2.6	7.7	9.6	0

Results of enzyme analysis (Skrabanja and Stams, 1989) showed that in cells of A. hydrogenoformans, grown on glutamate in the presence of a methanogen, high activities were found of the enzymes isocitrate dehydrogenase and citrate lyase, whereas no fumarate reductase and fumarase was found; whereas in glutamate grown cells high activities were found of β-methylaspartase and citramalate lyase, giving strong evidence for the mesaconate pathway.

Enzymes for the hydroxyglutarate pathway could not be detected. The absence fumarate reductase activity is consistent with an oxidative way of propionate formation from glutamate. The presence of this enzyme is necessary in a reductive process of propionate formation, and was found to be present in Selenomonas acidaminophila (Nanninga et al, 1987).

Enzyme analysis of cell extracts of Su 883 revealed that the same pathways could be involved in the formation of acetate and propionate as in A. hydrogenoformans.

Incubation of granular sludge

To get information on the importance of oxidative and reductive propionate formation; granular sludge was incubated with biotrypticase, a source of aminoacids.If methanogenesis is inhibited by the presence of bromoethane sulfonic acid, less propionate is formed compared with the control (table 3). Bacause at 37°C, but not at 55°C acetate and propionate are rapidly converted, and bres does not completely inhibit methanogenesis, nothing can be said about the importance of oxidative propionate formation at mesophilic temperatures. Under thermophilic conditions however, the inhibition of methane formation resulted in less propionate formation. This observation is a clear indication that propionate is indeed formed in oxidative process.

CONCLUDING REMARKS

A. hydrogenoformans and the termophilic isolate Su 883 form propionate oxidatively from glutamate, as demonstrated by shift in product formation at different H_2-concentrations and the absence of fumarate reductase, a key enzyme in reductive propionate formation.

Incubations of granular sludge with biotrypticase indicated that oxidative propionate formation may be more important than generally thought.

REFERENCES

Barker, H.A., 1981, Amino acid degradation by anaerobic bacteria, Ann. Rev. Biochem., 50:23

Bradford, M.M., 1976, A rapid and sensitive method for the quantitation of microgram quantities of protein utilizing the principle of protein-dye binding, Anal. Biochem., 72:248

Brandis-Heep, A., Gebhardt, N.A., Thauer, R.K., Widdel, F., and Pfennig, N., 1983, Anaerobic acetate oxidation to CO_2 by Desulfobacter postgatei. 1. Demonstration of all enzymes required for the operation of the citric acid cycle, Arch. Microbiol., 136:222

Buckel, W., and Barker, H.A., 1974, Two pathways of glutamate fermentation by anaerobic bacteria, J. Bact., 117:1248

Cheng Guansheng, Skrabanja, A.T.P., and Stams, A.J.M., 1989, Isolation and characterization of a thermophilic anaerobic bacterium converting succinate to propionate, In preparation

Hansen, T.A., and Stams, A.J.M., 1989, A rod-shaped, Gram-negative, propiogenic bacterium with a wide substrate range and the ability to fix nitrogen. Arch. Microbiol., submitted

Hsiang, M.W., and Bright, H.J., 1969, β-methylaspartase from Clostridium tetanomorphum, in : Lowenstein, J.M., (ed), Methods in Enzymology, 13:347

Huser, B., 1981, Methanbildung aus Acetat, PhD thesis, Zurich

Laanbroek, H.J., Abee, T., and Voogd, I.L., 1982, Alcohol conversions by Desulfobulbus propionicus Lindhorst in the presence and absence of sulfate and hydrogen, Arch. Microbiol., 133:178

Lerud, R.F., and Whiteley, H.R., 1971, Purification and properties of α-ketoglutarate reductase from Micrococcus aerogenes, J. Bact. 106:571

Naaninga, H.J., and Gotschal, J.C., 1985, Aminoacid fermentation and hydrogen transfer in mixed cultures, FEMS Microb. Ecol., 31:261

Nanninga, H.J., Drent, W.J., and Gotschall J.C., 1987, Fermentation of glutamate by Selenomonas acidaminophila sp. nov., Arch. Microbiol., 147;152

Odom, J.M., and Peck, H.P., 1981, Localization of dehydrogenases, reductases and electron transfer components in the sulfate reducing bacterium Desulfovibrio gigas, J. Bacteriol., 147:161

Schink, B., 1984, Fermentation of 2,3-butanol by Pelobacter carbinolycus sp. nov. and Pelobacter propionicus and evidence for propionate formation from C_2-compounds, Arch. microbiol.,137:33

Schweiger, G., and Buckel, W., 1984, On the degradation of (R)-lactate in the fermentation of alanine to propionate by Clostridium propionicum, FEBS Lett., 171:79

Skrabanja, A.T.P., and Stams, A.J.M., 1989, Pathways of glutamate conversion in Acidaminobacter hydrogenoformans, In preparation

Stams, A.J.M., Veenhuis, M., Weenk, G.H., Hansen, T.A., 1983, Occurence of polyglucose as a storage polymer in Desulfovibrio species and Desulfobulbus propionicus, Arch. Microbiol., 136:54

Stams, A.J.M., and Hansen, T.A., 1984, Fermentation of glutamate and other compounds by Acidaminobacter hydrogenoformans gen. nov. sp. nov., an obligate anaerobe isolated from black mud. Studies with pure cultures and mixed cultures with sulfate-reducing and methanogenic bacteria, Arch. Microbiol., 137:329

Winnacker, E.L., and Barker, H.A., 1970, Purification and properties of a NAD-dependent glutamate dehydrogenase from Clostridium SB 4, Biochem. Biophys. Acta, 212:225

ANAEROBIC DEGRADATION OF FURFURAL BY DEFINED MIXED CULTURES

Siegfried M. Schoberth, Ulrich Ney, and
Hermann Sahm

Institut für Biotechnologie
Kernforschungsanlage Jülich
D-5170 Jülich, Federal Republic of Germany

INTRODUCTION

The heteroaromatic compound furfural (2-furaldehyde) is a constituent of sulfite evaporator condensate (SEC) as seen in Table 1. SEC is a waste water from pulp and paper industry and arises in bulk quantities in paper mills that operate the acidic sulfite cooking process (Brune et al., 1982). Although acetic acid is the main compound (Table 1), furfural is the substance most toxic towards anaerobic bacteria (Morris et al, 1978; Folkerts et al., 1989).

It has been shown that SEC can be successfully treated through anaerobic digestion using mixed enrichment cultures when either a few trace elements are added to SEC (Brune et al., 1982), or a commercial fertilizer (Aivasidis, 1985; Ney et al., in preparation). When furfural serves as sole carbon and energy source, it is converted according to equation 1. From these cultures, a novel species of sulfate reducing bacteria, *Desulfovibrio furfuralis* was isolated (Brune et al., 1983; Folkerts et al., 1989). In the presence of sulfate, this organism carries out an anaerobic degradation of furfural to acetic acid (equ. 2).

$$C_5H_4O_2 + 3\ H_2O \longrightarrow 2.5\ CH_4 + 2.5\ CO_2 \qquad (1)$$

$$C_5H_4O_2 + 0.5\ SO_4^{2-} + 2\ H_2O \longrightarrow 2\ CH_3COO^- + H^+ \\ + CO_2 + 0.5\ H_2S \qquad (2)$$

More recently, we studied a 20 l-high rate fixed-bed anaerobic reactor treating SEC (Macario et al., 1989; Ney et al., in preparation). Both microbiological and immunological methods were employed. These investigations were conducted over a period of 14 months. The results show that despite shifts in the methanogenic subpopulations, the performance of the reactor remained fairly constant, as did the number of furfural degrading organisms (*D. furfuralis*).

Microbiology and Biochemistry of Strict Anaerobes Involved in Interspecies Transfer
Edited by J.-P. Bélaich *et al.*
Plenum Press, New York, 1990

141

Table 1. Composition of Sulfite Evaporator Condensate (SEC)

Compound	Concentration	% of Total Carbon
Acetic acid	300 – 600 mM	81.6 – 78.7%
Furfural	21 – 49 mM	14.3 – 16.1%
Methanol	30 – 80 mM	4.1 – 5.2%
Total sulfur*	20 – 37 mM	2.7 – 2.4%
COD**	24000 – 60000 mg/l	
pH	2	

* : Sulfate and sulfite
**: Chemical oxygen demand

Here we report on the stepwise construction of a defined coculture that is able to degrade furfural in a continuously operated fixed bed loop reactor. This work is a prerequisite to develop starter cultures that can be used in anaerobic digestion of furfural containing effluents.

DEGRADATION OF FURFURAL BY DEFINED BATCH COCULTURES

It was suggested that a complete degradation of furfural to methane (equ. 1) necessitated three steps (Brune et al., 1982; equ. 3 - 5), the conversion of H_2 and acetate being carried out by methanogenic bacteria:

$$C_5H_4O_2 + 4\ H_2O \longrightarrow 2\ CH_3COO^- + 2\ H^+ + CO_2 + 2\ H_2 \qquad (3)$$

$$4\ H_2 + CO_2 \longrightarrow CH_4 + 2\ H_2O \qquad (4)$$

$$CH_3COO^- + H^+ \longrightarrow CH_4 + CO_2 \qquad (5)$$

This is supported by the findings that in mixed batch cultures a transient formation of acetate occurs and that the chief amount of CH_4 is produced towards the end of the fermentation (Brune et al., 1982). The reaction according to equ. 3 does not support growth of *D. furfuralis* per se: either sulfate (equ. 2), sulfite or nitrate have to be added as electron acceptors (Folkerts et al., 1989). Therefore it had first to be proved that the electron acceptor sulfate could be replaced with a hydrogenotrophic methanogen.

Diculture

Methanobacterium bryantii strain MoHG was chosen as hydrogen accepting organism (equ. 4) because its immunotype was abundant throughout the longterm study of the anaerobic reactor mentioned above (Macario et al., 1989). Figure 1 shows that in the presence of *M. bryantii* both growth and formation of methane and acetic acid did occur in the absence of sulfate, sulfite or nitrate.
Only small amounts of furfural could be added stepwise since this compound is very toxic to methanogenic bacteria (Table

142

2). In addition, the growth rate of *D. furfuralis* has a narrow maximum (0.1 h^{-1}) at furfural concentrations between 4 and 5 mM (Folkerts et al., 1989). Figure 2 demonstrates that added excess sulfate was preferred as an electron acceptor (equ. 2) over *M. bryantii* as a hydrogen acceptor (equ.7).

Triculture

Both *Methanosarcina* spec. and *Methanothrix* spec. are able to convert acetate to methane (equ. 5), and morpho- and immunotypes of strains of both genera have been identified in SEC digesting reactors (Brune et al. 1982; Macario et al. 1989). *Methanosarcina* was the first choice of an acetotrophic partner for *D. furfuralis* since it is also able to degrade methanol (equ. 6) which is also a constituent of SEC. Table 2 lists the

$$CH_3OH \longrightarrow 0.75\ CH_4 + 0.25\ CO_2 + 0.75\ H_2O \qquad (6)$$

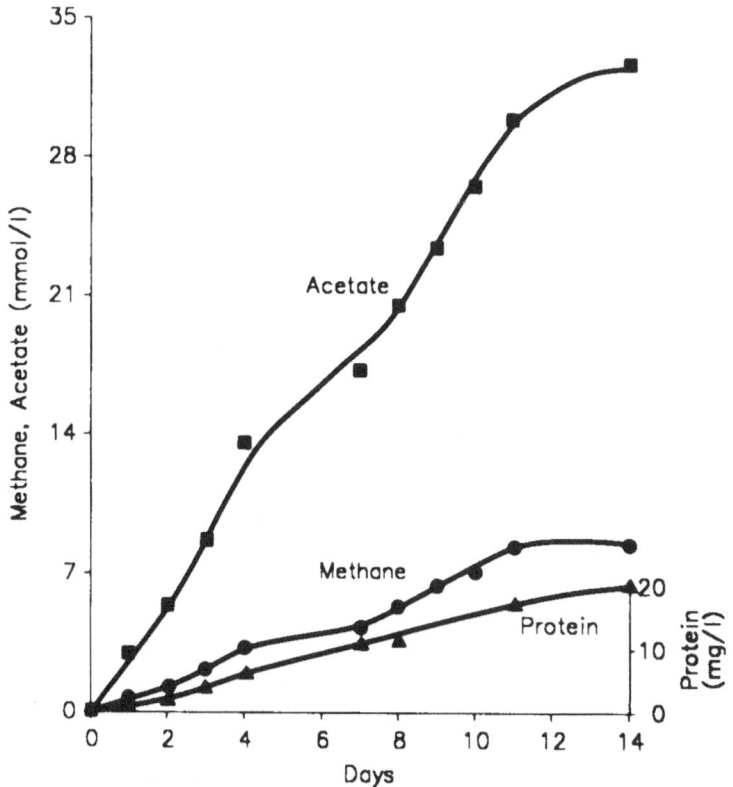

Figure 1. Degradation of Furfural by a Batch Co-culture of *Desulfovibio furfuralis* and *Methanobacterium bryantii* in the absence of sulfate. At each sampling point 1 - 2 mmol/l furfural were added. Growth is represented by cellular protein. The medium was similar as described in Folkerts et al., 1989.

strains that were tested with respect to their sensitivity towards furfural (Table 2). *M. barkeri* DSM 804 was chosen because its inhibitory constants (K_i) for furfural were slightly
higher than those of the other strains tested. Figure 3 shows
growth of a triculture consisting of *D. furfuralis*, *M. bryantii* and *M. barkeri* on furfural. The formation of methane is
accompanied by intermediate formation of acetic acid. Disappearance of acetate was not complete after 7 days. This has
been found already with enrichment cultures (Brune et al.,
1982). Using those enrichment cultures in continuous operation, it has also been found that the efficiency and kinetics
of immobilized systems are far superior over conventional
stirred tank reactors (Brune, G., 1982, Ph.D. thesis, University of Düsseldorf; Avasidis, 1985). Therefore we set up similar experiments as described above, yet employing continously
operated fixed-bed loop reactors.

DEGRADATION OF FURFURAL BY DEFINED COCULTURES IN FIXED-BED
LOOP REACTORS

The setup of the reactors was a small scale replica of
the fixed-bed loop (FBL) reactor used in SEC treatment (Aiva-

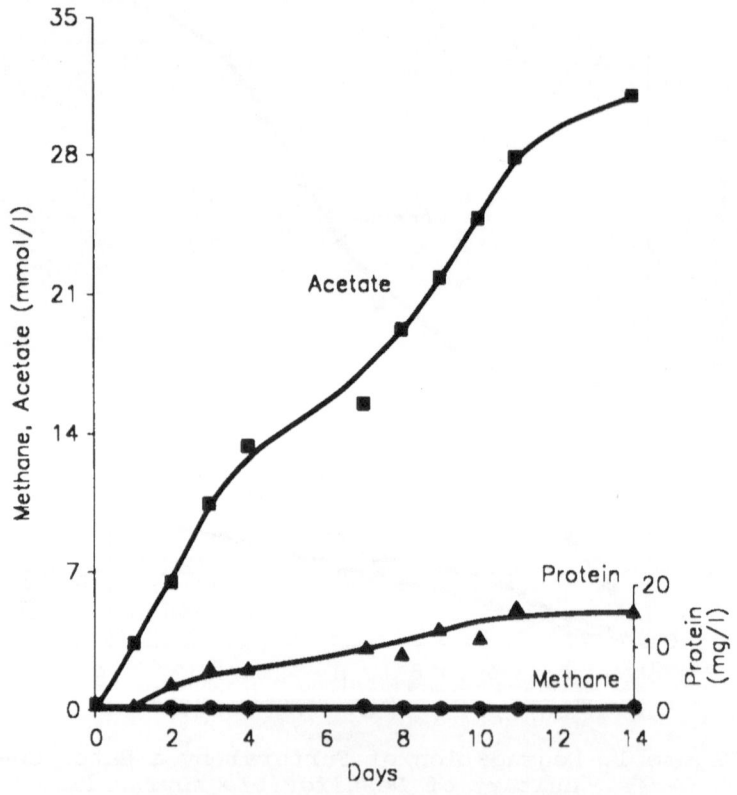

Figure 2. Degradation of Furfural by a Batch Coculture of *Desulfovibrio furfuralis*
and *Methanobacterium bryantii* with
10 mM sulfate. For further details see Fig. 1.

Table 2. Inhibition of Methanogens (K_i-values)
by Furfural

Organism	Substrates		
	Acetate	H_2/CO_2	Methanol
M. bryantii MoHG	-	1.78	-
M. barkeri DSM 800	1.25	1.17	1.74
M. barkeri DSM 804	1.83	1.02	2.05
M. barkeri DSM 1538	1.41	0.98	1.88
M. concilii DSM 3671	2.4	-	-

-: Not used as substrate for growth.

sidis, 1985; Macario et al., 1989; Ney et al, in preparation).
The working volume was 530 ml, including 69 ml of porous sinterglass fillings (SIRAN[TR], SCHOTT GmbH, Mainz; Aivasidis,
1985; Aivasidis and Wandrey, 1988). The organisms were the
same as used already in the batch experiments (see above).

Diculture

Table 3 shows that furfural was completely degraded and
its conversion to acetic acid was complete 90 - 93% according
to equ. 2 and 7 at retention times of less than 2 days. This

$$C_5H_4O_2 + 3\ H_2O \longrightarrow 2\ CH_3COO^- + H^+ \\ + 0.5\ CO_2 + 0.5\ CH_4 \qquad (7)$$

has to be compared with the long reaction times necessary in
batch (Fig. 1 and 2). Effective interspecies hydrogen transfer
between *D. furfuralis* and *M. bryantii* was indicated by very
low concentrations of H_2 in the gaseous phase of 200 ppm (20
Pa). When sulfate was added, sulfate reduction in *D. furfuralis* outcompeted interspecies hydrogen transfer to *M. bryantii*,
and formation of methane was almost nil (Table 3, reactor B),
as found already in the batch experiments (Fig. 2).

Tri- and tetracultures

In addition to the strains already described above, *Methanobrevibacter arboriphilus* AZ DSM 744 and *Methanothrix concilii* DSM 3671 were used in the last experiment described in
Table 4. *M. arboriphilus* whose immunocounts increased considerably during treatment of SEC by enrichment cultures (Macario et al., 1989) replaced *M. bryantii*. *Methanothrix* was added
for three reasons: first, because both morpho- and immunotypes

Table 3.   Continuous degradation of furfural
           by a coculture of *D. furfuralis* and
           M. bryantii in 530 ml FBL reactors

Parameter		Value	
Reactor		A	B
Retention time	(h)	32	28
Sulfate added	(mM)	0	10
Furfural$_{in}$	(mM)	12	12
Furfural$_{out}$	(mM)	0.01	0.02
Acetate$_{out}$	(mM)	21.5	22.2
Gas$_{out}$	(l/l/d)	0.38	0.12
CH$_4$	(%)	54%	0.8%
CO$_2$	(%)	43%	66%
Yield (Acetate/Furfural)		1.79	1.85

pH 7; Temperature, 37° C. The medium was similar as described by Brune et al., 1982. For
details see text.

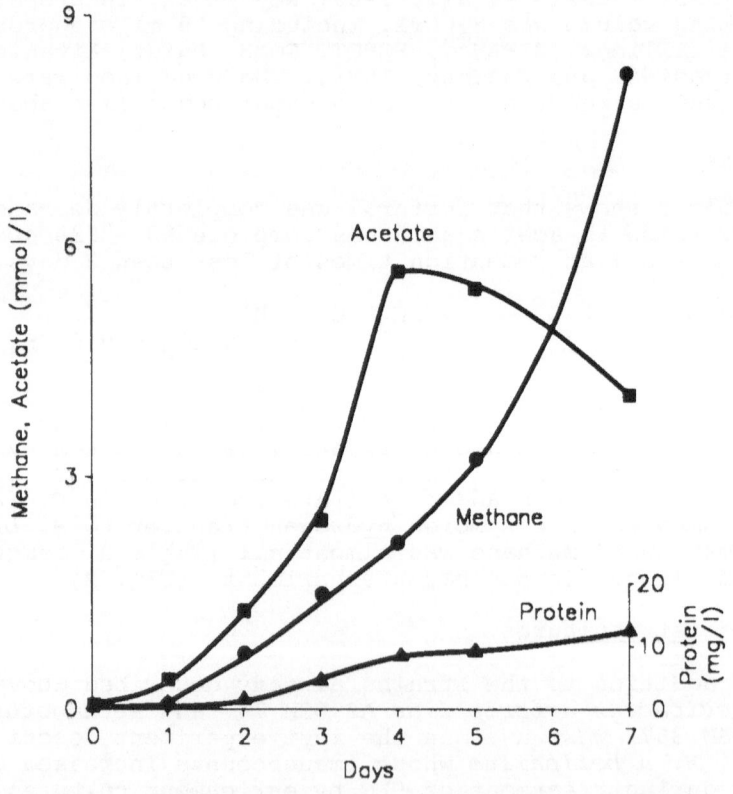

Figure 3. Degradation of Furfural by a Batch Co-
          culture of *D. furfuralis*, *M. bryantii*
          and *M. barkeri*. See Fig. 1.

Table 4. Continuous degradation of furfural by defined cocultures in 530 ml FBL reactors

Parameter		Value			
Reactor		A	B	C	D
Retention time	(h)	21	15.2	46.6	25.6
Sulfate added	(mM)	0	10	15	10
Sulfite added	(mM)	0	0	0	15
$Furfural_{in}$	(mM)	10	10	24	24
$Furfural_{out}$	(mM)	0.8	0.01	0	0
$Acetate_{in}$	(mM)	100	100	150	146
$Acetate_{out}$	(mM)	40	17	30	21
$Methanol_{in}$	(mM)	0	0	50	50
$Methanol_{out}$	(mM)	0	0	0	0
Gas_{out}	(l/l/d)	3.4	7.4	4.9	8.6
CH_4	(%)	57	51	52	54
Acetate					
converted*	(%)	67	86	85	89
Yield CH_4**	(%)	60	84	88	90
Yield CH_4***	(%)	90	97	99	98

The organisms used were: *D. furfuralis* (A - D); *M. bryantii*
(A - C); *M. barkeri* (A - D); *M. arboriphilus* and *M. concilii*
(D). pH was 6.7. *: $acetate_{out}$ per total $acetate_{in}$ including
acetate from furfural (equ. 2, 3, 7). **: yield per total
of incoming carbon substrates according to equ. 1, 5, 6;
***: per total of substrates converted. For other details
see Table 3 and text.

of *Methanothrix* are found along with those of *Methanosarcina*.
Second, we suggest that the extensive bundles of cellular filaments of *Methanothrix* seen in fixed bed loop SEC reactors
(Aivasidis, 1985; Macario et al., 1989) facilitate the adhesion of other organisms onto the carrier surface (Ney et al.,
in preparation), and third, *Methanothrix* has more favorable
kinetic constants for acetate degradation at low concentrations than *Methanosarcina* (Brune et al. 1982).

Table 4 summarizes the results of these experiments. The
stationary concentration of furfural is almost nil under all
experimental conditions. Although acetate (A - D) and methanol
(C, D) had been added to simulate more closely the composition
of SEC (Table 1), it was reduced by almost 90% (B - D). Only
in the absence of small amounts of sulfate or sulfite (A),
acetate reduction was less effective. Even better rated the
efficiency of methane production (yield CH_4 in Table 4). It
also improved when sulfate was added. As in the experiments
shown in Table 3, the very low concentration of H_2 in the
gaseous phase of 100 - 200 ppm (10 - 20 Pa) indicated effective interspecies hydrogen transfer.

Biomass was determined in separate experiments using the
same experimental setup and culture (*D. furfuralis*, *M. arbori-*

philus, *M. barkeri* and *M. concilii*) yet natural SEC as substrate. 65 g cellular dry weight/l was found in immobilized form between and on the surface of the sinterglass fillings, 31 g/l within the pores of the glass rings, and only 0.9 g/l in the free liquid phase. From these data it was calculated that 96 - 98% of the biomass was found in immobilized form. The specific activities of methane formation from furfural, acetate, H_2/CO_2, and methanol were (in μmol CH_4/g dry weight/h; potential, and actual in parentheses) furfural 284 (56), acetate 1411 (1336), methanol 376 (150) and H_2/CO_2 246 (56).

CONCLUSION

Axenic cultures of *D. furfuralis* do not grow on furfural in the absence of sulfate or other external electron acceptors (Folkerts et al., 1989). In this work it has been shown that under these conditions growth and conversion of furfural do occur when a hydrogenotrophic methanogen is added. The yield of acetate formed per furfural converted was 1.79 and 1.85 (Table 3), comparable to yields of 1.61 - 1.9 in axenic cultures with sulfate reported by Brune et al. (1983). Only traces of hydrogen (10 - 20 Pa) were detected in the culture gasphase (Fig. 1 and Table 3). Thus these experiments exemplify an obligate syntrophic relationship involving interspecies hydrogen transfer according to equations 3, 4 and 7. With the addition of acetotrophic methanogens to the diculture, a complete conversion of furfural to methane was accomplished (Fig. 3, Table 4). In the fixed-bed loop reactors employed in this study, furfural was degraded at influent concentrations of up to 24 mM though this compound is very toxic in batch cultures (Table 2; Folkerts et al., 1989; see also Morris et al., 1978).

The medium used in the present study (see Tables 3, 4 and text) did already closely resemble SEC as arising in paper mills (Table 1). In addition, using the defined tetraculture from this study (Table 4, D) we have shown that natural SEC with a COD of 35500 mg/l can be degraded with a yield of up to 84% at a hydraulic retention time of 17.9 h (Ney et al., 1989). The space-time yield of 39000 mg COD/l/d reached in these experiments corresponds to a turnover of furfural, acetate and methanol equivalent to the production of 25.4 mmol CH_4/l/h. These findings are promising and indicate that defined starter cultures for anaerobic treatment of furfural containing effluents can be developed.

Acknowledgements

This work was supported by a grant from the Deutsche Gesellschaft für Chemisches Apparatewesen, Chemische Technik und Biotechnologie e.V. (DECHEMA) to U.N.

REFERENCES

Aivasidis, A. 1985, <u>Water Sci. Tech.</u> 17:207

Aivasidis, A., and Wandrey, C., 1988. <u>Water Sci. Tech.</u>, 20:211.

Brune, G., Schoberth, S.M., and Sahm, H., 1982,<u>Process Bio-chem.</u>, 17(3):20.

Brune, G., Schoberth, S.M., and Sahm, H., 1983, <u>Appl. Envi-ronm. Microbiol.</u>, 46:1187.

Folkerts, M., Ney, U., Kneifel, H., Stackebrandt, E., Witte, E.G., Förstel, H., Schoberth, S.M., and Sahm, H., 1989, <u>Syst. Appl. Microbiol.</u>, 11:161.

Morris, J.A., Khettry, A., and Seitz, E.W., 1978, <u>J. Amer. Oil. Chem. Soc.</u>, 56:595.

Macario, A. J. L., Conway de Macario, E., Ney, U., Schoberth, S. M., and Sahm, H., 1989, <u>Appl. Environm Microbiol.</u> 55:1996.

Ney, U., Schoberth, S.M., and Sahm, H., Anaerobic degradation of sulfite evaporator condensate by defined bacterial mixed cultures, <u>in</u> "DECHEMA Monograph: Dechema-Biotechno-logy-Conferences", Verlag Chemie, Weinheim (1989), in press.

ENERGETICS AND KINETICS OF TWO COMPLEMENTARY HYDROGEN SINK REACTIONS IN A DEFINED 3-CHLOROBENZOATE DEGRADING METHANOGENIC CONSORTIUM

Jan Dolfing

Department of Biochemistry
University of Groningen
9747 AG Groningen, The Netherlands

INTRODUCTION

Recently a syntrophic microbial consortium was constructed that grows with 3-chlorobenzoate as the sole energy source (Dolfing and Tiedje 1986). It consists of three different bacteria: a hydrogenotrophic methanogen, an obligate syntrophic hydrogen-producing benzoate degrader, and strain DCB-1, an organism that reductively dehalogenates 3-chlorobenzoate to benzoate. The reducing equivalents for the latter reaction are derived from hydrogen (Figure 1). This consortium has two unique characteristics that distinguish it from the classical syntrophic co-cultures: (i) the presence of an organism that scavenges hydrogen as electron donor for a reductive dechlorination reaction and obtains energy for growth from this reaction (Dolfing and Tiedje 1987, Dolfing 1989), and (ii) the fact that hydrogen is scavenged by two coexisting organisms. Both characteristics will be discussed in the present paper.

REDUCTIVE DECHLORINATION AS AN ENERGY-YIELDING, HYDROGEN SCAVENGING REACTION

Thermodynamic calculations indicate that reductive dechlorination of 3-chlorobenzoate to benzoate is an energy-yielding reaction, not only under standard conditions, but also under environmentally relevant conditions (Table 1). Per mole of hydrogen consumed this reaction yields more energy than sulfate reduction or methanogenesis.
This consideration has led to the speculation that the dechlorinating organism in the consortium obtains energy for growth from the dechlorina-tion reaction. Tests with the consortium have supported this hypothesis (Dolfing and Tiedje 1987). The growth yield of the consortium on 3-chlorobenzoate was about 30 % higher than on benzoate, and this difference was matched by a specific increase in the presence of the dechlorinating organism in the 3-chlorobenzoate grown cultures. When the consortium was maintained on benzoate rather than on 3-chlorobenzoate for several generations the dechlorinating organism was diluted out. These observations make it unlikely that strain DCB-1 maintains itself solely by scavenging intermediates from the latter parts of the food chain. On the contrary, they suggest that this organism obtains energy from the

Microbiology and Biochemistry of Strict Anaerobes Involved in Interspecies Transfer
Edited by J.-P. Bélaich *et al.*
Plenum Press, New York, 1990

151

Substrates   Products

Dechlorinating organism (strain DCB-1)

$3\text{-Clorobenzoate}^- + H_2 \longrightarrow \text{Benzoate}^- + H^+ + Cl^-$

Benzoate oxidizer (strain BZ -2)

$\text{Benzoate}^- + 7\ H_2O \longrightarrow 3\ H_2 + HCO_3^- + 3\ \text{Acetate}^- + 3H^+$

Methanospirillum (strain PM-1)

$2\ H_2 + 0.5\ HCO_3^- \longrightarrow 0.5\ CH_4 + H_2O + 0.5\ OH^-$

Consortium

$3\text{-Chlorobenzoate}^- + 5.5\ H_2O \longrightarrow 0.5\ CH_4 + 3.5\ H^+ + Cl^- + 3\ \text{Acetate}^- + 0.5\ HCO_3^-$

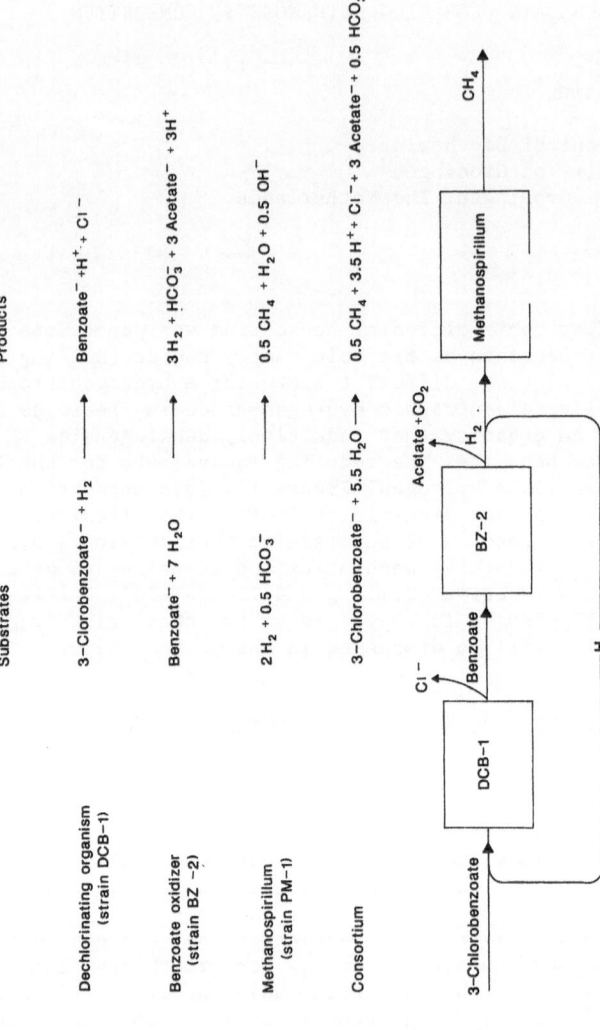

Figure 1. Organisms, reactions and interactions in a three-tiered syntrophic consortium growing on the methanogenic degradation of 3-chlorobenzoate (after Dolfing and Tiedje 1986).

Table 1. Change in free energy (Delta G°) values for various hydrogen scavenging reactions. Data are from Thauer et al. (1977), and from Dolfing and Tiedje (1987).

	kJ/mol H_2
Reductive dechlorination of 3-chlorobenzoate	-125
Nitrate reduction to nitrogen gas	-224
Nitrate reduction to ammonium	-150
Sulfate reduction	-39
Methanogenesis	-35

dechlorination reaction. Interestingly, the higher growth yield of the 3-chlorobenzoate grown consortium relative to the benzoate grown consortium would then imply that per mol of hydrogen consumed the cell yield of the dechlorinator is higher than the cell yield of the methanogen.

The second line of evidence in support of the energy-hypothesis is that the addition of 3-chlorobenzoate to energy-starved consortium cells resulted in a significantly higher increase in the ATP level of the cells than the addition of benzoate was able to effect (Dolfing and Tiedje 1987).

Recent studies with strain DCB-1 in pure culture (Dolfing 1989) have confirmed and extended these observations: the organism can be cultured in a defined medium with 3-chlorobenzoate as the only energy source. The growth yield of the organism was stoichiometric to the amount of 3-chlorobenzoate dechlorinated (Figure 2), while benzoate did not support growth. Also, the addition of 3-chlorobenzoate to starved cells resulted in a rapid increase in the ATP level of the cells, while benzoate had no effect (Figure 3).

The evidence that hydrogen is the source of reducing equivalents for reductive dechlorination in the consortium is based on the observation that growth of the consortium on four moles of benzoate resulted in the formation of three moles of methane, while growth of the consortium on four moles of 3-chlorobenzoate resulted in the formation of only two moles of methane (Dolfing and Tiedje 1986). Furthermore, experiments with D_2 and D_2O have indicated that the reducing equivalents for dechlorination in the

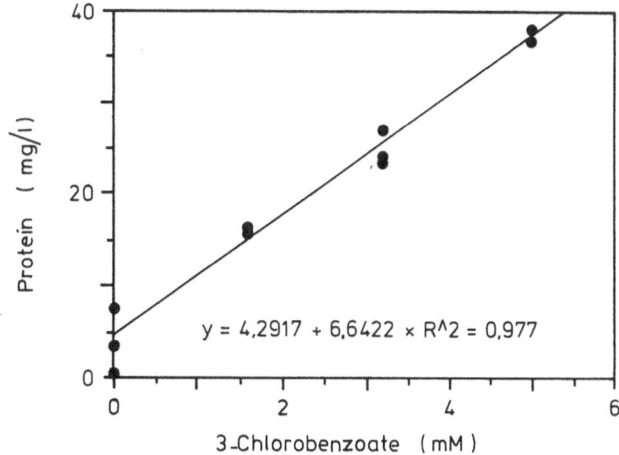

Figure 2. Growth yield of strain DCB-1 on 3-chlorobenzoate (after Dolfing 1989)

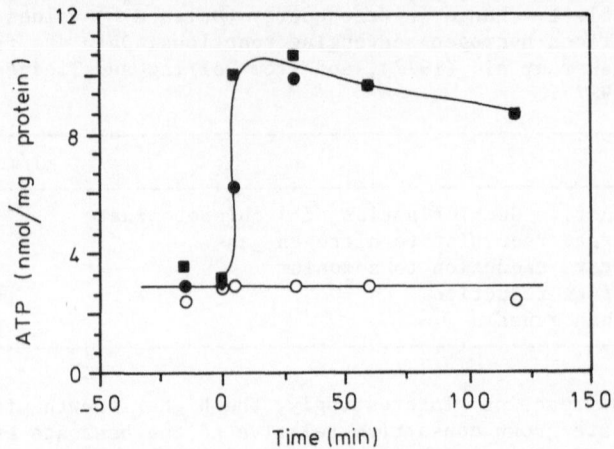

Figure 3. Effect of the addition of 3-chlorobenzoate on the
ATP concentration in resting cells of strain DCB-1.
3-Chlorobenzoate was added at t=0 to replicate vials (●,■),
a third vial (○) had only benzoate (after Dolfing 1989).

consortium originate from D_2 (H_2) and not from D_2O (H_2O) (Dolfing,
unpublished).
In view of its dechlorination-coupled hydrogen scavenging abilities and
its ability to grow with the dechlorination reaction as a "powerful"
energy source, it is not surprising that the dechlorinating organism can
act as an efficient hydrogen scavenger. It has been possible to cultivate
the benzoate degrader on benzoate with 3-chlorobenzoate as electron
acceptor, with strain DCB-1 as the hydrogen scavenger, i.e. in the absence
of sulfate reduction or methanogenesis. Growth of the benzoate degrader
with reductive dechlorination as the sole hydrogen sink reaction was also
possible when 3-chlorobenzoate was replaced by 2,5-dichlorobenzoate or
2-methyl,5-chlorobenzoate, i.e. other compounds that are specifically
dechlorinated by strain DCB-1 (Dolfing and Tiedje, in prep.), and whose
dechlorination is exergonic (Harrison, personal communication, manuscript
in preparation).

SCAVENGING OF HYDROGEN BY TWO CO-EXISTING ORGANISMS

The defined 3-chlorobenzoate degrading consortium grows as a well
balanced system. During growth benzoate is generally only present at
concentrations of about 50 µM, and the concentration of hydrogen is about
70 nM (Dolfing and Tiedje 1986). This plus the fact that one third of the
potential reservoir of hydrogen produced by the benzoate degrading
population is specifically available to the dechlorinating population,
even though hydrogen is scavenged by two co-existing organisms, raises the
question how this is achieved without one organism outcompeting the other
for its energy source, and how their hydrogen scavenging activities
influence the activity of the hydrogen producing partner.
Thermodynamic considerations indicate that the hydrogen concentration
influences the energetics of each organism in the consortium (Dolfing
1988a). Figure 4 depicts the relationships between the delta G values and
the hydrogen concentration. Formally Figure 4 gives an estimation of the
boundary conditions for the exergonic and endergonic domains of the
overall reactions. It seems reasonable to assume that also a relationship

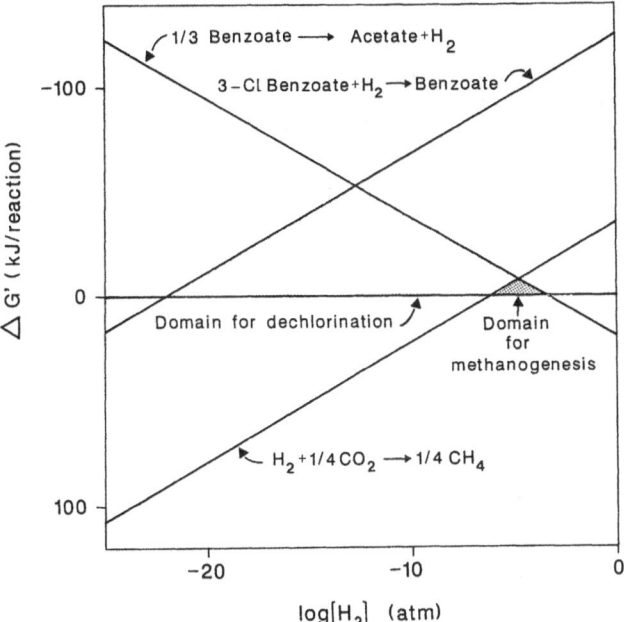

Figure 4. Effect of hydrogen partial pressure on the free
energy of conversion of 3-chlorobenzoate, benzoate, and
carbon dioxide.

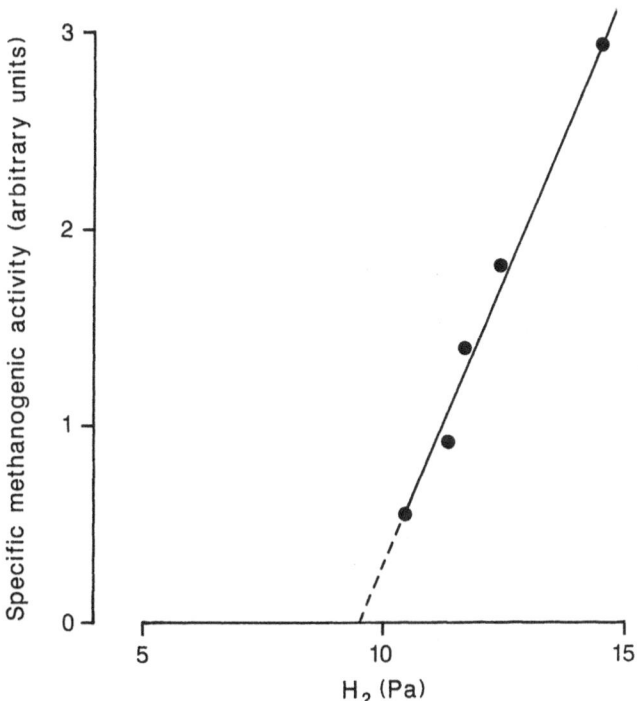

Figure 5. Effect of hydrogen partial pressure on the
specific methanogenic activity of Methanospirillum PM-1 in
a benzoate degrading syntrophic co-culture.

exists between the hydrogen concentration and the kinetics of these reactions. With respect to hydrogen as a substrate a Monod-type relationship would be expected. Experiments by e.g. Kristjansson et al. (1982) and Robinson and Tiedje (1984) have indeed shown that at hydrogen concentrations well below the half saturation constant the rates of hydrogen consumption and methanogenesis increase with increasing hydrogen concentrations. Similar results were obtained with the methanogen from the consortium, Methanospirillum PM-1 (Figure 5). This experiment also corroborated the existence of a minimum threshold concentration for hydrogen uptake in methanogens of about 10 Pa H_2 (Lovley 1985). Measured directly in the consortium the rate of methanogenesis increased with increasing hydrogen concentrations, but the rate of dechlorination did not increase with increasing hydrogen concentrations, possibly because the hydrogen concentration of about 70 nM prevailing in the consortium already allowed a maximal rate of dechlorination, which would imply that strain DCB-1 is a very good hydrogen scavenger indeed.

The following experimental set-up allowed testing of the hypothesis that the rate of benzoate degradation depends on the hydrogen concentration. The benzoate degrader was grown on benzoate in the presence of a hydrogenotrophic sulfate reducer plus a growth limiting amount of sulfate. Once sulfate had been depleted the hydrogen concentration in these co-cultures increased to about 2000 Pa, and benzoate degradation virtually stopped. By then adding different amounts of Methanospirillum cells different ratios between hydrogen producers and hydrogen consumers were established, and this resulted in a range of new equilibrium concentrations for hydrogen. The obtained correlation between the hydrogen concentration and the rate of benzoate degradation is shown in Figure 6. Based on these results a graphical scheme centered around hydrogen (Figure 7) can be formulated to evaluate the interactions between the three

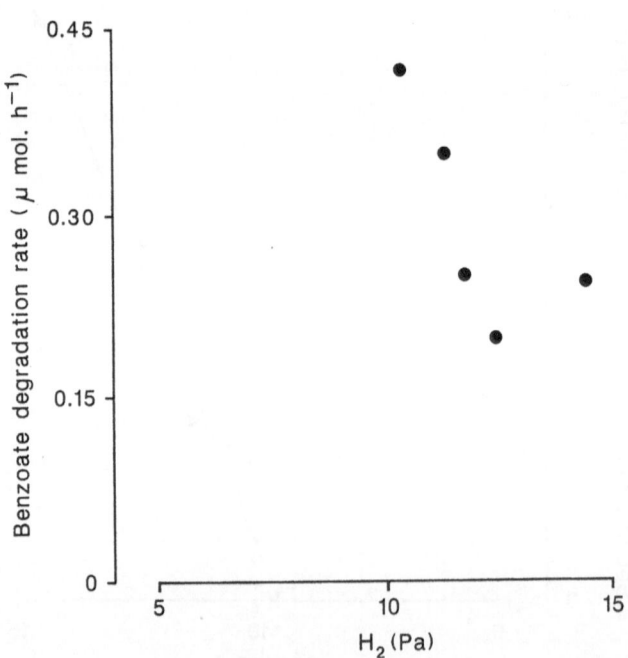

Figure 6. Relationship between hydrogen concentration and benzoate degradation rate in a methanogen-dependent co-culture.

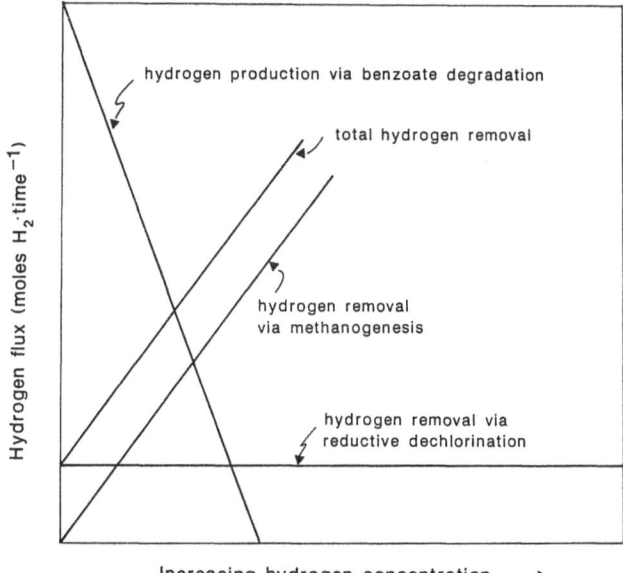

Figure 7. Outline of the hydrogen mediated kinetic
interdependency between hydrogen producers and hydrogen
consumers in 3-chlorobenzoate degrading consortium. The
potential hydrogen scavenging capacities are a function of
the hydrogen concentration.

organisms in the consortium. The rate of hydrogen production (in casu the
specific activity of the benzoate degrading population) decreases with
increasing hydrogen concentrations, while the rate of hydrogen consumption
via methanogenesis increases with increasing hydrogen concentrations.
Pursuant to our observations the rate of hydrogen consumption via
dechlorination is depicted as being independent of the hydrogen
concentration. This scheme predicts that in the consortium the rate of
benzoate degradation depends on the activities of both groups of hydrogen
consumers. If only one of the two groups is active the hydrogen
concentration will increase and consequently the rate of benzoate
degradation will decrease. Experimental results obtained with the
consortium support this hypothesis (Table 2). When in an experiment with
resting cells a mixture of benzoate plus 3-chlorobenzoate was added to the
consortium the rate of benzoate degradation was 50 % higher, at a slightly
lower hydrogen concentration, than when benzoate was added alone. The
enhanced benzoate degradation rate was triggered by the lower hydrogen
concentration. By offering a hydrogen sink additional to methanogenesis
the dechlorinating hydrogen scavenging population stimulated the rate of
benzoate degradation. The lowering of the hydrogen concentration was
hardly detectable (with our equipment), which is in agreement with the
observation that the rate of methanogenesis was hardly affected by the
activity of the dechlorinating population. This observation also implies
that the rate of benzoate degradation was strongly affected by small
changes in the hydrogen concentration in the concentration range of 1-4
Pa. Thus there was no significant competition for hydrogen between the two
hydrogen scavenging populations in the consortium as they practically
complemented each other's hydrogen scavenging potential at in situ
hydrogen concentrations during the degradation of 3-chlorobenzoate.

Table 2. Hydrogen fluxes in a defined 3-chlorobenzoate grown methanogenic consortium. The fluxes were measured over 24 hours after spiking of substrate-depleted cells with the indicated substrate.

Assay conditions	Hydrogen flux (μmoles/h)			Hydrogen (Pa)
	Benzoate degradation	Reductive dechlorination	Methanogenesis	
3-Chlorobenzoate	-*	1.3	--	--
Benzoate	2.3	--	2.3	4.0
3-Chlorobenzoate plus benzoate	3.7	1.3	2.4	2.5

*: -- = not determined.

The strong effect on the benzoate degradation rate of small changes in the hydrogen concentration suggested above would imply that the apparent K_i of the benzoate degrader for hydrogen is about a factor 10 lower than the range where these observations were made. It should be noted that the correlation between the hydrogen concentration and the rate of benzoate degradation as depicted in Figure 7 appears an oversimplification and can better be described by an inhibition term of the form $(1/(1+H_2/K_i)$. Such a term would also be in agreement with the observation that benzoate degradation still occurs at hydrogen concentrations up to 2000 Pa. Another reservation that should be made against the scheme depicted in Figure 7 is that the threshold concept was not included. The basic idea depicted here, however, should help in the design of future experiments aimed at a better quantitative understanding of the interdependencies between partner organisms in interspecies hydrogen based syntrophic co-cultures. The scheme has in a slightly modified format (Dolfing 1988b) already proved its usefulness, viz. in evaluating the effect of an inhibitor on one of the partners in a syntrophic co-culture. It was recently reported that acetate inhibits the rate of benzoate degradation in a syntrophic methanogenic coculture, and that higher acetate concentrations in the co-culture resulted in a decrease of the hydrogen concentration in the co-culture (Dolfing and Tiedje 1988). Such an effect is indeed predicted if the qualitative relationships between the partner organisms are as depicted in Figure 8. Figure 8 also indicates that specific inhibition of one of the partner organisms results in a new equilibrium concentration for hydrogen in such a way that, via the indirect effect on the other partner the inhibition is partially compensated (Dolfing and Tiedje 1988). Another important prediction made by this scheme is that doubling the population of e.g. methanogens in a syntrophic co-culture will not necessarily result in a doubling of the degradation rate of e.g. benzoate, not even when the co-culture consists of only a methanogen and a benzoate degrader. The stimulation of such an addition will depend on the ratio of the two partner organisms, and on the specific sensitivity of the two syntrophic organisms towards changes in the hydrogen concentration at the hydrogen concentration that is characteristic for a certain ratio of syntrophs.

It should be noted that the schemes presented in Figure 7 and 8 have been developed for specific activities, and not for growth parameters. This distinction may be of importance since the hydrogen concentration is expected to influence the energetics of the organisms, and thereby also their growth parameters (Dolfing 1988a). A direct link between energy generation and the rate of a reaction would translate into the use of log $[H_2]$ rather than $[H_2]$ in equations designed to model the dependency of

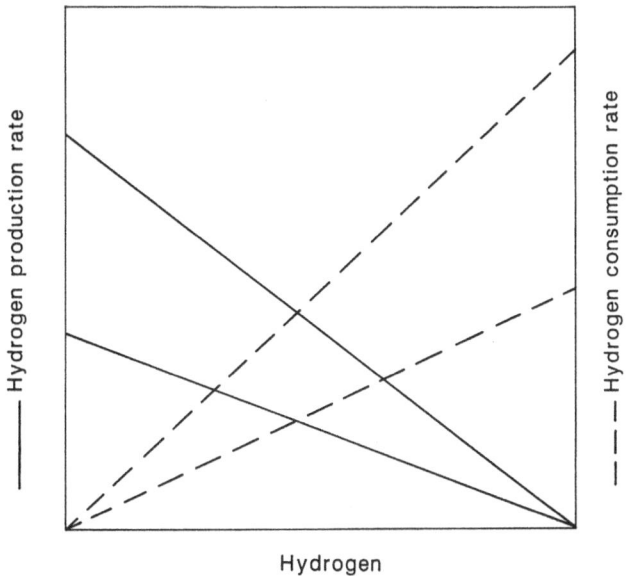

Figure 8. Modification of Figure 7, illustrating the effect of a specific inhibition of one of the partner organisms on the kinetics of the other partner as mediated via hydrogen.

growth related parameters on the hydrogen concentration in interspecies hydrogen transfer based co-cultures.

It is interesting to note that in the consortium methanogenesis from hydrogen could occur at hydrogen concentrations of 10–25 nM, i.e. well below the threshold value found for <u>Methanospirillum</u> after growth at high hydrogen concentrations. Apparently the culture history of the methanogen (cultivation under high or low partial pressures of hydrogen) influences the existence, or at least the value, of the threshold concentration for hydrogen uptake in <u>Methanospirillum</u>.

In the consortium the dechlorinating population was about 16 times the size of the methanogenic population (Dolfing and Tiedje 1987), but the hydrogen scavenging activity of the dechlorinating population was only half of that of the methanogens. Thus the specific hydrogen scavenging activity of the dechlorinating population was about 32 times less than that of the methanogenic population. This also implies that per mol of hydrogen consumed the cell yield of the dechlorinator was 32 times that of the methanogen, indicating that the dechlorinator is able to conserve the energy available from the dechlorination reaction very efficiently, and that the continued presence of the dechlorinating organism in the consortium, where long term competition for hydrogen seems apparent, is brought about by the relatively high growth yield of the dechlorinator, rather than by a high specific activity. An important conclusion of the present paper is that via the sensitivity of the benzoate degrader towards hydrogen this apparent long term competition for hydrogen is structured in such a way that the hydrogen scavenging capacities of the two groups of hydrogen consumers complement each other and together match the hydrogen production potential of the benzoate degrader. This results in co-existence rather than competition in this stable syntrophic co-culture.

Dolfing, J., 1988a, Acetogenesis, in: "Biology of anaerobic microorganisms", Zehnder, A. J. B., ed., John Wiley and Sons, Inc., New York.

Dolfing, J., 1988b, Hydrogen-based interactions in syntrophic propionate degrading microbial consortia, in: "Fifth international symposium on anaerobic digestion - Poster-papers", Tilche, A., and Rozzi, A., eds., Monduzzi Editore S.p.A., Bologna.

Dolfing, J., 1989, Reductive dechlorination of 3-chlorobenzoate is coupled to ATP production and growth in an anaerobic bacterium, strain DCB-1. Arch. Microbiol., in the press.

Dolfing, J., Tiedje, J. M., 1986, Hydrogen cycling in three-tiered food web growing on the methanogenic conversion of 3-chlorobenzoate. FEMS Microbiol. Ecol., 38:293.

Dolfing, J., Tiedje, J. M., 1987, Growth yield increase linked to reductive dechlorination in a defined 3-chlorobenzoate degrading methanogenic coculture. Arch. Microbiol., 149:102.

Dolfing, J., and Tiedje, J. M., 1988, Acetate inhibition of methanogenic, syntrophic benzoate degradation, Appl. Environ. Microbiol., 54:1871.

Lovley, D. R., 1985, Minimum threshold for hydrogen metabolism in methanogenic bacteria, Appl. Environ. Microbiol., 49:1530.

Robinson, J. A., and Tiedje, J. M., 1984, Competition between sulfate reducing and methanogenic bacteria for H_2 under resting and growing conditions, Arch. Microbiol., 137:26.

Thauer, R. K., Jungermann, K., and Decker, K., 1977, Energy conservation in chemotrophic anaerobic bacteria, Bacteriol. Rev., 41:100.

ROLE OF HYDROGEN IN THE GROWTH OF MUTUALISTIC METHANOGENIC COCULTURES

J. Benstead, D. B. Archer* and D. Lloyd

School of Pure and Applied Biology
University of Wales College of Cardiff
P.O. Box 915
Cardiff CF1 3TL

and

*AFRC Institute of Food Research
Colney Lane
Norwich NR4 7UA

Production of hydrogen and methane by a mutualistic methanogenic coculture (comprising *Desulfovibrio* strain FR-17 and *Methanobacterium* strain FR-2) during batch and continuous growth was monitored by membrane inlet mass spectrometry. This technique allows continuous non-invasive measurement of dissolved gases.

During batch growth in a sulphate-free medium containing ethanol (initially 0.05M), the concentration of dissolved hydrogen increased over the first 7.5 days; during this period no methane was detected. After 8 days the level of hydrogen decreased and an increase in the level of dissolved methane was observed. Maximum levels of dissolved hydrogen and methane were 93 and 875 μM, respectively. In continuous culture, the rates of methane and hydrogen production increased with increasing dilution rates over the range 0.39 to 0.83 day^{-1}.

INTRODUCTION

Natural populations of bacteria occurring in anaerobic digestion reactors are extremely complex, and in order to assess certain features of such systems, it is convenient and appropriate to study simplified laboratory systems. One such system consists of a mutualistic coculture of a *Desulfovibrio* sp. growing in the absence of sulphate on ethanol as sole carbon source and providing hydrogen for growth of a *Methanobacterium* sp.[1-3] This system serves as a model for interspecies hydrogen transfer between mutualistic bacterial species and may provide predictive information on the behaviour of more heterogeneous populations with respect to control phenomena and responses to perturbation.

Microbiology and Biochemistry of Strict Anaerobes Involved in Interspecies Transfer
Edited by J.-P. Bélaich *et al.*
Plenum Press, New York, 1990

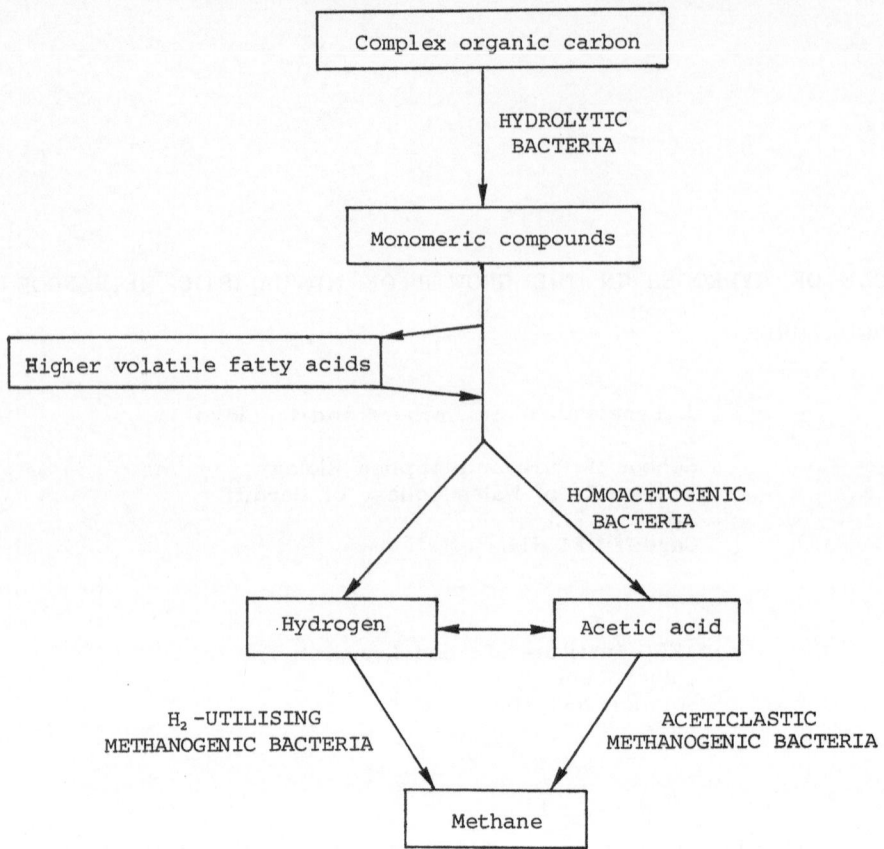

Fig. 1.   Major stages involved in the anaerobic degradation
of organic wastes.

The anaerobic degradation of organic wastes (Anaerobic Digestion) involves the participation of several groups of microorganisms[4] (Fig. 1). The hydrolytic bacteria ferment organic polymers to organic acids and alcohols, and these fermentation products are converted to acetate and hydrogen by the acetogenic bacteria. Removal of hydrogen and acetate by H_2-utilising and aceticlastic methanogenic bacteria, respectively, is essential to the whole process of waste mineralization.

Hydrogen is involved in most of the principal reactions of anaerobic digestion and has a controlling influence on both the rate and nature of the end products.   For example, the formation of acetic acid from glucose by the acid forming bacteria is coupled to the simultaneous production of large quantities of hydrogen.

$$C_6H_{12}O_6 + 2H_2O \longrightarrow 2CH_3COOH + 2CO_2 + 4H_2$$

This reaction is favourable providing the partial pressure of hydrogen is maintained at a low level by the hydrogen-utilising methanogenic bacteria.   At high levels of hydrogen or low pH, one response of the acid forming bacteria is to direct their metabolism to the production of higher fatty acids (butyric, propionic) instead of acetic acid to reduce the production of acid and free hydrogen.

Table 1.   Analytical methods available for hydrogen detection and their limits of sensitivity

Analytical method for measurement of hydrogen	Detection limit	Comments
Membrane inlet mass spectrometry	0.25μM (335 ppm)	Direct measurement of several dissolved gases Simultaneously
Gas-chromatography (Thermal conductivity detector)	20 ppm	Simultaneous quantification of methane
Exhaled hydrogen monitor (Electrochemical cell)	2 ppm	Portable unit with battery back up. Digital read out from 0-999 ppm hydrogen
Reduction gas detector (mercury vapour detector)	10 ppb	No interference from other gases. Can be used for long term monitoring or control

Under elevated levels of hydrogen, the oxidation of propionate and butyrate is inhibited[5-8], leading to a decline in the pH of the digester when the buffering capacity is exhausted. The methanogenic bacteria are sensitive to low pH and consequently acid conditions inhibit both the hydrogen and acetate-utilising methanogens. Unless immediate action is taken, the process of waste treatment may fail.

The effect of organic overloading on anaerobic systems results in hydrogen accumulation in the headspace.[9] Glucose, propionate or butyrate additions to mesophilic digester systems result in a rise in dissolved hydrogen levels.[10]

Methods which detect an imbalance in the microbial community at a very early stage (e.g. rise in hydrogen level in solution or in the digester gas) are highly desirable and may facilitate a more stringent monitoring and hence control of the process. Modulation of anaerobic digestion on a laboratory scale has been carried out by using the dissolved hydrogen signal from a membrane inlet mass spectrometer to control feed rates.[11,12]

Hydrogen monitoring in the digester gas also has the potential to be a useful indicator of process performance.[13-16] Levels of hydrogen in the gas for an established digester should be in the region of 10^{-4}–10^{-5} atm[16,17]; these correspond to dissolved gas concentrations of approximately 10^{-7}–10^{-8} mol/L. These levels are lower than the minimum detection limit by membrane inlet mass spectrometry (0.25μM). The technique finds application, however, where the hydrogen levels in digesters or microbial cultures are sufficient for detection. Some of the methods available for the detection and quantification of hydrogen are shown in Table 1.

In this study we describe the monitoring of dissolved hydrogen by membrane inlet mass spectrometry in batch and continuous methanogenic cocultures. We show that liquid phase measurements provide a more direct and useful assessment of batch culture status than those methods previously employed.

MATERIALS AND METHODS

Bacterial Strains and Their Maintenance

Methanobacterium strain FR-2 (DSM 2257) and *Desulfovibrio* strain FR-17 (NCIB 12086) were maintained at 37°C in Met 3 medium, under a head-space of H_2/CO_2 (4:1 v/v) for growth of *Methanobacterium* strain FR-2 or supplemented with 0.02M Na_2SO_4 and 0.05M ethanol for growth of *Desulfovibrio* strain FR-17. Anaerobic media and solutions were prepared following the procedures of Hungate.[16] Met 3 contained (g l^{-1}): Oxoid yeast extract (5), sodium formate (2), sodium acetate (2), K_2HPO_4 (0.45), NaCl (0.9), $(NH_4)_2SO_4$ (0.9), KH_2PO_4 (0.45), $MgSO_4.7H_2O$ (0.19), $FeSO_4. 7H_2O$ (0.01), resazurin (0.001) and 10 ml trace mineral solution.[19] The medium was purged with N_2/CO_2 (4:1 v/v) whilst being boiled and after cooling, an anaerobic solution of Na_2CO_3 (8% w/v) was added to give a final concentration of 0.2% (w/v), and the pH was adjusted to 6.8.

Medium was dispensed (4.5 ml) under N_2/CO_2 (4:1 v/v) into screw-capped glass tubes fitted with butyl rubber septa (Bellco, Vineland, N.J., U.S.A.), autoclaved and reduced by the addition of 0.1 ml of a solution containing $Na_2S.9H_2O$ (1.7% w/v) and cysteine hydrochloride (1.7% w/v). For the growth of *Methanobacterium* strain FR-2, the atmosphere was replaced with H_2/CO_2 (4:1 v/v) at 200kPa pressure.

Growth of Coculture

Cocultures of the *Desulfovibrio* and *Methanobacterium* spp. were grown at 37°C in sulphate-free medium containing 0.05M ethanol.[2] The medium was inoculated with pure logarithmic phase cultures of *Desulfovibrio* strain FR-17 (0.95% by vol.) and *Methanobacterium* strain FR-2 (0.95% by vol.).

A 950ml-working volume fermenter equipped with pH and temperature controllers was used (L. H. Engineering, Stoke Poges, U.K.). The rate of agitation was increased to 700 rpm during batch growth, with an initial slow rate of stirring (200 rpm) to allow the coculture to establish. For continuous culture, pre-reduced and sterilized medium was pumped from a 20L aspirator kept under a slight positive pressure of N_2/CO_2 (4:1 v/v) to the fermenter via two filters (a 3µm and 0.45µm) arranged in series. All tubing used was of butyl rubber.

Headspace Gas Analysis

A Pye series 104 chromatograph fitted with a Katharometer detector was used for the quantification of H_2 and CH_4. Samples (0.5 ml) were injected onto a Porapak Q column at ambient temperature.

Nitrogen was used as the carrier gas at a flow rate of 60 ml/min and the bridge current was 60 mA. Hydrogen at levels below 1,000ppm (101Pa) was assayed in a hydrogen monitor (Gas Measurement Instruments Ltd., Renfrew, United Kingdom).

Dissolved Gas Analysis by Mass Spectrometry

Levels of dissolved H_2 and CH_4 were continuously monitored during batch growth of methanogenic cocultures using a quadrupole mass spectrometer type SX200 with associated digital peak programmer (VG Gas Analysis, Aston, Middlewich, Cheshire, U.K.). The mass spectrometer was linked to the fermenter by means of a stainless steel probe with a silicon rubber membrane inlet. The probe was inserted into the fermenter and sterilized *in situ*. After autoclaving, the probe was attached to the mass spectrometer and liquid phase calibration was performed in the growth medium using standard gas mixtures equilibrated to saturation. The mass-to-charge ratios (m/z) used to measure the concentrations of H_2 and CH_4 were 2 and 15 respectively and the solubility of saturated aqueous solutions at 37°C were taken as 747µM and 1147µM respectively.

Ethanol and Acetate Determinations

Levels of ethanol and acetic acid in fermentation samples were assayed using enzymatic kits (BCL, Lewes, U.K.).

EXPERIMENTAL RESULTS

Continuous monitoring of the levels of dissolved hydrogen and methane by mass spectrometry during batch growth of a mutualistic methanogenic coculture:

Experiment 1. gas collection by the downward displacement method. The dissolved hydrogen level was shown to increase during the first 7.5 days of incubation and attained a maximum concentration of 93 µM; no

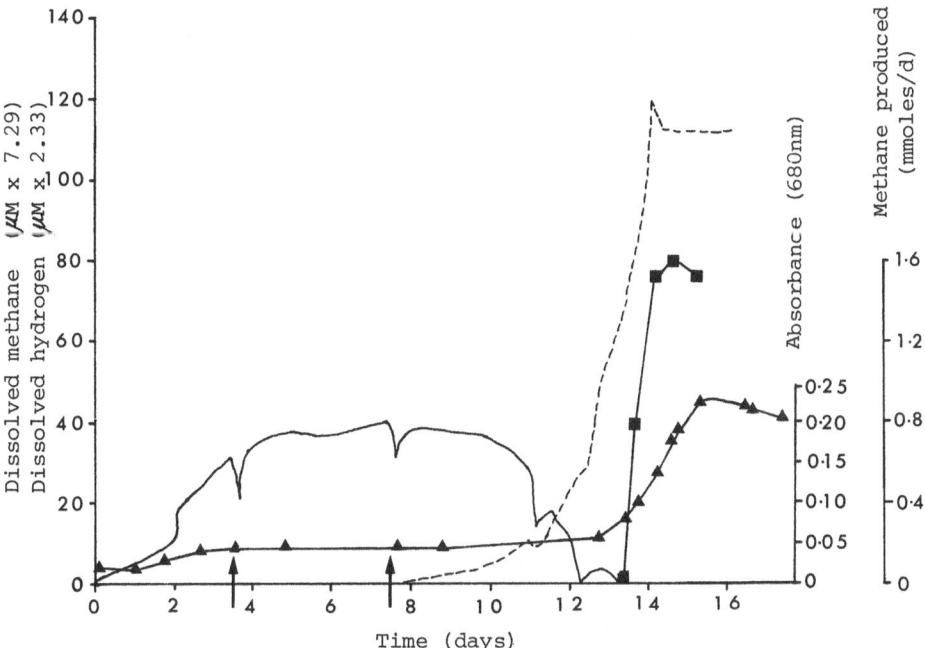

Fig. 2.   The progress of fermentation during batch growth of a coculture (comprising *Desulfovibrio* strain FR-17 and *Methanobacterium* strain FR-2) in sulphate-free medium containing ethanol (initially 0.05M). Symbols: ▲ , absorbance; ■ , methane produced; —— , dissolved hydrogen; ----, dissolved methane. A 950ml-working volume fermenter was used and levels of dissolved gases were continuously monitored using membrane inlet mass spectrometry. Arrows indicate the times at which the medium was reinoculated with pure cultures of both *Desulfovibrio* and *Methanobacterium* in order to establish growth of the coculture.

methane was detected during the same time interval (Fig. 2). Because the initial inoculum was not successful, at 3.6 and 7.6 days, the fermenter was reinoculated with pure cultures of both *Desulfovibrio* and *Methanobacterium.*

The level of dissolved hydrogen immediately after reinoculation decreased temporarily; this may result from the transfer of a low concentration of sulphate to the fermenter. In the presence of sulphate, hydrogen will be preferentially utilized by the *Desulfovibrio* sp. to produce H2S. Mass spectrometry showed that hydrogen was consumed by the coculture after 8 days with a corresponding increase in dissolved methane which attained a maximum concentration of 875μM.

Gas produced from the fermenter was collected over 4M NaH2PO4 (pH 2.0) and analysed by gas-chromatography. The estimated amount of methane produced (mmol/d) is shown in Fig. 2. Growth of the coculture was followed spectrophotometrically at 680nm; absorbance increased from 0.015 to 0.225 over a period of 15 days.

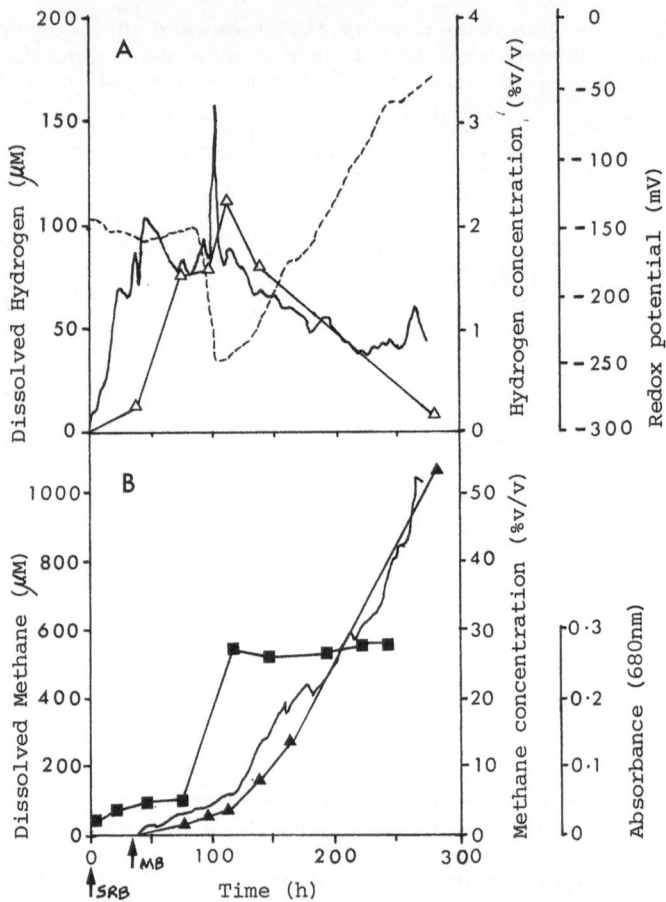

Fig.3. The progress of fermentation during batch growth of a coculture (comprising *Desulfovibrio* strain FR-17 and *Methanobacterium* strain FR-2) in sulphate-free medium containing ethanol (initially 0.05M). Symbols: Fig. 3A; ———, dissolved hydrogen; △ , headspace hydrogen; ---, redox potential. Fig. 3B; ——— , dissolved methane; ▲ , headspace methane; ■ , absorbance (680 nm). Arrows labelled SRB and MB indicate the time of addition of pure cultures of *Desulfovibrio* and *Methanobacterium* spp., respectively.

Experiment 2. analysis of gas samples taken directly from the culture headspace. Dissolved gas measurement by membrane-inlet mass spectrometry during the growth of a methanogenic coculture showed good agreement with the analysis of gases from the headspace by gas-chromatography (Fig. 3).

The medium was inoculated with a batch culture of the *Desulfovibrio* sp. at time 0 and dissolved hydrogen was detected after 3h incubation. The *Methanobacterium* sp. was introduced when the dissolved hydrogen concentration had increased to 76µM (Time 40h). Methane was first detected by mass spectrometry after 45h incubation, when the dissolved hydrogen concentration was 94µM.

A burst of hydrogen occurred which coincided with a decrease in the redox potential of the medium after 100h (Fig. 3).

Methanogenesis during continuous growth of the coculture

The rate of methane production and ethanol concentration during the growth of the coculture at various dilution rates has previously been examined.[3] The main conclusions drawn from this work showed the achievement of steady states for all dilution rates up to approximately 2.1 day^{-1}. The rate of methane production was shown to increase with increasing dilution rate, whereas, the concentration of ethanol remained low until high dilution rates were employed.

It was of interest in the present study to monitor hydrogen levels during continuous growth of the coculture. The steady-state hydrogen levels in the gas produced from the fermenter was shown to increase from 476ppm at a dilution rate of 0.39 day^{-1} to 756ppm at a dilution rate of 0.83 day^{-1} (results not shown).

Stability of the coculture to an increase in dilution rate

The effect of an increase in the dilution rate from 0.4 to 0.6 day^{-1} at time 0 is shown in Fig. 4. Increased ethanol in the fermenter gave an immediate stimulation in the rate of methane production. That increase in the level of ethanol will stimulate the growth of the *Desulfovibrio* sp. was predicted from mathematical analysis of the growth dynamics of mutualistic associations.[20]

The hydrogen level was shown to increase from a steady-state level of 476 ppm to over 3000 ppm. This observed increase in the level of hydrogen suggested that the rate of hydrogen production by the *Desulfovibrio* sp. exceeded the rate of hydrogen utilization by the *Methanobacterium* sp.

Effect of a new supply of medium on growth of the coculture

Connection of a new supply of media when the previous 20 L aspirator was exhausted resulted in a decrease in both the absorbance and rate of methane production of the coculture with a corresponding increase in both ethanol and hydrogen levels (Fig. 4). The stability of the coculture was restored after 150 hours. These observations were not considered to be caused by the introduction of air during the procedure since no increase in the redox potential of the media was observed. However, an increase in dissolved H_2S from reducing agents employed in the medium was detected by mass spectrometry (results not shown) and this may account for inhibition of growth.

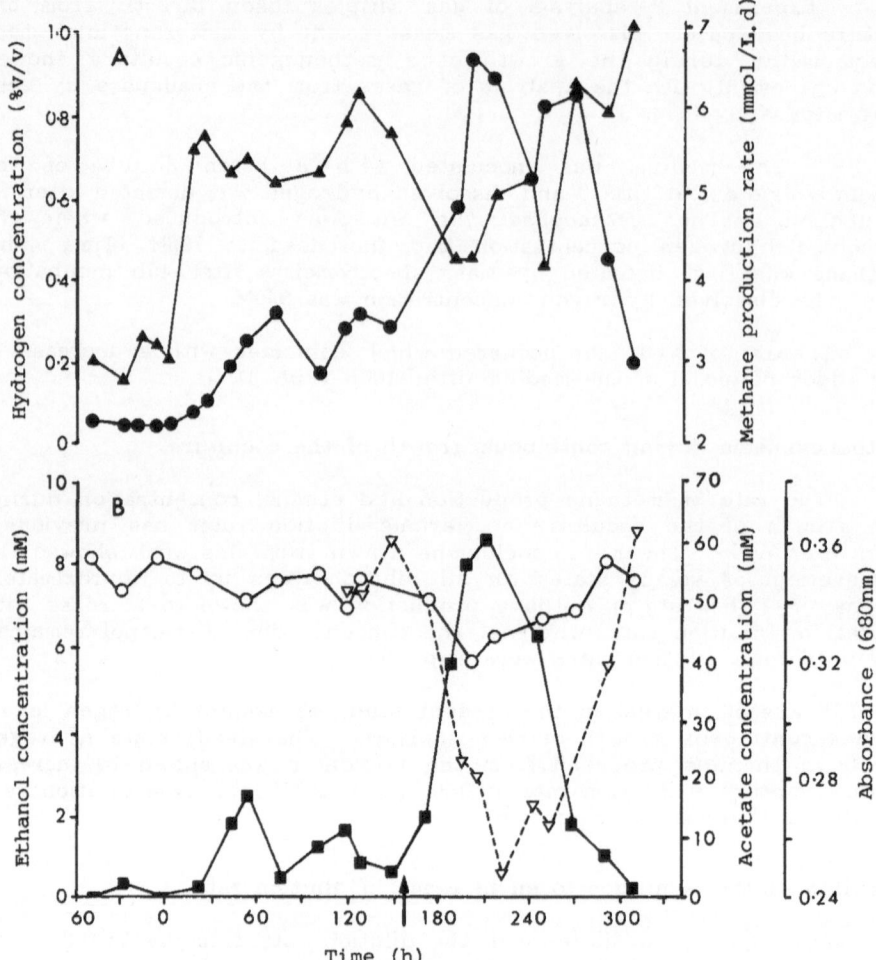

Fig. 4.   Progress of fermentation in a continuously-fed coculture
showing effects of increased dilution rate at time 0 (from
0.4-0.6 day⁻¹).   At the time indicated (arrow) a new batch of
growth medium was connected.   Symbols: Fig. 4A; ▲ ,
methane production rate; ● , hydrogen measured in tower;
Fig. 4B; ○ , acetate; ■ , ethanol; ▽ , absorbance (680nm).

DISCUSSION

        Mutualistic   interactions   are   essential   for   the   methanogenic
degradation   of   a   number   of   intermediates,   e.g.   propionic   and   butyric
acids, formed during the anaerobic digestion of wastes.   The oxidation of
propionate   by   the   obligate   proton-reducing   acetogenic   species   is
thermodynamically   favourable   only   when   coupled   to   methanogenesis   by
the H₂-utilising methanogenic bacteria.   The rates of both the hydrogen
producing and the hydrogen consuming reactions are influenced by the
hydrogen concentration.   In some systems, formate may replace hydrogen
as the linking intermediate between mutualistic species.[21,22]

        Direct measurements of numbers of methanogenic bacteria would be
extremely useful to assess process performance.   However, difficulties

have arisen in enumerating methanogens by conventional techniques. Acetate-utilising methanogens occur naturally in clumps (Methanosarcina barkeri, M.mazei) or in long filaments (Methanothrix soehgenii) so enumeration of these organisms as colony forming units is extremely inaccurate. Also the difficulties in culturing strict anaerobes on solid media may result in a considerable underestimate of numbers. These and other techniques for enumerating methanogens have recently been reviewed.[23]

Membrane inlet mass spectrometry has proved a useful technique for monitoring fermentative activity in situ.[24] This technique evidently provides a more direct measure of batch growth of defined methanogenic coculture conditions than gas phase measurements. Frequent sampling directly from the headspace of the fermenter via a butyl rubber septum should be avoided to prevent gas leakage. An alternative method of gas sampling by the downward displacement method has been described[25] and has enabled gas production by mutualistic methanogenic cocultures and other mixed cultures to be quantified. The major limitation of this method is that all measurements are time averaged because gas analysis is based on the accumulation of gas in the tower.

Rapid, transient changes in the concentrations of gases in solution which may go undetected by periodic sampling of the gas phase can be continuously monitored by membrane inlet mass spectrometry.

Hydrogen monitoring in anaerobic digesters estimates the hydrogen concentration in the bulk liquid which may not reflect the actual concentration present within microbial aggregates.[26] However, despite this limitation, hydrogen monitoring is an extremely useful indicator of process performance when used in conjunction with other conventional indicators (e.g. pH and volatile fatty acid levels).

ACKNOWLEDGEMENT

The financial support of a SERC CASE studentship is acknowledged by JB.

REFERENCES

1. M. P. Bryant, L. L. Campbell, C. A. Reddy and M. R. Crabill, Growth of *Desulfovibrio* in lactate or ethanol media low in sulphate in association with H_2-utilizing methanogenic bacteria, *Appl. Environ. Microbiol.* 33: 1162-1169 (1977).
2. D. B. Archer and G. E. Powell, Dependence of the specific growth rate of methanogenic mutualistic cocultures on the methanogen, *Arch. Microbiol.* 141: 133-137 (1985).
3. M. J. Tatton, D. B. Archer, G. E. Powell and M. L. Parker, Methanogenesis from ethanol by defined mixed continuous cultures, *Appl. Environ. Microbiol.* 55: 440-445 (1989).
4. M. P. Bryant, Microbial methane production - theoretical aspects, *J. Anim. Sci.* 48: 193-201 (1979).
5. S. H. Zinder, Microbiology of anaerobic conversion of organic wastes to methane: recent developments, *American Society for Microbiology News* 50: 294-298 (1984).
6. M. J. McInerney, M. P. Bryant and D. A. Stafford, Metabolic stages and energetics of microbial anaerobic digestion, *in:* "Anaerobic Digestion", pp.91-98, D. A. Stafford, B. I. Wheatley and D. E. Hughes, eds. (1980).

7. M. J. Wolin, Interactions between H_2-producing and methane-producing species, *in:* "Microbial Production and Utilization of Gases (H_2, CH_4, CO)", pp. 141-150, H. G. Schlegel, G. Gottschalk and N. Pfenning, eds., Golze, Göttingen, F.R.G. (1976).

8. H. F. Kaspar, and K. Wuhrmann, Product inhibition in sludge digestion, *Microb. Ecol.* 4: 241-248 (1978).

9. F. E. Mosey, New developments in the anaerobic treatment of industrial wastes, *Water Pollut. Cont.* 81: 540-552 (1982).

10. T. N. Whitmore, M. Lazzari and D. Lloyd, Comparative studies of methanogenesis in thermophilic and mesophilic anaerobic digesters using membrane inlet mass spectrometry, *Biotechnol. Lett.* 7: 283-288 (1985).

11. T. N. Whitmore, D. Lloyd, Mass spectrometric control of the thermophilic anaerobic digestion process based on levels of dissolved hydrogen, *Biotechnol. Lett.* 8: 203-208 (1986).

12. T. N. Whitmore, D. Lloyd, G. Jones and T. N. Williams, Hydrogen-dependent control of the continuous anaerobic digestion process, *Appl. Microbiol. Biotechnol.* 26: 383-388 (1987).

13. R. F. Hickey, The role of intermediate and product gases as regulators and indicators of anaerobic digestion, Ph.D. Dissertation, University of Massachusetts/Amherst, USA (1987).

14. F. E. Mosey and X. A. Fernandes, Mathematical modelling of methanogenesis in sewage sludge digestion, *in:* "Microbiological Methods for Environmental Biotechnology", pp. 159-168, J. M. Grainger and J. M. Lynch, eds., Academic Press, London, (1984).

15. D. B. Archer, M. G. Hilton, P. Adams, and H. Wiecko, Hydrogen as a process control index in a pilot scale anaerobic digester, *Biotechnol. Lett.* 8: 197-202 (1986).

16. F. E. Mosey, Mathematical modelling of the anaerobic digestion process: regulatory mechanisms for the formation of short-chain volatile acids from glucose, *Water Sci. Tech.* 15: 209-232 (1983).

17. D. B. Archer, The microbiological basis of process control in methanogenic fermentation of soluble wastes, *Enz. Microb. Technol.* 5: 162-170 (1983).

18. R. E. Hungate, A roll tube method for cultivation of strict anaerobes, *in:* "Methods in Microbiology", Vol. 3B, pp. 117-132, R. Norris and D. W. Ribbons, eds., Academic Press, Inc., New York.

19. D. B. Archer and N. R. King, A novel ultrastructural feature of a gas-vacuolated Methanosarcina, *FEMS Microbiology Letters* 16: 217-223.

20. G. E. Powell, Equalization of specific growth rates for syntrophic associations in batch culture, *J. Chem. Technol. Biotechnol.* 34B: 97-100 (1984).

21. J. H. Thiele and J. G. Zeikus, Control of interspecies electron flow during anaerobic digestion: significance of formate transfer versus hydrogen transfer during syntrophic methanogenesis in flocs, *Appl. Environ. Microbiol.* 54: 20-29 (1988).

22. D. R. Boone, R. L. Johnson and Y. Liu, Diffusion of the interspecies electron carrier H_2 and formate in methanogenic ecosystems and its implications in the measurement of Km for H_2 or formate uptake, *Appl. Environ. Microbiol.* 55: 1735-1741 (1989).

23. M. W. Peck and D. B. Archer, Methods for the quantification of methanogenic bacteria, *International Industrial Biotechnology* 9: 5-12, (1989).

24. D. Lloyd, R. I. Scott and T. N. Williams, Membrane-inlet mass spectrometry: measurement of dissolved gases in fermentation liquids, *Trends Biotechnol.* 1: 60-63 (1983).

25. B. H. Kirsop, M. G. Hilton, G. E. Powell and D. B. Archer, Methanogenesis in the anaerobic treatment of food-processing wastes, *in:* "Microbiological Methods for Environmental Biotechnology", pp. 139-158, J. M. Grainger and J. M. Lynch, eds., Academic Press, London (1984).

26. R. Conrad, T. J. Phelps and J. G. Zeikus, Gas metabolism evidence in support of the juxtaposition of hydrogen-producing and methanogenic bacteria in sewage sludge and lake sediments, *Appl. Environ. Microbiol.* 50: 595-601 (1985).

THERMODYNAMICAL AND MICROBIOLOGICAL EVIDENCE OF TROPHIC MICRONICHES FOR

PROPIONATE DEGRADATION IN A METHANOGENIC SLUDGE-BED REACTOR

Serge R. Guiot, F. Alexander MacLeod, and André Pauss

N.R.C., Biotechnology Research Institute
6100 Royalmount Avenue
Montréal, Qc. Canada H4P 2R2

INTRODUCTION

The upflow anaerobic sludge bed (UASB) concept, first proposed by Hemens et al. (1962), has been developed and extensively promoted by the Lettinga school, and applied to a large variety of wastewaters (Lettinga et al. 1980, 1983). The UBF reactor used in this study hybridized the UASB reactor with a packing filter (Guiot and van den Berg 1984). These reactors are characterized by the ability to accumulate a large amount of biomass due to adhesion of bacterial cells to each other. In the UASB and UBF reactors, the upflow velocity will select for organisms which can adhere to each other to form well settling granules which can be several millimeters in diameter. The granules accumulate in the reactor and then are exposed to the continuous feeding which is injected into the bottom of the reactor.

The complete anaerobic degradation of organic matter to CO_2 and CH_4 requires the concerted action of four major metabolic groups of bacteria (Mah 1982, Beaty et al. 1986). Fermentative bacteria hydrolyze the substrate polymers and ferment the products to volatile fatty acids (VFA), CO_2 and H_2. The obligate H_2-producing acetogenic (OHPA) bacteria degrade propionate and longer-chain VFAs and some aromatic compounds to acetate, CO_2 and H_2. A third group referred as the H_2-consuming acetogens reduce CO_2, CO, CH_3OH and methyloxy-groups of aromatic compounds into acetate and sometimes butyrate. The fourth group includes the H_2-using methanogenic bacteria which convert CO_2 and H_2 to CH_4 and the acetoclastic methanogens which cleave acetate to CH_4 and CO_2. Clearly, the aggregation of anaerobic microorganisms into granules would optimize the cooperation between the partner organisms, by reducing the diffusion distance for the transfer of metabolites, namely. Thus aggregation would create the close cell associations which are obligatory for the degradation of certain substrates. For example, the reduction degree of end-products of acidogenesis (Shink and Thauer 1988) and the degradation of propionate (Boone and Bryant 1980) and butyrate (Dwyer et al. 1988) are thermodynamically controlled by the H_2 concentration. These last reactions are exergonic only if the partial pressure of H_2 is lower than 10 and 100 Pa for propionate and butyrate, respectively (McInerney and Bryant 1980). These two substrates could not be oxidized unless the H_2 produced is scavenged by the H_2-consuming organisms. Aggregation of bacteria of different metabolic groups is thus of high importance for the energetics and kinetics of the overall substrate conversion in anaerobic digestion (Schink and Thauer 1988).

Microbiology and Biochemistry of Strict Anaerobes Involved in Interspecies Transfer
Edited by J.-P. Bélaich *et al.*
Plenum Press, New York, 1990

Using in situ metabolite concentrations, the in situ Gibbs free energy change of the reactions occurring in the process can be calculated. In most of the reports available, the gas phase H_2 partial pressure was used for thermodynamic calculations, with the assumption therefore that the liquid-to-gas H_2 transfer was not limited (Seitz et al. 1988, Dwyer et al. 1988). However this mass transfer rate was shown to be ineluctably limited in biomethanation process (Robinson and Tiedje 1982, Pauss et al. 1989a). Using punctual in situ concentrations of all the metabolites, including dissolved H_2, Conrad et al. (1986) reported that the propionate degradation was endergonic in various natural environments (ΔG from +1.8 to 14.1 kJ/reaction), while propionate degradation would be permitted in microbial clusters. However without the determination of the metabolite fluxes, the assumed degradations could not be effectively substantiated.

This paper presents results of three different experimental approaches which were conducted independently, very recently in our laboratory: the rate of propionate degradation within a granule bed, was compared to the ΔG calculated with dissolved H_2 concentration that was measured in the bulk liquid with a new hydrogen/air fuel cell detector (Pauss et al. 1989b); electron microscopy was used to study the ultrastructure of cross-cleaved granules (MacLeod et al. 1989); and specific metabolic activities were partitioned as a function of the depth of the granule. All results converge to substantiate the concept that defined intra-granular zones are particularly suited for interspecies H_2 transfer.

MATERIAL AND METHODS

Reactors

Bioreactors were upflow sludge bed and filter (UBF) reactors as described by Guiot and van den Berg (1984). The first reactor had a 20.5 L working volume. It was continuously fed a concentrated synthetic medium (in g/L, sucrose 380, yeast extract 3.8, KH_2PO_4 7.6, K_2HPO_4 10, $(NH_4)_2SO_4$ 19, NH_4HCO_3 76), diluted with bicarbonate buffered tap water (in g/L, $NaHCO_3$ 2.75, $KHCO_3$ 3.5) in a ratio of 1/127 (vol./vol.). During the period considered for propionate degradation (day 49 to 55), 158 g of Na-propionate was added per liter of synthetic medium. The recirculation pump (Jabsco, Ca.) insured an upflow velocity of 1.0 m/h, and an effluent recirculation to dilute feed ratio of 12/1. The reactor was maintained at 35°C. The gas production was recorded daily with a gas meter (Wet Tip Gas Meter Co, Wayne, Pa.).

The second reactor had a working volume of 13.5 L. The continuous feed stream contained, in g/L: sucrose 380, yeast extract 3.8, KH_2PO_4 7.6, K_2HPO_4 10, $(NH_4)_2SO_4$ 19, NH_4HCO_3 76. The dilution stream (tap water with 4 g/L of $NaHCO_3$ and 5 g/L of $KHCO_3$) was added to get a hydraulic residence time of 0.35 d, which provided a feed to dilution streams ratio of 1/87 (vol./vol.). The recirculation pump (Jabsco, Ca.) insured an upflow velocity of 0.9 m/h, and an effluent recirculation to dilute feed ratio of 10/1. The reactor was operated at a specific organic loading rate of 1.3 g COD/g VSS d with an 82% substrate removal efficiency. The reactor was in operation for one month before the collection of the granules.

Both reactors were inoculated with granular sludge obtained from an UASB reactor treating cheese whey wastewater (Agropur, Notre-Dame-du-Bon-Conseil, Qc, Canada). The reactors' performance was routinely assessed by daily determination of feed flow rate, gas production rate, gas composition, effluent flow rate, pH, temperature, effluent COD and VFA content.

Analytical methods

The analytical procedures are described in detail in Arcand et al. (1989) and Pauss et al. (1989a). Acetate and propionate were determined by gas-liquid chromatography of the free acids, obtained by addition of 1 volume of formic acid 6% (wt./vol.) to 1 volume of centrifuged sample. CH_4 and CO_2 were analyzed at 40°C via gas chromatography using a thermal conductivity detector. The hydrogen fraction in the gas phase was determined by gas chromatography using thermal conductivity detection and nitrogen gas as the carrier. With a 1 ml sample loop, the detection limit was about 3 Pa (30 ppm).

Dissolved hydrogen was continuously quantified in the recirculation stream with a new, sensitive and reliable hydrogen/air fuel cell detector (Syprotec, Pointe-Claire, Qc, Canada) (Pauss et al. 1989a). The detection limit of dissolved hydrogen was 80 nmol/l.

Glucose was determined by HPLC (Spectra-Physics SP8100) using a 3 cm RP-8 column (Brownlee Spheri 5 μm RP-8) combined with a 25 cm Polypore H resin column (Brownlee) at 40°C, maintained at a constant flow rate of 0.35 mL/min, using 0.01 N H_2SO_4 as solvent. Glucose was detected by differential refractometry (Spectra-Physics SP6040).

The pH was measured with a combined Radiometer electrode (TTT85 Radiometer, Copenhagen). The COD was determined colorimetrically by the method of Knechtel (1978). The Suspended Solids (SS) were determined by drying the sample at 105°C. The Volatile Suspended Solids (VSS) were determined by the weight loss of the sample between 105 and 600°C.

Specific metabolic activity

The specific activities of the sludge were determined by measuring the rate of uptake of a defined substrate (glucose, acetate, propionate, H_2/CO_2) (Guiot et al. 1986). Tests were conducted under anaerobic conditions in serum bottles shaken at 100 rpm and maintained at 35°C (G24 shaker, New Brunswick Scientific Co., Edison, N.J.). Sludge samples were pulsed with a not-limiting concentration of the defined substrate as the sole C-source. The linear substrate degradation curves were calculated using the least squares method. The specific rate was obtained by dividing the uptake rate by the VSS concentration in the serum bottle. The H_2 uptake activity was estimated similarly by measuring the methane converted from H_2/CO_2 as the sole substrates after 6 to 20 hrs of vigorous agitation at 35°C (G24 shaker at 350 rpm).

Granule abrasion and granulometry

A two step abrasion of granules was performed by upflow fluidization of a prewashed granular sludge sample, with O_2-free N_2 (10 L/min, 20 psi), that was injected through a glass column, for a few hours. The abraded particles were screened from the size-reduced granules, using a sieve of 500 μm mesh opening, under an O_2-free N_2 atmosphere.

Granulometry is a measure of the particle size distribution of a sample. A sludge sample was screened through a sieve of 500 μm mesh opening. The dry weight of the sieved portion was measured relative to total solids of the sample. An aliquot of the sieve-retained granules was introduced in a Petri dish to which buffer was added just enough to cover the granules. Photographs of the most representative parts of the dish (with a graduated transparency underneath) were taken with a Photoautomat MPS45 coupled to a stereomicroscope (WILD Heerbrugg, Switzerland) with a 3.3 magnification. The photographic negatives were analyzed using a

Quantimet Q520 Image Analysis System (Cambridge Instruments Ltd, Cambridge, UK) in order to obtain an automatic determination of the granule size distribution. The size was converted to sludge mass by considering the granule as a sphere and assuming all granules had equivalent density.

Scanning electron microscopy (SEM)

Washed granules were placed in sealed 50 mL serum bottles and were fixed overnight at 4°C in a solution of 5% glutaraldehyde in anaerobic cacodylate buffer (0.1 M, pH 7.2). Cleaved preparations were obtained by quick freezing the fixed samples in liquid nitrogen, followed by cleavage with a mortar and pestal. Dehydration was completed by passage of whole and cleaved granules through graded water-ethanol and ethanol-Freon 113 series. The samples were affixed to aluminium specimen mounts, coated with gold-palladium and examined with a JEOL T220 scanning electron microscope operated at an accelerating voltage of 15 kV.

Calculations

The standard Gibbs free energy changes (ΔG^o) were first calculated at 25°C, from the standard Gibbs free energy changes of formation of each component (Barrow 1981; Thauer et al. 1977). They were then calculated at 35°C with the Gibbs-Helmoltz's equation (equation 1) (Table 1). The standard enthalpy change used in the latter equation was calculated from the standard enthalpy changes of formation of each component.

$$\Delta G^o{}_{35°C} = 308.15 \left[\frac{\Delta G^o{}_{25°C}}{298.15} - \Delta H^o \frac{308.15 - 298.15}{308.15 \times 298.15} \right] \tag{1}$$

The Gibbs free energy change (ΔG) of the propionate degradation and of the methane-producing reactions were calculated according to equation (2):

$$\Delta G = \Delta G^o + 2.303 \ R \ T \log \frac{P[products]^n}{P[reactants]^m} \tag{2}$$

where R is the ideal gas constant ($8.314 \ J \ K^{-1} \ mol^{-1}$), T is the temperature (K), and P[products] and P[reactants] are respectively the product of the activities of the products and of the reactants, each of them having its stoichiometric coefficient (n and m) as exponent. The partial pressure of methane and the dissolved hydrogen concentration, as well as the ionized fraction of acetic and propionic acids (calculated from the measured total acid concentrations and the acidity constants: K_a of $1.73 \ 10^{-5}$ and $1.31 \ 10^{-5}$ mol/L respectively), were used in the calculations.

Table 1. Standard Gibbs free energy change and standard enthalpy changes of the biological reactions considered, at 25 and 35°C[a].

Reactions	ΔG^o		ΔH^o
	25°C	35°C	25-35°C
	(kJ/reaction)		
(1) Propionate + 3 $H_2O \longrightarrow$ 3 H_2 + HCO_3^- + Acetate + H^+	+ 169.2	+ 168.9	+ 178.2
(2) 4 H_2 + H^+ + $HCO_3^- \longrightarrow CH_4$ + 3 H_2O	- 245.8	- 246.6	- 224.5
(3) Acetate + $H_2O \longrightarrow HCO_3^-$ + CH_4	- 31.0	- 32.2	+ 5.5

(a) using the Gibbs-Helmoltz's equation.

RESULTS AND DISCUSSION

Thermodynamic considerations

The operation and performance data of the first reactor, as well as the major metabolite concentrations, namely acetic and propionic acids and the dissolved and gaseous hydrogen, were relatively constant over the two time periods that were considered (before and during propionate supplementation). Average values with standard deviations are provided in Table 2, for each period. Sodium propionate was added to the feed stream such that the final concentration was 1200 mg/L in the influent. Its concentration was measured in the effluent. Its degradation was undoubtedly effective and continuous (1.56 mmol/g VSS d, over 7 days i.e. 12 hydraulic residence times). The Gibbs free energy changes were calculated for each data point as previously described, and averaged for each period, in Table 2. The methane-producing reactions were exergonic, allowing sufficient energy change to lead to the formation of one or more ATP (Thauer and Morris 1984). In contrast, the propionate degradation into acetate, bicarbonate and hydrogen was endergonic along the overall reactor operation. However, despite of this unfavorable thermodynamic constraint, propionate did appear to be actually degraded in the reactor.

Table 2. Operating conditions, performance and Gibbs free energy changes in the first reactor before (day 6 to 42) and during (day 49 to 55) the propionate supplementation.

Parameters	Days 6-42		Days 49-55	
	Average value (std deviation)			
Influent COD (g/L)	4.14	(0.57)	4.51	(0.24)
Organic loading rate (g COD/L_{dig} d)	7.00	(0.42)	7.92	(0.41)
Hydraulic residence time (d)	0.59	(0.07)	0.57	(0.003)
Soluble COD removal efficiency (%)	99	(1.2)	99	(0.2)
Methane productivity (L(STP)/L_{dig} d)	1.47	(0.19)	1.57	(0.07)
Methane in the gas phase (%)	53	(1.3)	63	(0.3)
Sludge content (g VSS/L_{dig})	20.7	(3.35)	15.8	
Substrate specific removal rate (g COD/g VSS d)	0.34	(0.04)	0.51	(0.03)
pH	7.13	(0.10)	7.38	(0.05)
Propionate in influent (mM)	0		13.87	(0.73)
Acetate in effluent (mM)	0.44	(0.64)	0.12	(0.12)
Propionate in effluent (mM)	0.15	(0.10)	0.39	(0.05)
Dissolved H_2 (μM)	3.1	(0.5)	1.8	(0.074)
H_2 in gas phase (ppm)	137	(68)	68.1	(3.33)
Propionate specific removal rate (mmol/g VSS d)	-		$\geq$ 1.56	(0.085)
ΔG (kJ/react.) Propion. $\rightarrow$ Acet./H_2	+ 22.7	(1.7)	+ 14.65	(2.8)
ΔG (kJ/react.) $H_2/CO_2 \rightarrow CH_4$	- 71.0	(1.7)	- 63.9	(0.7)
ΔG (kJ/react.) Acetate $\rightarrow CH_4/CO_2$	- 25.5	(4.7)	- 23.0	(2.7)

The use of formate, instead of H_2, as the electron acceptor in the acetate-fermentation of propionate yielded a similar ΔG value. The low sulfate content of the medium excluded propionate oxidation, coupled with sulfate reduction. The reductive carboxylation of propionate into butyrate would be the only metabolic pathway by which propionate could be taken up with respect to the thermodynamics as defined in the bulk medium. This new pathway is not yet established as playing an important role in anaerobic digestors (Tholozan et al. 1988).

An alternative explanation for the absence of the thermodynamic inhibition of the propionate degradation, as defined in the bulk medium, could be provided by the biomass configuration. The reactor biomass was aggregated into granules. Particles with a diameter smaller than 0.5 mm represented at the most 17% of the total biomass (dry wt.) (Table 3). The average diameter of the granule was 1.10 mm ($\pm$ 0.46). Therefore the existence of intra-granular trophic microniches can be postulated. Syntrophic associations in such intra-granular microenvironments would sufficiently lower the dissolved H_2 concentration surrounding the propionate-degrading bacteria such that propionate degradation could occur within the granule. As a consequence of the highest power of H_2 in the logarithmic term of equation (2), the ΔG is most sensitive to its concentration. Hydrogen needed to be about 30 times less concentrated in the granule than in the bulk liquid to render the propionate degradation exergonic. The existence of such microbial microniches have been already postulated by Conrad et al. (1986) in anoxic lake sediments and Boone (1984) in a reactor degrading animal waste.

Table 3. Granule size distribution of the sludge bed of the first reactor (averaged on six samples)

Size class (equival. diameter, mm)	< 0.5	0.5 - 1	1 - 1.5	1.5 - 2	2 - 2.5
Relative sludge mass (%)					
Avg.	17.2	12.1	34.3	28.4	8.0
Std dev.	8.6	2.6	4.8	4.8	1.8

Ultrastructure of the granule

Recent findings in our laboratory showed that granules obtained from a mesophilic reactor fed with a sucrose medium (described here as the second reactor) exhibited a three-layered structure and that each layer possessed different distinguishing morphologies. Detailed results on granule morphologies and a proposed mechanism of formation are presented in a full paper under submission (MacLeod et al. 1989). SEM observations on a cleaved granule are presented in Fig. 1 and 2. The central layer i.e. the granule core, consisted of cavities encased by rod-shaped bacteria which possessed flat ends (Fig. 2c). They were of the same size and shape as previously described for Methanothrix species (Dubourguier et al. 1988). The external layer (ca. 10-20 µm thick) contained a variety of organisms (Fig. 2a), including Methanococcales-like organisms and Methanococcales-like filaments (Zehnder et al. 1982). The middle layer consisted of a large number of cocci and rod-shaped bacteria. Transmission electron microscopy (TEM) (not shown here) revealed that among these rods there was a very electron dense organism which resembled a Syntrophobacter species described by Dubourguier et al. (1988), and which was juxtapositioned to

178

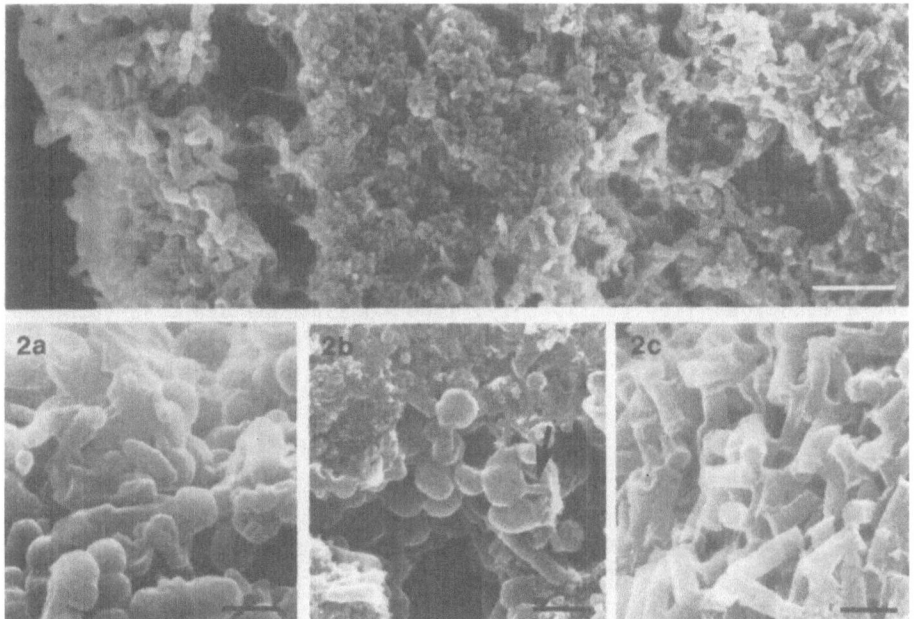

Fig. 1.   Scanning electron micrograph provides evidence of three morpholo-
          gically distinct layers in sucrose-fed granules. Bar, 5 μm.
Fig. 2a. The exterior layer contained a heterogenous population of rods,
          filaments and cocci. Bar, 1 μm.
Fig. 2b. Rods and cocci predominated in the second layer. The arrow iden-
          tifies a probable syntrophic association which is embedded in exo-
          polymeric material. Bar, 1 μm.
Fig. 2c. The central layer contained a large number of cavities which were
          surrounded principally by one bacterial morphology. Bar, 1 μm.

Syntrophobacter -like organisms. Such associations, embedded in exopoly-
mers, were also observed with SEM, as indicated by the arrow in the Fig.
2b.

    The juxtapositioning of different bacterial species in granular
material from anaerobic digestion has been reported earlier (Thiele et al.
1988, Dubourguier et al. 1988). Our model supports the development of such
a "lattice type" cell arrangement, in order to facilitate the interspecies
H_2 transfer. Furthermore, the results indicate that this arrangement is
particularly predominant in the middle layer of the granule. According to
this three-layer structure, H_2-consuming organisms present in the external
layer should utilize H_2 diffusing towards the second layer. The H_2-
consuming organisms present in the second layer would remove any remaining
H_2 produced by the acetogens present in this layer. This H_2 scouring would
insure that the level of dissolved H_2 would be sufficiently low in the
intra-granular space to permit the degradation of propionate.

Partition of metabolic activities

    Two abrasions were performed sequentially on the same sample of
granules, as described previously. The granules were sampled from the

first reactor at day 70. The specific activities were measured on the particles abraded by the first and the second treatment, as well as on granules remaining from the overall treatment. Results are given in Table 4. The first abrasion resulted in detachment of 14 % of the total biomass (dry wt.), and the second, 11 %. Assuming that the granules are spherical and are evenly dense, these successive mass reductions would correspond to a decrease in diameter of 5 and 4 %, respectively. This means that the actual abrasion technique was stripping out thin bacterial pellicles only. Analysis of the cumulative distribution of the granule volumes as a function of the size class indeed showed a small average reduction of the granules before and after each step-abrasion: average diameter (std. dev., granule count) as 1.09 (0.37, 151), 1.04 (0.35, 158) and 0.98 (0.40, 92), respectively. However since the stripping effect is in all probability uneven with respect to each individual granule, the results have to be interpreted cautiously and only in terms of tendencies. Namely the terminology of first and second pellicle, has to be taken as a figurative, rather than as a physical moiety.

Table 4. Partition of the metabolic specific activities as a function of the granule depth

Granule fraction		Specific activity			
Radius interval Avg. value, mm	Rel. dry wt. %	Avg. (st. dev.), mmol/g VSS d			
		Glucose	Acetate	Propionate	H_2
0.54–0.52	14	187 (24)	3.25 (0.2)	0.70 (0.76)	48 (2)
0.52–0.49	11	116 (5)	4.52 (0.4)	1.22 (0.23)	69 (6)
0.49–0	75	67 (0)	4.03 (0.5)	1.6 (0.1)	62 (12)

The specific metabolic activities as partitioned between the external pellicle, the second pellicle and the residual core of the granules, are given in Table 4. The acetoclastic activity is evenly distributed along the depth of the granule. This is not surprising, despite the evident dominance of Methanothrix-like bacteria in the granule core (Fig. 2c), for several reasons. Methanosarcina-like bacteria have indeed been observed in the external layer (Fig. 2a). Since the specific activity of Methano-sarcina-like microorganisms is recognized to be one order of magnitude larger than that of the Methanothrix-like bacteria, a certain number of the former group could produce an activity equivalent to a considerably larger number of the latter organisms. In addition it is likely that a large number of other species that were present in the first and second pellicles would be capable of utilizing acetate. Secondly a diffusional resistance to the substrate could have lowered the apparent activity in acetate tests on residual cores, which were still large aggregated configurations, in contrast to small particles present in the tests on the first and second pellicles.

In contrast Table 4 shows there is an obvious gradient distribution of the acidogenic (fermentative) bacteria and the propionate-degrading bacteria (H_2-producing acetogens). There was a clear predominance of fermentative bacteria in the external layers of the granules. Conversely propionate-degrading bacteria predominated further inside the granule. These observations confirm the ultrastructural model that was previously presented. Specifically, the partitioning of specific metabolic activities demonstrated that the outermost pellicle, containing H_2-consuming bacteria (Table 4, last column), can shield the propionate-degrading bacteria contained in inner layers, from the H_2 diffusing towards the granule interior.

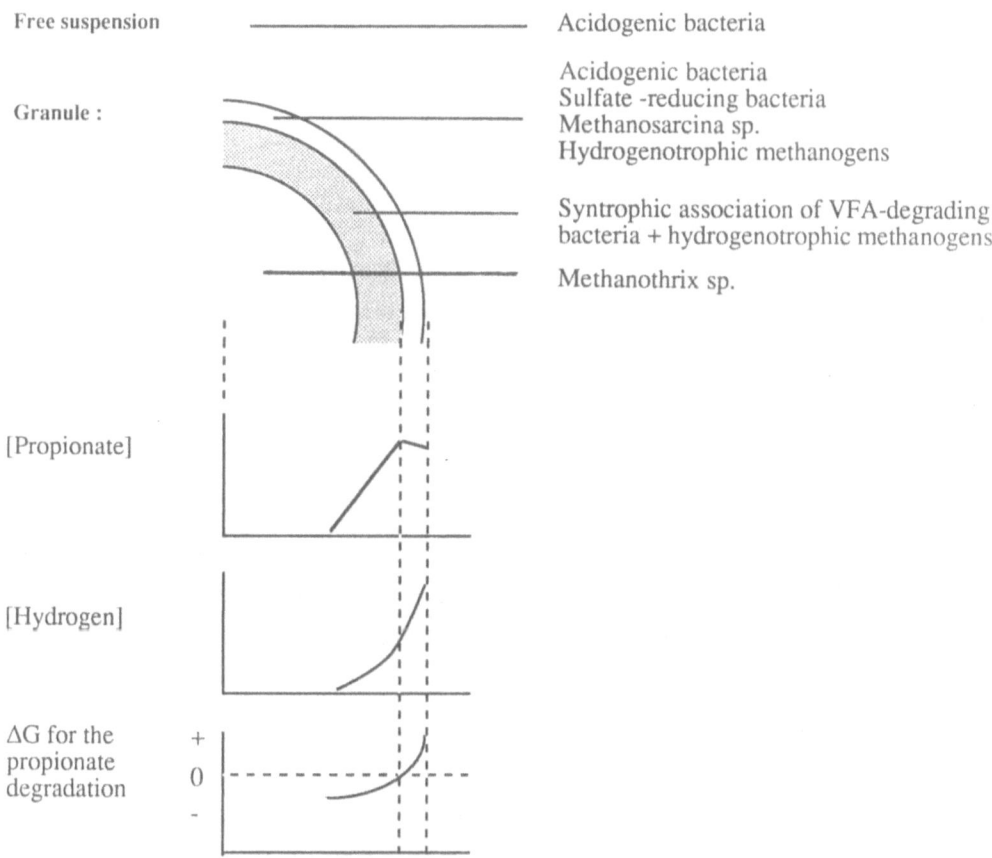

Free suspension ——————————— Acidogenic bacteria

Granule : ——————————— Acidogenic bacteria
Sulfate -reducing bacteria
Methanosarcina sp.
Hydrogenotrophic methanogens

——————————— Syntrophic association of VFA-degrading bacteria + hydrogenotrophic methanogens

——————————— Methanothrix sp.

[Propionate]

[Hydrogen]

ΔG for the propionate degradation

Fig. 3. Schematic representation of a sucrose-fed granule according to the three-layered model, and its expected effect on the propionate and H_2 concentration gradient

CONCLUSIONS

An effective and continuous degradation of propionate was measured in an anaerobic granular sludge-bed reactor, even though thermodynamic calculations showed that the H_2 concentrations in the environmental liquid would not permit such a reaction ($\Delta G > 0$). A three-layered structure of the sucrose-fed granule was observed with electron microscopy. Syntrophic bacteria associations were located between an external, presumably acidogenic layer, and an acetoclastic core. Measurement of acidogenic and propionoclastic activities as a function of the depth into the granule confirm the ultrastructural model. In such a structured organization (Fig. 3), the dissolved H_2 concentration surrounding the propionate-degrading bacteria would be reduced to a level much lower than that in the environmental liquid in order for propionate oxidation to occur inside the granule. Such an aggregate would be a stable metabolic arrangement that would create optimal niches for H2 interspecies transfer.

ACKNOWLEDGEMENTS

The authors thank Ms. C. Beaulieu, A. Dumont, S. Rocheleau and Mr. A. Corriveau for expert technical assistance.

REFERENCES

Arcand, Y., DesRochers, M., Chavarie, C., and Guiot, S. R., 1989, Effect of the hydraulic regime on the granule size distribution in an upflow anaerobic reactor, Proc. 21st Mid-Atlantic Industrial Waste Conference, June 25-27, Harrisburg, Pa., USA, 357.

Barrow, G. M., 1981, Physical chemistry for the life sciences, 2nd edition, McGraw-Hill Inc., New York.

Beaty, P. S., McInerney, M. J., and Wofford, N. Q., 1986, Energetics of H_2-producing syntrophic bacteria, in: "Biotechnological advances in processing municipal wastes for fuels and chemicals", A.A. Antonopoulos, ed., Argonne Natl Laboratory, Argonne, Il., 67.

Boone, D. R., 1984, Mixed-culture fermentor for simulating methanogenic digestor, Appl. Environ. Microbiol. 48:122.

Boone, D. R., and Bryant, M. P., 1980, Propionate-degrading bacterium, Syntrophobacter wolinii spp. nov. gen. nov., from methanogenic ecosystems, Appl. Environ. Microbiol. 40:626.

Conrad, R., Schink, B., Phelps, T. J., 1986, Thermodynamics of H_2-consuming and H_2-producing metabolic reactions in diverse methanogenic environments under in situ conditions, FEMS Microbiol. Ecol. 38:353.

Dubourguier, H. C., Prensier, G., and Albagnac, G., 1988, Structure and microbial activities of granular anaerobic sludge, p. 18-33, in: "Granular anaerobic sludge, microbiology and technology", G. Lettinga, A. J. B. Zehnder, J. T. C. Grotenhuis, and L. W. Hulshoff Pol (ed.), Pudoc, Wageningen.

Dwyer, D. F., Weeg-Aerssens, E., Shelton D. R., and Tiedje, J. M., 1988, Bionergetic conditions of butyrate metabolism by a syntrophic, anaerobic bacterium in coculture with hydrogen-oxidizing methanogenic and sulfidogenic bacteria, Appl. Environ. Microbiol. 54:1354.

Guiot, S. R., and van den Berg., L., 1984, Performance and biomass retention of an upflow anaerobic reactor combining a sludge blanket and a filter, Biotech. Lett. 6:161.

Guiot, S. R., Kennedy, K. J., and van den Berg, L., 1986, Comparison of the upflow anaerobic sludge blanket and sludge bed-filter concepts, Proc. EWPCA Conf. on Anaerobic Treatment, A Grown-Up Technology, Aquatec 1986, Amsterdam, The Netherlands, 533.

Hemens, J., Meiring, P. G., and Stander, G., 1962, Full-scale anaerobic digestion of effluents from the production of maize-starch, Water Waste Treat. 9:16.

Knechtel, J. R., 1978, A more economical method for the determination of chemical oxygen demand, Water Pollut. Control 116:25.

Lettinga, G., Hulshoff Pol, L. W., Wiegant, W., de Zeeuw, W., Hobma, S. W., Grin, P., Roersma, R., Sayed, S., and van Velsen, A. F. M., 1983, Upflow sludge blanket processes, in: Proceedings of the third international symposium on anaerobic digestion, 14-19 august 1983, Boston Massachusetts, USA, 139.

Lettinga, G., van Velsen, A. F. M., Hobma, W., de Zeeuw, W. J., and Klapwijk, A., 1980, Use of the upflow sludge blanket (USB) reactor concept for biological wastewater treatment, especially for anaerobic treatment, Biotechnol. Bioeng. 22:699.

MacLeod, F. A., Guiot, S. R., and Costerton, J. W., 1989, The ultrastructure of bacterial granules produced in an upflow anaerobic sludge blanket reactor, Appl. Environ. Microb., submitted.

Mah, R. A., 1982, Methanogenesis and methanogenic partnerships, Phil. Trans. R. Soc. London B 297:599.

McInerney, M. J., and Bryant, M. P., 1980, "Basic principle of biomass conversion processes for energy and fuels", S. Sofer and O. Zaborsky, eds, Plenum Press, New-York.

Pauss, A., Beauchemin, C., Samson, R., Guiot, S., 1989a, Continuous measurement of dissolved H_2 in an anaerobic reactor using a new hydrogen/air fuel cell detector, Biotechnol. Bioeng., in press.

Pauss, A., Samson, R., and Guiot, S. R., 1989b, Thermodynamic evidence of trophic microniches in methanogenic sludge-bed reactors. Appl. Microbiol. Biotechnol., submitted.

Robinson, J. A., and Tiedje, J. M., 1982, Kinetics of hydrogen consumption by rumen fluid, anaerobic digestor sludge, and sediment. Appl. Environ. Microb. 44:1374.

Schink, B., and Thauer, R. K., 1988, Energetics of syntrophic methane formation and the influence of aggregation, in: "Granular anaerobic sludge; microbiology and technology", G. Lettinga, A. J. B. Zehnder, J. T. C. Grotenhuis, and L. W. Hulshoff Pol, eds., Pudoc, Wageningen, The Netherlands, 5.

Seitz, H. J., Schink, B., and Conrad, R., 1988, Thermodynamics of hydrogen metabolism in methanogenic cocultures degrading ethanol or lactate, FEMS Microbiol. Lett. 55: 119.

Thauer, R. K., Jungermann, K., and Decker, K., 1977, Energy conservation in chemotrophic anaerobic bacteria, Bacteriol. Rev. 41: 100.

Thauer, R. K., and Morris, J. G., 1984, Metabolism of chemotrophic anaerobes: old views and new aspects, 36th Symposium of the Society for General Microbiology, The microbes, vol. 2, 123.

Thiele, J. H., Chartrain, M., and Zeikus, J. G., 1988, Control of interspecies electron flow during anaerobic digestion, role of floc formation in syntrophic methanogenesis, Appl. Environ. Microbiol. 54:10.

Tholozan, J. L., Samain, E., Grivet, J. P., Moletta, R., Dubourguier, H.C., and Albagnac, G., 1988, Reductive carboxylation of propionate to butyrate in methanogenic ecosystems, Appl. Environ. Microbiol., 54:441.

Zehnder, A. J. B., Ingvorsen, K., and Marti, T., 1982, Microbiology of methane bacteria, in: "Anaerobic digestion 1981", D. E. Hughes, D. A. Stafford, B. I. Wheatley, W. Baader, G. Lettinga, E. J. Nyns, W. Verstraete, and R. L. Wentworth, eds., Elsevier Biomedical Press, Amsterdam.

DISSIMILATION OF ETHANOL AND RELATED COMPOUNDS BY DESULFOVIBRIO STRAINS

T.A. Hansen and D.R. Kremer

Department of Microbiology, University of Groningen
Kerklaan 30, 9751 NN Haren, The Netherlands

INTRODUCTION

Strains of the genus Desulfovibrio are rod-shaped, vibroid or spirilloid sulfate-reducing bacteria that oxidize several organic and a few inorganic compounds. They lack a complete TCA cycle and an oxidative acetyl-CoA / carbon monoxide pathway and thus are unable to oxidize acetyl-CoA to CO_2; D. baarsii has the carbon monoxide pathway (Schauder et al. 1986) and does oxidize C_2-units to CO_2 but should be excluded from Desulfovibrio on the basis of phylogenetic data. One group of substrates is used by most or all Desulfovibrio strains. This group comprises hydrogen, formate, lactate, pyruvate, C_4-dicarboxylic acids such as malate and fumarate and short primary alcohols such as ethanol. A wide variety of compounds has been reported to be utilized by certain specific strains or groups of strains. These include choline, glycerol, dihydroxyacetone, 1,3-propanediol, fructose, furfural, amino acids, oxamate and oxalate (Widdel 1988; Hansen 1988).
The reducing equivalents liberated in the oxidation of the substrates are used in the reduction of sulfate to sulfide. Sulfate is reduced only after an activation to adenosine phosphosulfate (APS). The reduction of APS is catalyzed by APS reductase, a soluble protein which was recently shown to be a cytoplasmic enzyme with the aid of immunoelectron microscopy (Kremer et al. 1988a). The product, bisulfite, is reduced to sulfide by a bisulfite reductase located in the cytoplasm (Kremer et al. 1988a) or possibly via a pathway with two additional reductases and trithionate and thiosulfate as intermediates (LeGall and Fauque 1988). The natural electron donors of APS reductase and bisulfite reductase are not known.
For a proper understanding of the process of dissimilatory sulfate reduction the following questions have to be answered: i) how and where are the substrates oxidized ? ii) what are the carriers of the reducing equivalents ? iii) how are these reducing equivalents transferred to the reduction of APS and bisulfite ? iv) how is biologically useful energy conserved ? There are several factors that have complicated the task of answering the above questions. Unlike oxygen in aerobic respiring bacteria sulfate has to be transported into the cell in an energy-dependent process; therefore the mechanism and net costs of sulfate transport have to

Microbiology and Biochemistry of Strict Anaerobes Involved in Interspecies Transfer
Edited by J.-P. Bélaich *et al.*
Plenum Press, New York, 1990

185

be determined (Cypionka 1987; Thauer 1989). Sulfate-reducing bacteria possess an immensely diverse set of redox carriers with many species or strain-specific differences. Many redox proteins were purified but the function of some of them remains to be established (LeGall and Fauque 1988). Another complicating factor is that unlike e.g. cytochrome oxidases, the two main reductases of the sulfate reduction pathway are not membrane-bound or membrane-associated proteins.

Table 1. DEHYDROGENATION OF SUBSTRATES IN DESULFOVIBRIO

substrate	enzyme	acceptor	localization
hydrogen	hydrogenase	cytochrome c_3	periplasm
	hydrogenase	?	membrane
	(hydrogenase	?	cytoplasm)
formate	FDH	cytochrome c	periplasm
L-lactate	L-LDH	?	membrane
D-lactate	D-LDH	?	membrane/soluble
pyruvate	pyruvate DH	ferredoxin flavodoxin	cytoplasm
malate	malic enzyme	NADP	cytoplasm
glycerol	glycerol-3-P-DH	?	membrane
glycerol	glyceraldehyde-P-DH	NAD	cytoplasm
alanine	ala DH	NAD	cytoplasm

DEHYDROGENATION OF SUBSTRATES BY DESULFOVIBRIO

It has long been recognized that sulfate-reducing bacteria unlike aerobic heterotrophs make no or only a limited use of NADH as a general redox carrier in dissimilation processes (Thauer et al. 1977; Peck and Lissolo 1988). Reducing equivalents enter electron transport chain(s) at different redox levels and sites (periplasmic or cytoplasmic aspect of the membrane) depending on the substrate that is oxidized (Table 1). A role for NADH as redox carrier in the dissimilation of alanine and glycerol was demonstrated for two marine Desulfovibrio strains; these organisms, however, differ from well-studied strains by the unusually large number of compounds that are used by them (Stams and Hansen 1986; Kremer and Hansen 1987). Until recently no biochemical data were available on the oxidation of ethanol and other alcohols by Desulfovibrio.

OXIDATION OF ETHANOL BY DESULFOVIBRIO STRAINS

The oxidation of ethanol to acetate should easily allow a substrate-level phosphorylation if it would proceed via a reaction sequence involving acetyl phosphate (reactions a and b).

(a) ethanol ——→ acetaldehyde
        2H
                                       2H   ATP
                            acetyl-CoA ——→ acetyl phosphate ——→ acetate

(b) ethanol ——→ acetaldehyde ——→ acetyl phosphate ——→ acetate
        2H   2H   ATP

(c) ethanol ——→ acetaldehyde ——→ acetate
        2H   2H

We recently studied alcohol oxidation by four Desulfovibrio strains, namely D. gigas, a D. baculatus strain, D. carbinolicus (a strain which can grow slowly on methanol) and a marine Desulfovibrio with an unusually wide substrate range (Kremer et al. 1988b). All strains behaved similarly in that growth on ethanol led to the appearance of high NAD-dependent alcohol dehydrogenase activities. Aldehyde dehydrogenase activity was not NAD- or NADP-dependent and had highest activities when assayed with oxidized benzylviologen. This activity was not CoA- or phosphate-dependent and, in D. gigas, was stimulated fivefold by 20 mM K^+. The affinity for acetaldehyde in crude extracts was high with a Km of approximately 8 µM. The data show that the Desulfovibrio strains use reaction sequence (c). There is no substrate-level phosphorylation in the oxidation of ethanol to acetate. The oxidation of lactate and of ethanol to acetate are similar in that the first steps (lactate to pyruvate and ethanol to acetaldehyde) have standard redox potentials of approximately - 200 mV and that the resulting products are very strong reductants. However, ethanol is oxidized to acetaldehyde in the cytoplasm with NAD as acceptor whereas lactate is dehydrogenated by a membrane-bound lactate dehydrogenase with an unknown electron acceptor (possibly at the menaquinone / cytochrome b level) and unlike pyruvate oxidation acetaldehyde dehydrogenation does not result in the formation of acetyl-CoA. Molar growth yields on ethanol plus sulfate or sulfite were always lower than on lactate which is in agreement with a pathway without substrate phosphorylation. The use of NAD as electron acceptor in the ethanol dehydrogenase of "classical" Desulfovibrio species such as D. gigas shows that the role of NADH as a redox carrier in these sulfate reducers is not limited to atypical organisms such as the marine strains. NADH dehydrogenase activity which was assayed with the tetrazolium dye MTT was present in the soluble fraction of cell-free extracts of D. gigas. The marine Desulfovibrio, however, possesses a membrane-bound NADH dehydrogenase. The lack of CoA and phosphate dependence of the acetaldehyde dehydrogenase may seem wasteful at first sight since other non-sulfate-reducing anaerobes have been shown to use CoA-dependent acetaldehyde dehydrogenases (e.g. Schink et al. 1987). The reduction of sulfate with ethanol as a substrate consumes one ATP per ethanol oxidized to acetate in sulfate activation. Therefore, electron transport from NADH and from the reducing equivalents obtained in acetaldehyde oxidation to the reduction of APS and bisulfite in total must result in the production of more than one ATP.

SYNTROPHIC ETHANOL OXIDATION BY <u>DESULFOVIBRIO</u> AND METHANOGENS IN THE ABSENCE OF SULFATE

Bryant et al. (1977) demonstrated that <u>Desulfovibrio</u> strains can grow on lactate and also on ethanol in media without added sulfate provided the organisms are cocultured with hydrogen-utilizing methanogens. In a whey biomethanation system studied by Chartrain and Zeikus (1986) <u>Desulfovibrio</u> <u>vulgaris</u> was shown to play a major role in syntrophic lactate and ethanol degradation. Syntrophic ethanol degradation is summarized in the following equations:

<u>Desulfovibrio</u>:   2 ethanol $\longrightarrow$ 2 acetate + 4 H_2
methanogen: 4 H_2 + CO_2 $\longrightarrow$ CH_4 + 2 H_2O

If the conversion of ethanol into acetate and hydrogen is not accompanied by a substrate-level phosphorylation, how can the sulfate reducer grow ?   In fact, however in mixed cultures of <u>Desulfovibrio</u> <u>gigas</u> and <u>Methanospirillum</u> <u>hungatei</u> on ethanol in media without sulfate the growth of the sulfate reducer is negligible in comparison to that of the methanogen (Kremer et al. 1988b). Part of the growth of the <u>Desulfovibrio</u> may have been due to the use of a similarly rich medium as used by Bryant et al. (1977).   The oxidation of ethanol to acetate and hydrogen or formate most probably yields very little biologically useful energy.
Thauer (1989) studied the conversion of lactate to acetate, CO_2 and 2 H_2 by <u>Desulfovibrio</u>. The oxidation of lactate to pyruvate was found to be an energy-requiring process. This raises the question whether ethanol can be oxidized to acetaldehyde and hydrogen without energetic problems. A thermodynamic analysis shows that there is no reason to expect an energy requirement. The low Km of the acetaldehyde dehydrogenase and the high activities of the enzyme in cell-free extracts are indicative that the intracellular acetaldehyde concentrations are kept very low by the cell, possibly to avoid side-reactions of acetaldehyde with reactive sulfur species. Therefore, in a culture growing on ethanol the actual values of the redox potentials of ethanol/ acetaldehyde and of acetaldehyde/acetate differ significantly from the standard values. With the assumption of a thousandfold concentration difference between ethanol (e.g. 5 mM) and acetaldehyde (e.g. 5 µM) the redox potential becomes 90 mV lower and attains a value of approximately -290 mV. This is the same value as we have for the hydrogen/protons couple at a pH_2 of 10 Pa. It is therefore not necessary to assume that the reduction of protons requires energy in this system. With acetaldehyde the situation remains very favorable with a redox potential that will be 200 mV lower than the actual hydrogen redox potential.

METABOLISM OF ETHANEDIOL AND CHOLINE

Oxidation of ethylene glycol (ethanediol) and its oligomers up to tetraethylene glycol by a <u>Desulfovibrio</u> <u>desulfuricans</u> strain was discovered by Dwyer and Tiedje (1986). We recently found that <u>Deseulfovibrio</u> <u>carbinolicus</u> described by Nanninga and Gottschal (1987) grows rapidly on ethanediol. Ethanediol is usually dehydrated to acetaldehyde (e.g. Strass and Schink 1986). This would mean that ethanediol can be used as a non-toxic equivalent of acetaldehyde. <u>Desulfovibrio</u> <u>carbinolicus</u> did ferment ethanediol to acetate and ethanol upon the first transfer into a medium without sulfate from a culture that had been grown in the presence of sulfate. However, growth and fermentation did not occur after the second transfer. This result does not contradict our finding that the oxidation of acetaldehyde to acetate is not associated with a substrate-level phosphorylation.

Choline (trimethylethanolamine) is a growth substrate for Desulfovibrio desulfuricans, both in the presence and absence of sulfate. The first step in choline metabolism was reported to be a cleavage into trimethylamine and acetaldehyde (Hayward 1960). Thus, fermentation of choline would be similar to the fermentation of acetaldehyde. Fermentative growth on choline therefore means that the organism obtains energy in the dismutation of acetalde-hyde. Whether choline-fermenting Desulfovibrio strains do use a CoA- or phosphate-dependent acetaldehyde dehydrogenase or whether other energy-conserving processes are involved such as electron transport-associated pmf generation remains to be investigated.

REFERENCES

Bryant, M. P., Campbell, L. L., Reddy, C. A., and Crabill, M. R., 1977, Growth of Desulfovibrio in lactate or ethanol media low in sulfate in association with H_2-utilizing methanogenic bacteria, Appl. Environ. Microbiol., 33:1162.

Chartrain, M., and Zeikus, J. G., 1986, Microbial ecophysiology of whey biomethanation: characterization of bacterial trophic populations and prevalent species in continuous culture, Appl. Environ. Microbiol., 51:188.

Cypionka, H., 1987, Uptake of sulfate, sulfite, and thiosulfate by proton-anion symport in Desulfovibrio desulfuricans, Arch. Microbiol., 13:285.

Dwyer, D. F., and Tiedje, J. M., 1986, Metabolism of polyethylene glycol by two anaerobic bacteria, Desulfovibrio desulfuricans and a Bacteroides sp., Appl. Environ. Microbiol., 52:852.

Hansen, T. A., 1988, Physiology of sulphate-reducing bacteria, Microbiol. Sc., 5:81.

Hayward, H. R., 1960, Anaerobic degradation of choline iii. acetaldehyde as an intermediate in the fermentation of choline by extracts of Vibrio cholinicus, J. Biol. Chem., 235:3592.

Kremer, D. R., and Hansen, T. A., 1987, Glycerol and dihydroxy-acetone dissimilation in Desulfovibrio strains, Arch. Microbiol., 147:249.

Kremer, D. R., Veenhuis, M., Fauque, G., Peck Jr., H. D., LeGall, J., Lampreia, J., Moura, J. J. G., and Hansen, T. A., 1988a, Immunocytochemical localization of APS reductase and bisulfite reductase in three Desulfovibrio species, Arch. Microbiol., 150:296.

Kremer, D. R., Nienhuis-Kuiper, H. E., and Hansen, T. A., 1988b, Ethanol dissimilation in Desulfovibrio, Arch. Microbiol., 150:552.

LeGall, J., and Fauque, G., 1988, Dissimilatory reduction of sulfur compounds, in: "Biology of anaerobic microorga-nisms," A. J. B. Zehnder, ed., Wiley, New York, London, p. 587.

Nanninga, H. J., and Gottschal, J. C., 1987, Properties of Desulfovibrio carbinolicus sp. nov. and other sulfate-reducing bacteria isolated from an anaerobic-purification plant, Appl. Environ. Microbiol., 53:802.

Peck, H. D., and Lissolo, T., 1988, Assimilatory and dissimilatory sulphate reduction: enzymology and bioenergetics, in: ."The nitrogen and sulphur cycles," J. A. Cole and S. J. Ferguson, eds., Cambridge University Press, Cambridge.

Schauder, R., Eikmanns, B., Thauer, R. K., Widdel, F., and Fuchs, G., 1986, Acetate oxidation to CO_2 via novel pathway not involving reactions of the citric acid cycle, Arch. Microbiol., 145:162.

Schink, B., Kremer, D. R., and Hansen, T. A., 1987, Pathway of propionate formation from ethanol in Pelobacter propionicus, Arch. Microbiol., 147:321.

Stams, A. J. M., and Hansen, T. A., 1986, Metabolism of L-alanine in Desulfotomaculum ruminis and two marine Desulfovibrio strains, Arch. Microbiol., 145:277.

Strass, A., and Schink, B., 1986, Fermentation of polyethylene glycol via acetaldehyde in Pelobacter venetianus, Appl. Microbiol. Biotechnol., 25:37.

Thauer, R. K., 1989, Energy metabolism of sulfate-reducing bacteria, in: "Autotrophic bacteria," H. G. Schlegel and B. Bowien, eds., Science Tech Publishers, Madison, p. 397.

Thauer, R. K., Jungermann, K., and Decker, K., 1977, Energy conservation in chemotrophic anaerobic bacteria, Bacteriol. Rev., 41:100.

Widdel, F., 1988, Microbiology and ecology of sulfate and sulfur-reducing bacteria, in: "Biology of anaerobic microorganisms," A. J. B. Zehnder, ed., Wiley, New York London, p. 469.

FACTORS AFFECTING HYDROGEN UPTAKE BY BACTERIA GROWING IN THE HUMAN LARGE

INTESTINE

G.R. Gibson, J.H. Cummings and G.T. Macfarlane

MRC Dunn Clinical Nutrition Centre
100 Tennis Court Road
Cambridge. CB2 1QL
UK

INTRODUCTION

The human large intestine is a highly complex ecosystem that contains somewhere in the region of 400 different species of bacteria[1]. The vast majority of these bacteria are strict anaerobes and grow on a wide variety of substrates that have either escaped digestion in the small bowel or have been produced by the host[2]. In Western populations, between 10-60g of carbohydrate and 6-18g of proteinaceous material are potentially available for fermentation each day, producing a total bacterial mass of approximately 90g[3].

Hydrogen plays an important role in the anaerobic food chain in the large bowel. Many gut species form hydrogen to dispose of excess reducing power generated in reactions where large quantities of organic matter are being oxidized. Hydrogen removal can potentially be achieved by hydrogen utilizing species such as methanogenic, sulphate-reducing and acetogenic bacteria, or by excretion in breath or flatus. A survey of the literature shows excretion of hydrogen and methane to be in the region of 0.5-41 d^{-1}.[3] Disposal of hydrogen generated during colonic fermentation by either of these routes is important. For example, due to the stoichiometry of the reaction $4H_2 + CO_2 \longrightarrow CH_4 + 2H_2O$ utilization of hydrogen by terminal electron accepting species such as the methanogens converts 4 volumes of hydrogen to 1 volume of methane. Thus, from a purely physical point of view, this can be a significant factor affecting the host's health and well being.

Methane production in the colon has been studied for many years. For reasons hitherto unknown, only a proportion of the population has been found to excrete methane in breath[4]. Since human colonic methanogens have an obligate requirement for hydrogen, growing on either H_2/CO_2 or H_2/CH_3OH[5], it was considered that competition for this gas by other bacterial species could be a factor regulating their growth in the large bowel.

Competition for mutual substrates between methanogenic and sulphate-reducing bacteria occurs in many natural environments[6]. This suggested that similar processes could take place in the large intestine. In this paper we report studies on interactions between these two groups of bacteria and the factors that influence their growth and activites in the colon.

Microbiology and Biochemistry of Strict Anaerobes Involved in Interspecies Transfer
Edited by J.-P. Bélaich *et al.*
Plenum Press, New York, 1990

METHODS

Breath methane. End expiratory breath samples were collected in 20ml plastic syringes. The methane content was detected by GC[7].

Faecal slurries. Triplicate slurries (5% w/v) were prepared with and without the methanogenic inhibitor 2-bromoethanesulphonic acid (BES) and the sulphate reducer inhibitor sodium molybdate, as previously described[8]. The slurries were incubated at 37°C and headspace gas and liquid samples taken periodically. Hydrogen and methane were measured by GC[9] and H_2S was determined after precipitation of sulphides in 10% (w/v) zinc acetate [10]

Enumeration, isolation and characterization of sulphate-reducers. The agar shake dilution method was used. Media and conditions of cultivation were those described by Gibson et al.[11] Pure cultures were isolated and characterized using the criteria of Keith et al.[12]

Gut contents. These were obtained within 4h of death from 4 individuals had died suddenly, using the procedures of Cummings et al.[13] Samples were taken from the right colon (caecum and ascending colon), and the left colon (descending colon and sigmoid/rectum). Short chain fatty acids (SCFA) were measured by GC[14]. Sulphate reduction rates were determined using ^{35}S-labelled sodium sulphate as described by Jørgensen[15]. Dry weights were measured as outlined by Keith and Herbert[16].

Effect of sulphated polysaccharides on sulphate reduction and methanogenesis. These experiments were carried out with faecal slurries in batch culture using porcine gastric mucin and chondroitin sulphate as described by Gibson et al.[8]

Multichamber continuous culture system. A 3-stage continuous culture system was used to study the effect of mucin on interactions between methanogenic and sulphate-reducing bacteria. The system was designed to reproduce the different conditions of nutrient availability, growth rates and pH found in the right and left colon[2,17] Gases, sulphides, sulphate reduction rates, SCFA and bacterial populations were determined as described above.

RESULTS

Methane production and sulphate reduction in the colon

Breath methane concentrations, faecal sulphate reducer counts and rates of ^{35}S sulphate reduction were determined in two culturally and geographically diverse populations from the United Kingdom (Cambridge) and South Africa (Hekpoort Village, Western Transvaal). The rural black South Africans had a considerably higher individual carriage rate of methanogenic bacteria than their UK counterparts. A definite inverse relationship between methane production levels on one hand and faecal sulphate reducers and sulphate reduction rates on the other, was found (Table 1). Six of the British subjects and 17 of the rural Africans were methanogenic. These persons had undetectable or low sulphate reducer counts and trace levels of sulphate reduction. In contrast, the 14 non-methanogenic British and 3 non-methanogenic South Africans had high sulphate reducer populations and sulphate reduction rates. Some individuals had exceptionally high carriage levels of sulphate reducers ($> 10^{10}$ (g dry wt faeces)$^{-1}$, which corresponds to approximately 1% of the total gut microflora.

Table 1. Sulphate reducing and methanogenic activities in 20 British persons and 20 rural black South Africans.

Population	Methane producers			Non methane producers		
	Breath[a] methane	SRB count[b]	Sulphate[c] reduction	Breath[a] methane	SRB count[b]	Sulphate[c] reduction
British	2.9-47.2 (6)	ND	0.01-0.06 (6)	ND	6.7-10.2 (14)	7.6-81 (14)
Rural Black South African	3.2-50.8 (17)	5.3-6.8 (4)	0.01-0.08 (17)	ND	8.8-10.8 (3)	3.2-264 (3)

a = ppm; b = Log_{10} $(g\ dry\ wt.\ faeces)^{-1}$;c = nmol sulphate reduced h^{-1} $(g\ dry\ wt.\ faeces)^{-1}$. The number of positive cases are shown in parenthesis; ND = Not detected.

Characterization of colonic sulphate-reducing bacteria

Desulfovibrios were the numerically predominant sulphate reducers in faecal samples, constituting about 60% of the total count (Fig. 1). These bacteria utilized succinate, lactate, ethanol, amino acids (Glu/Ser/Ala) valerate and hydrogen as electron donors. Other sulphate reducing species were assigned to the genera Desulfobacter, Desulfomonas, Desulfobulbus and Desulfotomaculum. These sulphate reducers used a variety of electron donors including acetate, ethanol, pyruvate, lactate, propionate, hydrogen and butyrate.

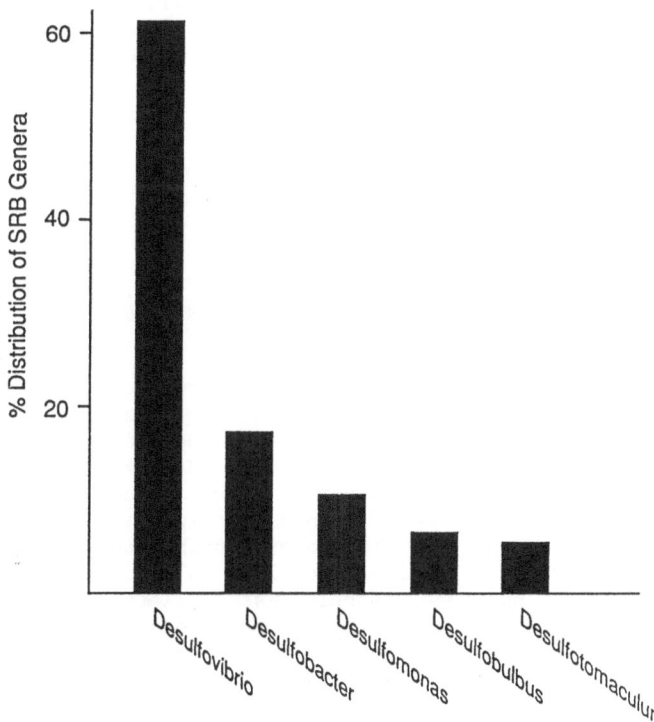

Fig. 1. Distribution of sulphate-reducing bacteria in faeces.

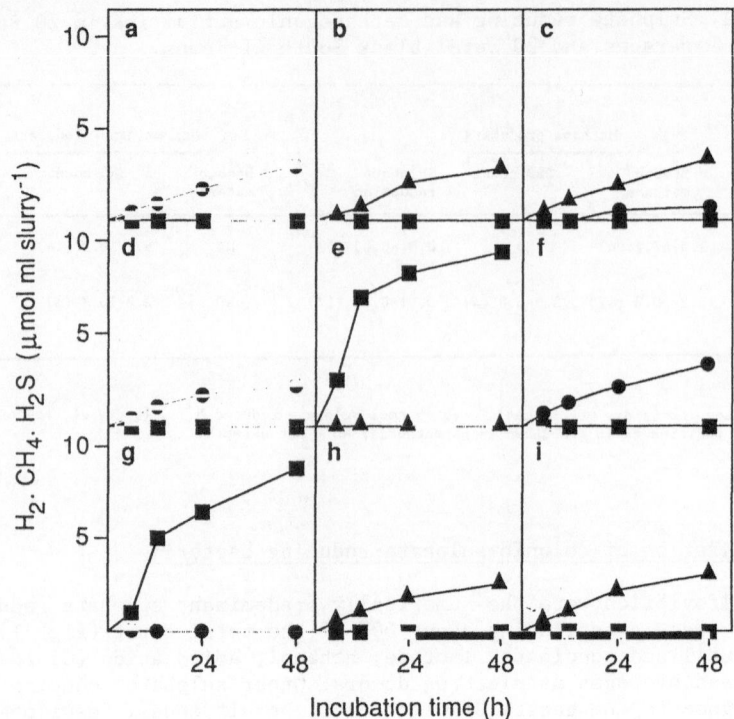

Fig. 2. Competition between methanogenic and sulphate-reducing bacteria for hydrogen in faecal slurries. a, methanogenic; b, sulphate reducing; c, mixed slurries; d,e,f equivalent slurries to a,b,c + 20mM sodium molybdate; g,h,i equivalent slurries to a,b,c + 20mM 2-bromoethane-sulphonic acid. (■) H_2; (●) CH_4; (▲) H_2S. Results are means from experiments with faeces from 3 methanogenic and 3 sulphate reducing persons.

Competition studies with sulphate reducers and methanogens

The pattern of distribution of methanogenic and sulphate-reducing species in different persons indicated that these bacteria competed for mutual substrates in the colon. This was confirmed in slurry experiments. Gas samples were taken for measurements of hydrogen and methane and liquid samples for H_2S. Sulphide production was used as a crude index of sulphate reducer activity. Fig. 2a,b shows that hydrogen did not accumulate in more than trace amounts in either methanogenic or sulphate reducer slurries. When the slurries were mixed however, sulphate-reducing bacteria outcompeted the methanogens for hydrogen (Fig. 2c). When 20mM molybdate was added to the slurries (Fig. 2d-f), sulphate reduction was completely inhibited. High levels of hydrogen accumulated in the sulphate reducer slurries (Fig. 2e) and methanogenic bacteria were able to convert all the available hydrogen to methane in the mixed slurries (Fig. 2f). Hydrogen accumulated in methanogenic slurries containing 20mM BES (Fig. 2g). The hydrogen concentration at 48h was approximately four times the corresponding methane value in the control slurries (Fig. 2a), as would be predicted if methanogenesis was the sole route of hydrogen uptake.

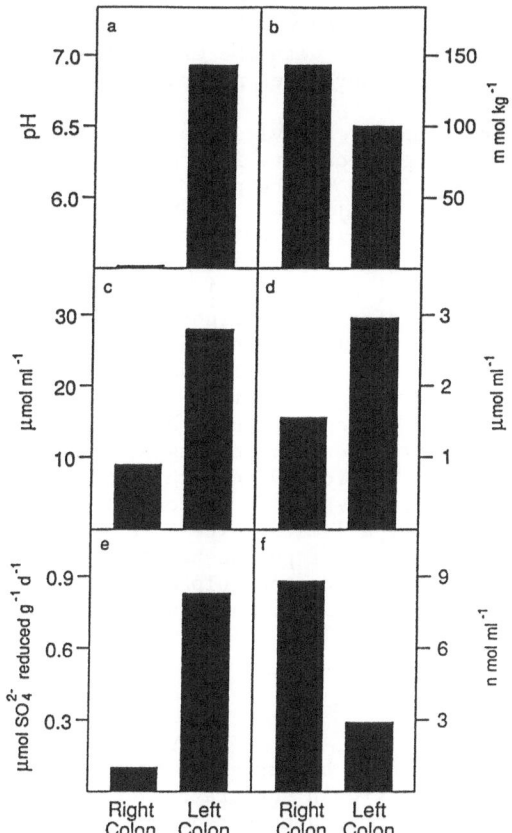

Fig. 3. Physicochemical characteristics and bacterial activities in different regions of the large intestine. a, pH; b, SCFA concentration; c, methane concentration; d, H_2S concentration; e, ^{35}S-SO_4^{2-} reduction rates; f, H_2 concentration.

Sudden death victim studies

The activities of the colonic microflora are influenced to a large extent by the physiology and anatomical architecture of the gut. The caecum and right colon receive digesta from the small intestine and are the major sites of fermentation in the large bowel. As a result, bacterial growth rates and activities are high. In contrast, the left colon is comparatively nutrient limited since readily fermentable substrates have been degraded in the right colon. Consequently, the use of faeces for microbiological studies gives little information on processes occurring in the proximal region of the gut. This problem can be largely overcome by using gut contents from sudden death victims[13]. We therefore obtained the colons from 4 persons, who, without any symptoms of ill health had died suddenly. Gut materials from different areas of the colon were obtained within 4h of death and pH, SCFA concentrations, methane concentrations, sulphate reducing activities and hydrogen concentrations determined.

The major chemical end products of fermentation in the large gut are the SCFA acetate, propionate and butyrate. SCFA concentrations were highest in the right colon confirming that the majority of fermentation occurred in the caecum and ascending colon (Fig. 3b). The high rates of SCFA production in the right colon lower gut pH with the effect that pH in this region is acidic (ca. pH 5.5) but approaches neutrality in the left colon (Fig 3a).

As in the faecal samples, activities of sulphate-reducing and methanogenic bacteria in gut contents from each of the sudden death victims were mutually exclusive. Three of the persons had significant concentrations of methane in all areas of the large bowel (Fig 3c), whereas this was replaced by high levels of sulphide and sulphate reducing activity in the remaining individual (Fig. 3d,e). These data show that sulphate reduction and methanogenesis occur in all areas of the large intestine. However, the activities differ in the right and left colon and this may be largely attributed to the physicochemical environment found in different parts of the gut. Methane, hydrogen sulphide concentrations and sulphate reduction rates were greater in the left compared to the right colon. This correlated inversely with hydrogen concentrations in all 4 individuals (Fig. 3f). Measurements of colonic gases suggest that hydrogen availability is probably a limiting factor for the growth of H_2 species in the gut, nevertheless unlike the rumen where H_2 is completely converted to CH_4, hydrogen does accumulate to a limited extent in the large intestine. These effects may be related to the gut pH. We have previously shown that the activities of colonic methanogens and sulphate reducers are strongly inhibited during growth at pH values below 6.5, with both processses occurring optimally at a neutral or slightly alkaline pH[18].

Multichamber system studies

Colonic sulphate reducing bacteria are able to outcompete gut methanogens for the mutual growth substrate hydrogen.
However, for this to occur sulphate is required. The question then arises as to the availability of this metabolite in the large bowel. Sulphate may be supplied to the colon in either dietary residues or in endogenous secretions. The sulphate content of the normal Western diet ranges from 2-16 mmol d^{-1} [19], and it is likely that this amount will show significant variability within a given population. However, a proportion of dietary sulphate is absorbed in the small intestine. Endogenous sources of sulphate are therefore of some interest. The sulphated polysaccharide mucin is constantly excreted by goblet cells in the colonic epithelium, whilst chondroitin sulphate enters the colon with the large quantities of epithelial cells that are shed daily. These substances are extensively degraded by gut bacteria[20,21]. Batch culture incubation of mixed faecal bacteria demonstrated that both mucin and chondroitin sulphate stimulated sulphate reducing activity when compared to starch, which is a sulphate free polysaccharide (Table 2). These substrates also sustained high rates of methanogenesis when sulphate-reducing bacteria were absent or inactive, showing that they were not inhibitory to the methanogens.

To further determine the role of methanogenic and sulphate-reducing bacteria in colonic hydrogen metabolism, we investigated the effect of mucin on sulphate reduction and methanogenesis in a 3-chambered continuous culture system. The multichamber system (MCS) used in this study attempted to reproduce in vitro some of the nutritional, physical and chemical characteristics found in different regions of the large intestine (Fig. 4).

A complex mixture of carbohydrates and proteins was fed to the system. The basal medium contained low levels of sulphate ($0.5gl^{-1}$). Vessel 1 was designed to reproduce the acidic, nutrient rich, fast growth conditions of the right colon. Due to bacterial growth, culture effluent entering vessels 2 and 3 became progressively more substrate limited. Vessels 2 and 3 were characterized by conditions of higher pH and slower growth rates to resemble

Table 2. Influence of sulphated polysaccharides on hydrogen uptake
processes in the large intestine.

Polysaccharide	Slurry type	Methanogenic rate[a]	Sulphide production rate[b]
Starch	Methanogenic	54 ± 2	6 ± 2
	Non methanogenic	0	60 ± 9
Mucin	Methanogenic	76 ± 9	8 ± 2
	Non methanogenic	0	163 ± 11
Chondroitin sulphate	Methanogenic	86 ± 8	13 ± 2
	Non methanogenic	0	118 ± 11

The slurries were obtained from 3 methanogenic and 3 non methanogenic persons.
Polysaccharide concentrations were 0.2% w/v. Results are means ±SEM.

a = nmol methane produced $ml^{-1}h^{-1}$; b = nmol sulphide produced $ml^{-1}h^{-1}$.

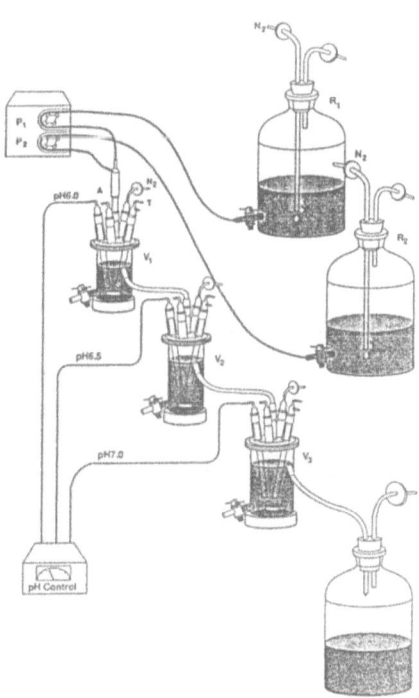

Fig. 4. The multichamber continuous culture system. Medium from reservoir
R1 was fed to vessel V1 by pump P1. Mucin (5.8 gd^{-1}) or
distilled water from reservoir R2 were added via pump P2. V1
sequentially fed vessels V2 and V3. T = temperature control,
N2 = nitrogen gas, A = alkali input. The operating volumes of V1,
V2, and V3 were 0.3, 0.5 and 0.81 with dilution rates of 0.08, 0.04
and $0.034h^{-1}$. Total retention time of the system was 62.7h. pH was
automatically controlled V1 (6.0), V2 (6.5), V3 (7.0).

conditions found in the left colon. The media reservoirs and vessels of the system were maintained, but not sparged, with an atmosphere of oxygen free nitrogen. Each vessel of the MCS was inoculated with faeces from a person who was essentially a methane producer but who, unusually, harboured low numbers of sulphate-reducing bacteria, which could be enriched for by the addition of sulphate. During the first 48 days of the experiment, distilled water was added to vessel 1. This was then replaced by mucin for a further 22 days. Subsequently, distilled water again replaced mucin and the experiment was terminated after a total of 120 days. The effect of mucin upon methanogenic and sulphate reducing activities was determined at 3-4 day time intervals.

Methane production occurred in all vessels of the MCS before mucin was fed to the system (Table 3). The higher amounts of methane produced in vessels 2 and 3 confirmed that methanogenic bacteria grew best under less acidic conditions and low dilution rates. During this time, sulphate reducing activity, as determined by H_2S levels, ^{35}S-SO_4^{2-} reduction rates and viable sulphate reducer counts was low. When mucin was added to the system, methanogenesis was strongly inhibited but hydrogen did not accumulate. Hydrogen sulphide production was stimulated in all vessels but most markedly in vessels 2 and 3. This correlated well with sulphate reduction rates and viable counts of sulphate-reducing bacteria (Table 3). Measurements of SCFA showed that the mucin was extensively fermented by mixed faecal bacteria growing in the MCS. Sulphate reduction was stimulated and methanogens in the MCS were competitively displaced as the major hydrogen utilizing bacteria. Methanogenic bacteria were not completely washed out of the MCS however, as shown by the resumption of methane production in vessels 2 and 3 when mucin addition was stopped. In the absence of mucin, sulphate reduction diminished towards the initial baseline levels.

Despite the fact that low levels of methanogenesis or sulphate reduction occurred in vessel 1, hydrogen only accumulated in trace amounts. This raises the possibility that hydrogen was being utilized by an alternative route, such as acetogenesis. Although the reduction of CO_2 to acetate by hydrogen in the reaction

$$4H_2 + 2CO_2 \longrightarrow CH_3COO^- + H^+ + 2H_2O \quad (\Delta G_o' = -95 \text{ kJ mol}^{-1})$$

is thermodynamically unfavourable compared to either methane production from H_2/CO_2

$$4H_2 + CO_2 \longrightarrow CH_4 + 2H_2O \quad (\Delta G_o' = -131 \text{ kJ mol}^{-1})$$

or dissimilatory sulphate reduction with hydrogen as the electron donor

$$4H_2 + SO_4^{2-} + 2H^+ \longrightarrow H_2S + 4H_2O \quad (\Delta G_o' = -152.2 \text{ kJmol}^{-1})$$

acetogenic bacteria have been found to compete with methanogens in acidic lake sediments[22]. Whilst we did not test for acetogenic activity, the possibility that acetogens were able to compete effectively against either methanogenic or sulphate-reducing bacteria in the acid cultures of vessel 1 cannot be excluded.

Table 3.   Effect of mucin on activities of methanogenic and sulphate-reducing bacteria grown in a 3-chambered continuous culture system.

Vessel	No mucin					Mucin					No mucin				
	H_2^a	CH_4^b	H_2S^c	SO_4^{2-d} redn	Total^e SRB	H_2^a	CH_4^b	H_2S^c	SO_4^{2-d} redn	Total^e SRB	H_2^a	CH_4^b	H_2S^c	SO_4^{2-d} redn	Total^e SRB
1	6.3 ±2.2	0.4 ±0.1	1.1 ±0.1	T	ND	3.1 ±0.2	ND	5.9 ±0.1	23 ±7.0	2.8 ±0.7	2.2 ±0.1	ND	2.1 ±0.3	2.0 ±1.0	ND
2	2.1 ±0.3	0.7 ±0.3	1.2 ±0	T	3.3 ±0.3	0.4 ±0.1	ND	8.4 ±0.4	118 ±16.0	7.2 ±0.2	ND	1.3 ±0.3	2.8 ±1.0	14.0 ±9.0	4.0 ±0.9
3	2.7 ±0.1	1.0 ±0.2	1.1 ±0.2	2.8 ±1.0	3.8 ±0.1	0.1 ±0.4	ND	8.4 ±0.4	112 ±23.0	7.5 ±0.1	ND	1.6 ±0.4	4.4 ±0.6	21.0 ±11.0	2.9 ±0.1

The multichamber chemostat was operated for 48 days in the absence of mucin. During this period, gas and liquid samples were taken on days 3, 9, 22, 30 and 40. Mucin was then added to the system from days 48 to 70, with samples taken on days 54, 60 and 69. The system was then operated for a further 50 days without mucin and with sampling on days 87, 102 and 120. Results are presented as means ± SD

a = nmol ml^{-1}; b = µmol ml^{-1}; c = µmol ml^{-1}; d = nmol SO_4^{2-} reduced h^{-1}; e = $\log_{10}$ ml^{-1}.

ND = Not detected; T = < 1.0 nmol ml^{-1}.

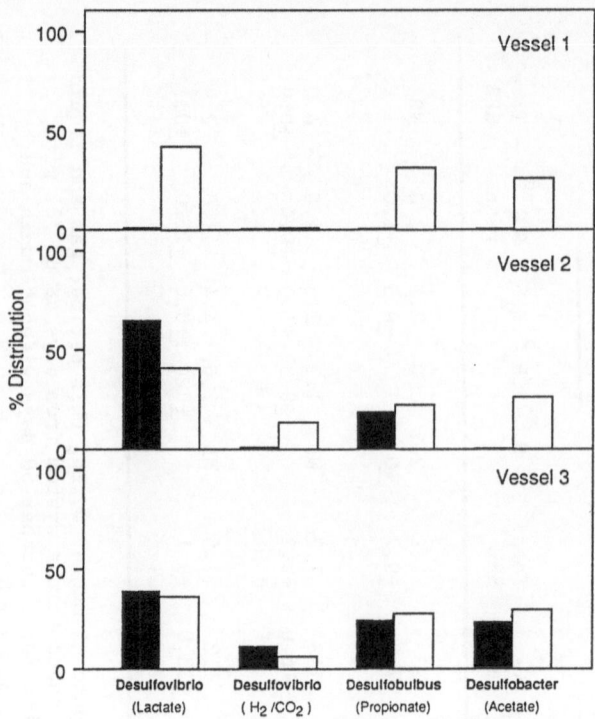

Fig. 5. Effect of mucin on the distribution of different sulphate-
reducing bacteria in the MCS. Shaded bars are non-mucin cultures
and open bars are mucin cultures. The electron donor used for
enumeration and isolation is given in parenthesis.

Sulphate-reducing bacteria in the MCS

Sulphate reducers were never detected in vessel 1 in the absence of
mucin (Fig. 5). Since small populations of sulphate reducing bacteria
occurred in vessels 2 and 3 and sulphate was present in the feed medium,
other factors such as pH and dilution rate must have affected their growth.
Mucin stimulated sulphate reducer growth overall in the MCS and many
different types were detected. Lactate utilizing Desulfovibrio spp. were
numerically predominant. This agrees with the faecal sulphate-reducer counts
shown in Fig. 2 and suggests that in the colonic environment, these bacteria
are the most sucessful species in competing for limiting amounts of
sulphate. The competition studies of Laanbroek et al.[23] support this
conclusion. When mucin was metabolised in the MCS the relative proportions
of Desulfobacter spp. and Desulfobulbus spp. increased as sulphate became
more available. Significant populations of hydrogen utilizing sulphate
reducers were only found in vessel 2 in the mucin cultures although they
were present in vessel 3, irrespective of the presence of mucin.

CONCLUSIONS

Our studies demonstrate that individuals can be differentiated on the
basis of whether methanogenic or sulphate-reducing bacteria predominate in
their colons and that intestinal sulphate reducers outcompete methanogens
for hydrogen. Kristjansson et al.[24] have shown that this results from the
greater affinity for hydrogen of the sulphate reducers (K_s 1 μmol l^{-1})
compared to the methanogens (K_s 6 μmol l^{-1}). Hydrogen utilizing sulphate

reducers constitute only a small proportion of total sulphate-reducing bacterial numbers in the large gut. Other species that are able to utilize a wide variety of electron donors and do not directly compete with hydrogen requiring methanogens were only found in non-methanogenic persons, suggesting that some factor was limiting their growth. Experiments in our laboratory have shown that in some, but not all, methanogenic individuals, feeding increased levels of sulphate stimulates the growth of sulphate-reducing bacteria in the colon within a few days and represses methane production (S. Christl and J.H. Cummings, unpublished results). Taken together, these observations provide good evidence that sulphate availability, whether from dietary or endogenous sources, is the principal factor that determines whether sulphate reducers or methanogenic bacteria colonise the large intestine.

REFERENCES

1. S.M. Finegold, V.L. Sutter and G.E. Mathisen, Normal indigenous microflora in: "Human Intestinal Microflora in Health and Disease," D.J. Hentges, ed., Academic Press, London (1983).
2. G.T. Macfarlane, S. Hay and G.R. Gibson, Influence of mucin on glycosidase, protease and arylamidase activities of human gut bacteria grown in a 3-stage continuous culture system, J. Appl. Bacteriol., 66: 407 (1989).
3. G.T. Macfarlane and J.H. Cummings, The colonic flora, fermentation and large bowel digestive function, in "The Large Intestine: Physiology, Pathophysiology and Diseases," S.F. Phillips, J.H. Pemberton and R.G. Shorter, eds., Raven Press, New York (1990), in press.
4. J.H. Bond, R.R. Engel and M.D. Levitt, Factors influencing pulmonary methane excretion in man, J. Exp. Med., 133: 572 (1971).
5. T.L. Miller and M.J. Wolin, Methanosphaera stadtmaniae gen. nov., sp. nov. a species that forms methane by reducing methanol with hydrogen, Arch. Microbiol., 141: 116 (1985).
6. R.S. Oremland and S. Polcin, Methanogenesis and sulfate reduction : competitive and non-competitive substrates in estuarine sediments. Appl. Env. Microbiol., 44: 1270 (1982).
7. I. Segal, A.R.P. Walker, S. Lord and J.H. Cummings, Breath methane and large bowel cancer risk in contrasting African populations, Gut, 29: 608 (1988).
8. G.R. Gibson, J.H. Cummings and G.T. Macfarlane, Competition for hydrogen between sulphate-reducing bacteria and methanogenic bacteria from the human large intestine, J. Appl. Bacteriol., 65: 241 (1988).
9. C. Allison and G. T. Macfarlane, Effect of nitrate upon methane production by slurries of human faecal bacteria, J. Gen. Microbiol., 134: 1397 (1988).
10. J.D. Cline, Spectrophotometric determination of H_2S in natural waters, Limnol. Oceanogr., 14: 454 (1969).
11. G.R. Gibson, G.T. Macfarlane and J.H. Cummings, Occurrence of sulphate-reducing bacteria in human faeces and the relationship of dissimilatory sulphate reduction to methanogenesis in the large gut, J. Appl. Bacteriol., 65: 103 (1988).
12. S.M. Keith, R.A. Herbert and C.G. Harfoot, Isolation of new types of sulphate-reducing bacteria from estuarine and marine sediments using chemostat enrichments, J. Appl. Bacteriol., 53: 29 (1982).
13. J.H. Cummings, E.W. Pomare, W.J. Branch, C.P.E. Naylor and G.T. Macfarlane, Short chain fatty acids in human large intestine, portal hepatic and venous blood, Gut, 28: 1221 (1987).
14. L.V. Holdeman, E.P. Cato and W.E.C. Moore (eds.), Anaerobe laboratory manual 4th Edn., Virginia Polytechnic and State University, Blacksburg (1977).
15. B.B. Jørgensen, A comparison of methods for the quantification of

bacterial sulfate reduction in coastal marine sediments. I. Measurement with radiotracer techniques Geomicrobiol. J., 1: 11 (1978).

16. S.M. Keith and R.A. Herbert, Dissimilatory nitrate reduction by a strain of Desulfovibrio desulfuricans, FEMS Microbiol. Ecol., 18: 55 (1983).

17. G.R. Gibson, J.H. Cummings and G.T. Macfarlane, Use of a three-stage continuous culture system to study the effect of mucin on dissimilatory sulfate reduction and methanogenesis by mixed populations of human gut bacteria, Appl. Env. Microbiol., 54: 2750 (1988).

18. G.R. Gibson, J.H. Cummings, G.T. Macfarlane, C. Allison, I. Segal, H.H. Vorster and A.R.P. Walker, Alternative pathways for hydrogen disposal during fermentation in the human colon, Gut, (1990), in press.

19. T.J.H. Florin, G. Neale and J.H. Cummings, Dietary and endogenous sulphate losses from the upper gastrointestinal tract, Proc. Med. Res. Soc. Summer Meeting (1989).

20. A.M. Roberton and R.A. Stanley, In vitro utilization of mucin by Bacteroides fragilis, Appl. Env. Microbiol., 43: 325 (1982).

21. A.A. Salyers and M.O'Brien, Cellular location of enzymes involved in chondroitin sulfate breakdown by Bacteroides thetaiotaomicron, J. Bacteriol., 43: 325 (1982).

22. T.J. Phelps and J.G. Zeikus, Influence of pH on terminal carbon metabolism in anoxic sediments from a mildly acidic lake. Appl. Env. Microbiol., 48: 1088 (1984).

23. H.J.Laanbroek, H.J. Geerlings, L. Sijtsma and H. Veldkamp, Competition for sulfate and ethanol among Desulfobacter, Desulfobulbus and Desulfovibrio species isolated from intertidal sediments, Appl. Env. Microbiol., 47: 329 (1984).

24. J.K. Kristjansson, P. Schönheit and R.K. Thauer, Different Ks values for hydrogen of methanogenic bacteria and sulfate reducing bacteria : an explanation for the apparent inhibition of methanogenesis by sulfate, Arch. Microbiol., 131: 278 (1982).

ISOLATION AND CHARACTERIZATION OF AN ANAEROBIC BACTERIUM

DEGRADING 4-CHLOROBUTYRATE

Kaoru Matsuda[*], Kazunori Nakamura, Yoichi Kamagata ánd
Eiichi Mikami

[*]Technical Institute, Kawasaki Heavy Industries
Kawasaki-cho, Akashi, Hyogo 673, Japan
 Fermentation Research Institute, Agency of Industrial
Science and Technology, Tsukuba, Ibaraki 305, Japan

ABSTRACT

An anaerobic rod-shaped bacterium (strain K-1) was isolated from
mesophilic digester sludge acclimatized with 4-chlorobutyrate as main
energy and carbon source. This bacterium was mesophilic, Gram-negative
and motile, and degraded 4-chlorobutyrate to butyrate, acetate and
hydrogen. Cell morphology varied from singles to pairs, and occasionally
long chains. When growth peaked at a pH exceeding 6.0, cells immediately
lysed, but not less than pH 6.0. The optimal conditions for growth were
an initial pH 6.5-6.8 and 37°C. Growth required yeast extract or
clarified rumen fluid, but neither peptone nor clarified sludge fluid
could be replaced. Fructose was utilized as a carbon source, but not
other sugars. The isolate could also utilize other chlorinated compounds
such as 3-chlorobutyrate and 3-chloropropionate, and the main products
were butyrate and acetate, propionate and acetate, respectively.
However, butyrate could not be utilized. Unsaturated fatty acids such as
crotonate or vinyl acetate, or 3-hydroxybutyrate could be degraded and
the same products from the degradation of 4-chlorobutyrate were observed.
Reductive dechlorination (or saturation of unsaturated bond) may thus
possibly serve as a regeneration system for a reduced coenzyme, which may
be formed in an oxidation pathway of such compounds to acetate.

INTRODUCTION

Anaerobic degradation of organic compounds has been extensively investigated for application to anaerobic treatment of wastewaters. Of
various organic compounds, chlorinated compounds such as
tetrachloroethylene[1], trichloroethylene[1,2] and chlorophenol[3,4] have
been the focus of interest because of their toxicity or recalcitrance.
Attention has also been paid to syntrophic bacteria which degrade
short chain fatty acids. In the last several years, these bacteria have
been identified and their degradation pathways studied[5-9].
Syntrophomonas wolfei is a representative syntrophic bacterium which
catabolizes C_4 to C_8 saturated fatty acids to acetate or acetate and
propionate[5]. S. wolfei can be grown in coculture with an H_2-utilizing
bacterium, since the anaerobic oxidation of saturated fatty acids

Microbiology and Biochemistry of Strict Anaerobes Involved in Interspecies Transfer
Edited by J.-P. Bélaich et al.
Plenum Press New York 1990

requires thermodynamically favorable conditions[10] acquired by maintaining of a low H_2 concentration in the culture. However, less attention has been paid to anaerobic degradation of chlorinated fatty acids and related compounds.

In the present study, 4-chlorobutyrate was used as a model substrate to examine the anaerobic degradation of chlorinated fatty acids, and an anaerobic bacterium which degrades 4-chlorobutyrate was isolated. We describe some characteristics of our isolate and discuss the role of dechlorination in anaerobic metabolism.

MATERIALS AND METHODS

Sources of organisms

The 4-chlorobutyrate degrading bacterium, strain K-1, was isolated from an enrichment culture inoculated with mesophilic digester sludge of a sewage treatment plant. A mesophilic H_2-utilizing methanogen was isolated from an enrichment culture inoculated with mesophilic digester sludge of the same plant. It utilized H_2/CO_2 and formate as substrates for growth. _Desulfovibrio vulgaris_ (DSM 2119) was obtained as an H_2 consuming sulfate-reducer from the Deutsche Sammlung von Mikroorganismen. _S. wolfei_ (DSM 2245B) cocultured with _Methanospirillum hungatei_ was obtained from the Deutsche Sammlung von Mikroorganismen. It utilized C_4 to C_8 saturated fatty acids and crotonate as substrates for growth.

Media and conditions of cultivation

The preparation of media and isolation were conducted by the methods of Hungate[11], Bryant[12], and Balch and Wolfe[13]. Cultivation was performed in 120ml serum vials each containing 50ml of medium or 26ml test tubes each containing 10ml of medium and closed with butyl rubber stoppers and aluminum crimps (Sanshin Corp.). The basal medium(A) contained the following components (per liter) : K_2HPO_4 0.75g ; KH_2PO_4 0.75g ; $MgCl_2 \cdot 6H_2O$ 0.36g ; NH_4Cl 1.0g ; yeast extract 0.5g ; $Na_2S \cdot 9H_2O$ 0.5g ; $NaHCO_3$ 2.5g ; resazurin 2mg ; mineral solution[14] 9ml ; vitamin solution[15] 10ml. The basal medium(B) used for several growth tests had the same composition except that K_2HPO_4 0.75g and KH_2PO_4 0.75g were replaced by Na_2HPO_4 4.73g and KH_2PO_4 4.54g. The pH of each medium was adjusted to 6.8 by the addition of 10% KOH solution. Preparation of media was made under a 100% N_2 gas phase. They were autoclaved at 121°C for 15min. The $NaHCO_3$, Na_2S solution and each substrate at the indicated concentrations were added to each medium through a 0.45-μm membrane filter (Millipore Corp.) just before inoculation. The gas phase was replaced with N_2/CO_2(80/20,v/v), and pressurized to 1 atm(101.3KPa). Unless otherwise stated, all incubations were conducted at 37°C.

For mixed culture experiments, either an H_2-utilizing methanogen or _D. vulgaris_ was added to the cultures. In the experiment on butyrate(2mM) degradation, 2mM Na_2SO_4 was added to the medium through a 0.45-μm membrane filter just before inoculation.

Cultivation of _S. wolfei_ was done as in the case of strain K-1. Incubation was conducted at 36°C.

Enrichment and isolation

The enrichment culture was made on the basal medium(A) containing 0.25g/l of 4-chlorobutyrate in a 120ml serum vial and inoculated with 10ml of mesophilic digester sludge. Following the complete degradation of 4-chlorobutyrate, 10ml of culture were transferred to fresh medium. During enrichment, the substrate concentration was increased to a final level of 2.3g/l.

A pure culture was obtained by repeated application of roll-tube method, using the basal medium(A) containing 17g/l of purified agar (Difco) and 1.5g/l of 4-chlorobutyrate.

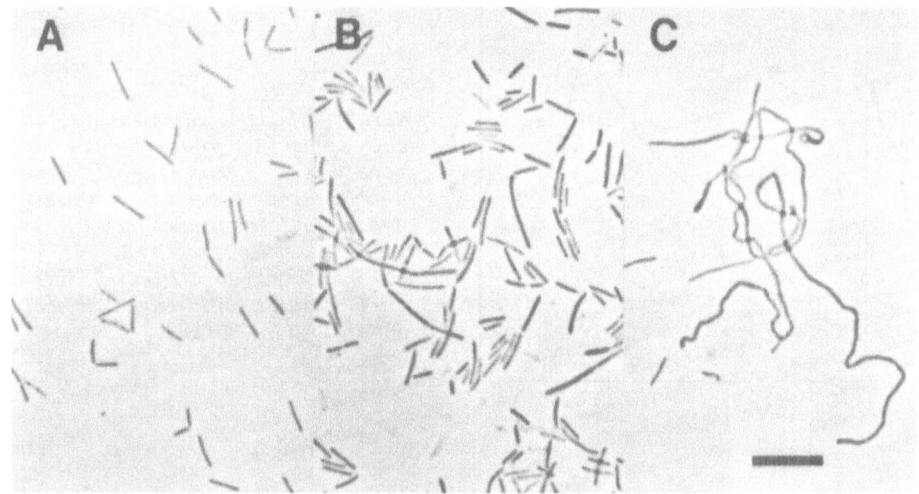

Fig.1.Phase-contrast photomicrographs of strain K-1 under various conditions. (A)Growth in medium containing 19mM 4-chlorobutyrate(4CB)
and 500mg/1 yeast extract(YE) at 30°C. (B)Growth in medium containing 19mM 4CB and 500mg/1 YE at 37°C. (C)Growth in medium containing
19mM 4CB and 100mg/1 YE at 37°C. Bar represents 10µm.

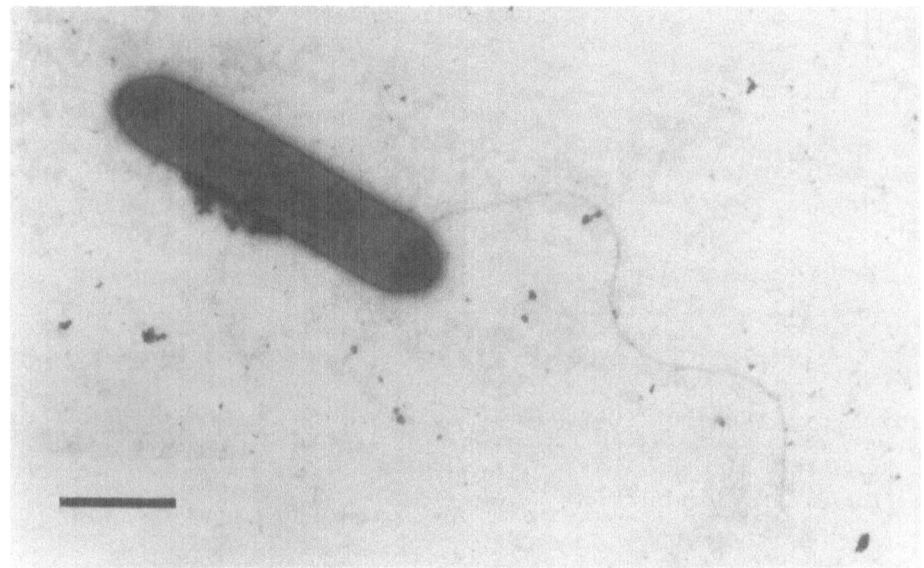

Fig.2. Electron micrograph of strain K-1 negatively stained with uranyl
acetate. Bar represents 1µm.

Electron microscopy

Cells were treated by negative staining using 1%(v/v) uranyl acetate
and observed under a transmission electron microscope operated at 80 kV
(Japan Electron Optics Laboratory JEM 1200-EX).

Analysis

Optical density of the culture was determined at 600nm using a
Hitachi 150-20 spectrophotometer. Fatty acids were measured using a gas

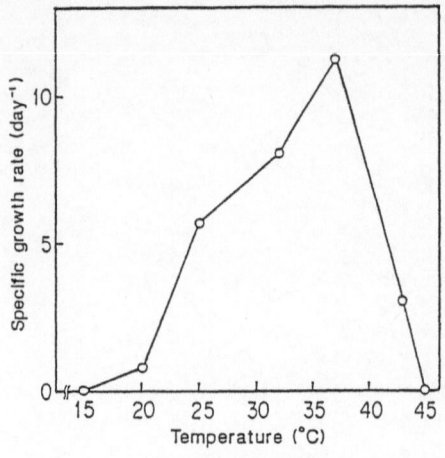

Fig.3.

Effect of the temperature on the specific growth rate of strain K-1. Experiments were performed on the basal medium(B) containing 0.05% yeast extract and 18.8mM 4-chlorobutyrate. Initial culture conditions were pH 6.7 under $N_2/CO_2(80/20)$.

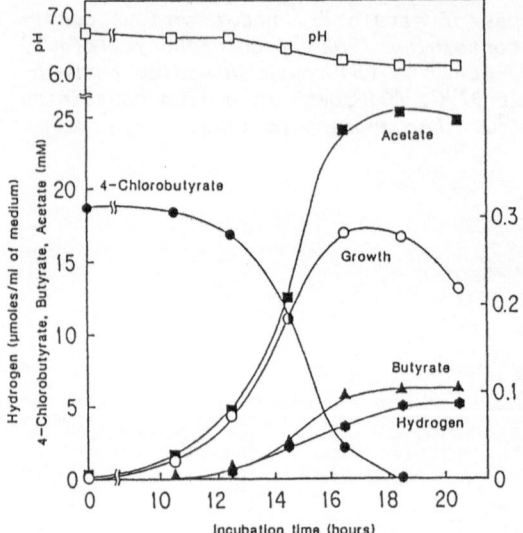

Fig.4.

Time course of growth and 4-chlorobutyrate degradation by strain K-1. Experiments were performed on the basal medium(B) containing 0.05% yeast extract and 18.8mM 4-chlorobutyrate. Initial culture conditions were pH 6.7 and 37^oC under $N_2/CO_2(80/20)$.

chromatograph (Shimadzu GC-5A) equipped with a flame ionization detector and a column of Shimalite PEG-6000. H_2 and CH_4 were measured using a gas chromatograph (Shimadzu GC-8AIT) equipped with a thermal conductivity detector and a molecular sieve column (60-80 mesh).

RESULTS

Morphology

The cells of strain K-1 were straight or slightly curved rods, averaging 1.5-6.5 µm in length and 0.4-0.8 µm in width (Fig.1A). They were slightly motile, and stained Gram-negative and the Gram-type[16] was negative. Spores were observed and growth occurred after pasteurization (30min at 70^oC). Electron microscopy indicated one flagellum (Fig.2). Cells appeared in singles or pairs when grown at 30^oC, and somewhat as chains when grown at 37^oC (Fig.1A,B). However, they appeared as helical long chains when the amount of yeast extract in the medium decreased from 500mg/l to 100mg/l (Fig.1C). Growth required yeast extract or rumen fluid. If the medium contained only 100mg/l of yeast extract, growth was delayed and cultures did not continue after transfer.

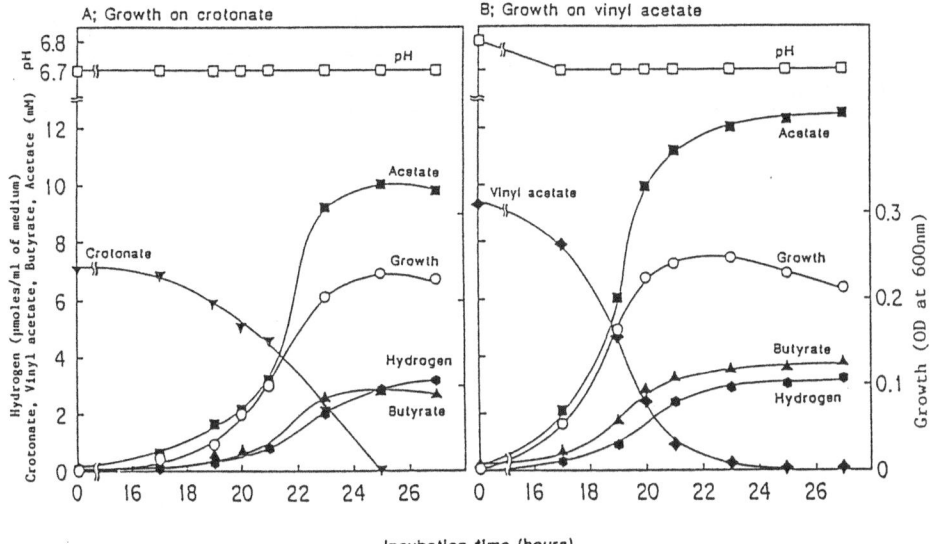

Fig.5. *Time course of growth and unsaturated compound degradation by strain K-1. (A)Growth on crotonate;(B)Growth on vinyl acetate. Experiments were performed on the basal medium(B) containing 0.05% yeast extract and 7.0mM crotonate or 9.1mM vinyl acetate under optimal conditions.*

Table 1. *Amounts of products from 4-chlorobutyrate or unsaturated compounds by strain K-1.*

Substrate	Growth (ΔOD_{600})	Amount of substrate consumed ($\Delta \mu mol$)	Amount of products ($\Delta \mu mol$) Acetate	Butyrate	H_2	Carbon recovery (%)
4-Chlorobutyrate	0.28	978	1286	315	265	98.0
Crotonate	0.23	366	501	129	168	103.7
Vinyl acetate	0.25	474	636	181	167	105.3
None	0	—	7	0	3	—

The basal medium(B) contained 0.05% yeast extract and 0.25% $NaHCO_3$. Initial culture conditions were pH 6.7-6.8 and 37^oC under $N_2/CO_2(80/20)$.

Growth properties on 4–chlorobutyrate

Optimal growth conditions were examined on the basal medium(B). Temperature range for growth was 20^oC to 43^oC (Fig.3). Active motility on the culture at less than 30^oC was observed. The initial pH for growth was 6.2 to 7.6. Optimal pH and temperature were 6.5-6.8 and 37^oC, respectively. Degradation of 4–chlorobutyrate and product formation during growth under optimal conditions are shown in Fig.4. Growth was accompanied by complete degradation of 4–chlorobutyrate to butyrate, acetate and hydrogen (Table 1). Under an H_2/CO_2 atmosphere, the growth of strain K-1 was not affected and the degradation pattern of 4–chlorobutyrate did not change (data not shown). Under optimal conditions, the maximum specific growth rate was $0.47h^{-1}$(t_d=1.5h). No growth occurred with yeast extract alone (Table 1). Cells immediately lysed when growth peaked at over pH6.0, but not at less than pH6.0.

Complete inhibition of growth and 4–chlorobutyrate utilization occurred by the addition(0.1g/l) of antibiotics such as penicillin G, vancomycin and streptomycin. Electron acceptors such as sulfur(excess), sulfate(10mM), sulfite(1mM), thiosulfate(10mM), nitrate(2mM) and nitrite(2mM) were not utilized for growth.

Table 2. Anaerobic degradation of chlorinated compounds by strain K-1.

Chlorinated compound	Degradation	Fermentation products
Chloroacetate(C_2)	–	
2-Chloropropionate(C_3)	–	
3-Chloropropionate(C_3)	+	Propionate, Acetate
2-Chlorobutyrate(C_4)	–	
3-Chlorobutyrate(C_4)	+	Butyrate, Acetate, Hydrogen
4-Chlorobutyrate(C_4)	+	Butyrate, Acetate, Hydrogen
5-Chlorovalerate(C_5)	–	

Symbols: +, degradation; -, no degradation
Experiments were performed in test tubes containing 10ml of the basal
medium(A) containing 0.05% yeast extract and 5mM substrate. Initial
culture conditions were pH 6.8 and 37°C under N_2/CO_2(80/20). Degradation was determined after a 1-week incubation period.

Table 3. Anaerobic degradation of organic compounds by strain K-1.

Organic compound	Degradation	Fermentation products
Acetate(C_2)	–	
Propionate(C_3)	–	
n-Butyrate(C_4)	–	
n-Valerate(C_5)	–	
n-Caproate(C_6)	–	
Crotonate($C_{4:1}$)	+	Butyrate, Acetate, Hydrogen
Vinyl acetate($C_{4:1}$)	+	Butyrate, Acetate, Hydrogen
Acrylate($C_{3:1}$)	+	Propionate, Acetate, Hydrogen
3-Hydroxybutyrate(C_4)	+	Butyrate, Acetate, Hydrogen
Lactate(C_3)	+	Propionate, Acetate, Hydrogen
Pyruvate(C_3)	+	Butyrate, Acetate, Hydrogen

Symbols: +, degradation; -, no degradation
Experiments were performed in test tubes containing 10ml of the basal
medium(A) containing 0.05% yeast extract and 5mM substrate. Initial
culture conditions were pH 6.8 and 37°C under N_2/CO_2(80/20). Degradation was determined after a 1-week incubation period.

Degradation of organic compounds

Fermentation products on various organic compounds are summarized in
Tables 2 and 3. 3-Chloropropionate was degraded to propionate and
acetate, and 3-chlorobutyrate to butyrate, acetate and a small amount of
hydrogen. However, chloroacetate, 2-chloropropionate, 2-chlorobutyrate
and 5-chlorovalerate were not utilized.
Crotonate, vinyl acetate, 3-hydroxybutyrate, and pyruvate were also
utilized and produced the same products as the degradation of 4-
chlorobutyrate. Acrylate and lactate were degraded to propionate,
acetate and hydrogen. However, none of saturated fatty acids(C_{2-6}) could
be utilized. Growth curves on crotonate and vinyl acetate under optimal
conditions are shown in Fig.5. The specific growth rates on crotonate
and vinyl acetate were $0.39h^{-1}$(t_d=1.8h) and $0.42h^{-1}$(t_d=1.7h), respectively.

Table 4. *Utilization of various sugars by strain K-1.*

Substrate	Final absorbance of the culture (OD_{600})
Fructose	0.60
Other sugars	< 0.10

The following sugars were not utilized for growth :
adonitol, cellobiose, dulcitol, glucose, lactose,
maltose, mannose, maltotriose, melibiose, raffionse,
ribose, sucrose, sorbitol, trehalose, and xylose.
Cultivations were performed on the basal medium(A)
containing 0.5% substrate for 1-week. Initial cul-
ture conditions were pH 6.7 and 37°C under
N_2/CO_2(80/20). Growth on the basal medium(A) in the
absence of substrate was reached an absorbance <
0.10.

Sugars examined for being utilized are listed in Table 4. Fructose was utilized as a substrate, but no others. Other organic compounds which could not be utilized for growth as follows: amyl alcohol, butanol, ethanol, glycerol, methanol; i-propanol, alanine, glutamate, glycine, citrate, formate, fumarate, gluconate, succinate, chitin, and starch.

Mixed culture experiments

Strain K-1 cocultured with an H_2-utilizing methanogen, which was isolated in our laboratory, could degrade 4-chlorobutyrate without any change of the fermentation product ratio (butyrate:acetate), although all formed hydrogen was converted to methane (data not shown). Strain K-1 cocultured with D. vulgaris could not degrade butyrate. S. wolfei cocultured with M. hungatei degraded 4-chlorobutyrate. However, the degradation of 4-chlorobutyrate by strain K-1 was much faster than that by S. wolfei with M. hungatei (data not shown).

DISCUSSION

Taxonomical location of strain K-1

Strain K-1 is a spore-forming, Gram-negative, motile, straight or slightly curved rod, 1.5-6.5 μm in length and 0.4-0.8 μm in width. Cells possess one flagellum, appear in singles or pairs, and occasionally long chains.

Based on these characteristics, the isolate in the present study may be classified with the genus Clostridium, since it is anaerobic and a spore-former but not a sulfate-reducer. This strain K-1 may be class-ified with a novel species of the genus Clostridium, since it utilizes only fructose among sugars.

Physiology

Strain K-1 could degrade several chlorinated fatty acids other than 4-chlorobutyrate. The susceptibility of dechlorination may be related to the distance of the chlorinated carbon from the carboxyl group, since degradation occurred with 3-chloropropionate, 3-chlorobutyrate and 4-chlorobutyrate, but not with 2-chloropropionate and 2-chlorobutyrate.

Strain K-1 fermented not only chlorinated fatty acids but also un-saturated and hydroxyl fatty acids such as crotonate and 3-

hydroxybutyrate. Comparing the growth rate of strain K-1 to that of other crotonate-utilizing anaerobic bacteria, strain K-1 ($\mu=0.39h^{-1}$) grows much faster than Ilyobacter polytropus[17] ($\mu=0.28h^{-1}$), Clostridium kluyveri[18] ($\mu=0.058h^{-1}$) and other Clostridium species[19] ($\mu=0.15h^{-1}$). These organisms ferment 3-hydroxybutyrate and crotonate to butyrate and acetate.

Stieb and Schink discussed the pathways of the fermentation of crotonate and 3-hydroxybutyrate by I. polytropus as follows[17]; a portion of each of these compounds is finally converted to acetate via beta-oxidation. First, crotonyl-CoA is generated by an initial activation to the CoA-derivative and it is hydrated to 3-hydroxybutyryl-CoA. Next, 3-hydroxybutyryl-CoA is dehydrogenated to acetoacetyl-CoA, which is finally converted to two molecules of acetate via acetyl-CoA. The reducing equivalents are released by these reactions and are consumed as electron donors for reducing crotonyl-CoA to butyryl-CoA. These overall reactions are accompanied by the synthesis of one molecule ATP per two molecules of substrate through fermentation. Thus, two molecules of substrate are stoichiometrically fermented to one molecule of butyrate and two molecules of acetate. However, in our experiments, the production ratio of butyrate to acetate was lower than that theoretically expected, and hydrogen formation was also recognized (Table 1).

The fermentation of 3 and 4-chlorobutyrate by strain K-1 may take a pathway similar to that of crotonate or 3-hydroxybutyrate. The degradation pathway of chlorobutyrate may be as follows; a portion of chlorobutyrate is finally oxidized to two molecules of acetate, probably via beta-oxidation by which ATP is generated through substrate-level phosphorylation. In this pathway, reduced coenzymes are released, and are consumed as electron donors for the reductive dechlorination of the other portion of chlorobutyrate to butyrate. Thus, two molecules of chlorobutyrate are stoichiometrically fermented to one molecule of butyrate and two molecules of acetate. However, in our experiments, the production ratio of butyrate to acetate was lower than that theoretically expected, and hydrogen formation was also recognized as being the same as in the case of unsaturated compounds (Table.1). This production ratio was not affected under an H_2/CO_2 atmosphere, or cocultivation with an H_2-utilizing methanogen.

The degradation pathways of acrylate and lactate may be considered as follows: a portion of each compound is oxidized to acetate with ATP generation via lactate and pyruvate. Reduced coenzymes are released in these reactions, and they are consumed as electron donors for reducing of another substrate to propionate. Thus, each of these compounds is fermented to propionate and acetate. An different fermentation pattern was observed with pyruvate degradation. Pyruvate was fermented to butyrate, acetate and hydrogen. In this case, butyrate formation may have occurred through the coupling of two molecules of acetyl-CoA.

The fermentation of 3-chloropropionate may take a pathway similar to that of acrylate or lactate. The degradation pathway of chloropropionate may be as follows; a portion of chloropropionate is finally oxidized to acetate via acrylate, lactate and pyruvate. In this pathway, reduced coenzymes are released, and are consumed as electron donors for the reductive dechlorination of other portion of chloropropionate to propionate. Thus, chloropropionate is fermented to propionate and acetate.

As mentioned above, under anaerobic conditions, a certain system of regeneration of the reduced coenzyme to the oxidized form is essential for overall metabolism. Thus, an oxidized coenzyme may be regenerated by saturation of the unsaturated bond and reductive dechlorination. These reactions involve the conversion of a portion of each substrate, such as halogenated fatty acids and unsaturated fatty acids, to saturated fatty acids. However, it is unclear why hydrogen was generated from the degradation of chlorobutyrate, crotonate, vinyl acetate, acrylate, 3-hydroxybutyrate, lactate and pyruvate. It may be caused by excessive oxidation of substrate to acetate.

ACKNOWLEDGMENTS

This research was supported by the grant of New Energy and Industrial Technology Development Organization(R.& D. on the New Wastewater Treatment System), and performed in cooperation by Aqua Renaissance Research Association and Fermentation Research Institute.
We thank Miss Akemi Ohmiya for making the electron micrographs(Fruit Tree Research Station, Ministry of Agriculture Forestry and Fisheries).

REFERENCES

1. T.M. Vogel and P.L. McCarty, Biotransformation of Tetrachloroethylene to Trichloroethylene, Dichloroethylene, Vinyl Chloride, and Carbon Dioxide under Methanogenic Conditions, Appl. Environ. Microbiol. 49:1080-1083 (1985)
2. E.J. Bouwer and P.L. McCarty, Transformations of Halogenated Organic Compounds under Denitrification Conditions, Appl. Environ. Microbiol. 45:1295-1299 (1983)
3. R. Hakulinen, S. Woods, J. Ferguson and M. Benjamin, The Role of Facultative Anaerobic Micro-organisms in Anaerobic Biodegradation of Chlorophenols, Wat. Sci. Tech. 17:289-301 (1985)
4. S.A. Boyd and D.R. Shelton, Anaerobic Biodegradation of Chlorophenols in Fresh and Acclimated Sludge, Appl. Environ. Microbiol. 47:272-277 (1984)
5. M.J. McInerney, M.P. Bryant, R.B. Hespell and J.W. Costerton, Syntrophomonas wolfei gen. nov. sp. nov., an Anaerobic, Syntrophic, Fatty Acid-Oxidizing Bacterium, Appl. Environ. Microbiol. 41:1029-1039 (1981)
6. D.R. Boone and M.P. Bryant, Propionate-Degrading Bacterium, Syntrophobacter wolinii sp. nov. gen. nov., from Methanogenic Ecosystems, Appl. Environ. Microbiol. 40:626-632 (1980)
7. F. Roy, E. Samain, H.C. Dubourguier and G. Albagnac, Synthrophomonas sapovorans sp. nov., a new obligately proton reducing anaerobe oxidizing saturated and unsaturated long chain fatty acids, Arch. Microbiol. 145:142-147 (1986)
8. F. Widdel and N. Pfennig, Studies on dissimilatory sulfate-reducing bacteria that decompose fatty acids I, Isolation of new sulfate-reducing bacteria enriched with acetate from saline environments. Description of Desulfobacter postgatei gen. nov., sp. nov., Arch. Microbiol. 129:395-400 (1981)
9. M. Stieb and B. Schink, Anaerobic oxidation of fatty acids by Clostridium bryantii sp. nov., a sporeforming, obligately syntrophic bacterium, Arch. Microbiol. 140:387-390 (1985)
10. R.K. Thauer, K. Jungermann and K. Decker, Energy conservation in chemotrophic anaerobic bacteria, Bacteriol Rev. 41:100-180 (1977)
11. R.E. Hungate, A roll tube method for cultivation of strict anaerobes, in: Methods microbiology. Vol.3B, J.R.Norris, D.W.Ribbons, eds., Academic Press, New York (1969)
12. M.P. Bryant, Commentary on the Hungate technique for culture of anaerobic bacteria, Am. J. Clin. Nutr. 25:1324-1328 (1972)
13. W.E. Balch and R.S. Wolfe, New approach to the cultivation of methanogenic bacteria:2-mercaptoethanesulfonic acid (HS-CoM)-dependent growth of Methanobacterium ruminantium in a pressurized atmosphere, Appl. Environ. Microbiol. 32:781-791 (1976)

14. H. Morii, M. Nishihara and Y. Koga, Isolation, characterization and physiology of a new formate-assimilable methanogenic strain (A2) of Methanobrevibacter arboriphilus, Agric. Biol. Chem. 47:2781-2789 (1983)
15. W.E. Balch, G.E. Fox, L.J. Magrum, C.R. Woese and R.S.Wolfe, Methanogens: reevaluation of a unique biological group, Microbiol. Rev. 43:260-296 (1979)
16. T. Gregersen, Rapid method for distinction of Gram-negative from Gram-positive bacteria, Eur. J. Appl. Microbiol. Biotechnol. 5:123-127 (1978)
17. M. Stieb and B. Schink, A new 3-hydroxybutyrate fermenting anaerobe, Ilyobacter polytropus, gen. nov. sp. nov., possessing various fermentation pathways, Arch. Microbiol. 140:139-146 (1984)
18. R.K. Thauer, K. Jungermann, H. Henninger, J. Wenning and K. Decker, The energy metabolism of Clostridium kluyveri, Eur. J. Biochem. 4:173-180 (1968)
19. J. Bader, H. Günther, E. Schleicher, H. Simon, S. Pohl and W. Mannheim, Utilization of (E)-2-butenoate (crotonate) by Clostridium kluyveri and some other Clostridium species, Arch. Microbiol. 125:159-165 (1980)

CARBON AND ENERGY FLOW DURING ACETOGENIC METABOLISM

OF UNICARBON AND MULTICARBON SUBSTRATES

N.D.Lindley E.Gros P.LeBloas M.Cocaign and P.Loubière

Dept Genie Biochimique, INSA-CTBM
Avenue de Rangueil, 31077 Toulouse Cedex
France

In anaerobic environments carbon and energy flow is even more closely inter-related than in aerobic habitats. This is due to the high proportion of carbon substrate transformed to fermentation end-products in order to create the necessary energetic balance for efficient cell growth. Some of these fermentation metabolites play an essential role in energy (ATP) production via substrate level phosphorylation reactions, while others can be viewed as electron sinks to avoid accumulation of excess reducing equivalents (as reduced co-enzymes). In an established multi-species population the product of one species fermentative metabolism will contribute to the substrate requirements of co-existing species. Thus, an essential interspecies carbon flow will occur involving various different metabolic types. In addition, some species that overcome their excess reducing equivalent yield via hydrogen gas production will almost certainly be present. This source of reducing power is essential for homoacetogenic and methanogenic autotrophs that fix carbon dioxide as principle substrate.

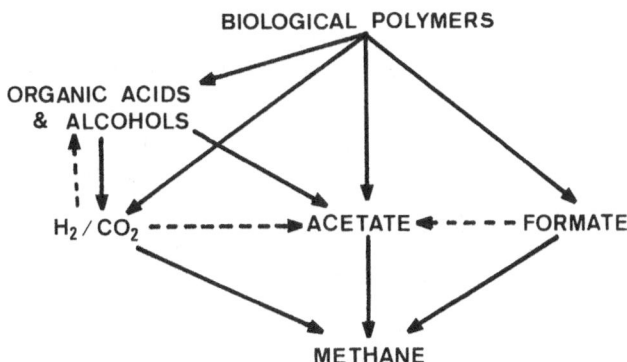

Figure 1. Network of carbon flow within natural environments showing the major reactions and intermediary products

Microbiology and Biochemistry of Strict Anaerobes Involved in Interspecies Transfer
Edited by J.-P. Bélaich *et al.*
Plenum Press, New York, 1990

A typical metabolic network representing potential carbon and energy flow in natural environments can be established (1), and has often been portrayed in diagrams similar to figure 1. The component groups of bacteria involved in such habitats can be described as follows:

1) extracellular enzyme-secreting hydrolytic bacteria that ferment complex biological polymers (polysaccharides, lipids, proteins, etc) into organic acids (lactate, butyrate, propionate, etc) and alcohols (butanol, ethanol, methanol, etc), carbon dioxide and hydrogen.

2) hydrogen-producing acidogens which ferment multi-carbon alcohols and organic acids of at least three carbon atoms to hydrogen gas and acetate.

3) acetogenic bacteria which ferment multi-carbon and/or unicarbon compounds into organic acids via acetyl-CoA, carbon dioxide and hydrogen.

4) methanogenic archaebacteria fermenting acetate, single carbon compounds and hydrogen/carbon dioxide mixtures to methane and in some cases carbon dioxide.

Such a network of metabolic types avoids to any great extent many of the inhibitory phenomena associated with the build-up of toxic concentrations of fermentation end-products often encountered in single-species laboratory cultures. The final products, methane and carbon dioxide are gases which will readily diffuse out of the anaerobic environment, to be consumed by the methane-oxidising methanotrophs in adjacent aerobic environments. The relative importance of each metabolic group within any given environment will be dependent on the nature of the initial substrate and the predominating physico-chemical conditions.

In this paper the influence of available substrate on the growth and acetogenic metabolism of Eubacterium limosum will be discussed. This bacterium and others of similar metabolic flexibility can be seen as potential regulators of interspecies hydrogen transfer since they can exploit a wide range of carbon substrates, some of which will lead to production of hydrogen. However, they can also use hydrogen as source of reducing power to fix carbon dioxide.

GROWTH ON SINGLE SUBSTRATES

E. limosum can metabolise many simple carbon substrates, including hexose and pentose sugars, lactate, reduced single-carbon compounds (methanol, formate, carbon monoxide, betaine, methoxyl-substituted aromatics) and carbon dioxide/hydrogen mixtures. Such versatile metabolic potential places these organisms in a somewhat unique position: their metabolism of heterotrophic substrates involves hydrogen production, while autotrophic carbon dioxide fixation requires hydrogen as source of reducing power for conversion of substrate to methyl-level intermediate for acetyl-CoA synthesis. When growing methylotrophically on methanol or methoxylated aromatics, hydrogen is neither produced nor an essential co-substrate (table 1). Addition of hydrogen to the gas phase does however lead to an altered acidogenic metabolism in that acetate tends to be produced at the expense of butyrate.

Table 1 Growth rates and fermentation products of batch cultures of <u>Eubacterium limosum</u> on various substrates.

SUBSTRATE	μ	YIELDS (% CARBON) ACETATE	BUTYRATE	CO_2	H_2
glucose	0.31	30.1	6.0	32.9	+
fructose	0.32	34.8	12.1	21.0	+
mannitol	0.32	36.0	19.8	16.3	+
ribose	0.21	45.5	3.0	40.0	+
dihydroxyacetone	0.28	40.2	24.0	9.1	
lactate	0.16	16.7	62.6	5.1	
methanol	0.12	35.2	81.4	-32.2	

Differences in growth rate can be explained if the carbon flow through intermediary metabolism is examined for each substrate. Sugars and sugar alcohols convert all the carbon substrate to anabolic precursors and acidogenic end-products via glycolysis, while during methylotrophic growth (and also for lactate) gluconeogenesis is involved (figure 2). A substantial part of the carbon flow (more than 75%) is associated with organic acid synthesis and is distinct from precursor synthesis during growth on methanol. Biomass synthesis and hence growth rate on unicarbon substrates is fixed by the rate of gluconeogenesis. The major bottleknecks would appear to be located within certain enzymic reactions whose kinetics are such that glycolysis is the preferred direction (LeBloas & Lindley, unpublished results). This seeming absence of gluconeogenic-specific enzymes leads to problems in ensuring the transfer of carbon flux at rates adequate to support growth rates comparable with glycolysis. Some of the enzymes which might be expected to be induced during growth on methanol can not be measured at significant levels. Work in progress in our laboratory should enable this aspect to be better understood within a short time.

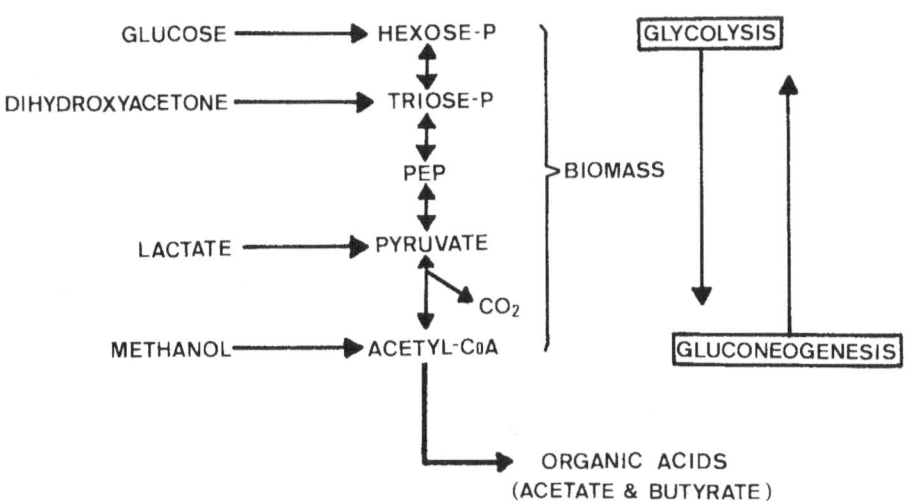

Figure 2. Entry of various substrates into intermediary metabolism (biomass precursor formation and acidogenesis) of <u>Eubacterium limosum</u>.

Table 2 Stoichiometric equations for acidogenic fermentations.

GLUCOSE

100 Glucose $\longrightarrow$ 31 CO_2 + 159 Acetate + 21 Butyrate + 13 H_2

METHANOL

100 Methanol + 28 CO_2 $\longrightarrow$ 9 Acetate + 22 Butyrate

HYDROGEN/CARBON DIOXIDE

100 Hydrogen + 52 CO_2 $\longrightarrow$ 22 Acetate

During chemostat growth of E. limosum on methanol, the kinetics of substrate consumption suggest that the rate at which carbon dioxide can be fixed may also be a limiting factor for this organism's metabolism. Whilst specific rates of methanol consumption increases proportionately to the growth rate, carbon dioxide consumption reaches a maximum at $\mu=0.06$/h and thereafter remains constant. Thus, in order to achieve an adequate energetic balance the bacterium must produce ever-increasing proportions of butyrate. Washout occurs when the relative rates of consumption of each sustrate attain the theoretical value for homobutyric fermentation (2). Growth at higher dilution rates can only be achieved if a further substrate of a less reduced status than methanol eg. acetate, is added to the feed mixture (see figure 3).

Although growth rates obtained on various substrates can be explained, albeit as yet, somewhat speculatively, the total absence of hydrogen gas production during growth on methanol cannot be satisfactorily accounted for. In general, the end-products will be synthesised according to the substrate's level

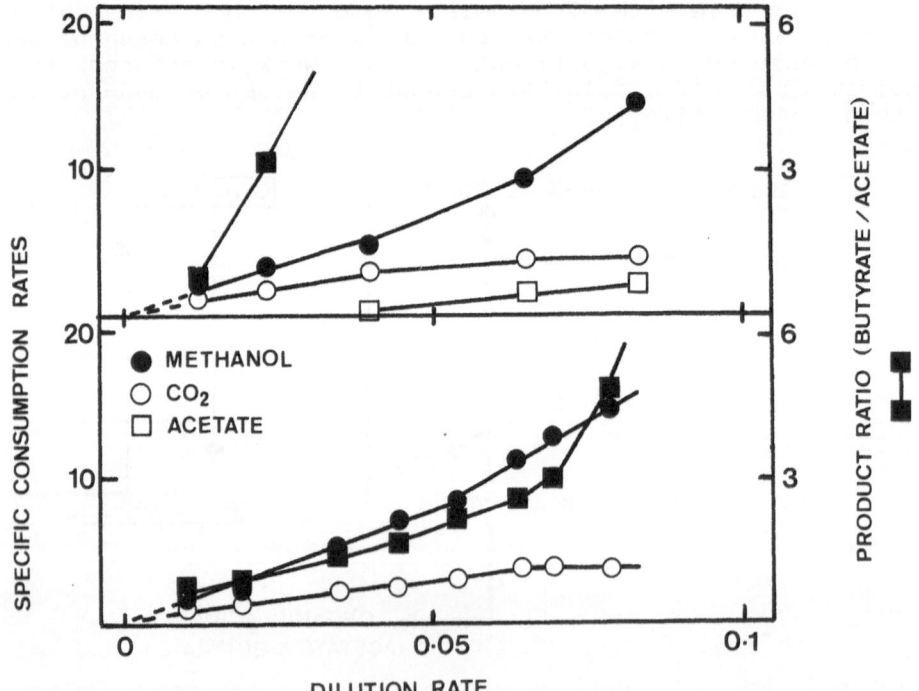

Figure 3. Effect of dilution rate on the specific rates of substrate consumption and the product ratio

of reduction, and since methanol/carbon dioxide is a substrate
of a similarly reduced status to glucose it follows that the
level of reduction of the products should also be similar.
This is achieved during methylotrophic growth by a greatly
enhanced synthesis of butyrate which compensates for the lack
of hydrogen production (see table 2). However, the inability
to waste reducing equivalents via hydrogen gas production
during methylotrophic growth necessitates a metabolism which
can either guarantee a continued flux towards butyrate or
vary the amount of carbon dioxide as co-substrate as a function
of the butyrate accumulated. Both metabolic adaptations are
present in E.limosum; the methanol/carbon dioxide consumption
ratio does vary within fairly narrow limits (3), and recent
evidence has shown that intracellular butyrate concentrations
are maintained lower that those of the culture broth due to
the operation of a non-passive mechanism of butyrate excretion
(4). Although such an efflux of butyrate against a concentration gradient prolongs the period of growth during which maximum specific rates may be maintained this involves an ever-
increasing ATP expenditure ultimately leading to the collapse
of the cross-membrane pH-gradient and a concommitant fall
in the specific rates of growth and substrate consumption.

The absence of hydrogen gas production during methylotrophic growth though not the metabolic functions compensating
for the lack of versatility are difficult to explain from
the evolutionnary point of view. There appears to be no definite advantage to the organism whose metabolism would be
endowed with increased flexibility if some of the excess reducing equivalents could be wasted via hydrogen gas. It remains
to be seen whether hydrogenase-transferase enzymes are absent
during methylotrophic growth, or merely inactive due to the
kinetic constants of the various dehydrogenase enzymes. It
is worth noting that both mannitol and glycerol partially
repress hydrogenase activity during growth of Clostridium
acetobutylicum with glucose as principal carbon source and
that this phenomenon is related to partial repression of the
hydrogen-producing hydrogenase (Soucaille, pers.comm.).

Whatever the actual mechanism controlling hydrogen production, it is clear that acidogenic bacteria such as E. limosum
have three distinct modes of metabolism dependent on the substrate presented and that as both a producer and consumer of
hydrogen the influence on the global energy budget within
anaerobic environments will be significant. Such conclusions
are however based upon growth on simple substrates and little
information exists regarding the growth of such micro-organisms
when presented with a mixture of substrates. Indeed, in a
recent symposium, the organising committee included in the
closing statement a regret that the manner in which anaerobic
species utilise defined mixtures seems not to have received
much attention (5). Before examining some of our recent findings related to the manner in which E.limosum uses mixtures
of sugars and methanol, it is first important to assess whether
such experiments are relevant to natural environments.

The initial substrate for anaerobic environments is usually portrayed as a variety of biological polymers. Of particular interest in the context of this presentation, are those
polymers susceptable to be degraded by hydrolytic species
to yield either sugars or single carbon substrates. One of the

major polymers will be lignocellulose material whose attack will generate not only hexose and pentose sugars, but also methoxylated aromatic compounds. These lignin-derived methyl esters are analagous with free methanol as regards consumption by E. limosum (6). They are attacked by a limited number of acetogens by a cleavage reaction producing a single carbon intermediate (unpublished work by Cocaign & Lindley suggests this may be formaldehyde) and a phenolic residue. Growth rates on syringate and vanillate do not significantly differ from those on methanol though ferulate is more slowly metabolised. In addition to these sources of reduced unicarbon substrates, free methanol might also be anticipated in environments in which pectin is to be found. To date, none of the pectinolytic bacteria isolated have been able to use the methanol liberated during degradation of the pectin (7). These two sources of methanol equivalents will of course be supplemented with formate a major end-product of acidogenic metabolism. Thus, it is quite reasonable to anticipate some methylotrophic acidogenesis in natural environments, though the physico-chemical conditions will determine whether or not acidogens can compete with methanogens for such substrates.

BATCH FERMENTATIONS OF GLUCOSE/METHANOL MIXTURES

When presented with substrate mixtures (8.5mM glucose/ 50mM methanol/50mM carbon dioxide) under batch conditions glucose is the preferred substrate of E.limosum, though the pattern of both growth and substrate consumption cannot be adequately described as a classical diauxic pattern of fermentation. Following a lag phase during which the redox potential is modified, growth is initiated with consumption of glucose. A specific growth rate of 0,31/h is established and a mixed acidogenesis occurs with both acetate and butyrate being produced (fig.4). During this period some hydrogen production takes place, though it is interesting to note that less

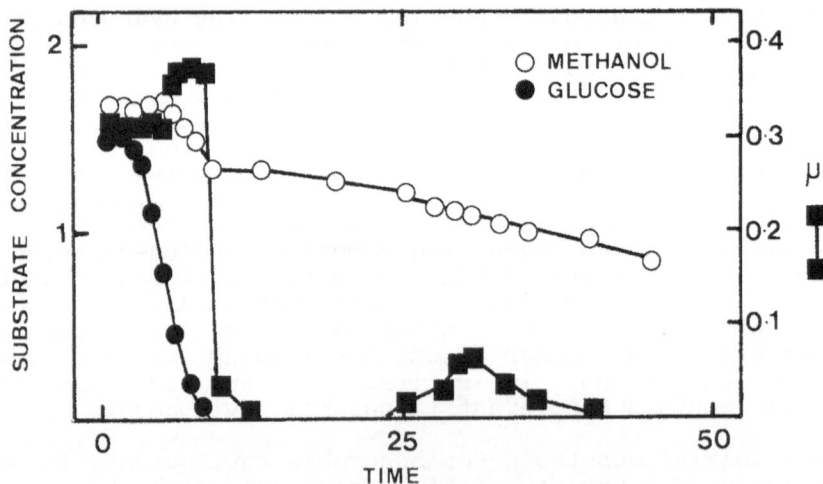

Figure 4. Batch growth of Eubacterium limosum on methanol/ glucose substrate mixtures using a methanol-grown inoculum. Substrate concentrations are shown in g/l.

hydrogen and more butyrate are produced, compared with cultures
grown on glucose alone. When the residual glucose falls below
5mM consumption of methanol begins, simultaneous with an
increase in the specific growth rate to 0.38-0.40/h which
is maintained until all the glucose has been utilized. Growth
then halts and a prolonged lag phase occurs before renewed
consumption of methanol takes place. During this phase the
metabolism is homobutyric, though growth rates are considerably
lower than those normally associated with methylotrophic
growth. This can be explained by the changes in both the
physiological state of the micro-organism and the accumulated
organic acid concentrations within the fermentation broth.
Rates of growth are in fact similar to those normally obtained
in methanolic cultures following the consumption of equivalent
substrate levels. Of interest is the finding that hydrogen
produced during the initial phase of growth on glucose is
rapidly re-consumed during the phase in which simultaneous
consumption of glucose and methanol occurs. This profile
was achieved only when using inocula grown on methanol. More
classic diauxic growth with no simultaneous consumption occurs
for cultures for which the inoculum was pre-grown on glucose.
If the outflowing culture of a double-carbon-limited chemostat
growing on the same mixture was used, the growth profile was
somewhat different. Simultaneous consumption of both substrates takes place directly growth begins with a specific growth
rate of 0.36/h being maintained up until complete depletion
of the glucose. Following a brief lag phase, the residual
methanol is utilized. Throughout this culture, no significant
concentration of hydrogen could be detected in the gas phase
(i.e. less than 1%).

Similar work using a different strain of E. limosum described a typical diauxic growth curve for glucose/methanol
mixtures (6), though these authors based their conclusions
on biomass analysis alone, with no measurements of substrate
consumption. These experiments were performed with extremely
low glucose concentrations and in our opinion, the initial
period of rapid growth attributed to glucose consumption,
was most likely a mixed substrate growth. In order to verify
whether or not this was possible, we repeated our work with
mixtures of substrates in which the initial glucose levels
were below those at which simultaneous substrate utilisation
might be expected (i.e. 4mM glucose/50mM methanol). Simultaneous consumption of both substrates was observed from the
onset, though the specific rate of methanol consumption was
never comparable to rates found during growth on methanol
alone.

The finding that hydrogen could be used from the point
at which methanol was consumed is to be expected from previous
work. It has been observed in our laboratory that cultures
of E. limosum grown on glucose maintain a metabolic activity
leading to acetate production from hydrogen/carbon dioxide
(or from formate) following the complete depletion of glucose,
though no growth seems possible on these substrates (8).
This finding may well be dependent on the strain used, since
there appears to be considerable variation concerning the
ability to support autotrophic growth. The absence of the
recently discovered mechanism coupling the methylene-THF dehydrogenase activity to a sodium ion gradient to generate
biochemical energy in autotrophic acidogens and methanogens

may explain strain differences regarding capacity to exploit hydrogen/carbon dioxide (9).

CHEMOSTAT GROWTH ON GLUCOSE/METHANOL MIXTURES

The results obtained with substrate mixtures under batch conditions imply that regulation of substrate consumption involves both repression-derepression and induction mechanisms. Kinetic behaviour under carbon-limitation in chemostat cultures was thus investigated. Two complementary approaches have been undertaken involving either the use of a fixed growth rate and varying mixture composition, or a defined mixture under a range of growth rates.

Various Methanol/Glucose Mixtures with Fixed Growth Rate

A relatively low dilution rate (0.05/h) was fixed to investigate the behaviour of E. limosum when presented with a substrate mixture of constant carbon concentration, but varying proportions of glucose and methanol. Throughout this experiment both acetate (25mM) and carbon dioxide (50mM) were included in the medium at constant values. The substrate consumption data shows that glucose was always entirely consumed, but that residual methanol was present in the outflow at high initial methanol concentrations. Indeed at all steady-states a maximum of 55mM methanol was able to be metabolised, this value being fixed by end-product inhibition phenomena

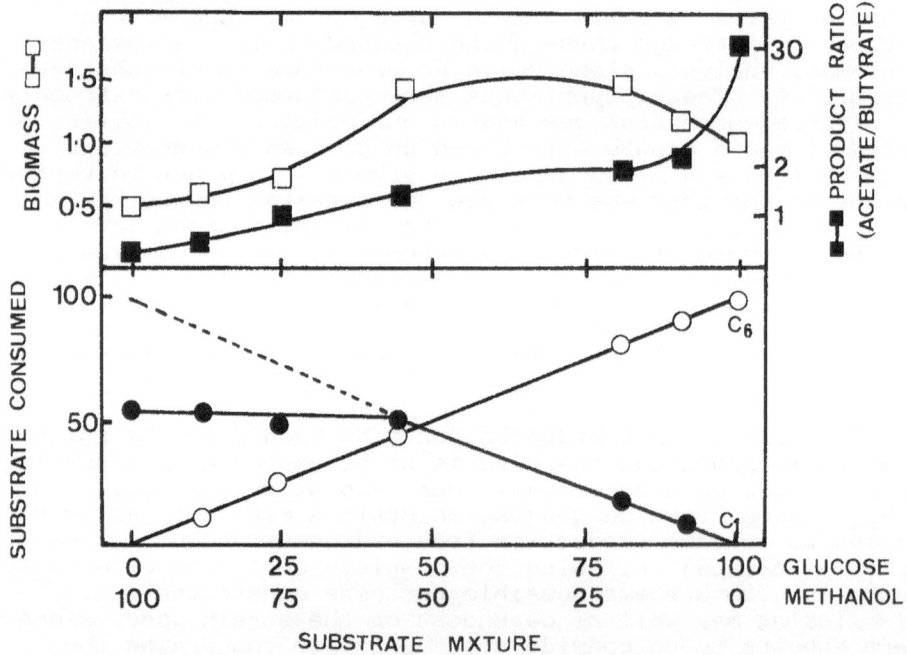

Figure 5. Use of various glucose/methanol mixtures in chemostat cultures with D=0.05/h. Substrate concentrations are shown as mM carbon, whilst products are the ratio of molar concentrations.

rather than nutritional limitations. As a consequence biomass
values were not constant, but proportionate to the total carbon
consumption, i.e. maximal at 50/50 when double-substrate carbon
limitation was established (figure 5). Hydrogen gas was absent
under all mixtures, other than at trace amounts during growth
on glucose alone. The organic acids produced showed a net
shift towards acetate as the proportion of glucose in the
medium increased with a significant fall in production of
butyrate during growth on glucose alone. Growth of E. limosum,
even on glucose alone under carbon-limiting growth conditions
does not produce significant hydrogen, due presumably to the
excess carbon dioxide in the medium. However, when a fault
in pH-regulation occurred leading to the transient build-up
of residual glucose, a rapid change in gas-phase composition
took place with hydrogen accumulating. Upon the restoration
of glucose-limitation this hydrogen was rapidly removed.

Effect of Various Growth Rates on Substrate Utilisation of a Defined Mixture

One of the commonly observed phenomena reported for
chemostat cultures grown on substrate mixtures is that accumulation of the substrate supporting the lower growth rate is
often displaced towards higher growth rates in the presence
of other substrates. This has been explained by the modified
position of the growth-rate determining reaction. In other
words, the rate at which an anabolic precursor can be supplied
tends to fix a potential growth rate, but if this rate-deter-
mining metabolite can be supplied by an alternative route
(due to the metabolism of a co-substrate) the biochemical
bottleneck will be removed. The use of a wide range of growth
rates for E. limosum and a fixed medium composition (8mM glu-
cose/50mM methanol) has confirmed that this phenomena supplies
also to an anaerobic bacterium. An important variation was
however observed in that the buid-up of residual methanol
at growth rates for in excess of the maximum growth rate this

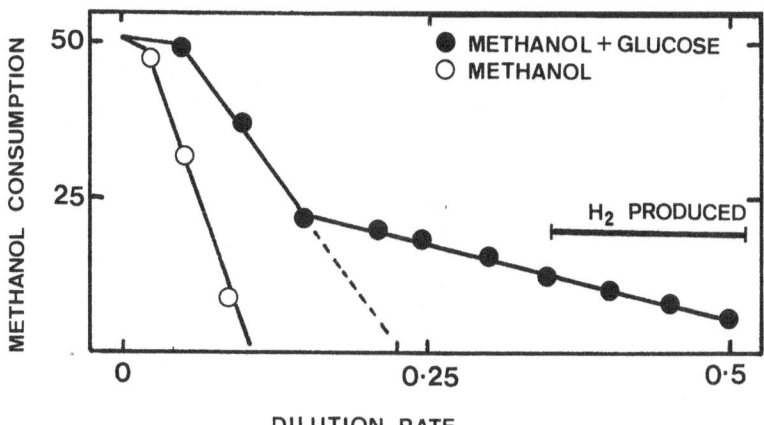

Figure 6. The effect of dilution rate on the substrate con-
           sumption kinetics for Eubacterium limosum grown
           on either methanol alone, or methanol/glucose
           (50/50) mixtures.

substrate can support was atypical (figure 6). Even at growth
rates five times higher than for methanol alone, some methanol
was consumed. Not only this, but growth rates in excess of
the batch-mode maximum for growth on glucose were established
in which no residual glucose occurred. It would appear, therefore, that the growth of E. limosum on substrates mixtures
is more efficient than on either substrate alone. Without
the necessary enzyme activities and/or isotope labelling
patterns (work currently in progress) it is difficult to
explain why such improved growth occurs. It is reasonable
however to speculate that methanol functions principally as
a source of energy and thus allows the glucose to be more
efficiently metabolised.

The presence of various substrates can be seen to have a
profound effect on the growth and acidogenic metabolism of
E. limosum. Perhaps the most interesting of these phenomena
is this organism's capacity to grow more rapidly in the presence of low concentrations of methanol/glucose than on either
substrate when present as sole carbon source. It is also
important to note that with an excess of carbon dioxide very
little production of hydrogen gas takes place, even when grown
on glucose alone. This is all the more noticeable if the
bacterium has previously been grown on unicarbon substrates.
From the carbon balance equations it is clear that the reducing
equivalents are re-used to transform carbon dioxide to acetate.
The enzymes specific to unicarbon metabolism are subject to
multiple control, some under strict derepression/induction
control of enzyme synthesis, whilst others, most likely the
constituative methanol dissimilating enzymes enable some
methanol to be utilised when glucose is present. In natural
environments, sugars are most likely to be in limiting concentrations and thus bacteria such as E. limosum can be expected
to exert a mixed metabolism and hence hydrogen available for
co-existing species will be lower than might be expected from
simple extrapolation of experimentation using glucose as sole
carbon source.

While the results obtained to date are far from complete,
it is already clear that our understanding of the physiology
of acetogens is far from complete. This is particularly true
as regards their potential role in the complex carbon and
energy flow found in natural environments. In this presentation an attempt has been made to situate mixed substrate
effects on the probable inter-species interactions within
natural environments, similar reasoning has also proved to
be useful in realising the full biotechnological potential
of these versatile microorganisms (10). The ability to direct
carbon and energy flow along desired metabolic pathways at the
expense of those normally employed should enable both yields
and specific productivities of industrially important chemicals
to be significantly improved.

REFERENCES

1) J.G. Zeikus . Metabolic communication between biodegradative populations in nature.
    In : " Microbes in their Natural Environments ". J.H. Slater , R. Whittenbury
    & J.W.T. Wimpenny (eds .) , p.p. 423 - 462 . Cambridge University Press ,
    Cambridge , U.K. (1983) .

2) N.D. Lindley , P. Loubière and G. Goma . Chemostat growth of *Eubacterium limosum* on unicarbon substrate mixtures . In :" Mixed and Multiple Substrates and Feedstocks " . G. Hamer & T. Egli (eds.) . In press . (1989).

3) S. Pacaud , P. Loubière , G. Goma and N.D. Lindley . Organic acid production during methylotrophic growth of *Eubacterium limosum* : displacement towards increased butyric acid yields by supplementing with acetate . Appl. Microbiology. Biotechn. , 23 : 330 - 335 . (1986).

4) P. Loubière , G. Goma and N.D. Lindley . A non - passive mechanism of butyrate excretion operates during acidogenic fermentation of methanol by *Eubacterium limosum* . Antonie van Leeuwenhoek . In press . (1989) .

5) G. Hamer and T. Egli . Concluding remarks . In : " Mixed and Multiple Substrates and Feedstocks ". G. Hamer & T. Egli . (eds.) . In press . (1989) .

6) B.R. Sharak - Genthner and M.P. Bryant . Additional characteristics of one - carbon - compound utilization by *Eubacterium limosum* and *Acetobacterium woodii* . Appl. Environ. Microbiol. , 53 : 471 - 476 . (1987) .

7) B. Schink. Ecology of C1-utilizing anaerobes . In : " Microbial Growth on C1 - Compounds ; 5th International Symposium ". H.W. van Verseveld & J.A. Duine (eds.) , p.p. 81-88 , Martinus Nijhoff (Pub.) , Dordrecht , The Netherlands . (1987) .

8) P. Loubière , S. Pacaud , G. Goma and N.D. Lindley . The effect of formate on the acidogenic fermentation of methanol by *Eubacterium limosum* . J. Gen. Appl. Microbiol. , 33 : 463 - 470 . (1987) .

9) V. Müller. Sodium bioenergetics in methanogens and acetogens . In : " Microbial Growth on C1 - compounds : 6th International Symposium " to be published (1989) .

10) N.D. Lindley and P. Soucaille . Control of carbon flow in anaerobes by mixed substrate feeding strategies . In : " Stratégie d'Utilisation des Substrats pour la production des Métabolites Microbiens ". J.M. Le Beault & J.G. Pan . (eds.) . Société Française de Microbiologie , Paris . In press . (1989) .

ENRICHMENT OF A MESOPHILIC, SYNTROPHIC BACTERIAL CONSORTIUM

CONVERTING ACETATE TO METHANE AT HIGH AMMONIUM CONCENTRATIONS

Anna Blomgren, Anda Hansen, and Bo H. Svensson

Department of Microbiology
Swedish University of Agricultural Sciences
Box 7025, S-750 07 UPPSALA, Sweden

INTRODUCTION

Nitrogen-rich organic materials, such as swine and poultry manure, slaughterhouse and fish industry waste, etc., generally produce smaller amounts of biogas than similar substrates with lower nitrogen contents. In addition, methane formation rates are lower during fermentation of the nitrogen-rich materials (McCarty & McKinney, 1961; van Velsen, 1979; 1981; Wiegant, 1986). These differences seem to be due to the fact that ammonium, which at high concentrations inhibits methane formation (van Velsen, 1979; 1981; Wiegant, 1986), is formed during the anaerobic degradation of nitrogen-rich substrate and, as in the case of manure, is also added via the urine (cf Fig. 1). Elevated ammonium concentrations can prolong the period necessary for starting up a bioreactor. In addition, longer retention times are necessary in order to obtain a given reduction in COD in nitrogen-rich material as compared with organic matter with a lower nitrogen content. Research within this field has thus far been performed mainly in laboratory reactor systems, where effects of temperature, pH, loading rates and adaptation phenomena have been studied. Present know-ledge within this field is briefly summarized below.

In biogas systems not adapted to high ammonium levels, concentrations above 1.7 $NH_4^+-N^{-1}$ can inhibit biogas processes (Albertson, 1961; McCarthy & MacKinney, 1961; Koster & Lettinga, 1984). These studies have mostly been carried out using anaerobic domestic sewage sludge, in which NH_4^+-N levels are typically below 1 g l^{-1}, as a seeding material. Van Velsen (1979; 1981) emphasized that to ensure the rapid and safe start-up of biogas reactors in which ammonium concentrations are expected to become high, it is essential that a proper inoculum be chosen: The time needed to achieve an acceptable fermentation has been reported to increase with increasing ammonium concentrations, when anaerobic domestic sewage sludge is used as

Microbiology and Biochemistry of Strict Anaerobes Involved in Interspecies Transfer
Edited by J.-P. Bélaich *et al.*
Plenum Press, New York, 1990

the inoculum in reactors fed volatile fatty acids (C-1 - C-5) (van Velsen, 1981). However, when sludge from a swine-manure reactor already at 2.4 g NH_4^+-N l^{-1}, was used, the lag period was shorter or nonexistent, at considerably higher ammonium levels. These investigations also indicated that it was mainly the methanogenic bacteria that were inhibited, since volatile fatty acids were degraded as soon as biogas formation commenced. The observed accumulation of volatile fatty acids in reactors where ammonium concentrations increased without any prior adaptation confirmed this result (c.f. van Velsen, 1979; Wiegant, 1986; Zeeman et al., 1985). Similar results have been reported for granular sludge, where total inhibition was observed for a potato juice-fed reactor at 1.9 g NH_4^+-N l^{-1} (Koster, 1986). After adaptation of the same sludge, methane formation still occurred at 11.8 g NH_4^+-N l^{-1} although at a much lower rate (Koster & Lettinga, 1988).

Biogas reactors running under thermophilic conditions appear to be affected at lower ammonium concentrations than those working at mesophilic temperatures, and an increase in pH seems to enhance the effect of ammonium. Because free ammonia increases exponentially, both as result of a rise in temperature and an increase in pH, it is believed that ammonia rather than ammonium is the inhibitor (Wiegant, 1986; Zeeman et al., 1985; Mathiesen, 1986).

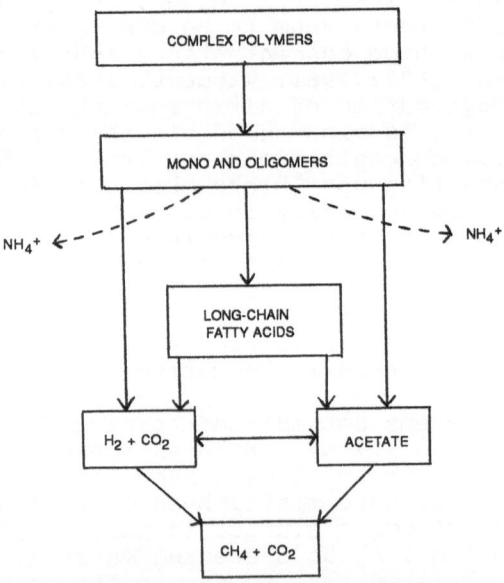

Figure 1. Anaerobic degradation of nitrogenous organic matter to biogas and ammonium.

Pure-culture studies have shown that the ammonium ion may affect methanogenic bacteria in two ways (Sprott & Patel, 1986). Firstly the methane synthesizing enzyme system may be inhibited by ammonium. Secondly the hydrophobic ammonia molecule probably diffuses passively into the cell. Inside the cell ammonia will be rapidly converted into ammonium owing to the intracellular pH conditions (Fig. 2). To avoid depletion of cytoplasmic protons the cell pumps protons into the cytoplasm via a hypothesized potassium antiporter. That this sequence of events actually occurs is supported by the results of short-term (20-80 min) experiments, i.e. an increase in ammonium concentrations occurred along with a concomitant decrease in potassium concentrations inside the cell, when ammonium chloride was added to a buffered suspension of bacteria (Sprott & Patel, 1986). This effect is more pronounced at alkaline pH's (Sprott et al., 1984) supporting the hypothesis that a passive diffusion of ammonia occurs. This phenomenon is not specific to methanogens. Both Escherichia coli and Bacillus polymyxa lose cytoplasmic potassium ions when treated with buffer containing ammonium hydroxide (Sprott et al., 1984). Other weak bases such as ethanolamine, diethanolamine and methylamine can induce a potassium efflux in Vibrio alginolyticus and E. coli, especially at alkaline pH's, (Nakamura et al., 1982).

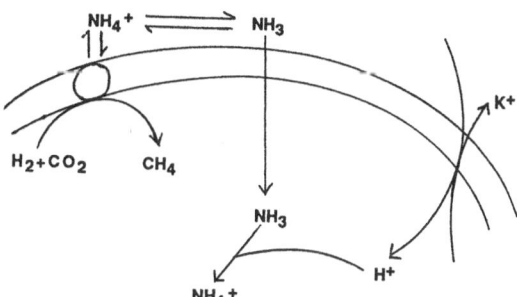

Figure 2. Proposed mechanisms by which ammonium affects methanogens (modified and redrawn from Spott and Patel, 1986).

Aceticlastic methanogens seem to be more sensitive to ammonium, at concentrations above the threshold level of 1.7 g NH_4^+-N l^{-1} mentioned above, than are the hydrogenotrophic methanogens (Koster & Lettinga, 1984). In the tests conducted by Sprott and Patel (1986) with pure cultures, referred to above, the aceticlastic methanogens (Methanosarsina barkeri and Methanotrix conclii) also seemed to be more sensitive than the hydrogenotrophs to increasing ammonium concentrations (Table 1). Methanospirillum hungatei, which needs acetate as a carbon source even though it is a hydrogenotrophic bacterium, was very sensitive to ammonium. However, Parkin and Miller (1982) have shown that in enrichment cultures adapted to elevated ammonium concentrations which contained acetate as the only substrate, the degradation of acetate to methane proceeed unabatedly up to around 8 g NH_4^+-N l^{-1}, and some degradation even occurred at concentrations as high as 17 g NH_4^+-N l^{-1}.

The purpose of our experiments was to elucidate the microbial mechanisms by which acetate is converted to methane at high ammonium concentrations in order to clear up contradictions in the literature. Enrichment cultures of methanogenic bacteria on acetate from different reactors, running at ammonium concentrations between 6-10 g NH_4^+-N l^{-1}, were started. The results from one series of these enrichments, indicating that a syntrophic fermentation of acetate to methane by a co-culture occurs under mesophilic conditions, are discussed in this article.

Table 1. Eight methanogens classified as to their sensitivity/tolerance to ammonium according to Sprott and Patel (1986).

NH_4^+-sensitive	NH_4^+-tolerant
Methanospirillum hungatei	Methanobrevibacter aboriphilus
Methanosarcina bakeri	Methanobrevcibacter smithii
Methanothrix concillii	Methanobacterium strain G2R
Methanobacterium bryantii	
Methanobacterium formicicum	

MATERIALS AND METHODS

Cultivation procedures

We used the mineral medium described by Zehnder et al. (1980), but with ferrous chloride instead of ferric chloride. In the cultures used to enrich bacteria converting acetate to methane at high ammonium concentrations, 26.7 g NH_4Cl (7 g NH_4^+-N l^{-1}) was added per liter of mineral medium.

Basically the anaerobic techniques of Hungate (1950) as modified by Balch and Wolfe (1976), were used. To enhance the effect of ammonia, oxygen-free N_2 was used as headspace, which resulted in a pH of 8.

The medium used for the dilution series under N_2 was supplemented with acetate to give an initial concentration of 50 mM, and roll tubes were prepared by including 2% agar (Difco). No extra NH_4Cl was added to the agar medium, since such an addition prevented the agar from solidifying.

To enrich and purify for hydrogen-utilizing methanogens and 0.1 g l^{-1} vancomycin (SIGMA) was added.

The medium used for further enrichment and purification of acetate-oxidizing organisms was kept under N_2 and supplemented with 1 g l^{-1} yeast extract (Oxoid). 2-Bromoethanesulfonic acid (BES; MERCK) was added to reach a final concentration of 10 g l^{-1} in order to prevent growth of methanogens.

Enrichments

To isolate acetate-utilizing methanogens, enrichments were started, both as batch cultures and as a continuous culture. Seeding material was taken from a lab-scale mesophilic sludge digestor. The digestor was fed daily with swine manure, resulting in a retention time of 19 days, and operated at 5.5-6.0 g NH_4^+-N l^{-1} and pH = 7.5.

Table 2. Culture parameters used for methanogenic enrichment on acetate at high ammononium concentrations (7 g NH_4^+-N l^{-1}).

	Continuous culture	Batch culture
Temperature	37°C	37°C
pH	8.0	8.0
Initial gas phase	N_2	N_2
Sodium acetate	5 mM	50 mM
Culture volume	500 ml	400 ml
NH_4^+-N	7 g l^{-1}	7 g l^{-1}
Flow rate	24 ml day^{-1}	–
Retention time	20 days	–

The continuous culture was started by adding 100 ml of digestor sludge to 400 ml of reduced mineral medium, supplied with 7 g NH_4^+-N l^{-1} and kept under N_2 in a 1 l serum bottle. The bottle was connected to a pump, a gas collection bag and an out-flow vial and incubated at 37°C. By continuously adding reduced mineral medium and a 0.46 M sodium acetate solution, the NH_4^+-N and the sodium acetate concentrations in the culture were maintained at 7 g l^{-1} and 5 mM respectively. The flow rate of the culture was 1 ml h^{-1}, giving a retention time of 20 days, i.e. identical to that of the reactor supplying the seeding material and pH = 8.0.

Enrichments in batch cultures were started in 500-ml serum bottles. A 50 ml portion of digestor sludge was transferred to 350 ml of reduced mineral medium. Two series of bottles were started: one with and one without an extra addition of NH_4Cl (7 g NH_4^+-N l^{-1}). Sodium acetate was added to a final concentration of 50 mM and the cultures were incubated at 37°C. In the cultures with 7 g NH_4^+-N l^{-1} the pH was originally 8.0, while in cultures without the extra addition of NH_4Cl it was initially 7.5 and thereafter adjusted to 8.0 by adding Na_2CO_3. The conditions for the two ways of enrichment are summarized in Table 2.

Analytical methods

Methane and acetate concentrations were quantified by gas chromatography (Packard Model 428) with flame ionization detection and nitrogen as carrier gas. Methane was quantified on a Porapak T column (2 m long and with a 2 mm inner diameter; i.d.) working isothermically at 80°C and with a carrier gas flow of 30 ml min^{-1}. Injector and detector temperatures were both 150°C. Gas samples (0.3 ml) were removed from the culture headspaces with a 1-ml syringe. Acetate was chromatographed on a 25-m wide bore column (CP Sil 5CB; di 5 μ; Chrompack, Holland), with an i.d. of 0.53 mm, working over a temperature gradient of 80-130°C at a rate of 5°C min. A pre-column (10 m length, i.d. 0.53 mm) was placed in front of the CP Sil column. The carrier gas flow was 8 ml min^{-1}. Samples were removed from the culture liquid with a syringe and centrifuged. The supernatant was acidified by adding formic acid to a final concentration of 2M prior to injection (0.2 μl).

Microscopy

Phase-contrast and epifluorescence microscopy were performed using a Zeiss Microscope equipped with a mercury lamp for epifluorescence and a camera.

RESULTS

Characteristic methane-forming bacteria populations had developed after about 9 months of incubation. In batch cultures without extra NH_4Cl a typical Methanosarcina, forming large packets of coccoid cell units, dominated. In both the batch and the continuous cultures at 7 g NH_4^+-N l^{-1}, three bacteria dominated: one irregular coccus and two rods differing in thickness. Both rods were able to form long chains, and the thicker one showed differentiation during growth, such as budding-like division and swelling (Fig. 3). The irregular coccus showed autofluorescence at 420 nm, indicating the presence of coenzyme F_{420}, which is characteristic for methanogens.

Repeated dilution in liquid acetate media under N_2 succeded in removing all bacteria except for the three dominant types, i.e. the irregular coccus and the two rods. Attempts to separate them by dilution in roll tubes with acetate-agar medium under N_2 repeatedly failed; although colonies grew in the tubes, growth was never obtained after transfer to the liquid media.

Dilutions in mineral medium under H_2/CO_2 promoted growth and methanogenesis of the fluorescent coccus (strain MAB1). Acetate was not used for methanogenesis by this bacterium.

When BES was added to the dilutions of the triculture in mineral medium under H_2/CO_2, to inhibit methanogenesis, growth of the thick rod (strain SAR1) was promoted. In these cultures acetate was formed.

The complete degradation of sodium acetate (50 mM) was followed in batch cultures with and without extra ammonium. Acetate was stoichiometrically converted to methane in both cases. The rate of acetate degradation (about 0.5 µmole acetate day^{-1}) corresponded to the rate of methane formation. The degradation of acetate to methane in the continuous culture was also stoichiometrical.

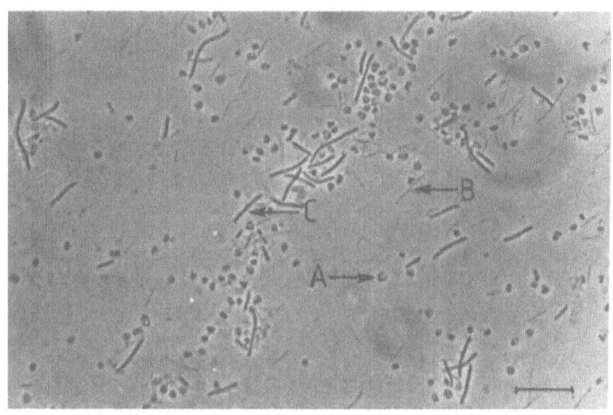

Figure 3. Phase contrast view (1000 x) of the mesophilic enrichment culture converting acetate syntrophically to
methane: A is the fluorescent strain MAB1, B the thin
rod (strain TRX1) and C the thick rod (strain SAR1)
which is probably a homoacetogen. The bar corresponds
to 10 µm.

DISCUSSION

Only two genera of mesophilic methanogens, Methanosarcina
and Methanothrix, have thus far been shown to stoichiometrically
convert acetate to CH_4 and CO_2 (Zinder, 1988). The morphology of
both these bacteria is very characteristic; therefore they are
easy to recognize using microscopy. Methanosarcina forms large
packets of coccoid cell units, while Methanotrix is a rod-shaped
bacterium normally interconnected end to end in long filaments.

In the batch culture without extra ammonium Methanosarcina predominated. In the batch culture and in the continuous culture at 7 g NH_4^+-N l^{-1} neither Methanosarcina nor Methanotrix was observed. Instead we found an irregular fluorescent coccus (Strain MAB1) and two different rods. Its morphological dissimilarities to the two genera mentioned above and its ability to form methane from H_2/CO_2 show that Strain MAB1 is a hydrogenotrophic methanogen. The thick rod (Strain SAR1) in the co-culture was not able to convert acetate to methane; instead, it utilized H_2/CO_2 for growth while excreting acetate, as is characteristic of homoacetogenetic bacteria. Such bacteria have been shown to be capable of oxidizing acetate to hydrogen and carbon dioxide under thermophilic conditions (Zinder & Koch, 1984). The presence of the hydrogenotrophic methanogen and the acetogen in our culture therefore suggests that an interspecies hydrogen transfer is involved in the acetate degradation.

Two mechanisms are known by which methane and carbon dioxide can be formed from acetate. The aceticlastic mechanism involves the conversion of the methyl group on the acetate to CH_4, with the carboxyl group giving rise to carbon dioxide. Both Methanosarcina and Methanothrix use this reaction to obtain energy for growth. The second mechanism involves two steps. In the first step acetate is oxidized to H_2 and CO_2, while in the second, CO_2 is reduced to CH_4 by H_2. The net energy gained is the same for the two reactions. In the two step conversion, however, at least two types of bacteria have to share the energy available.

The second mechanism for conversion of acetate to carbon dioxide and methane has been described for a thermophilic enrichment culture (Zinder and Koch, 1984; Lee and Zinder, 1988). In this enrichment culture methanogenesis from acetate was found to require two organisms coupled via interspecies hydrogen transfer. One member of the co-culture was a non-methanogenic bacterium oxidizing acetate to CO_2 and reducing protons to H_2. The other member was a metanogen that reduced CO_2 to CH_4 by using the H_2 formed. Without the methanogen acetate oxidation was not possible. The reason being that this reaction is thermodynamically unfavourable, unless the hydrogen partial pressure is kept low by the hydrogen-consuming organism.

Based on the results discussed above, i.e. the degradation of acetate and the formation of stoichiometrical amounts of methane combined with the fact that Strain MAB1 is a hydrogen-utilizing methanogen, unable to grow on acetate alone, we suspect that the same mechanism for methanogenesis from acetate was operative in the enrichment cultures at high ammonium concentrations. The difficulties encountered when trying to separate the three dominant organisms growing on acetate by dilution in liquid media and in roll tubes can be taken as further evidence that a two-step mechanism is involved.

Thus different populations of bacteria seem to have developed, depending on whether concentrations of ammonium were high or low. By drastically lowering the NH_4^+-concentration (to 0.3 g NH_4^+-N l^{-1}) below that of the reactor used (5.5-6 g NH_4^+-N l^{-1}) for seeding resulted in aceticlastic methane formation, as indicated by the resulting highly enriched Methanosarcina-culture. At higher concentrations, acetate is most likely converted to CH_4 by means of the two-step mechanism, which requires that a syntrophic relationship exists between an acetateoxidizing bacterium and a hydrogenotrophic methanogen. To our knowledge this is the first evidence indicating that the syntrophic degradation of acetate to CH_4 and CO_2 can occur under mesophilic conditions. Experiments now in progress utilizing ^{14}C-labelled acetate seem to confirm these results.

Thus the prolongation of the adaptation period required for the stable degradation of organic material to methane at high levels of ammonium may be a result of a shift in the acetate-utilizing bacterial population.

ACKNOWLEDGEMENTS

This work was jointly supported by the Council of Nordic Ministers and the Swedish Energy Agency (Contract No. 266 099-0 BIOGAS/SLU). The authors want to thank Ms Ann-Cristine Lundquist for an excellent job typing the manuscript and Dr. David Tilles for improving it linguistically.

REFERENCES

Albertson, O. E., 1961, Ammonia nitrogen and the anaerobic environment, J. Wat. Pollut. Contr. Fed., 33:978.
Balch, W. E. and Wolfe, R. S., 1976, New approach to the cultivation of methanogenic bacteria: 2-mercaptoethane-sulfonic acid (HS-CoM)-dependent growth of Methano-bacterium ruminantium in a pressurized atmosphere, Appl. Environ. Microbiol., 32:781.
Hungate, R. E., 1950, The anaerobic mesophilic cellolytic bacteria, Bact. Rev., 14:1.
Koster, I. W., 1986, Characteristics of the pH-influenced adaptation of methanogenic sludge to ammonia toxicity, J. Chem. Techn. Biotechnol., 36:445.
Koster, I. W. and Lettinga, G., 1984, The influence of ammonium-nitrogen on the specific activity of pelletized methano-genic sludge, Agr. Wastes, 9:205.
Koster, I. W. and Lettinga, G., 1988, Anaerobic digestion at extreme ammonia concentrations, Bio. Wastes, 25:51.
Lee, M. J. and Zinder, S.H., 1988, Isolation and characerization of a thermophilic bacterium which oxidizes acetate in syntrophic association with a methanogen and which grows acetogenically on H_2-CO_2, Appl. Environ. Microbiol., 54:124.
McCarty, P. L. and McKinney, R. E., 1961, Salt toxicity in an-aerobic digestion, J. Wat. Pollut. Contr. Fed., 33:399.
Mathisen, B., 1987, Production of biogas - influence of high ammonium concentrations, SLU, Rapp. Allm. 108(6):1 (In Swedish).
Nakamura, T., Tokuda, H. and Unemoto, T., 1982, Effects of pH and monovalent cations on potassium exit from the marine bacterium, Vibrio alganolyticus, and the manipulation of cellular cation content, Biochem. Biophys. Acta, 692:389.
Parkin, G. F. and Miller, S. W., 1982, Response of methane fermentation to continous additions of selected

industrial toxicants, in "Proceedings of the 37th Industrial Waste Conference", Purdue University, Lafayette IN, p. 729.

Sprott, G. D. and Patel, G. B., 1986, Ammonium toxicity in pure culture of methanogenic bacteria, System. Appl. Microbiol., 7:358.

Sprott, G. D., Shaw, K. M. and Jarrell, K. F., 1984, Ammonia/potassium exchange in methanogenic bacteria, J. Biol. Chem., 259:12602.

Velsen, A. F. M. van, 1979, Adaptation of methanogenic sludge to high ammonia nitrogen concentrations, Water Res., 13:995.

Velsen, A. F. M. van, 1981, "Anaerobic Digestion of Piggery Waste" (Ph. D. Thesis), Wageningen.

Wiegant, W. M., 1986, "Thermophilic Anaerobic Digestion of Waste and Wastewater Treatment" (Ph. D. Thesis), Wageningen.

Zeeman, G., Wiegant, W. M., Koster-Treffers, M. E. and Lettinga, G., 1985, The influence of total ammonia concentration on thermophilic digestion of cow manure, Agr. Wastes, 14:19.

Zehnder, A. J. B., Huser, B. A., Brock, T. D. and Wuhrman, K., 1980, Characterization of an acetate-decarbocylating, non-hydrogen oxidizing methane bacterium, Arch. Microbiol., 124:1.

Zinder, S. H., 1988, Conversion of acetic acid to methane by thermophiles, in: "Anaerobic Digestion 1988", E. R. Hall and P. N. Hobson eds, Pergamon Press, Oxford, p. 1.

Zinder, S. H. and Koch, M., 1984, Non-acetoclastic methanogenesis from acetate: acetate oxidation by a thermophilic syntrophic coculture, Arch. Microbiol., 138:263.

CONFERENCES - B : BIOCHEMISTR

ELECTRON CARRIER PROTEINS IN

DESULFOVIBRIO VULGARIS MIYAZAKI

Tatsuhiko Yagi and Mari Ogata

Department of Chemistry
Shizuoka University
836 Oya, Shizuoka 422, Japan

INTRODUCTION

Desulfovibrio vulgaris Miyazaki, a sulfate-reducing bacterium, resembles the type strain, D. vulgaris Hildenborough, in the amino acid sequences of cytochrome c_3 (Shinkai et al., 1980), cytochrome c-553 (Nakano et al., 1983; Van Rooijen et al., 1989), and rubredoxin (Shimizu et al., 1989), but differs in morphology (Kobayashi and Skyring, 1982) and in the characteristics of periplasmic hydrogenase (Yagi et al., 1978). D. vulgaris Miyazaki lives on lactate and sulfate. The overall reaction to yield energy is the oxidation of lactate with sulfate.

$$2 \ CH_3CHOHCOO^- + SO_4^{2-} \rightarrow 2 \ CH_3COO^- + 2 \ HCO_3^- + H_2S \qquad \Delta G=-160kJ$$

This reaction is composed of two reaction paths, the lactate degradation path and the sulfate reduction path.

The lactate degradation path is composed of four reactions, and results in the production of ATP by the substrate level phosphorylation (Ogata et al., 1981; Ogata and Yagi, 1986; Ogata et al., 1988).
Lactate dehydrogenase (2e-transfer)
lactate + ferricytochrome c-553 → pyruvate + ferrocytochrome c-553
Pyruvate dehydrogenase (2e-transfer)
pyruvate + CoA + ferredoxin
→ acetyl-CoA + bicarbonate + reduced ferredoxin
Phosphate acetyltransferase
acetyl-CoA + phosphate → acetyl phosphate + CoA
Acetate kinase
acetyl phosphate + ADP → acetate + ATP

The sulfate reduction path is composed of the following reactions.
Sulfate adenylyltransferase
sulfate + ATP → adenosine phosphosulfate + pyrophosphate
Inorganic pyrophosphatase
pyrophosphate + water → 2 phosphate
Adenosine phosphosulfate reductase (2e-transfer)
adenosine phosphosultate + reduced carrier
→ AMP + sulfite + oxidized carrier
Sulfite reductase (desulfoviridin) (6e-transfer)
sulfite + reduced ferredoxin → hydrogen sulfide + ferredoxin

Microbiology and Biochemistry of Strict Anaerobes Involved in Interspecies Transfer
Edited by J.-P. Bélaich *et al.*
Plenum Press, New York, 1990

237

Since 2 ATP molecules produced from 2 lactate molecules in the former path are consumed to convert sulfate to adenosine phosphosulfate in the latter path, the electron transfer from the lactate degradation path to the sulfate reduction path must be coupled to the phosphorylation of ADP to yield ATP. D. vulgaris Miyazaki also contains hydrogenase and produces hydrogen transiently during growth (Tsuji and Yagi, 1980).

In spite of a rather simple electron transfer system, D. vulgaris Miyazaki is a rich source of electron carrier proteins. So far, cytochrome c_3, cytochrome c-553, high molecular-weight cytochrome (h.m.cytochrome) (Tsuji and Yagi, 1980), two ferredoxins, and rubredoxin (Ogata et al., 1988) have been detected. In some cultural conditions flavodoxin is produced instead of ferredoxin (Ogata and Yagi, 1986).

Much effort has been made to elucidate structural, electrochemical, and biochemical characteristics of these electron carrier proteins, which are supposed to constitute an electron transfer network in D. vulgaris Miyazaki cells. However, our knowledge is still insufficient, and the synthetic pathway of heme, one of the most important prosthetic groups of electron carrier proteins, has not yet been elucidated. This paper focuses on the purification and characterization of h.m.cytochrome, some electron transfer reactions among carrier proteins, and the path of anaerobic heme synthesis in D. vulgaris Miyazaki.

MATERIALS AND METHODS

Purification of electron carrier proteins. All procedures were carried out under pure nitrogen (99.9999%). The Tris-HCl buffers used were pH 7.4 and had been deaerated by bubbling with pure nitrogen for at least 60 min before use. Cells of D. vulgaris Miyazaki (100 g wet weight) were suspended in 200 ml of deaerated 10 mM Tris-HCl containing 10 mg DNase I (Sigma), and 4 mg benzylsulfonyl fluoride was added. The mixture was subjected to sonic disintegration for 12 min and centrifuged for 60 min at 80000g. The supernatant (sonic sup, 260 ml) was mixed with 8.0 g streptomycin sulfate, left to stand overnight under nitrogen at 0°C, and centrifuged to remove the precipitate. The supernatant (streptomycin sup, 265ml) was passed through a column (22 × 400 mm) of DE32 (Whatman) which had been thoroughly equilibrated with 10 mM Tris-HCl, and washed with deaerated 10 mM Tris-HCl. The filtrate and washings from the DE32 column (DE filtrate, 280 ml) contained cytochromes. The dark brown portion at the top of the DE32 in the column (DE top) was taken out and layered on a short column of prewashed DE32, and washed successively with deaerated 50 mM Tris-HCl and 150 mM Tris-HCl, then rubredoxin and ferredoxins were eluted with 150 mM Tris-HCl containing 0.4 M KCl. The eluate was 70% saturated with ammonium sulfate under nitrogen and centrifuged to remove the precipitate. The supernatant was then made 80% saturated and passed through a short column of Sepharose CL-6B, and the adsorbed carrier proteins were eluted with deaerated Tris-NaCl (50 mM Tris-HCl containing 0.2 M NaCl). The eluate (DE top-CL eluate, 18 ml) was dialyzed thoroughly against deaerated 10 mM Tris-HCl; and rubredoxin, ferredoxin I, and ferredoxin II were separated and purified by means of the DE32 column chromatography (Ogata et al., 1988; Shimizu et al., 1989).

DE filtrate was concentrated under nitrogen, and the precipitate formed was removed by centrifugation (concd DE filtrate, 40 ml). It was then diluted 2-fold with deaerated 10 mM Tris-HCl and passed through a column of DE32 again. The filtrate and washings (2nd DE filtrate, 90 ml) contained cytochromes. Ammonium sulfate was added to 85% saturation under nitrogen, and centrifuged to separate the precipitate (am.sulf.ppt) from the supernatant (am.sulf.sup, 104 ml). Am.sulf.ppt, which contained h.m. cytochrome, was dissolved in deaerated Tris-NaCl, and chromatographed on a column (22 × 2000 mm) of Sephadex G-50 (fine) with deaerated Tris-NaCl as the elution buffer. The h.m.cytochrome eluted from the column (G50 eluate,

64 ml) was dialyzed against deaerated distilled water under nitrogen, and passed through a column of CM-cellulose (ammonium form) to adsorb h.m.cytochrome, which was eluted by the gradient of deaerated aqueous ammonia from 0 mM to 20 mM to obtain the final preparation of h.m.cytochrome.

Am.sulf.sup (hereafter, no precaution was made to exclude the atmospheric oxygen) was passed through a short column of Sepharose CL-6B to adsorb cytochrome c_3 and cytochrome c-553. The cytochromes were eluted with Tris-NaCl, chromatographed on a Sephadex G-50 (fine) column (22 × 2000 mm) to separate cytochrome c_3 and cytochrome c-553, and were purified by means of the CM-cellulose column chromatography (Gayda et al., 1987; Yagi, 1979).

Properties of high molecular-weight cytochrome. The millimolar absorbancy of cytochrome was estimated by comparing the absorbance of the native cytochrome with that of its pyridine ferrohemochrome, whose millimolar absorbancy at 550 nm is 29.1. The standard redox potential of h.m.cytochrome was calculated from the redox equilibrium between the cytochrome (3 μM on the basis of heme concentration) and FMN (0.05 mM) in deaerated 20 mM Tris-HCl. After each addition of sodium dithionite solution (40 mg in 50 ml 0.5 M Tris-HCl) to the h.m.cytochrome-FMN mixture under the stream of nitrogen, the concentration of ferro-h.m.cytochrome was estimated from the net alpha-peak height, (A553-(A568+A538)/2), and the concentration of FMN in the oxidized form was estimated from the absorbance difference between 440 and 490 nm, which is independent of the redox state of h.m.cytochrome. Enzymic reaction of h.m.cytochrome was observed spectrophotometrically in a quartz-made long-necked reaction cell (optical path: 10 mm) under a stream of nitrogen (or hydrogen in some experiments).

Molecular weight markers. The molecular weight marker proteins used were α_2-macroglobulin, reduced (Mr: 170000), rabbit muscle phosphorylase b (Mr: 97400), bovine liver glutamate dehydrogenase (Mr: 55400), porcine muscle lactate dehydrogenase (Mr: 36500), and soybean trypsin inhibitor (Mr: 20100), contained in Combithek (Boehringer Mannheim).

Assays of the enzymes involved in the synthesis of protoporphyrin. Ultracentrifugal supernatant of the bacterial sonicate (sonic sup) obtained from the cell suspension of D. vulgaris Miyazaki (5 g wet cells in 15 ml of 10 mM Tris-HCl, pH 7.4), was used in these experiments.

5-Aminolevulinate synthase: A reaction mixture containing 0.1 mmol glycine, 0.1 mmol succinate, 10 μmol $MgCl_2$, 8.45 μmol ATP, 0.37 μmol CoA, 8 μmol EDTA, 0.29 μmol pyridoxal phosphate and 0.25 ml sonic sup in 1.0 ml of 0.1 M Tris-HCl, was shaken for 30 min at 37°C, and the reaction was terminated by the addition of 0.5 ml of 10% trichloroacetic acid. The amount of 5-aminolevulinate in the deproteinized solution was determined spectrophotometrically at 552 nm with the modified Ehrlich's reagent (Urata and Granick, 1963), and corrected for the control determined at time 0 of the reaction.

Porphobilinogen synthase: A reaction mixture containing 10 μmol 5-aminolevulinate, 15 μmol 2-mercaptoethanol, 0.15 mmol KCl, and 0.9 ml sonic sup in 3.0 ml of 0.1 M Tris-HCl, was shaken for 30 min at 37°C. The reaction was terminated by the addition of 1.0 ml of 20% trichloroacetic acid containing 0.1 M $HgCl_2$. The amount of porphobilinogen in the deproteinized solution was determined spectrophotometrically with Ehrlich's reagent (Shemin, 1970), and corrected for the control taken at time 0.

Porphobilinogen deaminase: A reaction mixture containing 1.6 μmol porphobilinogen, 10 μmol EDTA, and 0.5 ml sonic sup in 3.0 ml of 0.1 M Tris-HCl, pH 8.2, was incubated for 45 min at 37°C. Two 0.05-ml aliquots of the reaction mixture were taken out, and diluted to 2 ml. One of them received 2 ml of Ehrlich's reagent, and the other, 2 ml of Ehrlich's blank reagent (no N,N-dimethylaminobenzaldehyde contained). Then the absorbance at 552 nm corrected for the blank was read to estimate the amount of porphobilinogen. A similar measurement was made at time 0, and the amount of porphobilinogen consumed during the reaction was calculated (Bogorad, 1962).

Protoporphyrinogen IX was prepared by reducing protoporphyrin IX (0.24 mM in 0.01 M KOH) by vigorous shaking with 1% sodium amalgam at 80°C for 3 min (Sano and Granick, 1961), followed by filtration to remove the insolubles. Enzymic oxidation of protoporphyrinogen IX to protoporphyrin IX will be described in the RESULTS section.

RESULTS

Characterization of High Molecular-Weight Cytochrome

Purification of high molecular-weight cytochrome. The summary of the purification of h.m.cytochrome, as well as other electron carrier proteins, is given in Table 1. One of the remarkable features of h.m.cytochrome is its instability against atmospheric oxygen. When purification was conducted under the atmospheric oxygen, the elution profile of the cytochrome from the Sephadex G-50 column was distorted as shown in Fig. 1. The purity index (A553 of the ferro form/A280 of the ferri form) of the final preparation of h.m.cytochrome was 2.4. Preparations of h.m.cytochrome with a lower purity index were obtained occasionally. Since the recovery of h.m.cytochrome from sonic sup was very low at this stage, and it is not stable even under nitrogen, these preparations were used in some experiments without being further purified.

Molecular weight. The molecular weight of h.m.cytochrome was determined to be 67000 by slab-SDS gel electrophoresis, run with marker proteins (Fig. 2). The purified h.m.cytochrome preparation contained a proteinaceous impurity (Mr: 48000). The content of the impurity in a typical preparation was determined to be 4% by densitometric analysis after being stained with Coomassie Brilliant Blue G (Fig. 2). Heme was detected only in the major protein band by peroxidase-staining technique (Thomas et al., 1976).

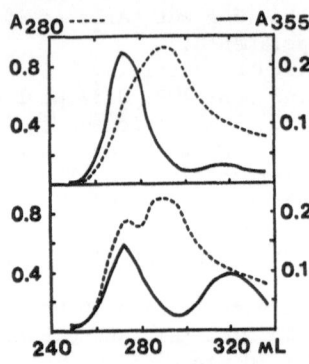

Elution volume

Fig. 1. Elution pattern of h.m.cytochrome from a column (22 × 2000 mm) of Sephadex G-50 (fine). Upper frame: under the anaerobic conditions. Lower frame: under atmospheric oxygen.

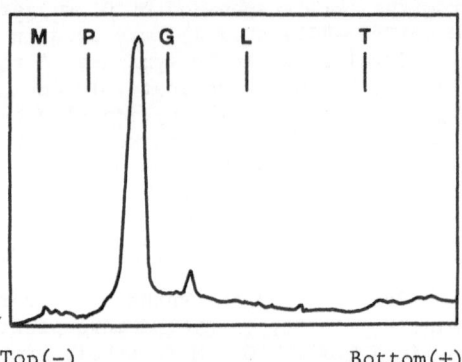

Top(−)   Bottom(+)

Fig. 2. Densitogram of h.m.cytochrome, SDS gel electrophoresed and stained. The position of the marker proteins are indicated in the figure. M: α_2-macroglobulin, reduced, P: phosphorylase b, G: glutamate dehydrogenase, L: L-lactate dehydrogenase, and T: trypsin inhibitor.

Table 1. Summary of Purification of Cytochromes, Ferredoxins, and
Rubredoxin from Desulfovibrio vulgaris Miyazaki

	vol × A260	vol × A280	vol × A400	vol × α-peak	
Sonic sup (from 100g wet cells)	36300	23400	2770	190	c₃ : 90
Streptomycin sup	31600	23100	2730	190	c-553:10
┌DE filtrate (cytochromes)	6900	7180	2050	157	h.m.c:90
Concd DE filtrate	2530	2950	1090	144	
2nd DE filtrate		1360		144	
Am.sulf.ppt (h.m.cyt)		550		38.6	
G50 eluate (h.m.cyt)		340		34.0	
H.m.cytochrome		5.2		12.3(Recov:14%)	
Am.sulf.sup (cyt c₃,cyt c-553)		200		105	
G50 eluate (cyt c₃)		21.5		67.6	
Cytochrome c₃		17.4		59.0(Recov:66%)	
G50 eluate (cyt c-553)		3.8		4.1	
Cytochrome c-553		3.1		3.8(Recov:38%)	
└DE top-CL eluate	740	430	34.9		
Ferredoxin I	21.2	19.4	9.6		
Ferredoxin II	14.5	14.7	6.0		
Rubredoxin	1.0	1.9	(vol×A490: 0.85)		

Spectra. The spectra of the ferri and ferro forms of h.m.cytochrome
are typical c-type (Table 2). As there is no distinct peak at 695 nm, a
methionyl side chain is not a ligand to the heme. The millimolar absorbancy
of ferro-h.m.cytochrome at 553.5 nm (alpha-peak) was estimated to be 26.3
per heme, or 290 per protein assuming the heme content to be 11.

Table 2. Properties of Cytochromes from D. vulgaris Miyazaki

	Cyt c₃	Cyt c-553	H.m.cyt
Mol.wt. (holoprotein)	13994	9013	67000
Mol.wt. (apoprotein)	11528	8397	
No. of AA residues	107	79	
Hemes/mol protein	4	1	11
Heme ligands	His	His + Met	His(not Met)
pI	10.5	10.5	9.1
E°' (mV)	-230,-320,	+26	Avrg: -174
	-330,-360		-80 ~ -300
CO complex	yes	no	yes
Spectral properties	nm: ε$_{mM}$	nm: ε$_{mM}$	nm: ε$_{mM}$
ferri form			
(Met-ligand)		695: 0.9	
alpha-beta	532: 38.5	526: 10.4	530 : 110
gamma	410:440	410:111	409.5:1300
delta	350: 88.0	360: 29.1	355 : 280
(protein)	280: 34	280: 20	280 : 120
ferro form			
alpha	552:110	553: 24.7	553.5: 290
beta	524: 57.2	524: 16.4	523.5: 160
gamma	419:693	418:153	419.5:1970
delta	323:126.5	317: 33.1	325 : 360

Properties of h.m.cytochrome were determined in this study.
Properties of cytochrome c₃ and cytochrome c-553 described in
several papers were collected and summarized here.

Heme content. A h.m.cytochrome sample of 27% impurity was dissolved in water and thoroughly dialyzed. An 8.20 ml portion of this solution (heme concentration: 0.0514 mM) was lyophilized to give 3.52 mg dry protein. The impurity in this preparation has been proven to have no heme by the heme-staining technique (Thomas et al., 1976). The heme content of h.m.cytochrome is thus 11 per molecule.

Standard redox potential of h.m.cytochrome. Redox equilibrium was attained between h.m.cytochrome and FMN as shown in Fig. 3. From the slope and the intersect of the line, the n-value of the Nernst equation was determined to be 0.47, and the standard redox potential, −174 mV, at pH 7.

$$E^{\circ\,\prime}_{\text{h.m.cyt}} \ (mV) = E^{\circ\,\prime}_{\text{FMN}} + \frac{59}{2}\log\frac{\text{FMN}}{\text{FMNH}_2} + \frac{59}{0.47}\log\frac{\text{ferro-h.m.cyt}}{\text{ferri-h.m.cyt}}$$
$$= -174$$

The properties of h.m.cytochrome, cytochrome c_3, and cytochrome c-553 from D. vulgaris Miyazaki are summarized in Table 2.

Redox reactions of high molecular-weight cytochrome. Hydrogenase: Each of 2 quartz-made long-necked optical cells received 0.7 μmol cytochrome c_3 or 0.3 μmol h.m.cytochrome in 3.0 ml of deaerated 20 mM Tris-HCl, and were kept under a stream of hydrogen for 1h. Then 0.2 ml of 8 μM hydrogenase solution was added and incubated at 30°C under a stream of hydrogen. Only cytochrome c_3 was reduced, in accordance with the established carrier specificity of Desulfovibrio hydrogenase (Yagi et al., 1968). When the reaction mixture containing hydrogenase-reduced cytochrome c_3 (0.3 ml) was added to the mixture containing h.m.cytochrome under a stream of hydrogen, reduction of h.m.cytochrome was observed.

Lactate dehydrogenase: The reduction rate of electron carrier proteins by partially purified D-lactate dehydrogenase (Shimizu et al., 1989) was measured spectrophotometrically and the results are illustrated in Table 3.

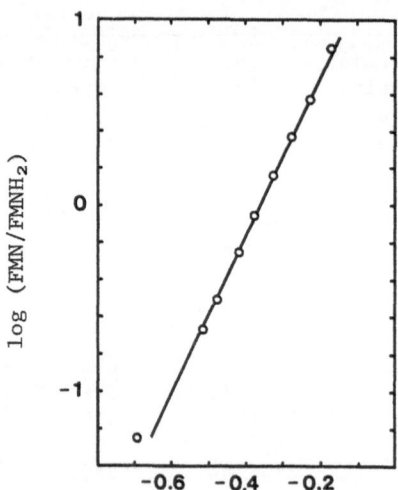

$$\log \frac{\text{ferri-h.m.cytochrome}}{\text{ferro-h.m.cytochrome}}$$

Fig. 3. Nernst plot for the redox equilibrium between h.m.cyto- chrome and FMN.

Table 3. Efficiency of Some Electron Carrier Proteins for
D-Lactate Dehydrogenase from D. vulgaris Miyazaki

Electron acceptor, tested	Concentration (μM)	Reaction rate (nmol lactate oxidized/min)
Nitrotetrazolium blue[a]	600	12
Cyt c-553	10	0.014
H.m.cyt	2	0.0055
Rubredoxin	24	0.037
Cyt c-553 (+Q)[b]	10	1.84
H.m.cyt (+Q)	2	0.30
Rubredoxin (+Q)	24	2.73
Cyt c-553 + rubredxn (+Q)	10 + 24	2.49
H.m.cyt + rubredxn (+Q)	2 + 24	3.28

[a] Standard assay conditions (Ogata et al., 1981)
[b] +Q: 2-Methyl-1,4-naphthoquinone was added to the reaction mixture to give a final concentration of 20 μM.

Anaerobic Synthesis of Protoporphyrin in D. vulgaris Miyazaki

Non-oxidative enzymes involved in the synthesis of protoporphyrin.
The activities of 5-aminolevulinate synthase, porphobilinogen synthase, and
porphobilinogen deaminase in the sonic sup of D. vulgaris Miyazaki were measured and the results are shown in Table 4.

An oxidative step in the synthesis of protoporphyrin. Anoxygenic oxidation of protoporphyrinogen IX to protoporphyrin IX in the presence of
sonic sup of D. vulgaris Miyazaki was observed by spectrophotometry: A deaerated reaction mixture (3.0 ml) containing 0.12 μmol protoporphyrinogen
IX, 3 μmol EDTA, and 0.5 ml sonic sup in 50 mM Tris-HCl, pH 7.4, was placed
in a long-necked optical cell and kept under a stream of nitrogen. The spectrum of the reaction mixture did not change when 30 μmol sulfite was added
to the cell, but changed slightly when 15 μmol sulfate and 30 μmol ATP were
added. The spectrum of protoporphyrin IX, if produced in the reaction mixture, has characteristic peaks at 540, 578, and 633 nm, but the desulfoviridin (alpha and beta peaks at 630 and 585 nm, respectively) and cytochromes in the sonic sup interfered with the spectrophotometric detection
of protoporphyrin in the reaction mixture. In some experiments, the addition of protoporphyrinogen to sonic sup resulted in the reduction of a cytochrome. The following experiments were carried out to specify the substance
which is reduced by the addition of protoporphyrinogen in the bacterial
extract of D. vulgaris Miyazaki.

Table 4. Activities of Non-Oxidative Enzymes Involved in the
Synthesis of Protoporphyrin in D. vulgaris Miyazaki

Enzyme	Reaction observed	Activity (nmol/min) per ml of sonic sup
5-Aminolevulinate synthase	5-Aminolevulinate produced	0.24
Porphobilinogen synthase	Porphobilinogen produced	4.56
Porphobilinogen deaminase	Porphobilinogen consumed	1.66

Sonic sup (6 ml) was passed through a column (18 × 97 mm) of Sephadex G-25 which had been thoroughly washed with deaerated water to remove endogenous substrates such as lactate and sulfate. The filtrate was then passed through a short column of CM-cellulose (ammonium form) to remove cytochrome c_3 (cytochrome c-553 and h.m.cytochrome were not removed; Yagi, 1979), and finally through a short column of DE32 which had been equilibrated with deaerated 10 mM Tris-HCl to remove ferredoxins and rubredoxin. The final filtrate (crude enzyme) was used to demonstrate anaerobic oxidation of protoporphyrinogen.

The reaction mixture containing 0.25 ml crude enzyme, 3 µmol EDTA, and 0.12 µmol protoporphyrinogen IX in 3.0 ml of deaerated 50 mM Tris-HCl was placed in a reaction cell under a stream of nitrogen and the spectrum was recorded. Figure 4 shows the results. Curves 0 (time 0, 2h, and 4h from the bottom in each set of curves) show that little spectral change was observed even after 4h. Curves 1 show that the spectrum characteristic of protoporphyrin appeared when sulfate and ATP were added to the reaction mixture. Curves 2 show that externally added cytochrome c-553 was reduced in the ab-

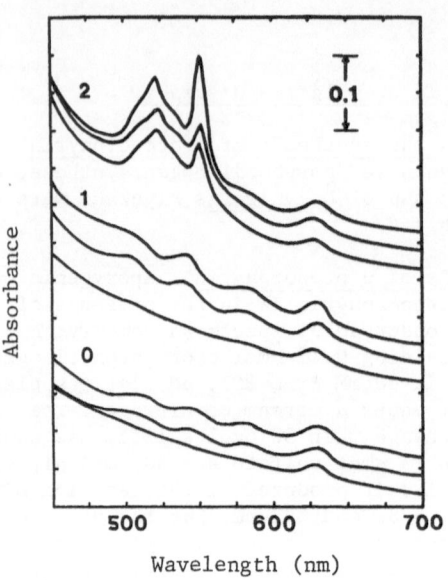

Fig. 4. Spectral change of the reaction mixture containing crude enzyme and protoporphyrinogen. Each cell contained 0.25 ml crude enzyme, 3 µmol EDTA, and 0.12 µmol protoporphyrinogen IX in 3.0 ml of deaerated 50 mM Tris-HCl. Curves 0: No addition, curves 1: sodium sulfate (15 µmol) and ATP (30 µmol) added, and curves 2: cytochrome c-553 (20 nmol) added. In each set of curves, time 0, 2h, and 4h, from the bottom.

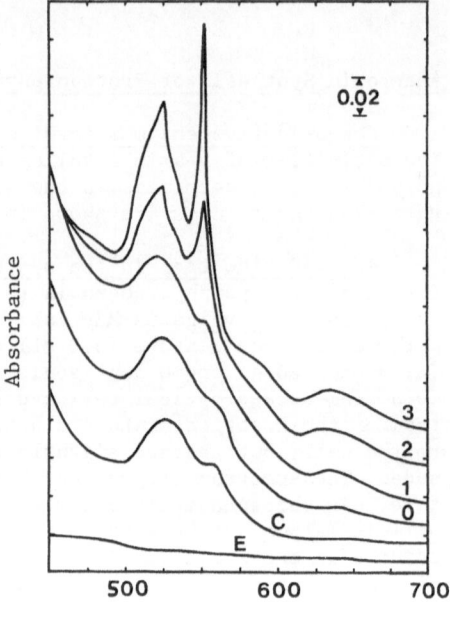

Fig. 5. Formation of protoporphyrin from protoporphyrinogen coupled to the reduction of cytochrome c-553. Curve E: Spectrum of the enzyme preparation, curve C: cytochrome c-553 was added (final concentration: 6.6 µM), curves 0, 1, 2, and 3: time 0, 1h, 2h, and 3h after the addition of protoporphyrinogen (final concentration: 43 µM).

sence of sulfate and ATP. These lines of evidence suggest that cytochrome c-553 is an electron acceptor for the anaerobic oxidation of protoporphyrinogen to protoporphyrin, but the presence of desulfoviridin in the crude enzyme was obstructive for identifying the protoporphyrin produced during the reaction.

Crude enzyme (12 ml) was, therefore, chromatographed on a column (22 × 390 mm) of Sephacryl S-200 which had been thoroughly washed with deaerated 75 mM Tris-HCl containing 1.5 mM EDTA, and the eluate was collected in 8-ml fractions. Desulfoviridin was eluted in fractions nos. 8 and 9, whereas the activity to reduce cytochrome c-553 in the presence of protoporphyrinogen was detected in fraction no. 11 (the elution position of a protein of Mr between 30000 and 40000). Figure 5 shows the results of experiments using this fraction as an enzyme preparation. The reaction cell, under a stream of nitrogen, received 2.45 ml enzyme (curve E), then 0.05 ml of 0.4 mM cytochrome c-553 was added (curve C). Protoporphyrinogen (0.5 ml, 0.26 mM) was introduced to the cell (curve 0) and kept anaerobic for 1h (curve 1), 2h (curve 2), or 3h (curve 3) in the dark. Reduction of cytochrome c-553 and increase of 633 nm peak occurred concomitantly. The concentration of protoporphyrin produced, calculated from the increase of 633-nm peak height from time 0 to 3h (Jacob and Jacob, 1976), was 2.7 μM, whereas the concentration of cytochrome c-553 reduced was 6.4 μM. In the control system without the enzyme preparation, cytochrome c-553 was reduced, but more slowly, with the addition of proto-porphyrinogen.

DISCUSSION

Several electron carrier proteins isolated from D. vulgaris Miyazaki have been studied extensively. The tertiary structures of cytochrome c_3 (Higuchi et al., 1984) and cytochrome c-553 (Nakagawa, A., Higuchi, Y., Yasuoka, N., Katsube, Y., and Yagi, T., to be published), and the primary structures of ferredoxin I, ferredoxin II (Okawara et al., 1988a,b), and rubredoxin (Shimizu et al., 1989) were established. The redox behaviors of cytochrome c_3 have been analyzed by means of spectrophotometry (Tabushi et al., 1983; Yagi, 1984), electrochemistry (Niki et al., 1979), EPR (Gayda et al., 1987; Benosman et al., 1989), and Raman spectrometry (Verma et al., 1988). Cytochrome c_3 is a natural electron carrier for hydrogenase (Yagi et al., 1968). Cytochrome c-553 is a natural electron carrier for formate dehydrogenase (Yagi, 1979) and lactate dehydrogenase (Ogata et al., 1981). Ferredoxin I and ferredoxin II work as electron carriers for pyruvate dehydrogenase, but the latter is only 40% as efficient as the former (Okawara et al., 1988b). A reconstructed reaction system containing pyruvate dehydrogenase, ferredoxin I, cytochrome c_3, and hydrogenase was found to catalyze decomposition of pyruvate to yield hydrogen in the presence of CoA (Ogata et al., 1988). This proves the electron transfer between ferredoxin I and cytochrome c_3. Ferredoxin I is also an electron mediator between sulfite reductase (desulfoviridin) and hydrogenase/cytochrome c_3 system (Ogata et al., 1988). However, little is known about the structural, electrochemical, and biochemical characteristics of h.m.cytochrome.

In this study, h.m.cytochrome was purified and characterized. It is a monomeric multihemoprotein of Mr 67000 containing 11 hemes. The methionyl side chain is not a ligand to the heme. Unlike the other cytochromes (c_3 and c-553) from D. vulgaris Miyazaki, h.m.cytochrome is unstable against atmospheric oxygen. Under atmospheric oxygen, the gel chromatographic elution pattern of h.m.cytochrome is distorted by unknown reason.

Higuchi et al. (1987) reported crystallization of a h.m.cytochrome from D. vulgaris Hildenborough. This cytochrome contains 16 hemes in a molecule of Mr 75000 (69000 by SDS gel electrophoresis and 81000 by gel filtration chromatography on a TSK G3000W column). Since the relative molecular mass per heme of our h.m.cytochrome (67000/11) is not much different from theirs (75000/16), both proteins may be homologous. Another polyhemic cyto-

chrome isolated from D. vulgaris Hildenborough (Loutfi et al., 1989) contains 8 hemes in a dimeric molecule of Mr 26000, and is called cytochrome
c_3 (Mr 26000). However, the purified protein of cytochrome c_3 (Mr 26000)
from D. vulgaris Hildenborough was reported to give a major band of Mr
70000 on SDS polyacrylamide gel electrophoresis, and its molecular weight
estimated by ultracentrifugation was 43300 (Loutfi et al., 1989). Moreover,
cytochrome c_3 (Mr 26000) and Higuchi's h.m.cytochrome from the same organism are very similar in amino acid composition. These lines of evidence
suggest that Higuchi et al. and Loutfi et al. were dealing with the same
protein, which is homologous to h.m.cytochrome of D. vulgaris Miyazaki.

The results of the redox equilibrium measurement between h.m.cyto-
chrome and FMN show that the midpoint potential of h.m.cytochrome is -174
mV, a value much more negative than the ordinary monohemic cytochromes c
with histidyl and methionyl ligands. The n-value of the Nernst equation for
h.m.cytochrome is 0.47, i.e., the apparent number of electrons transferred
in the redox reaction is less than unity. This can be explained if 11 hemes
of h.m.cytochrome have their midpoint potentials distributed between -80
and -300 mV. It is noteworthy that the potential range of h.m.cytochrome
overlaps that of cytochrome c_3 (Table 2). H.m.cytochrome can, therefore,
be regarded as "a cytochrome c_3," a polyhemic low-potential cytochrome with
histidyl ligands.

Whereas cytochrome c_3 (Mr 26000) was reported to be an electron carrier for hydrogenase from D. vulgaris Hildenborough (Loutfi et al., 1989),
no specific function has been assigned to Higuchi's h.m.cytochrome purified
without any attempt to exclude atmospheric oxygen. H.m.cytochrome from D.
vulgaris Miyazaki purified under anaerobic conditions does not accept electrons directly from hydrogen/hydrogenase, but is reduced in the presence of
cytochrome c_3. This means that h.m.cytochrome could mediate electron transfer between cytochrome c_3 and other electron transfer systems. H.m.cyto-
chrome also accepts electrons from lactate/lactate dehydrogenase, and the
reaction rate is enhanced in the presence of 2-methyl-1,4-naphthoquinone,
whose naturally occurring homolog is probably an essential component for
this enzyme.

During the growth of D. vulgaris, protoporphyrin must be synthesized
to supply heme for the synthesis of cytochromes. The synthetic pathway of
protoporphyrin in aerobes involves non oxidative steps and oxidative steps.
In this study, the presence of three non-oxidative enzymes, 5-aminolevuli-
nate synthase, porphobilinogen synthase, and porphobilinogen deaminase,
were detected in D. vulgaris Miyazaki. The rather low activity of 5-amino-
levulinate synthase may mean that this is the rate-limiting enzyme for the
synthesis of protoporphyrin, but possibility remains that hydrolytic breakdown of succinyl-CoA occurred during the assay.

Two oxidative enzymes involved in the aerobic synthesis of protoporphyrin are coproporphyrinogen oxidase and protoporphyrinogen oxidase. Both
enzymes require molecular oxygen which cannot be substituted by other oxidizing agents (Sano and Granick, 1961; Batlle et al., 1965; Poulson, 1976).
In anaerobic bacterium, these steps must be bypassed to produce heme.
In the bacterial extract prepared from anaerobically grown E. coli cells,
nitrate and fumarate work as electron acceptors for the oxidative steps of
heme synthesis (Jacob and Jacob, 1975; 1976). In the present study, formation of protopoyphyrin from protoporphyrinogen was observed in crude enzyme
preparation obtained from D. vulgaris Miyazaki when supplimented with sulfate and ATP, but not with sulfite. Since the enzyme preparation used was
deprived of cytochrome c_3, ferredoxins, and rubredoxin, the electron acceptor for protoporphyrinogen dehydrogenation must be different from these
carrier proteins. Using partially purified enzyme which was free from
desulfoviridin, the externally added cytochrome c-553 was reduced upon the
addition of protoporphyrinogen, and the formation of protoporphyrin was
observed. The ratio of the cytochrome c-553 reduced to the protoporphyrin
produced was 6.4/2.7, or 2.4, which was less than the expected ratio, 6,

for the reaction. This is probably due to overestimation of the broad 633-nm peak height. Although slow non-enzymic reduction of cytochrome c-553 occurred in the presence of protoporphyrinogen, our results strongly suggest that cytochrome c-553 is a natural electron acceptor for one of the oxidative steps of protoporphyrin synthesis in D. vulgaris Miyazaki. Sulfate and ATP stimulate the production of protoporphyrin in the crude extract from which cytochrome c_3, ferredoxins and rubredoxin had been removed; and cytochrome c-553 is assumed to be the natural electron acceptor for the dehydrogenation of protoporphyrinogen. This probably means that ferrocytochrome c-553 is donating electrons to adenosine phosphosulfate reductase in this organism.

This study proves that there is electron transfer between cytochrome c_3 and h.m.cytochrome, but the relationship between cytochrome c-553 and other cytochromes has not been elucidated. Spontaneous electron transfer between cytochrome c_3 and ferredoxin has already been reported (Ogata et al., 1988). Presumably, cytochrome c-553, as well as rubredoxin, functions in redox systems of rather positive potential, which is insulated from a very negative ferredoxin/cytochrome c_3/h.m.cytochrome system. Further study will be necessary to elucidate the entire profile of the electron transfer network in D. vulgaris Miyazaki.

Acknowledgments. This research was supported by a grant from the Ministry of Education, Science, and Culture of Japan. We are indebted to Akemi Oishi, Kazumi Hayama, and Chigusa Tasaka for skillful technical assistance.

REFERENCES

Batlle, A. M. del C., Benson, A., and Rimington, C., 1965, Purification and properties of coproporphyrinogenase, Biochem. J., 97:731.
Benosman, H., Asso, M., Bertrand, P., Yagi, T., and Gayda, J.-P., 1989, EPR study of the redox interactions in cytochrome c_3 from Desulfovibrio vulgaris Miyazaki, Eur. J. Biochem., 182:51.
Bogorad, L., 1962, Porphyrin synthesis, Methods Enzymol., 5:885.
Gayda, J. P., Yagi, T., Benosman, H., and Bertrand, P., 1987, EPR redox study of cytochrome c_3 from Desulfovibrio vulgaris Miyazaki, FEBS Lett., 217:57.
Higuchi, Y., Kusunoki, M., Matsuura, Y., Yasuoka, N., and Kakudo, M., 1984, Refined structure of cytochrome c_3 at 1.8Å resolution, J. Mol. Biol., 172:109.
Higuchi, Y., Inaka, K., Yasuoka, N., and Yagi, T., 1987, Isolation and crystallization of high molecular weight cytochrome from Desulfovibrio vulgaris Hildenborough, Biochim. Biophys. Acta, 911:341.
Jacobs, N. J., and Jacobs, J. M., 1975, Fumarate as alternate electron acceptor for the late steps of anaerobic heme synthesis in Escherichia coli, Biochem. Biophys. Res. Commun., 65:435.
Jacobs, N. J., and Jacobs, J. M., 1976, Nitrate, fumarate, and oxygen as electron acceptors for a late step in microbial heme synthesis, Biochim. Biophys. Acta, 449:1.
Kobayashi, K., and Skyring, G. W., 1982, Ultrastructural and biochemical characterization of Miyazaki strains of Desulfovibrio vulgaris, J. Gen. Appl. Microbiol., 28:45.
Loutfi, M., Guerlesquin, F., Bianco, P., Haladjian, J., and Bruschi, M., 1989, Comparative studies of polyhemic cytochromes c isolated from Desulfovibrio vulgaris (Hildenborough) and Desulfovibrio desulfuricans (Norway), Biochem. Biophys. Res. Commun., 159:670.
Nakano, K., Kikumoto, Y., and Yagi, T., 1983, Amino acid sequence of cytochrome c-553 from Desulfovibrio vulgaris Miyazaki, J. Biol. Chem., 258:12409.

Niki, K., Yagi, T., Inokuchi, H., and Kimura, K., 1979, Electrochemical behavior of cytochrome c_3 of Desulfovibrio vulgaris, strain Miyazaki, on the mercury electrode, J. Am. Chem. Soc., 101:3335.

Ogata, M., Arihara, K., and Yagi, T., 1981, D-Lactate dehydrogenase of Desulfovibrio vulgaris, J. Biochem., 89:1423.

Ogata, M., and Yagi, T., 1986, Pyruvate dehydrogenase and the path of lactate degradation in Desulfovibrio vulgaris Miyazaki F, J. Biochem., 100:311.

Ogata, M., Kondo, S., Okawara, N., and Yagi, T., 1988, Purification and characterization of ferredoxin from Desulfovibrio vulgaris Miyazaki, J. Biochem., 103:121.

Okawara, N., Ogata, M., Yagi, T., Wakabayashi, S., and Matsubara, H., 1988a, Amino acid sequence of ferredoxin I from Desulfovibrio vulgaris Miyazaki, J. Biochem., 104:196.

Okawara, N., Ogata, M., Yagi, T., Wakabayashi, S., and Matsubara, H., 1988b, Characterization and complete amino acid sequence of ferredoxin II from Desulfovibrio vulgaris Miyazaki, Biochimie, 70:1815.

Poulson, R., 1976, The enzymic conversion of protoporphyrinogen IX to protoporphyrin IX in mammalian mitochondria, J. Biol. Chem., 251:3730.

Sano, S., and Granick, S., 1961, Mitochondrial coproporphyrinogen oxidase and protoporphyrin formation, J. Biol. Chem., 236:1173.

Shemin, D., 1970, δ-Aminolevulinic acid dehydratase, Methods Enzymol., 17A:205.

Shimizu, F., Ogata, M., Yagi, T., Wakabayashi, S., and Matsubara, H., 1989, Amino acid sequence and function of rubredoxin from Desulfovibrio vulgaris Miyazaki, Biochimie, 71: in press.

Shinkai, W., Hase, T., Yagi, T., and Matsubara, H., 1980, Amino acid sequence of cytochrome c_3 from Desulfovibrio vulgaris Miyazaki, J. Biochem., 87:1747.

Tabushi, I., Nishiya, T., Yagi, T., and Inokuchi, H., 1983, Kinetic study on the successive four-step reduction of cytochrome c_3. J. Biochem., 94:1375.

Thomas, P. E., Ryan, D., and Levin, W., 1976, An improved staining procedure for the detection of the peroxidase activity of cytochrome P-450 on sodium dodecyl sulfate polyacrylamide gels, Anal. Biochem., 75:168.

Tsuji, K., and Yagi, T., 1980, Significance of hydrogen burst from growing cultures of Desulfovibrio vulgaris, Miyazaki, and the role of hydrogenase and cytochrome c_3 in energy production system, Arch. Microbiol., 125:35.

Urata, G., and Granick, S., 1963, Biosynthesis of α-aminoketones and the metabolism of aminoacetone, J. Biol. Chem., 238:811.

Van Rooijen, G. J. H., Bruschi, M., and Voordouw, G., 1989, Cloning and sequencing of the gene encoding cytochrome c_{553} from Desulfovibrio vulgaris Hildenborough, J. Bacteriol., 171:3575.

Verma, A. L., Kimura, K., Nakamura, A., Yagi, T., Inokuchi, H., and Kitagawa, T., 1988, Resonance Raman studies of hydrogenase-catalyzed reduction of cytochrome c_3 by hydrogen. Evidence for heme-heme interactions. J. Am. Chem. Soc., 110:6617.

Yagi, T., 1979, Purification and properties of cytochrome c-553, an electron acceptor for formate dehydrogenase of Desulfovibrio vulgaris, Miyazaki, Biochim. Biophys. Acta, 548:96.

Yagi, T., 1984, Spectral and kinetic abnormality during the reduction of cytochrome c_3 catalyzed by hydrogenase with hydrogen, Biochim. Biophys. Acta, 767:288.

Yagi, T., Honya, M., and Tamiya, N., 1968, Purification and properties of hydrogenases of different origins, Biochim. Biophys. Acta, 153:699.

Yagi, T., Endo, A., and Tsuji, K., 1978, Properties of hydrogenase from particulate fraction of Desulfovibrio vulgaris, in: "Hydrogenases: Their Catalytic Activity, Structure and Function," H. G. Schlegel and K. Schneider, eds., p. 107, Erich Goltze, Göttingen.

INTERACTION STUDIES BETWEEN REDOX PROTEINS, CYTOCHROME C_3, FERREDOXIN

AND HYDROGENASE FROM SULFATE REDUCING BACTERIA

A. DOLLA[*], F. GUERLESQUIN[*], M. BRUSCHI[*] and R. HASER[**]

[*]L.C.B/C.N.R.S. B.P.71 – 13277 Marseille Cedex 9-F. [**]LCCMB

CNRS, Faculté de Médecine,secteur Nord–13326Marseille Cdex15

Sulfate reducing bacteria, which are all obligate anaerobes, have in common their ability to utilize the oxidized forms of sulfur as electron acceptor for the oxidation of organic substrates. This reduction of inorganic compounds known as dissimulatory reduction of sulfates is linked to energy conservation. Lactate is the most common energy source for the genus Desulfovibrio and the reduction of two lactate produces eight electron pairs and two ATP by substrate-level phosphorylation exactly balancing the amounts needed for the reduction of one sulfate.

The genus Desulfovibrio is characterized by the requirement of sulfate as the terminal acceptor of the electron transport chain which uses either organic compounds or H_2 as its initial electron donor. The metabolism of hydrogen is regulated by reversible hydrogenase which is specifically reduced or oxidized by the low potential cytochrome c_3. The electron transport chain contains in addition to ferredoxin and cytochrome c_3, b type cytochromes, flavodoxin and rubredoxin (1).

Three different types of hydrogenases

have been characterized in Desulfovibrio. All catalyze the same reaction, are composed of two different subunits and contain non heme iron, however they exhibit different metal contents and redox center clusters. (Fe) hydrogenase, containing only iron sulfur centers (two (4Fe-4S) clusters plus an atypical (4Fe-4S) or (3Fe-4S)) has been characterized in Desulfovibrio vulgaris (2).
(NiFe) hydrogenase containing a redox active nickel, two (4Fe-4S) clusters and a three iron center has been described in D. gigas (3) and (NiFeSe) hydrogenase, containing nickel, selenium and two (4Fe-4S) clusters has been found in D. baculatus (4).

Polyhemic cytochromes c_3

Cytochrome c_3 (Mr 13000) and cytochrome c_3 (Mr 26000) are distributed in all species of Desulfovibrio examined so far and constitute a new class of polyhemic low potential cytochrome c.

Cytochrome c_3 (Mr 13000) contains four hemes for a molecular weight comparable to eucaryotic cytochrome c. The iron axial ligands are two histidinyl side chains as with b type cytochromes and the four hemes exhibit non identical negative redox potentials (5). It acts as electron donor and acceptor for hydrogenase and other redox partners : ferredoxin, flavodoxin and rubredoxin (6). The amino acid sequences of cytochrome c_3 from six different species of Desulfovibrio have been

Microbiology and Biochemistry of Strict Anaerobes Involved in Interspecies Transfer   249
Edited by J.-P. Bélaich et al.
Plenum Press, New York, 1990

determined (7 and ref. there in). They are extremely variable from one cytochrome c_3 to another and no more than 25% of the total residues are kept in the same positions. These invariant amino acids are essentially located in the cys-his clusters which bind each heme. (fig. 1).

The three dimensional structure of the cytochrome c_3 from D. desulfuricans Norway has been solved at 2.5 Å resolution (8) and compared to that of cytochrome c_3 from D. vulgaris Miyazaki (9). The two structures show the same folding of the molecule with a core of non parallel hemes presenting a relatively high exposure to the solvent with short iron to iron distances (11 to 17 Å).

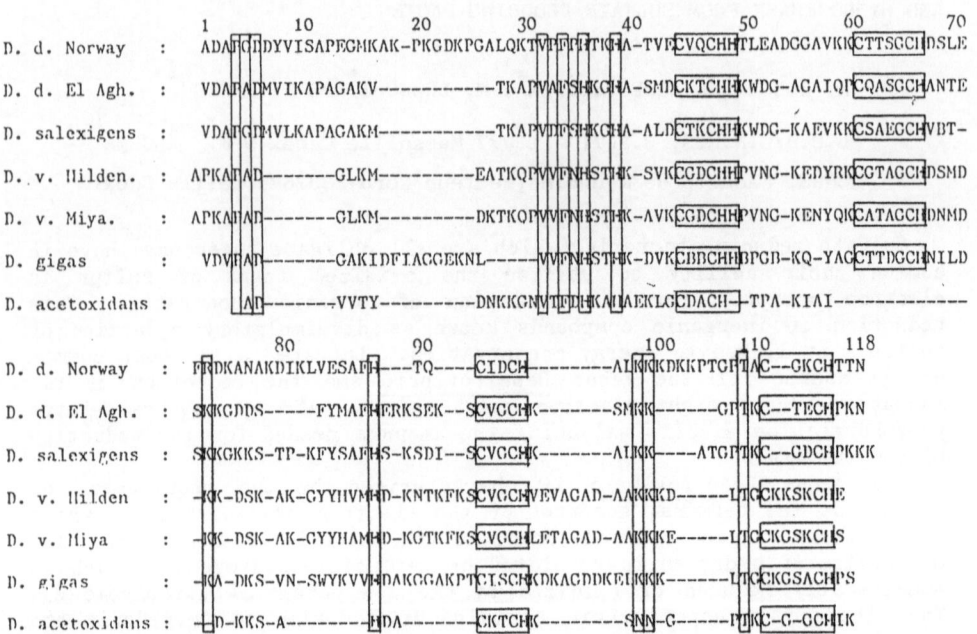

Fig. 1 Alignment of the amino acid sequences of cytochromes c_3 from D. desulfuricans Norway, D. desulfuricans El Agheila Z, D. salexigens, D. vulgaris Hildenborough, D. vulgaris Miyazaki, D. gigas and D. acetoxidans

Ferredoxins

Desulfovibrio ferredoxins are isolated as polymers of basic unit of Mr 6000. Presence of one or two (4Fe-4S) clusters and/or (3Fe-xS) cluster has been described in these proteins (10). In Desulfovibrio it has been demonstrated that the tetrahemic cytochrome c_3 is an obligate intermediate between hydrogenase and ferredoxin (11)(12). This electron transport chain can be used in the reverse direction when coupled to a sulfite reductase (12), or when Desulfovibrio grows autotrophically under a H_2 atmosphere (13). In sulfate reducing bacteria, the requirement of cytochrome c_3 for electron transfer between ferredoxin and hydrogenase (fig. 2) instead of direct coupling as observed in clostridia, reveals high specificity between these oxidoreduction partners. To improve the understanding of the electron transfer mechanism between hemes and iron sulfur clusters, we have studied the electron transfer complex between cytochrome c_3 and ferredoxin, and cytochrome c_3 and hydrogenase isolated from Desulfovibrio desulfuricans Norway strain.

The cytochrome c₃-ferredoxin I complex

In D. d. Norway cytochrome c_3, each heme exhibits a distinct redox potential (-165mV, -305mV, -365mV and -400mV) (14). Ferredoxin I from the same organism, contains one (4Fe-4S) cluster with a mid point reduction potential of -374mV (15).

Rapid kinetic experiments (16) for studying the electron exchange reaction between D. desulfuricans Norway cytochrome c_3 and ferredoxin I have shown the formation of an intermediate complex, followed by an intramolecular electron exchange within the cytochrome c_3-ferredoxin complex.

The complex formation between oxidized ferredoxin and cytochrome c_3 has been shown by NMR experiments (17). Presence of D. d. Norway ferredoxin produces ferricytochrome c_3 ^{1}H-NMR spectrum modifications. The chemical shift of perturbated heme methyl resonances has been used to determine the stoichiometry of the complex (1:1). Two of the four hemes, namely (-165mV and -305mV), are affected by the presence of ferredoxin I. The heme methyl resonances are average resonances of free and bound cytochrome c_3, indicating a fast exchange process on the NMR time scale. Thermodynamic parameters of the complex formation have been established by microcalorimetric measurements (18).

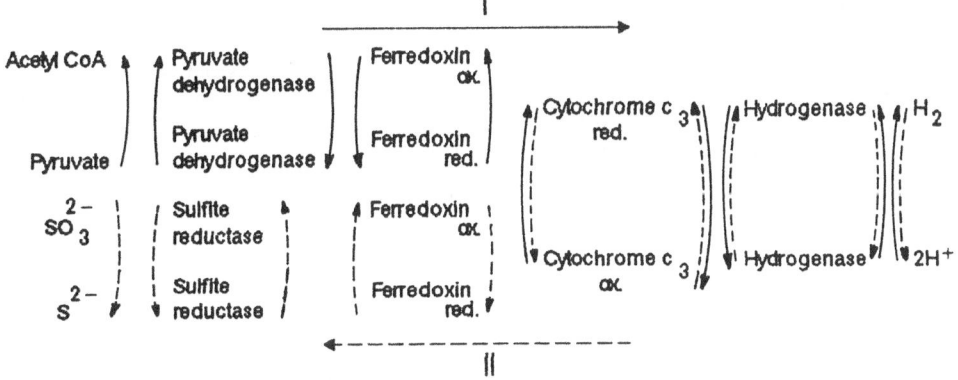

Fig. 2 Electron transport chain from Desulfovibrio.
I. The phosphoroclastic reaction.
II. The sulfite reduction.

The stoichiometry of one molecule of cytochrome c_3 per monomer of ferredoxin I was also found from these experiments. The association constant (K_A = 1.3 x 10^{6}M^{-1} at 283 K, Tris-HCl buffer 10^{-2}M and pH = 7.7) is highly dependent of the ionic strength, exhibiting an important electrostatic effect on the association, though the enthalpy (ΔH = 19 Kj.mol^{-1}) and entropy (ΔS = 183 J.K^{-1}.mol^{-1}) were positive and consistent with a hydrophobic process involved in the interaction. The use of two buffers (Tris-HCl and phosphate) during microcalorimetric experiments has shown a proton release during the complex formation. A pH-stat measurement of this proton release has indicated that one of the charged groups involved in the interacting site undergoes a pK shift during the association process from 7.35 to 6.05. Protein association processes are sometimes accompanied by proton release (19), but the identity of the proton-dissociating groups is up to now poorly described. Perutz et al. (20) reported an alkaline Bohr effect in hemoglobin, ascribed to a decrease in pK of conjugate bases from 7.7 in deoxyhemoglobin to 6.2 in oxyhemoglobin. By analogy in cytochrome c_3-ferredoxin I complex the proton release could possibly be associated to histidine 9 in ferredoxin or one of the α-NH_2 groups of cytochrome c_3 (alanine 1) or ferredoxin I (threonine 1).[2] On the other hand, Mathews (21) has reported a heme propionate group pK shift (7.3 to 6.2) between the oxidized and reduced forms of Pseudomonas aeruginosa cytochrome c_{551}. A possible role of D. d. Norway cytochrome c_3 heme propionates in the binding site may also be considered when describing this proton release.

On the basis of the thermodynamic parameters of D. d. Norway cytochrome c_3-ferredoxin I complex formation, four different models were generated with the ferredoxin iron-sulfur cluster facing each heme of cytochrome c_3 successively (22).

The models were obtained on the basis of known X-ray structure of D. d. Norway cytochrome c_3 and simulation of the ferredoxin structure derived from X-ray structure of Peptococcus aerogenes ferredoxin and ferredoxin I sequence.

One of the models was considered as the best structure of the complex in terms of charge interactions and complementarity of the topology of the contact surfaces. The distance between the heme 4 (sequential numbering) iron atom and the (Fe-S) cluster is 11.8 Å. Five ion-pair including cytochrome c_3 arginine or lysine residues and ferredoxin acidic residues, are involved in the interacting site.

In order to confirm this model, a cross-linked complex has been synthetized with the use of 1-cyclohexyl-3-(2-morpholinoethyl) carbodiimide metho-p-toluenosulfonate as bifunctional agent. The study of physiological properties (23) and EPR spectra (24) shows that the covalent complex is a valid model of the native one. To localize the cross-linking sites, the purified complex is hydrolyzed by specific proteases, and the covalently bound peptides corresponding to the interacting site of each of the two proteins have been obtained and their amino acid sequence analyzed. (Table I).

Table I. Enzymatic hydrolysis of the cytochrome c_3-ferredoxin covalent complex.

Enzyme	cross-linked peptides	Proposed cross-links	Hypothetical salt bridges
chymotrypsin	100 118 c3 Lys -- Asn 24 31 Fd Ala -- Lys and 1 23 Fd Thr -- Phe	c3 Lys100,101,103 104 and 113 Fd Asp 5 and 27 Glu 7,13,17,21, 27,29 and 30	c3 Fd 100 29 Lys -- Glu 101 5 Lys -- Asp 101 7 Lys -- Glu
S. aureus alkalin proteasis	86 118 c3 Asn -- Asn 1 7 Fd Thr -- Glu 72 85 c3 Phe -- Glu 43 54 Fd Cys -- Glu	c3 Lys 97,100,101, 103,104 and 113 Fd Asp 5 c3 Arg 73 Lys 75,79 and 82 Fd Asp 46 and 49	101 5 Lys -- Asp 73 49 Arg -- Asp

Ferredoxin electron transfer site on cytochrome c_3

Single chemical modification of D. d. Norway cytochrome c_3 arginine 73 residue gives evidence of the assignment of the highest redox potential (-165mV) to heme 4 in the structure. ^{1}H-NMR 1D spectra were used to establish the selectivity of the modification effects (25). Significant shift of heme (-165mV) and heme (-305mV) methyl lines were observed. These effects are similar to those induced by presence of ferredoxin I on the native protein, indicating that the modification is localized on the ferredoxin I interacting domain on cytochrome c_3. This result is in agreement with modelization of the cytochrome c_3 -ferredoxin complex where arginine 73 is involved in the electrostatic process of the association (22). Redox properties of the cytochrome c_3 derivative were investigated by electrochemistry (26). A significant decrease of the heme (-165mV) redox potential of 50mV was observed. Such a redox potential decrease may be interpreted by a polarity change of the heme crevice or by a hydrogen bond breaking after arginine chemical modification. Modelization of arginine modified cytochrome c_3 demonstrated that modified arginine was closer to heme 4 than to the three other hemes in the native structure (26).

The assignment of the highest redox potential to heme 4 in D. d. Norway cytochrome c_3 and the conclusion that it is the ferredoxin interacting site asks the question of the functional meaning of the three other hemes in the molecule. One of the possible role would be a specificity for various redox-partners.

Cytochrome c_3-hydrogenase interactions

Among the hydrogenase gene sequences up to now elucidated only (Fe) hydrogenase from D. vulgaris Hildenborough exhibit a 2(4Fe-4S) ferredoxin like sequence (27). From amino-acid sequence comparison with P. aerogenes ferredoxin, two (4Fe-4S) clusters might be inserted in the protein as in ferredoxins. However ten more cysteine residues are present in the sequence of the high molecular weight subunit and one cysteine has been found in the small subunit sequence.

D. gigas (Ni Fe) and D. baculatus (Ni Fe Se) hydrogenases genes sequences (28) do not contain any ferredoxin like sequence, but ten cysteine residues are conserved in all known (Ni Fe) hydrogenases sequences. Consideration of ferredoxins as a simple example of hydrogenases is now not so evident, as the (Fe-S) clusters insertion mode in hydrogenases is apparently similar to ferredoxins process but the cysteine position is different. The requirement of cytochrome c_3 for electron transfer between ferredoxin and hydrogenase has underlined the high specificity between these redox partners. With the cytochrome c_3-ferredoxin complex we have established the various thermodynamic and architectural parameters of the complex formation (18) (22).

The periplasmic (Ni Fe Se) hydrogenase from D. d. Norway consists of two subunits of Mr 59000 and Mr 29000 containing two (4Fe 4S) clusters (34). The rate constant of electron transfer in D.d.Norway hydrogenase-cytochrome c_3 complex determined by electrochemistry (3.10^7 $M^{-1} S^{-1}$) (29) and D. d. Norway cytochrome c_3 -ferredoxin complex by rapid kinetics (7.10^7 $M^{-1} S^{-1}$) (16), reveal a similar affinity of hydrogenase and ferredoxin to cytochrome c_3. However sequence comparisons between P. aerogenes ferredoxin, D. vulgaris hydrogenase ferredoxin like sequence and D. d. Norway ferredoxin I have been done (Fig. 3). The amino-acids involved in the salt bridges of the cytochrome c_3-ferredoxin complex interacting site are not found in P. aerogenes ferredoxin nor D. vulgaris hydrogenase while all Desulfovibrio ferredoxins have conserved most of them.

As the (Fe-S) cysteinyl ligands positions seem to be different in (Ni Fe) hydrogenase sequences one can expect an important structure variation in the cluster vicinity in (Ni Fe) hydrogenases and then an important change in the interacting residues positions. One goal of our studies will be to elucidate thermodynamic and structural parameters which govern the complex formation between hydrogenase and cytochrome c_3.

Intramolecular electron transfer in cytochrome c_3

In view to obtain structural informations on the architecture of the binding site of cytochrome c_3-ferredoxin and cytochrome c_3-hydrogenase complexes, covalent cross-linked complexes were prepared. The interacting site between ferredoxin and cytochrome c_3 was covalently bound by carbodiimide (23) and then the interacting heme (heme 4 at -165mV) was not anymore accessible for an other redox partner.

```
P.a    Fd      AYVIN-DSCIACGACKPECP-VNIIQG--SIYAID--A
D.d.N.FdI      TIVIHEECIGCESCVELCP-EVFAMIDGEEKAMV-TA
D.g.   FdI     PIEV-NDDCMACEACVEICP-DVFEMNEEGDKAVV-IN
D.a.   FdI     ARKFYVQQDECIACESCVEIAP-GAFAMDPEIEKAYVKDV
D.d.N.FdII     MGYSVIVQSQKCIGCGECVQVCPVEVYELQN--GKAVP-VN
D.a    FdIII   GYKITIQTQKCTGDGECVQVCPVEVYELQD--GKAVA-VN
D.v.H Hase        VQIDEAKCIGCDTCSQYCP--TAAIFGEMGEPHSIPH


P.a    Fd      -DSCIDCGS-CASVCPVGAPNPED
D.d.N.FdI      PDSTAECAQDAIQACPVEAISKE
D.g.   FdI     PDSDLDCVEEAIDSCPAEAIRS
D.a.   FdI     EGASQEEVEEAMDTCPVQSIEE
D.d.N.FdII     EEECLGC-ESCIEVCPQNAIVE
D.a    FdIII   EDECLGC-ESCVEVCEQDALTVEEN
D.v.H.Hase     IEACINCGQ-CLTHCPENAIYEA
```

Fig. 3. Sequence comparison of ferredoxins isolated from P. aerogenes (P.a.), D. desulfuricans Norway (D.d.N.), D. gigas (D.g.), D. africanus (D.a.) and ferredoxin like sequence from D. vulgaris hydrogenase (D.v.Hase).

The final reduction of ferredoxin after incubation of the covalent complex in presence of hydrogenase under H_2 atmosphere necessitates an intramolecular electron process within cytochrome c_3 (24).
Structural considerations of this intramolecular electron process have been done on the basis of D. d. Norway and D. v. Miyazaki cytochrome c_3 structures (8) (9) (37) (38). As up to now no information on the hydrogenase interacting heme on cytochrome c_3 is available, the model has been established on the reverse reaction, namely cytochrome c_3 reduction by ferredoxin. The electron enters the system propably through direct transfer between the exposed edge of heme 4 and the (4Fe-4S) cluster of ferredoxin.

Heme 4 is significantly closer to heme 3 than to heme 2 and heme 1 (Fig. 4). The inter-heme helical structure (84-101) is highly conserved in all the cytochrome c_3 known sequences. This helix provides the axial ligands His 89 and His 96 in such arrangement that may favor an electronic coupling between heme 4 and heme 3. From heme 3, the electron could propagate to heme 1, which is very close (10.9 A), an aromatic intervening group (Phe 34) favouring most probably the transfer.

The electron propagation could then proceed from heme 1 to heme 2. However the lack of heme 2 in the trihemic cytochrome c_7, leads to consider a privileged role for hemes 4,3 and 1 in the control the intramolecular electron exchange (fig. 4).

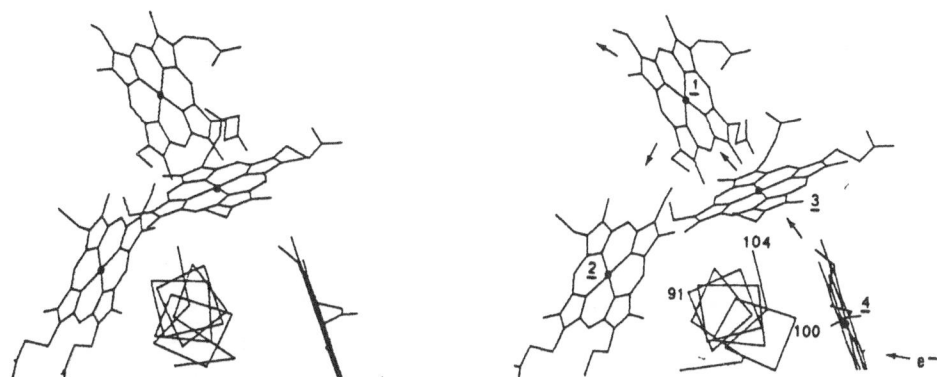

Fig. 4 . The heme cluster and the inter-heme α-helix (only α-carbons are shown) in D.d.N. cytochromes c_3-heme 4 (4) is the electron entrance gate when cytochrome c_3 is reduced by ferredoxin. The arrows point to the next hemes suggested to be involved in the intramolecular electron transfer. In the crystal structure (37) heme 2, with its two propionate groups (*) facing the external medium, serves to anchor an other cytochrome c_3 molecule in such a way that heme 1 and heme 4 come close together. This intermolecular organization might be related to the high conductivity of the protein, as established by the electrochemical studies of the hydrogenase cytochrome c_3 system (30).

It is therefore tempting to propose that heme 1 is the electron exit site when the electron enters the system via heme 4 and reversibly. Considering heme 4 as the ferredoxin interacting site we suggest that either the hydrogenase interacting site is the same that ferredoxin, namely heme 4, or the hydrogenase interacting heme would be heme 1.

Cellular localization of electron transfer proteins in Desulfovibrio

Up to now phosphoroclastic reaction and sulfite reduction (fig.2) were described as involving cytochrome c_3 (Mr 13000), ferredoxins or flavodoxins and hydrogenase. Recently, results on cellular localization of these redox proteins (31) (32) were determinant in the knowledge of physiological meaning of various species representation for one of these classes of proteins.

In Desulfovibrio desulfuricans Norway two cytochromes c_3 have been characterized : a cytochrome c_3 (Mr 13000) and a cytochrome c_3 (Mr 26000) (33). The cell localization where the various physiological redox pathways would be determinant to understand the similar physiological activities of this class of proteins. From NH_2-amino acid sequence, Le Gall and Peck (3) have shown that cytochrome c_3^2 (Mr 13000) would act as electron transfer protein in the periplasm, while cytochrome c_3 (Mr 26000) would function in the cytoplasm.

Two classes of hydrogenases have been described in Desulfovibrio desulfuricans Norway one (NiFeSe) hydrogenase has been found in the periplasmic part of the cell (34), whereas a (NiFe) hydrogenase has been extracted from Desulfovibrio desulfuricans Norway membrane (35).

The third class of hydrogenase namely, (Fe) hydrogenase has not been characterized in this bacteria up to now, but has been found in various Desulfovibrio species (2).

Ferredoxins cell localization is expected to be cytoplasmic as flavodoxins (36). However the structural homologies between the various classes of cytochrome c_3 (7) and of ferredoxins can explain the high affinity of the "in vitro" complexes described in Desulfovibrio desulfuricans Norway whatever the cell localization of the various partners. Further studies in cell localization and characterization of these redox proteins should promote progress in this area.

PUBLICATIONS

1. **J.M. Odom and H.D. Peck Jr.** Hydrogenase, electron transfer proteins and energy coupling in the sulfate reducing bacteria Desufovibrio, Ann. Rev. Microbiol. **38** : 551-592. (1984).

2. **H.M. Van der Westen, S.G. Mayhew and C. Veeger.** Separation of hydrogenase from intact cells of Desulfovibrio vulgaris. Purification and Properties, FEBS Lett. **86** : 122-126. (1978).

3. **R. Cammack, D. Patil, R. Aguire and E.C. Hatchikian.** Redox properties of the ESR - detectable nickel in hydrogenase from Desulfovibrio gigas, FEBS Lett. **142** : 289-292. (1982).

4. **M. Teixeira, G. Fauque, I. Moura, P.A. Lespinat, Y. Berlier, B. Pickril, H.D. Peck Jr., A.V. Xavier, J. Le Gall and J.J.G. Moura.** Nickel - (iron sulfur) - Selenium containing hydrogenases from Desulfovibrio baculatus (DSM) 1743, Eur. J. Biochem. **167** : 47-58. (1987).

5. **M. Bruschi, M. Loutfi, P. Bianco and J. Haladjian.** Correlation studies between structural and redox properties of cytochromes C_3, Biochem. Biophys. Res. Comm. **120** : 384-389. (1984).

6. **G.R. Bell, J.P. Lee, H.D. Peck Jr and J. Le Gall.** Reactivity of D. gigas hydrogenase toward artificial and natural electron donor or acceptors, Biochimie **60** : 315-320. (1978).

7. **M. Bruschi.** The primary structure of the tetrahaem cytochrome c_3 from Desulfovibrio desulfuricans (strain Norway 4). Description of a new class of low potential cytochrome c, Biochim. Biophys. Acta. **671** : 219-226. (1981).

8. **R. Haser, M. Pierrot, M. Frey, F. Payan, J.P. Astier, M. Bruschi and J. Le Gall.** Structure and sequence of cytochrome c_3, a multihaem cytochrome, **282** : 806-810. (1979).

9. **Y. Higuchi, M. Kusunoki, Y. Matsuura, N. Yasuoka, M. Kakudo.** Refined structure of cytochrome c_3 from D. vulgaris Miyazaki at 1.8 A resolution, J. Mol. Biol. **172** : 109-139 (1984).

10. **M. Bruschi and F. Guerlesquin.** Structure, function and evolution of bacterial ferredoxins, FEMS Microbiology Reviews. **54** : 155-176. (1988).

11. **J.M. Akagi.** Electron carriers for the phosphoroclastic reaction of Desulfovibrio desulfuricans, J. Biol. Chem. **242** : 2478-2483. (1967).

12. **B. Suh and J.M. Akagi.** Formation of thiosulfate from sulfite by Desulfovibrio vulgaris, J. Bacteriol. **99** : 210-215. (1969).

13. **J. Le Gall and J.R. Postgate.** The physiology of sulfate reducing bacteria, Adv. Microbiol. Physiol. **10** : 81-133. (1973).

14. **P. Bianco and J. Haladjian.** Current-potential responses for a tetrahemic protein : a method of determining the individual half-wave potentials of cytochrome c_3 from Desulfovibrio desulfuricans strain Norway, Electrochim. Acta. **26** : 1001-1004. (1981).

15. **F. Guerlesquin, J.J.G. Moura and R. Cammack.** Iron–sulphur cluster compostion and redox properties of two ferredoxins from Desulfovibrio desulfuricans Norway, Biochim. Biophys. Acta. **679** : 422–427. (1982).

16. **C. Capillère–Blandin, F. Guerlesquin and M. Bruschi.** Rapid kinetic studies of the electron exchange reaction between cytochrome c_3 and ferredoxin from D. desulfuricans Norway strain and their individual reactions with dithionite, Biochim. Biophys. Acta. **848** : 279–293. (1986).

17. **F. Guerlesquin, M. Noailly and M. Bruschi.** Preliminary ^{1}H NMR studies of the interaction between cytochrome c_3 and ferredoxin I from Desulfovibrio desulfuricans Norway, Biochem. Biophys. Res. Comm. **130** : 1102–1108. (1985).

18. **F. Guerlesquin, J.C. Sari and M. Bruschi.** Thermodynamic parameters of the cytochrome c_3 – ferredoxin complex formation. Biochemistry, **26** : 7438–7443. (1987).

19. **P.D. Ross and S. Subramanian.** Thermodynamics of protein association reactions : Forces contributing to stability, Biochemistry. **20** : 3096–3102. (1981).

20. **M.F. Perutz, H. Murihead, L. Mazzarella, R.A. Grawtha, J. Greer and J.V. Kilmartin.** Identification of residues responsible for the alkaline Bohr effect in haemoglobin, Nature (London). **222** : 1240–1243. (1969).

21. **S. Mathews.** The structure, function and evolution of cytochromes, Prog. Biophys. Mol. Biol. **45** : 1–56. (1985).

22. **C. Cambillau, M. Frey, J. Mosse, F. Guerlesquin and M. Bruschi.** Model of a complex between the tetraheme cytochrome c_3 and the ferredoxin I from Desulfovibrio desulfuricans Norway, Proteins. **4** : 63–70. (1988).

23. **A. Dolla and M. Bruschi.** The cytochrome c_3–ferredoxin electron transfer complex : cross linking studies, Biochim. Biophys. Acta. **932** : 26–32. (1988).

24. **A. Dolla, F. Guerlesquin, M. Bruschi, B. Guigliarelli, M. Asso, P. Bertrand and J.P. Gayda.** Cytochrome C_3–ferredoxin I covalent complex : evidence for an intramolecular electron exchange in cytochrome c_3, Biochim. Biophys. Acta. **975** : 395–398 (1989).

25. **A. Dolla, F. Guerlesquin, M. Noailly and M. Bruschi.** Chemical modification of arginine 73 of cytochrome c_3 from Desulfovibrio desulfuricans Norway, Biochem. (Life Sci. Adv.) **6** : 253–258. (1987).

26. **A. Dolla, C. Cambillau, P. Bianco, J. Haladjian and M. Bruschi.** Structural assignment of the heme potentials of cytochrome c_3, using a specifically modified argine, Biochem. Biophys. Res. Comm. **147** : 818–823.(1987).

27. **G. Voordouw and S. Brenner.** Nucleotide sequence of the gene encoding the hydrogenase from Desulfovibrio vulgaris (Hildenborough), Eur. J. Biochem. **148** : 515–520. (1985).

28. **G. Voordouw, N.K. Menon, J. Le Gall, E. Choi, H.D. Peck and A. Przybyla.** Analysis and comparison of nucleotide sequences encoding the genes for (Ni Fe) and (Ni Fe Se) hydrogenases from Desulfovibrio gigas and Desulfovibrio baculatus. Journal of Bacteriology. **171** : 2894–2899. (1989).

29. **J. Haladjian, P. Bianco, F. Guerlesquin and M. Bruschi.** Electrochemical study of electron exchange between cytochrome c_3 and hydrogenase from Desulfovibrio desulfuricans Norway, Biochim. Biophys. Res. Comm. **147** : 1289–1294. (1987).

30. **K. Kimura, A. Suzuki, H. Inokushi and T. Yagi.** Hydrogenase activity in the dry state. Isotope exchange and reversible oxidoreduction of cytochrome c_3, Biochim. Biophys. Acta. **567** : 96–105. (1979).

31. **J. Le Gall and H.D. Peck.** Amino-terminal amino acid sequences of electron transfer proteins from Gram-negative bacteria as indicators of their cellular localization : the sulfate-reducing bacteria, FEMS Microbiology Reviews **46** : 35-40. (1987).

32. **G. Voordouw, H.M. Kent and J.R. Postgate.** Identification of the genes for hydrogenase and cytochrome c_3 in Desulfovibrio, Can. J. Microbiol. **33** : 1006-1010. (1987).

33. **F. Guerlesquin, G. Bovier-Lapierre and M. Bruschi.** Purification and characterization of cytochrome c_3 (Mr 26000) isolated from Desulfovibrio desulfuricans Norway, Biochem. Biophys. Res. Comm. **105** : 530-538. (1982).

34. **R. Rieder, R. Cammack and D.O. Hall.** Purification and properties of the soluble hydrogenase from Desulfovibrio desulfuricans Norway, Eur. J. Biochim. **145** : 637-643. (1984).

35. **W.V. Lalla-Maharajh, D.O. Hall, R. Cammack and K.K. Rao.** Purification and properties of the membrane bound hydrogenase from Desulfovibrio desulfuricans, Biochem. J. **209** : 445-454. (1983).

36. **P.G. Curley and G. Voordouw.** Cloning and sequencing of the gene encoding flavodoxin from Desulfovibrio vulgaris Hildenborough, FEMS Microbiol Lett. **49** : 295-299. (1988).

37. **M. Pierrot, R. Haser, M. Frey, F. Payan and J.P. Astier.** Crystal structure and electron transfer properties of cytochrome c_3. J. Biol. Chem. **257** : 14341-14348 (1982).

38. **R. Haser and J. Mosse.** Heme cluster structures and electron transfer in multiheme cytochromes c_3. In : cytochrome systems, eds. S. Papa, B. Chance and L. Ernster, Plenum Publishing Corporation, pp. 423-430 (1987).

BIOCHEMISTRY OF THE METHYLCOENZYME M METHYLREDUCTASE SYSTEM

P. E. ROUVIERE, C. H. KUHNER and R. S. WOLFE

Microbiology Department, University of Illinois
407 S. Goodwin Urbana, IL. 61801 USA

ABSTRACT

The reduction of 2-(methylthio)ethanesulfonate (CH_3-S-CoM) is catalyzed by the methylreductase, also known as component C. The prosthetic group of component C is a Ni tetrapyrrole. In its active form, the methylreductase can reductively demethylate CH_3-S-CoM by itself using N-7-mercapto-heptanoyl-O^3-phospho-L-threonine (HS-HTP) as a reductant and produce CH_4 and CoM-S-S-HTP, the heterodisulfide of 2-mercaptoethanesulfonate (HS-CoM) and HS-HTP. However, such active preparations of component C are unstable, and "inactive" methylreductase needs other enzymatic fractions as well as ATP. The number of enzymatic fractions needed varies with the nature of the electron donor. When H_2 is used as the sole source of electrons, at least four enzymatic fractions (A1, A2, A3a and A3b) are required in addition to component C. In this complex system, HS-HTP cannot act as the sole source of electrons and still requires the presence of H_2 to reductively reactivate the methylreductase. If H_2 is used in conjunction with substrate amounts of HS-HTP, the requirement for A1 is bypassed, indicating that A1 is involved in the regeneration of HS-HTP from CoM-S-S-HTP, probably through the F_{420}-reducing hydrogenase. If titanium citrate is used as the sole source of electrons, the requirements for H_2 and for A3b, which contains the methylviologen hydrogenase, are bypassed. Our current model for the reactivation of the methylreductase proposes that A2 catalyzes an ATP-dependent allosteric modification of A3a, an FeS protein, lowering the mid-point potential of its FeS centers to values enabling it to reduce the nickel atoms in the methylreductase from Ni^{II} to Ni^{I}.

INTRODUCTION: THE METHYLREDUCTASE REACTION

In all methanogenic bacteria, the last step of methane formation is the reductive demethylation of 2-(methylthio)ethanesulfonate (methylcoenzyme M, CH_3-S-CoM). This reaction has been extensively studied in *Methanobacterium thermoautotrophicum*, but still remains poorly understood (Rouvière and Wolfe, 1988). In

Microbiology and Biochemistry of Strict Anaerobes Involved in Interspecies Transfer
Edited by J.-P. Bélaich *et al.*
Plenum Press, New York, 1990

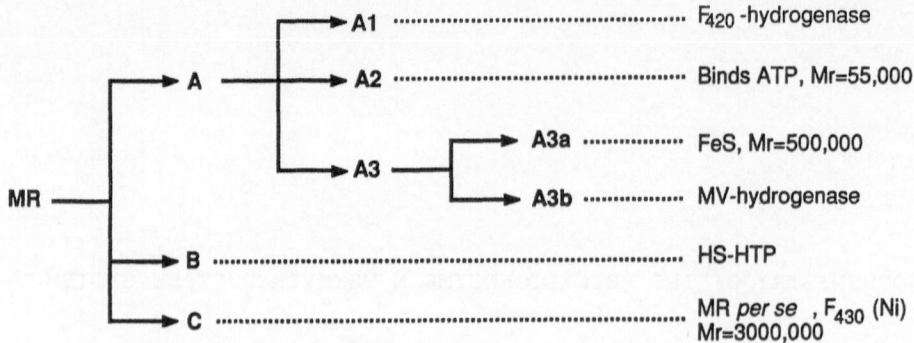

Fig. 1. Scheme for the resolution of the methylreductase system components.

the last few years two discoveries have increased our understanding of the functioning of the methylreductase system. The first one was the elucidation of the structure of component B (Fig.1), a cofactor absolutely required for the reaction, which was found to be N-7-mercapto-heptanoyl-O^3-phospho-L-threonine (HS-HTP) (Noll et al., 1986). It was shown to be the actual electron donor for the demethylation of CH_3-S-CoM when added in substrate amounts to cell free extracts under a N_2 atmosphere (Noll et al., 1987). The methylreductase reaction was then shown to proceed as follows:

$$CH_3\text{-}S\text{-}CoM + HS\text{-}HTP \longrightarrow CH_4 + CoM\text{-}S\text{-}S\text{-}HTP$$

where CoM-S-S-HTP is the heterodisulfide of 2-mercaptoethane sulfonate (coenzyme M, HS-CoM) and of HS-HTP (Bobik et al., 1987; Ellermann et al., 1988).

Another key observation was made when it was shown that in strain marburg this reaction could be catalyzed by only one protein, the methylreductase *per se,* also named component C, using cob(I)alamin (B_{12s}) as electron source (Ankel-Fuchs et al., 1986). Component C is a large protein with a molecular weight of 300,000 (Ellefson and Wolfe, 1981). It binds HS-HTP (Noll and Wolfe, 1986), two molecules of HS-CoM (Hartzell et al., 1987), and two molecules of factor F_{430}, a Ni tetrahydro-corphin (Hausinger et al., 1984). By analogy with vitamin B_{12}, factor F_{430} had long been suspected to be the site of the demethylation of CH_3-S-CoM. Electron paramagnetic resonance (EPR) studies of the Ni have shown that the activity of component C correlates with the Ni of F_{430} in its most reduced state, Ni^I (Albrecht et al., 1988). This active form of component C (C_a) is unstable.

In *Methanobacterium thermoautotrophicum* strain ΔH, although highly purified preparations of component C could demethylate CH_3-S-CoM, attempts to isolate active preparations of homogeneous component C were unsuccessful (Hartzell et al., 1988). When inactive component C (C_i) was used to produce methane from CH_3-S-CoM using H_2 as the electron donor, three other protein fractions were required, namely fractions A1, A2 and A3 (Nagle and Wolfe, 1983). In addition, catalytic amounts of ATP were required to prime the reaction. Once activated, the system could produce CH_4 in the absence of ATP. (Whitman and Wolfe, 1983). The overall reaction is shown below:

$$CH_3-S-CoM + H_2 \xrightarrow[\text{HS-HTP, ATP}]{\text{C, A1, A2, A3}} CH_4 + HS-CoM$$

Recently progress has been made in elucidation of the role of the additional A fractions. We present here a summary of the data which led us to propose our most detailed model for the functioning of the methylreductase.

FORMATION OF CH_4 WITH H_2 AS THE SOLE SOURCE OF ELECTRONS

The strategy used in our laboratory consists of fractionating cell free extract into components each of which is required to reconstitute the methylreductase activity with H_2 as the sole electron source. A major impediment to this method is that the fractions have no known functions and cannot be assayed independently. At least three functions would have to be fulfilled by the methylreductase system: (i) the catalysis of the demethylation of $CH_3-S-CoM$, (ii) the transfer of electrons to the methyl group, (iii) the utilization of ATP to activate the system.

Only the first function has been assigned unambiguously to component C as mentioned above. However it was not possible to determine which fractions included the necessary hydrogenase function since extracts of *Methanobacterium* possess two hydrogenase activities, one of which reduces the deazaflavin of methanogens, factor F_{420}, as well as methylviologen (F_{420}-hydrogenase), while a second reduces methylviologen only (MV-hydrogenase). The F_{420}-hydrogenase was contained in fraction A1, whereas the MV-hydrogenase was contained in fraction A3. Similarly, the mechanism of the activation by ATP could not be investigated since many fractions contained non specific ATPase activity. Component A2 was purified to homogeneity and was shown to bind ATP-agarose, suggesting that it played a role in the activation of the system by ATP (Rouvière and Wolfe, 1985). However it had no ATPase activity by itself and at that stage no hypothesis could be made on the functioning of the methylreductase system. Progress in the study of the A fractions was made by the utilization of alternative electron donors.

SIMPLIFICATION OF THE METHYLREDUCTASE COMPLEX

Titanium (III) citrate (Ti^{III}) was shown to replace H_2 in driving methanogenesis from CO_2 in crude cell free extract (Bobik and Wolfe, 1989). It was tested in the four component system described by Nagle and Wolfe. Ti^{III} could replace H_2 in the methylreductase system and also bypass fraction A1, indicating that it was involved in the transfer of electrons from H_2 to $CH_3-S-CoM$. At that time it had also been shown that the actual electron donor to the methyl group was HS-HTP as mentioned above (Bobik and Wolfe, 1987; Ellermann et al., 1988).

$$CH_3-S-CoM + HS-HTP \longrightarrow CH_4 + CoM-S-S-HTP$$

Active fractions of component C (C_a), with Ni in the Ni^I state (Albracht et al., 1988) were able to catalyze this reaction by themselves, if substrate amounts of HS-HTP were provided (Ellermann et al., 1988). This reaction proceeded in the absence of ATP. When substrate amounts of HS-HTP were added to the three component methylreductase system (i.e. A2, A3 and C) under a N_2 atmosphere no methane was produced. Activity was restored when H_2 was supplied in addition to HS-HTP, although H_2 by itself could not drive the reaction in the absence of component A1 (Rouvière et al., 1988). This observation showed that the methylreductase system had two distinct requirements for electrons: one to provide the electrons for the reduction of the methyl group, and the other presumably to reductively activate component C. This and other observations led us to propose the following model for the functioning of the methylreductase system (Rouvière et al., 1988):

$$C_i (Ni^{II}) \xrightarrow{\text{A2, A3, ATP, } Ti^{III} \text{ or } H_2} C_a (Ni^I)$$

$$CH_3-S-CoM + HS-HTP \xrightarrow{C_a} CH_4 + CoM-S-S-HTP$$

$$CoM-S-S-HTP \xrightarrow{Ti^{III} + B_{12} \text{ or } A1 + H_2} HS-CoM + HS-HTP$$

This model implied that in the absence of component A1 which contained the F_{420}-hydrogenase, the MV-hydrogenase in component A3 was involved in transferring electrons from H_2 for the reduction of the Ni of component C.

RESOLUTION OF COMPONENT A3 INTO TWO FRACTIONS

Using Ti^{III} as the sole source of electrons in a simplified three component system, namely A2, A3 and C, helped the study of component A3. Since components A2 and C were easily amenable to purification, one could purify component A3 on the basis of its ability to complement components A2 and C. When extracts were subjected to chromatography on Phenyl-sepharose, such a fraction was isolated. However when it was used to complement components A1, A2 and C in a H_2-driven assay no CH_4 was produced. A fifth fraction was necessary to reconstitute the methylreductase activity. This fraction was also required when the system was driven by H_2 and HS-HTP (Table 1). This indicated that fraction A3 had now been resolved into two fractions: component A3a, capable of complementing components A2 and C in a Ti^{III}-driven assay, and component A3b, required in addition to A3a when electrons for the reactivation of component C were provided by H_2 (Rouvière and Wolfe, 1989). Component A3b contained most of the MV-hydrogenase.

Component A3a was further characterized. Its molecular weight was estimated by gel filtration chromatography to be 500,000. Such a large molecular weight was confirmed by the very low migration of component A3a on 5% polyacrylamide gel electrophoresis. Component A3a is extremely oxygen sensitive and is inhibited by bathophenanthroline disulfonate, a chelator

Table 1. Required protein components of the methylreductase system according to the electron source used.

required components	electron sources
C_a	HS-HTP
C_i, A2, A3a	Ti^{III} + HS-HTP, Ti^{III} + B_{12}
C_i, A2, A3a, A3b	HS-HTP + H_2
C_i, A1, A2, A3a, A3b	H_2

C_a and C_i represent respectively the active and inactive forms of component C.

specific for Fe^{2+}, suggesting, along with its characteristic brown color, that component A3a was an FeS protein aggregate (Rouvière and Wolfe, 1989). Inhibition of our most purified component A3a, obtained by preparative polyacrylamide gel electrophoresis, indicated that it contained about one iron-sulfur center per 5,000 D which is compatible with FeS proteins such as ferredoxins. Homogeneous component A3a has not been obtained yet (Rouvière and Wolfe, 1989). Recently Reeve et al. (1989) have found a gene adjacent to those of the MV-hydrogenase which codes for a polyferredoxin with six FeS clusters. Component A3a would be a good candidate for the protein encoded by this gene.

PROPOSED ROLE FOR COMPONENT A3a

We propose that component A3a is involved in the transfer of electrons between the MV-hydrogenase in A3b and component C for its reductive activation. Electron paramagnetic studies have shown that the active form of component C had a Ni in the Ni^I state whereas in inactive component C it was in the Ni^{II} state (Albracht et al., 1988). The reduction of a Ni atom tetracoordinated between four N takes place at very low potentials. Jaun and Pfaltz (1986) reported a mid point potential of -600 mV (versus the H_2 electrode) for the couple Ni^I/Ni^{II} of the F_{430} pentamethyl-ester. Similarly in Ni tetracoordinated model compounds, mid-point potentials for the Ni^I/Ni^{II} couple have been shown to range from -1,500 mV to -600 mV (Lovecchio et al., 1974). Although it cannot be excluded, it seems unlikely that electrons from H_2 (E'^o = -420 mV) could reduce the Ni from Ni^{II} to Ni^I. We propose that ATP would be used to bridge the potential gap between H_2 and Ni^I. Two mechanisms could be envisioned (Fig. 2). In one case, ATP could induce an allosteric modification of component C, increasing the mid-point potential for the Ni of F_{430}, making it reducible by H_2 (Fig. 2B). In an alternative model, component A3a would be the site of the ATP utilization. The effect of ATP could be to lower the mid-point potential of the FeS centers on A3a below -420 mV, to the level of the Ni^I/Ni^{II} couple (Fig. 2A). We favor this later hypothesis for two reasons: (i) a similar phenomenon takes place at the level of the Fe protein of the nitrogenase system, where the binding of two ATP molecules lowers the mid-point potential of its FeS centers from −250 mV to −400 mV (Mortensen and Thornley, 1979), (ii) crude component A3 is inhibited irreversibly by 2',3'-dialdehyde of ATP, an analogue of ATP (Rouvière and Wolfe, 1987). A similar inhibitory effect of the dialdehyde of ATP has also been observed for the ATP-dependent reductive activation of the methyltetrahydro-methanopterin : HS-CoM methyltransferase (Kengen et al., 1988). In addition, it was reported that the addition of Mg-ATP plus CH₃-S-CoM induced a shift in g-values in the EPR signal of an FeS center in crude extracts of *Methanobacterium bryantii* (Rogers et al., 1988). Unfortunately, the

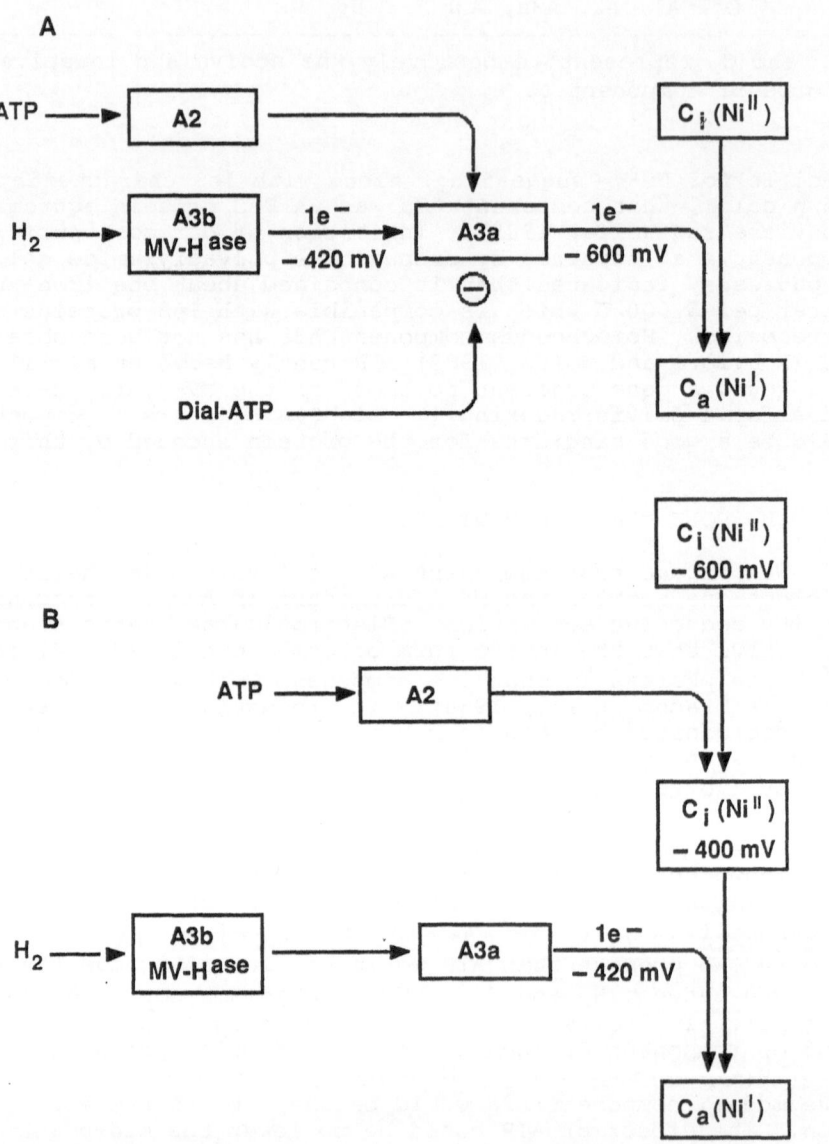

Fig.2. Two models for the involvement of ATP in the reductive reactivation of component C.

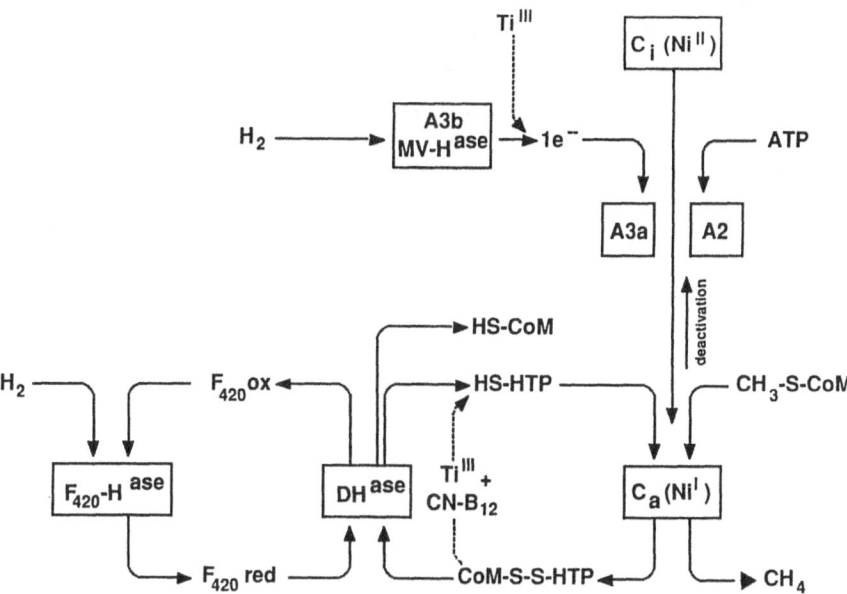

Fig.3. Model for the functioning of the methylreductase system.

effects of CH_3-S-CoM and ATP were not investigated independently. One could envision the system composed of A2, A3a and A3b to be a general system for providing electrons at extremely low potential through the utilization of ATP.

The reactivation process is probably more complex than the simple reduction of the Ni of F_{430}. It might in particular include the reorganization of the ligands of the F_{430}. Albracht et al. (1988) showed that there are at least two different EPR Ni^I signals for F_{430} in intact cells, one of them indicating a strong axial ligand, the other having none. Albracht et al. suggest that it might be due to a bond between Ni^I and HS-HTP. When the incorporation of HS-CoM into component C was studied, it was shown that it could only be incorporated as CH_3-S-CoM. The incorporation preceded the beginning of the production of CH_4 (reactivation), occurred about every 150 turnovers and required the presence of the A fractions as well as ATP (Hartzell et al., 1987). These observations too might reflect another aspect of the reactivation of component C.

ROLE OF COMPONENT A1: REGENERATION OF HS-COM AND HS-HTP

The involvement of component A1 in the reduction of CoM-S-S-HTP by H_2 was shown by using alternative electron donors. However its nature has not been elucidated yet. Several observations suggest that component A1 includes the F_{420}-hydrogenase: (i) the F_{420}-hydrogenase was always found to comigrate with component A1 (Nagle and Wolfe, 1983), (ii) with H_2 as the sole source of electrons, F_{420} was found to be stimulatory (Nagle and Wolfe, 1983; Whitman and Wolfe, 1983; Rouvière and Wolfe, 1989),

(iii) FAD, the prosthetic group of the F_{420}-hydrogenase, was found to be absolutely required in some highly resolved systems (Nagle and Wolfe, 1983), (iv) anti-F_{420}-hydrogenase antibodies inhibited the production of methane (Fox et al., 1987). The existence of a specific F_{420}-dependent : CoM-S-S-HTP dehydrogenase (DHase) would therefore be necessary. Recently an H_2-linked CoM-S-S-HTP dehydrogenase activity was detected in cell extracts. Neither NADH nor NADPH could provide the electrons for the reaction and unfortunately reduced F_{420} was not tested (Hedderich and Thauer, 1988). The involvement of the F_{420}-hydrogenase in the reduction of CoM-S-S-HTP would also be attractive since it is a membrane associated enzyme which could explain the link between the demethylation of CH_3-S-CoM and the production of ATP by the cell (Baron et al., 1987).

MODEL FOR THE FUNCTIONING OF THE METHYLREDUCTASE COMPLEX

In summary we present the following model (Fig. 3): first, inactive component C (C_i) must be reductively reactivated. Electrons from H_2 are transferred by the MV-hydrogenase (MV-Hase) in component A3b to the Ni of component C_i via component A3a, a large FeS aggregate. We propose that this reaction is made possible by the utilization of ATP, catalyzed by component A2, the ATP-binding protein, to modify the midpoint potential of the FeS center of A3a, allowing it to reduce the Ni^{II} of component C at a very low potential. Once activated, component C with a Ni^I, can demethylate CH_3-S-CoM by itself with HS-HTP as the electron donor. CoM-S-S-HTP, the product of the methylreductase reaction with CH_4 is then reduced with H_2 by component A1, which most likely includes the F_{420}-hydrogenase (F_{420}-Hase).

This model takes into account all of the observations made so far on the methylreductase system. It may serve as a reference for future investigations into the biochemistry of the methylreductase system.

ACKNOWLEDGEMENTS

This work was supported by grant DMB 8613679 from the National Science Foundation and Grant DE-FG02-86ER13651 from the Department of Energy.

REFERENCES

Albracht S. P. J., Ankel-Fuchs D., van der Zwaan J. W., Fontijn R. D. and Thauer R. K. (1988) Biochim. Biophys. Acta. 955, 86-102.

Ankel-Fuchs D., Hüster R., Mörschel E., Albracht S. P. J., and Thauer R.K. (1986) System. Appl. Microbiol. 7, 383-387.

Baron S. F., Brown D. P.and Ferry J. G. (1987) J. Bacteriol. 169, 3823-3825.

Bobik T. A., Olson K. D., Noll K. M. and Wolfe R. S. (1987) Biochem. Biophys. Res. Commun. 149, 455-460.

Bobik T. A. and Wolfe R. S. (1989) J. Bacteriol. 171, 1423-1427.

Ellefson W.L. and Wolfe R. S. (1981) J. Biol. Chem. 256, 4259-4262.

Ellermann J. R., Hedderich R., Böcher R. and Thauer R. K. (1988) Eur. J. Biochem. 172, 669-677.

Fox J. A., Livingston D. J., Orme-Johnson W. H. and Walsh C. T. (1987) Biochemistry 26, 4219-4227.

Hartzell P. L., Donnelly M. I. and Wolfe R. S. (1987) J. Biol. Chem. 262, 5581-5586.

Hartzell P. L., Escalante-Semerena J. C., Bobik T. A. and Wolfe R. S. (1988) J. Bacteriol. 170, 2711-2715.

Hausinger R. P., Orme-Johnson W. H.and Walsh C.T. (1984) Biochemistry 23, 801-804.

Hedderich R. and Thauer R. K. (1988) FEBS Lett. 234, 223-227.

Jaun B. and Pfaltz A. (1986) J. Chem. Soc. Chem. Commun. 17, 1327-1329.

Kengen S. W. F., Mosterd J. J., Nelissen R. L. H., Keltjens J. T., van der Drift C. and Vogels G. D. (1988) Arch. Microbiol. 150, 405-412.

Lovecchio, F. V., Gore E. S. and Bush D. S. (1974) J. Amer. Chem. Soc. 96, 3109-3118.

Mortenson L.E. and Thornley R. N. F. (1979) Ann. Rev. Biochem. 48, 387-418.

Nagle D. P. Jr. and Wolfe R. S. (1983) Proc. Natl. Acad. Sci. USA 80, 2151-2155.

Noll K. M., Rinehart K. L. Jr., Tanner R. S. and Wolfe R. S. (1986) Proc. Natl. Acad. Sci. U.S.A. 83, 4238-4242.

Noll K. M.and Wolfe R. S. (1986) Biochem. Biophys. Res. Commun. 139, 889-895.

Noll K. M., Donnelly M. I. and Wolfe R. S. (1987) J. Biol. Chem. 262, 513-515.

Reeve J.N., Beckler G.S., Cram D.S., Hamilton P. T., Brown J. W., Krzycki J. A., Kolodziej A. F., Alex L., Orme-Johnson W. H. and Walsh C. T. (1989) Proc. Natl. Acad. Sci. U.S.A. 86, 3031-3035.

Rogers K. R., Gillies K. and Lancaster J. R. (1988) Biochem. Biophys. Res. Commun. 153, 87-95.

Rouvière P. E., Escalante-Semerena J. C. and Wolfe R. S. (1985) J. Bacteriol. 162, 61-66.

Rouvière P. E. and Wolfe R. S. (1987) J. Bacteriol. 169, 1737-1739.

Rouvière P. E., Bobik T. A., and Wolfe R. S. (1987) J. Bacteriol. 170, 3946-3952.

Rouvière P. E. and Wolfe R. S. (1988) J. Biol. Chem. 263, 7913-7916.

Rouvière P. E. and Wolfe R. S. (1989) J. Bacteriol. 171, 4556-4562.

Whitman W. B. and Wolfe R. S. (1983) J. Bacteriol. 154, 640-649.

ENERGETICS OF THE GROWTH OF A NEW
SYNTROPHIC BENZOATE DEGRADING BACTERIUM

J.P. Bélaich[1], P. Heitz[1], M. Rousset[1] and J.L. Garcia[2]

[1] Laboratoire de Chimie Bactérienne CNRS , BP 71
F - 13277 Marseille Cédex 9 , France
[2] Laboratoire de Microbiologie ORSTOM , Case87
Université de Provence , 3 , Place V. Hugo
F - 13331 Marseille Cédex 3

INTRODUCTION

Benzoate can be anaerobically dissimilated through four distinct pathways (Evans and Fuchs 1988) including the photometabolism of *Rhodopseudomonas palustris* (Dutton and Evans 1969), nitrate respiration by *Pseudomonas sp.* (Taylor *et al.* 1970) or *Moraxella sp.* (Williams and Evans 1975), sulfate respiration by *Desulfococcus multivorans* , *Desulfosarcina variabilis* , *Desulfonema magnum* (Widdel 1980 , 1987), *Desulfotomaculum sapomandens* (Cord-Ruwisch and Garcia 1985), and *Desulfobacterium indolicum* (Bak and Widdel 1986), and syntrophic associations of hydrogen consumers with obligate hydrogen producing acetogenic (OHPA) bacteria (Mountfort and Bryant 1982) . The latter pathway was first studied with methanogenic consortia (Fina and Fiskin 1960 ; Ferry and Wolfe 1976 ; Keith *et al.* 1978) before new strains of syntrophic bacteria were isolated . Mountfort and Bryant (1982) isolated the first bacterium of this group , in syntrophic association with a H_2 - consuming bacterium . This was further characterized by Mountfort *et al.* (1984) as *Syntrophus buswellii* . Similar bacteria , possessing the characteristic undulating outer membrane and able to ferment benzoate , have been obtained by Shelton and Tiedje (1984) .

Using enrichment cultures on phenyl acetate or phenol , Barik *et al.* (1985) have succeeded in isolating two original benzoate degrading strains in coculture with *Wolinella succinogenes* . One of them (strain PA1) was successfully grown in pure culture on succinate . Finally Tschech and Schink (1986) isolated two new strains which are able to degrade monohydroxybenzoates or benzoate in syntrophic association with either a sulfate - reducing bacterium or a methanogen .

Various anaerobic bacteria , including *Streptococcus bovis* and *Coprococcus sp.* (Tsai and Jones 1975), *Pelobacter acidigallici* (Schink and Pfennig 1982), *Enterobacter cloacae* (Grbic - Galic and Pat - Polasko 1985), and *Eubacterium oxidoreducens* (Krumholz and Bryant 1986), have been found to ferment trihydroxy - monobenzenoid compounds in pure culture without requiring exogenous electron acceptors such as sulfate or nitrate . However none of these bacteria were found to ferment benzoate , so it was suggested that the ring clivage mechanisms in trihydroxy - , certain dihydroxy - or methoxy - substituted benzene and benzoate were different (Sleat and

Microbiology and Biochemistry of Strict Anaerobes Involved in Interspecies Transfer
Edited by J.-P. Bélaich *et al.*
Plenum Press, New York, 1990

269

Robinson 1984) . Mud from a polluted river was used for enrichments to isolate other species of anaerobic bacteria that might catabolize benzoate .This paper describes the isolation procedure and some features of a syntrophic benzoate degrader (S) grown in coculture with H_2 - utilizing sulfate reducer or methanogen.

MATERIALS AND METHODS

Source of organisms : *Desulfovibrio fructosovorans* (DSM 3604 , Ollivier *et al.* 1988) was obtained from our culture collection . *Methanospirillum hungatei* was isolated from the defined syntrophic association with *Syntrophus buswellii* (DSM 2612 TB). These H_2 - utilizing bacteria are unable to dissimilate benzoate or fatty acids other than formate . Both of the cocultures studied here have been deposited with the DSM (Deutsche Sammlung von Mikroorganismen und Zellkulturen GmbH), as strain 4156 A in the case of SF and 4156 B in the case of SH .

Media and culture techniques : the anaerobic Hungate technique (Hungate 1969) modified as to the use of syringes (Macy *et al.* 1972) was used throughout this study . The anaerobic bicarbonated pH 7 buffered sulfide - reduced medium contained benzoate (7 mM) and vitamins as sole organic substances . The composition of this medium has been described previously (Widdel and Pfennig 1984) . The medium was adjusted to pH 7.0 with 10 M KOH and boiled for 10 min under oxygen - free nitrogen .

After cooling , the medium was placed in an anaerobic glove box (La Calhène , Bezons , France) and dispensed in 20 - ml portions into 60 ml serum bottles (Wheaton Scientific Co , Milleville ,NJ , USA) . The bottles were stoppered with butyl rubber closures (Bellco Glass , Inc., Vineland , NJ , USA), flushed with a $N_2 : CO_2$ gas mixture (4 : 1), and sealed with aluminium crimp seals . The serum vials were autoclaved at 110°C for 30 min . A 0.4 ml portion of 1% (w/v) sterile Na2S.9H2O was added to each vessel prior to inoculation . For roll tube media , 20 g /l of agar was added after the medium was brought to a boil . The medium was dispensed in 4.5 ml portions into serum tubes using the anaerobic glove box . The tubes were then prepared in the same way as the serum bottles .

Enrichments were prepared using 30% inocula from mud of the river Huveaune in Marseilles . Ten percent transfers of the active methane producing enrichments were made at about 2-3 week intervals . The hydrogen utilizing bacteria , i.e. *D. fructosovorans* and *M. hungatei* , were maintained in serum bottles in basal medium without benzoate , but with fructose and formate , respectively , as carbon sources. The medium for the sulfate - reducing bacteria also contained 20 mM Na_2SO_4 . The cultures of the methanogen were grown under N_2 .

Cocultures of the benzoate degrading consortia were maintained by transfer (20% v/v) at 3 - week intervals . All incubations were at 30°C , with bottles held in a vertical position except for liquid cultures under $H_2 : CO_2$; these were incubated in a slanted position on a reciprocal shaker. Culture purity was checked by examining wet mounts and by inoculating the basal medium with 1% glucose , 1% yeast extract and 1% Biotrypcase (Biomérieux , Craponne , France).

Continuous culture experiments were carried out using a 5 liter chemostat (Spapex , La Seyne , France) equipped with an electronic level control (CNRS patent) ; both instruments were specially designed for use under strictly anaerobic conditions . The volume of the culture was 4 l ; the culture was stirred by means of a mechanical rotating blade stirrer at 200 rpm . The pH was maintained at 7.0 by the buffer , and the vessel was continuously flushed with a mixture of $N_2 : CO_2$ (80 / 20) at 2 l / h .

Analytical methods : growth was determined by optical density measurements at 530 nm , by cell dry weight determination or by direct microscopic counts using a counting chamber . The growth yields (Yb) were determined by measuring the quantities of cell mass (g dry weight) per mol of degraded benzoate .

Under optimal growth conditions it was observed that 79 mg dry weight per l of culture of SF cells corresponded to a concentration of 1.8×10^{11} cells /l . As the benzoate degrader and *D. fructosovorans* cannot be distinguished by light microscopy , it was assumed that the weight of one cell was equal to $79/1.8 = 43.9 \times 10^{11}$ mg .

This value was used to calculate the molar growth yields from bacterial enumerations . In some cases, specific *D. fructosovorans* enumerations were carried out . For this purpose an aliquot of the culture to be counted was inoculated into a gelose fructose sulfate medium and the colonies were counted after a 10 - day incubation . Methane and acetate were determined by gas chromatography and benzoate was analysed by high performance liquid chromatography (HPLC) using an Amin ion - exclusion HPX-87H Biorad column . G+C percents were determined using the equilibrium centrifugation technique . The DNA (5mg) and the cesium chloride solution were pipetted into the Beckman analytical centrifuge cell with a volume of 400 ml . Centrifugation was performed at 44,000 rpm , at 20°C for 48 h . Estimation of the G+C % values were done using the value of 50.0 % for the *E. coli* K12 DNA as standard .

Electron microscopy : 10 ml of culture were centifuged at 5,000 x g for 20 min and resuspended in 0.5 ml of basal medium . A drop of the culture was then placed on a Formvar-coated , carbon reinforced grid (200 mesh), and excess fluid drawn off with filter paper . Cells were stained with 1% (w/v) phosphotungstic acid (pH 7) for 5 sec . Preparations were examined using a TEM Hitachi H 600 electron microscope .

RESULTS

Enrichment of the benzoate degrading mixed culture Huv : after four transfers of the first enrichment culture onto benzoate medium a stable bacterial population (Huv) was found to exist . Benzoate was fermented in acetate and methane . At least four morphologically dominant species seem to have been present (**Fig.1**).

When observed under epifluorescence microscopy at 420 nm , the hydrogen - consuming methanogens resembled *M. hungatei* ; the acetate - utilizing methanogens had the appearance of a large filamentous rods of the *Methanothrix* type . The predominantly non - methanogenic bacteria were very motile coccoids ; they probably consisted of syntrophic benzoate degrading bacteria . The fourth strain was a spore forming short rod ,

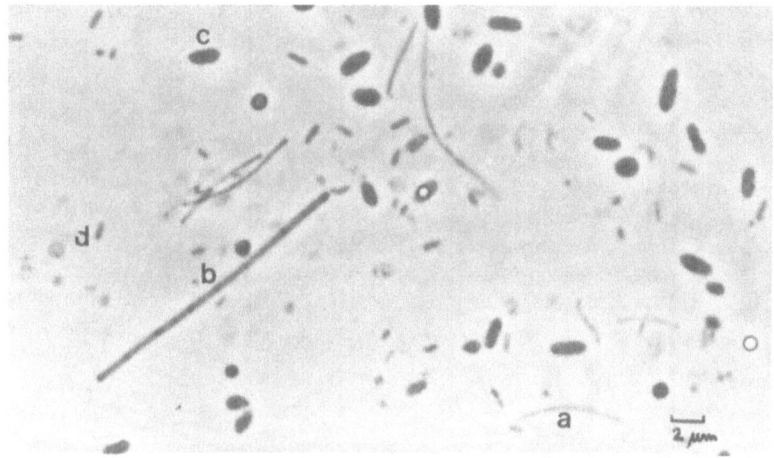

Fig.1. Light micrograph of stable enrichment culture Huv.
a = *Methanospirillum* - like methanogen ;
b = *Methanothrix* - like methanogen ;
c = sporulated presumed sulfate reducer ;
d = presumed syntrophic benzoate degrader .

probably a sulfate reducer , as indicated by its outgrowth into an anaerobic sulfate - lactate medium which resulted in H_2S production ; it was assumed that it might play an homoacetogenic role in this mixed culture on $H_2 : CO_2$ resulting from benzoate oxidation by the syntroph .

The kinetics of benzoate fermentation by the Huv mixed culture are shown in **Fig. 2** . An intermediary acetate accumulation can be seen to have occured , which had completely cleaved into methane and CO_2 by the end of the fermentation .

Isolation of a syntrophic benzoate degrader : it was attempted to isolate the benzoate degrader in coculture with *D. fructosovorans* by diluting the Huv culture into a suspension of the sulfate reducer and on a benzoate - sulfate agar medium in roll tubes . Few colonies had developed by the last dilution step after three months of incubation . Four of these colonies were transferred into a benzoate - sulfate liquid medium and the isolation procedure was repeated to ascertain the purity of the syntrophic culture .The latter operation gave isolates which were physiologically identical , named strain SF, containing exclusively a benzoate degrader (S) associated with *D. fructosovorans* (F).

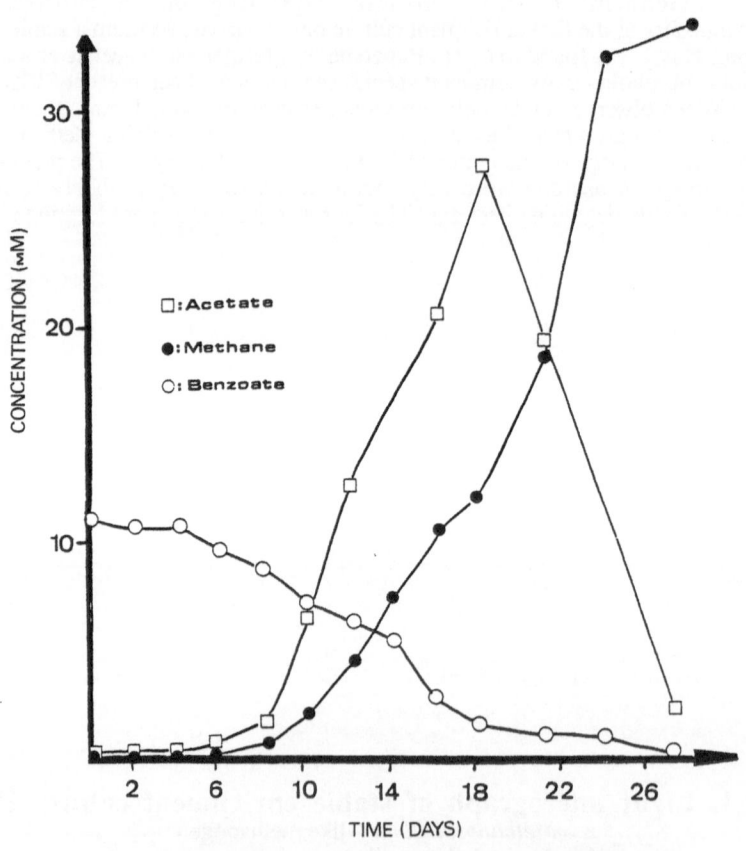

Fig.2. Kinetics of benzoate degradation
by the Huv enrichment culture .

Fig.3 shows that the final optical densities of Huv cultures were proportional to the initial benzoate concentrations in the medium . The doubling time of Huv was found to be 48 h .

Characterization of the benzoate degrader strain S : the benzoate degrader in coculture with *D. fructosovorans* did not use hydroxycinnamate (phenyl - 3 - propionate) , salycilate (2 - hydroxybenzoate), 3 - and 4 - hydroxybenzoate , protocatechuate , phenol or adipate . Only benzoate was found to be fermented . The coculture SF did not grow without sulfate which indicates the existence of an obligatory syntrophic dependence . When *M. hungatei* , an active hydrogen scavenger , was added to a sulfate - free medium inoculated with SF , growth occurred , accompanied by CH_4 production . After several transfers of this methanogenic coculture (SH) into a sulfate - free medium , SH was still contaminated by *D. fructosovorans* .

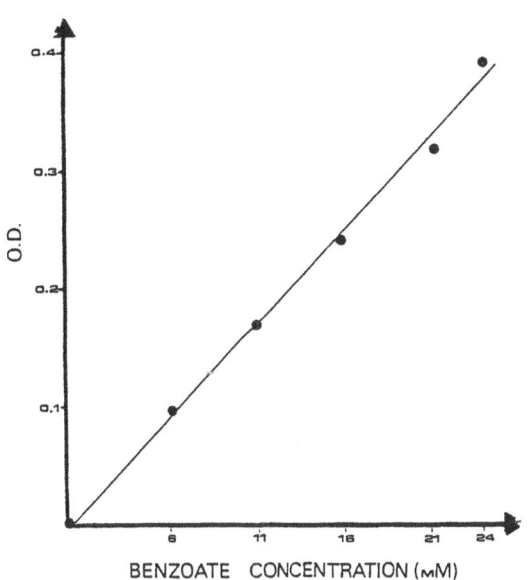

Fig.3. Final optical densities (OD) of Huv enrichment , versus the initial benzoate concentrations .

The cells of the benzoate degrader occurred singly , in pairs or in clusters, and stained Gram negative . No spore formation was ever observed . However, it was impossible to detect any differences among the cells of the SF culture , electron micrographs revealed clearly that this co - culture contained two quite different morphologies .

One , which appeared as slightly to strongly curved , rather clear cells measuring 1.5 to 3 µm in length and 0.6 to 1.2 µm in width with a few long dark wrinkles , corresponded to *D. fructosovorans* (Fig. 4A) .

The second kind , which appeared as a short dark rod - shaped bacterium with rounded ends and an undulatory outer membrane , usually measuring about 0.7 by 1 to 2.2 µm , corresponded to the syntroph . The cells were motile and possessed a single polar flagellum (Fig. 4B). Microphotographs obtained with the SH coculture confirmed the morphology of the syntroph (Fig. 4C) .

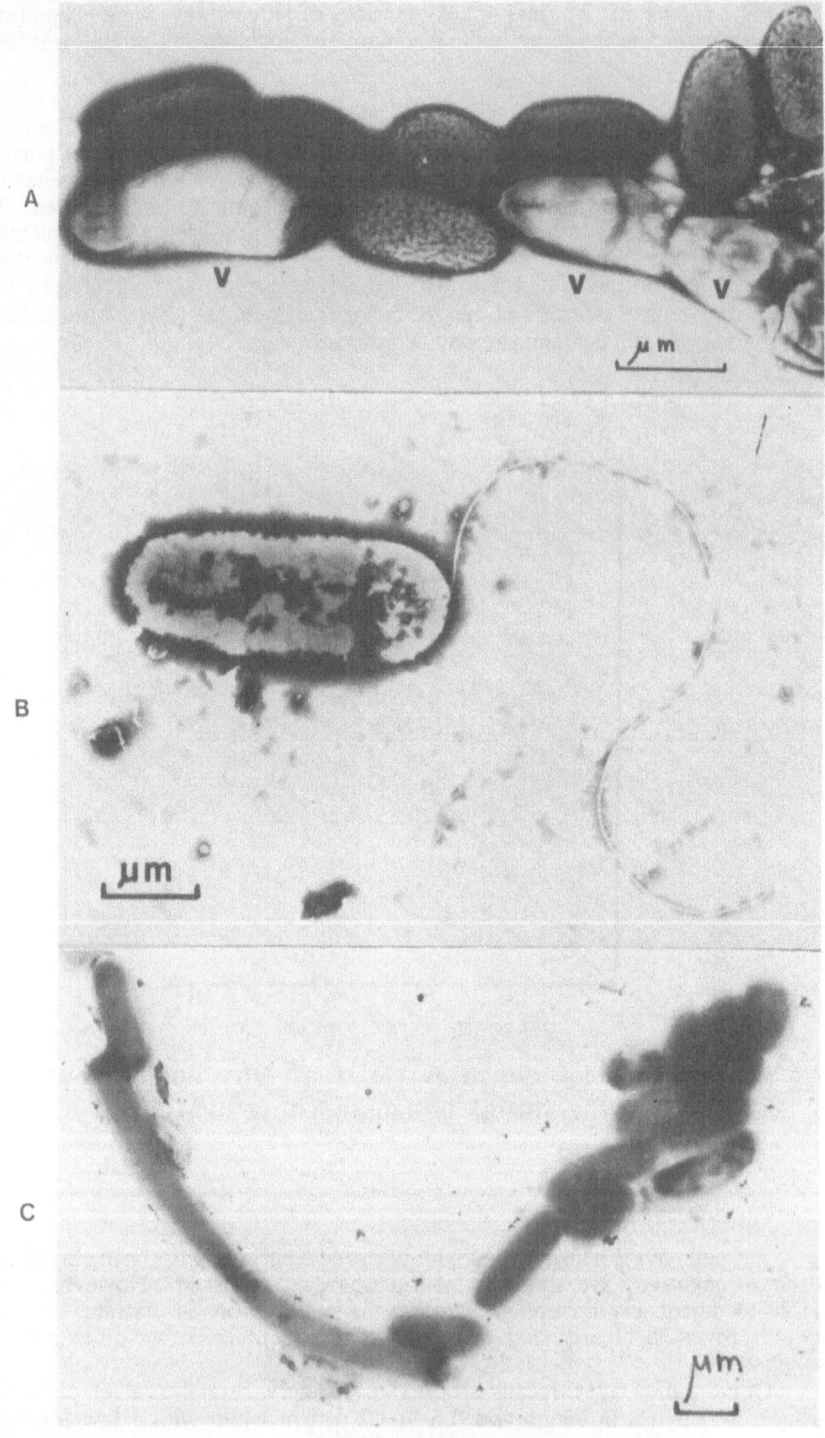

Fig.4. Electron micrographs of the syntrophic cocultures SF and SH .

A = coculture SF ;
B = monotrichous benzoate degrader ;
C = coculture SH .

DNA base composition : a mixture of pure DNAs of *Escherichia coli* K12 and *D. fructosovorans* was centrifuged to check the procedure . At equilibrium , two distinct peaks were obtained , which agree with results published earlier on *D. fructosovorans* (Ollivier *et al.* 1988). In a second set of determinations , the DNA extracted from a pellet obtained from the SF coculture was analyzed . Two peaks were observed , one of which corresponded to a mol % G + C of 65.4 , and consequently to the DNA of *D. fructosovorans* . The second corresponded to a value of 55.7% , and was attributed to the syntrophic bacterium .

Growth studies in batch cultures : since the SF coculture produced large amounts of sulfides , it was impossible to determine the growth by performing optical density measurements . Bacterial cell counts were used to estimate the growth and the biomass was estimated as described in Materials and Methods . **Table 1** shows that , in both cultures , with initial benzoate concentrations of up to 10 mM , the final bacterial concentrations were proportional to the amount of benzoate fermented .

At higher initial concentrations , the dissimilation of benzoate occurred up to about 18 mM in both cases , but the growth was not proportionately higher . In the case of the SH consortium , it was possible to distinguish the benzoate degrader S from *M. hungatei* because of the considerable morphological differences between the two . It can be seen from the table that SH culture contained about six times more S than *M.hungatei* cells . The growth rates of SF and SH were found to be equal to 0.0032 and 0.0036 h^{-1} , respectively . Addition of yeast extract to the medium did not improve the coculture growth . Only acetate was detected in both cases of fermentation . No adipate , butyrate , cyclohexane carboxylate , formate , heptanoate , hexanoate , octanoate , pimelate , propionate , succinate or valerate, which are all putative fermentation products of benzoate were detected . CO_2 was produced by both cultures and sulfide and methane were produced by SF and SH , respectively .

The benzoate fermentation relationship proposed by Mountfort and Bryant

(1982) is : $C_7H_6O_2 + 6 H_2O \longrightarrow 3 C_2H_4O_2 + CO_2 + 6 H^+ + 6 e^-$ (**equ .1**)

Table 1. Benzoate fermentation by batch cultures

	Benzoate (mM) initial	consumed	Acetate (mM)	$\mathbb{R}$ (x 10^7)	cells / ml	
	5	4.9	15	3.0	2.8	
	10	8.95	18	2.0	5.7	
SF	15	11.8	20	1.7	7.5	
	20	12.9	17	1.3	4.9	
	30	18.4	13	0.7	3.9	
					M.hungatei	Syntroph
	5	4.98	16	3.2	0.4	2.5
	10	9.95	19	1.9	0.6	6.0
SH	15	10.4	20	1.9	0.9	7.7
	20	12.2	18	1.4	0.5	4.4
	30	18.7	13	0.6	0.5	3.4

$\mathbb{R}$ = ratio mol acetate per mol benzoate ;
Culture tubes containing 10 ml medium were inoculated
with 2 ml of coculture and incubated for at least 25 days .
Values are means of triplicate tubes .

In this fermentation process , the ratio of acetate produced / benzoate fermented must therefore be equal to 3 whatever the hydrogen scavenger used . Surprisingly , it was observed that with both of the cultures studied here , this ratio was 3 only when the initial benzoate concentration was 5 mM (**Table 1**). The ratio decreased greatly when the initial benzoate concentrations were higher .

Growth studies in continuous cultures : from batch inocula , were obtained (with great difficulty) stable continuous cultures of SF and SH . **Fig. 5A and Table 2** show the results of a series of steady states obtained with SF at various benzoate concentrations .

Table 2 . Benzoate fermentation by continuous SF cultures

D^a (h^{-1})	Benzoate (mM) initial	consumed	Acetate (mM)	$\mathbb{R}$	cells / ml ($\times 10^8$)	Y_{coc}^b (g / mol)
0.004	6	4.6	15.0	3.26	2.22	20.9
0.005	6	5.2	18.5	3.55	2.40	20.0
0.004	10.2	5.4	20.0	3.7	1.60	13.0
0.005	10.1	6.0	16.5	2.8	2.25	16.3
0.007	10.0	7.0	20.0	2.9	1.35	8.5
0.010	10.0	7.8	22.0	2.9	1.36	7.9
0.0035	17.0	7.4	14.0	1.6	1.66	9.8
0.0044	17.0	8.1	13.9	1.7	1.80	9.8
0.0050	18.1	8.4	15.7	1.9	2.30	12.0
0.0071	17.0	10.9	16.9	1.6	1.25	5.0
0.0100	17.0	13.4	17.0	1.3	1.03	3.4

$\mathbb{R}$ = ratio mol acetate per mol benzoate ;

[a] D= dilution rate ; [b] Y_{coc}= overall growth yields of the coculture .

Measurements of relevant parameters were performed when the steady states were reached , i.e. after three culture volume changes .

In all cases , the maximum biomass was obtained at a dilution rate (D) of 0.005 h^{-1}. This biomass was unexpectedly significantly higher than that obtained in batch cultures . The shape of the curve giving the bacterial concentration versus the dilution rate was complex , possibly due to an inhibition phenomenon .

This hypothesis is strongly supported by the data in **Table 2** , which show that the dissimilation of benzoate decreased as the initial concentration of the substrate in the medium increased . Moreover, the ratio of acetate produced to benzoate fermented decreased with the increase in fermented benzoate . Similar results were obtained with SH culture (**Fig. 5B and Table 3**).

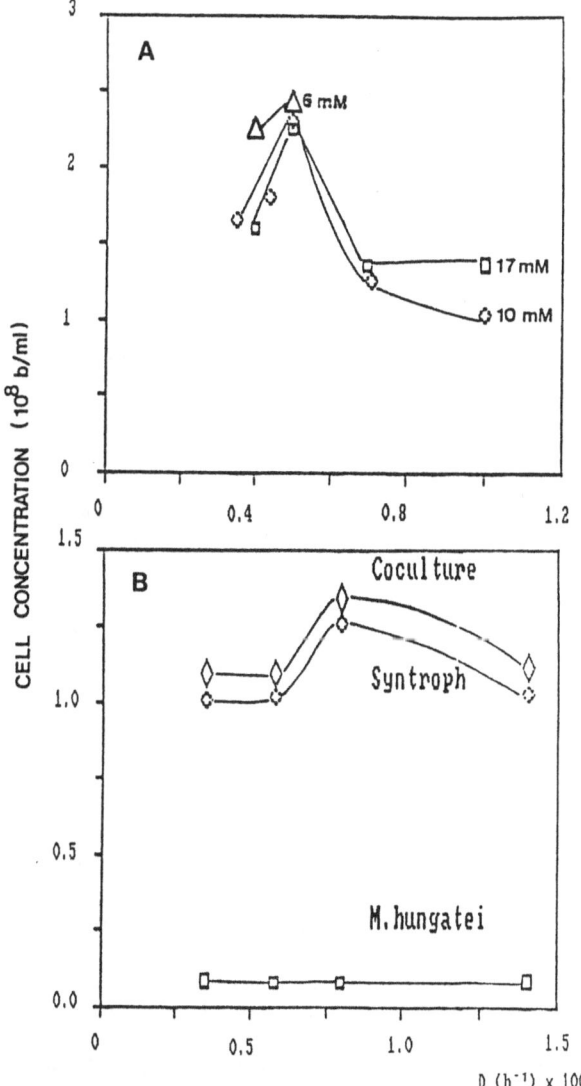

Fig. 5. Productivity of the syntrophic cocultures with 10 mM initial benzoate concentration.
A = coculture SF ; B = coculture SH .

Table 3 . Benzoate fermentation by continuous SH cultures

D^a (h^{-1})	Benzoate (mM) initial	Benzoate (mM) consumed	Acetate (mM)	R	cells (x 10^8 / ml) M.hungatei	cells (x 10^8 / ml) Syntroph	Y_{syn}^b (g/mol)
0.0058	6.0	5.0	14.8	3.00	0.07	1.08	9.4
0.0035	10.0	7.1	18.0	2.54	0.09	1.01	6.1
0.0058	10.0	8.0	15.0	1.88	0.08	1.02	5.5
0.0080	10.0	7.7	18.0	2.33	0.08	1.26	7.0
0.0140	10.0	9.2	21.0	2.30	0.09	1.08	4.8
0.0030	20.0	9.8	15.7	1.60	0.04	1.65	7.2

R = ratio mol acetate per mol benzoate ;

[a] D : dilution rate ; [b] Y_{syn} : syntrophic bacteria growth yields .

The maximum biomass concentration was obtained at 0.008 h^{-1} D . Molar benzoate growth yields of both cocultures were calculated (**Tables 2 and 3**). Unfortunately , as it was impossible to distinguish between the two kinds of bacteria in SF , it was not possible to estimate the contribution of each species to the overall growth yields ; whereas with SH , the two morphologies are quite different , so that the syntrophic molar growth yields could be obtained .

DISCUSSION

In the present paper, a new strain of anaerobic benzoate degrader was isolated from the Huveaune river . This strain was morphologically similar to Syntrophus buswellii which is the sole anaerobic degrader identified to date (Mountfort et al 1984). However our strain seemed to differ from the latter as regards some of its nutritional and physiological properties . It could be grown on the simple synthetic medium of Widdel and Pfennig (1984) containing p - aminobenzoic acid , biotin , vitamin B12 and thiamine as the only organic compounds other than benzoic acid .

Addition of complex organic substrates such as yeast extract did not improve the growth . On the other hand , *S. buswellii* growth seems to be dependent on rumen fluid (Mountfort and Bryant 1982). The mol % G+C corresponds to the DNA of the Bacteroidaceae family , to which most of the syntrophic bacteria are now thought to belong . A re - arrangement of syntrophs , or at least a re - classification as a new family is being undertaken at present (Zhao , Yang , Woese and Bryant , papers in preparation).

From the technical point of view , studies on the metabolism of the benzoate degrader are hampered by the need for the hydrogen scavenger to be present in the culture . Moreover the very low growth rate of the benzoate degrader increases the difficulty of gaining sufficient quantities of cells for biochemical studies using classical batch growth techniques . The use of a chemostat solves certain major problems encountered in the production of bacterial biomass . For example , the difficult problem of starting the cultures is overcome once the continuous syntrophic culture has reached the first steady state . We observed that both cocultures (SF and SH) became more and more stable in the course of cultivation . Moreover , and this was not predictible , the maximum bacterial densities of both cultures were found to be significantly higher in continuous than in batch cultures. An interesting point is that , in the case of both cocultures , the fermentation balance of benzoate depended on the amount of benzoate fermented under batch or continuous conditions .

When the fermented benzoate concentrations were low, the fermentation balance was in agreement with the relationship (**equ.1**).When the benzoate concentration was greater than 10 mM , the acetate quantities recovered from the medium were lower than predicted . This fact suggests that benzoate fermentation may be partially inhibited by high substrate concentrations .

As shown in **Tables 2 and 3** , the highest benzoate growth yield of SF was 20 g/mol.The growth yield of the syntroph in SH was evaluated at 9 g/mol, the SH yield being roughly 10 g/mol . The order of magnitude of these values was similar to that of the anaerobic growth yields obtained on substrates with approximatively the same molecular weight , such as hexoses , for example . However if we do not take into consideration the molar weight of the energy and carbon source but only the energy associated with the fermentation reaction , some interesting features appear. The relationships in **equ.2 and equ.3** show the theoretical benzoate fermentation reactions and the associated free energies corresponding to SH and SF, respectively :

$$4\ C_7H_5O_2^- + 19\ H_2O \longrightarrow 3\ CH_4 + 12\ CH_3CO_2^- + HCO_3^- + 9\ H^+$$
$\Delta G'^\circ = $ - 48.2 KJ / mole of benzoate
(Mountfort and Bryant , 1982) (**equ.2**)

$$4\ C_7H_5O_2^- + 16\ H_2O + 3\ SO_4^{2-} \longrightarrow 3\ S\ H^- + 12\ CH_3CO_2^- + 4\ HCO_3^- + 9\ H^+$$
$\Delta G'^\circ = $ - 59.8 KJ / mole of benzoate (our calculation) (**equ.3**)

In view of the two overall SH and SF benzoate growth yields , it can therefore be calculated that the ratios : (biomass) / (free energy associated with the catabolic reaction) which actually represent the efficiency of energy utilization were 0.21 and 0.33 g / KJ for SH and SF respectively . These values are significantly higher than those obtained with other microorganisms (see Fardeau and Belaich , 1986) and seem to suggest that some bacterial species growing under very low energy availability conditions , as in the case of interspecies hydrogen transfer , might develop energy coupling abilities which are much more efficient than those of species growing under energy - rich conditions .

Under continuous culture , relatively large quantities of cells were obtained , which served to determine the DNA mol % G+C and were sufficiently abundant to begin investigating the biochemistry and molecular biology of the syntroph and to purify the benzoate catabolism and hydrogenase enzymes involved in hydrogen interspecies transfer .

Acknowledgements . This work was partially supported by " Gaz de France ". We thank Dr. Jessica Blanc and Dr. V.A. Jacq for revising the manuscript .

REFERENCES

Bak , F. and Widdel , F. , 1986 . Anaerobic degradation of indolic compounds by sulfate - reducing enrichment cultures , and description of *Desulfobacterium indolicum* gen. nov. sp. nov.. Arch. Microbiol. 146 : 170 - 176.
Barik , S. , Brulla , W.J. and Bryant , M.P. , 1985 . PA-1, a versatile anaerobe obtained in pure culture , catabolizes benzenoids and other compounds in syntrophy with hydrogenotrophs, and P-2 plus *Wolinella* sp. degrades benzenoids . Appl . Environ. Microbiol .50 : 304 - 310 .

Cord-Ruwisch , R. , Ollivier , B. and Garcia , J.L. , 1986 . Fructose degradation by *Desulfovibrio* sp. in pure culture and in coculture with *Methanospirillum hungatei* . Curr. Microbiol. 13 : 285 -289.

Cord-Ruwisch , R. and Garcia , J.L., 1986 . Isolation and characterization of an anaerobic benzoate degrading sporeforming , sulfate - reducing bacterium , *Desulfotomaculum sapomandens* sp. nov. FEMS Microbiol. Lett. 29 : 325 - 330 .

Dutton , P. L. and Evans , W. C. ,1969 . The metabolism of aromatic compounds by *Rhodopseudomonas palustris* . Biochem. J. 133 : 525 - 535 .

Dwyer , D. F. , Krumme , M. L. , Boyd , S. A. and Tiedje , J. M . ,1986. Kinetics of phenol biodegradation by an immobilized methanogen consortium . Appl. Environ. Microbiol . 52 : 345 -351 .

Fardeau , M . L. and Belaich , J. P., 1986 . Energetics of the growth of *Methanococcus thermolithotrophicus* . Arch. Microbiol. 144 : 381 - 385 .

Ferry , J . G . and Wolfe , R. S . , 1976 . Anaerobic degradation of benzoate to methane by a microbial consortium . Arch. Microbiol. 107 : 33 - 40 .

Fina , L . R . and Fiskin , A. M . ,1960 . The anaerobic decomposition of benzoic acid during methane fermentation . II. Fate of carbon one and seven . Arch . Biochem. Biophys. 91: 163 - 165 .

Grbic - Galic , D. and Pat - Polasko , L . L . , 1985 . *Enterobacter cloacae* DG-6 : a strain that transforms methoxylated aromatics under aerobic and anaerobic conditions . Curr. Microbiol. 12 : 321 - 324 .

Guyot , J . P. , Traoré , I. and Garcia , J. L . , 1985 . Methane production from propionate by methanogenic mixed culture. FEMS Microbiol. Lett. 26 : 329 - 332 .

Hungate , R. E . , 1969 . A roll tube method for cultivation of strict anaerobes . In : " Methods in Microbiology ", vol 3B , pp . 117-132 . Norris , J.R. and Ribbons , D. W . (eds .) , Academic Press , London .

Keith , C. L. , Bridges , R. L., Fina , L. R ., Iverson , K. L. and Cloran , J. A. , 1978 . The anaerobic decomposition of benzoic acid during methane fermentation . Arch. Microbiol. 118 : 173 - 176 .

Krumholz , L . R . and Bryant , M. P. , 1985 . *Syntrophococcus sucromutans* sp. nov. gen. nov. uses carbohydrates as electron donors and formate , methoxymonobenzenoids or *Methanobrevibacter* as electron acceptor systems . Arch. Microbiol. 143 : 313 - 318 .

Macy , J. M. , Snellen , J. E . and Hungate, R. E . , 1972 . Use of syringe methods for anaerobiosis . Am J. Clin. Nutr. 25 : 1318 - 1323 .

Mountfort , D . O . and Bryant , M . P. , 1982 . Isolation and characterization of an anaerobic syntrophic benzoate - degrading bacterium from sewage sludge . Arch. Microbiol. 133 : 249 - 256 .

Mountfort , D . O ., Brulla , J . W . , Krumholz , L . R . and Bryant , M . P., 1984 . *Syntrophus buswellii* gen. nov. sp. nov. : a benzoate cataboliser from methanogenic ecosystems . Int. J. Syst. Bacteriol. 34 : 216 - 217.

Ollivier , B . , Cord-Ruwisch , R . , Hatchikian , E. C. and Garcia J.L. , 1988 . Characterization of *Desulfovibrio fructosovorans* sp. nov. Arch. Microbiol. 149 : 447 - 450 .

Schink , B. and Pfennig , N . , 1982 . Fermentation of tri-hydroxybenzenes by *Pelobacter acidigallici* gen. nov. sp. nov., a new strictly anaerobic, non spore - forming bacterium . Arch. Microbiol. 133 : 195 - 201 .

Tschech , A . and Schink , B. , 1986 . Fermentative degradation of mono-hydroxy-benzoates by defined syntrophic cocultures . Arch. Microbiol. 145 : 396 - 402 .

Widdel , F. , 1980 . Anaerobacter Abbau von Fettsaueren und Benzoesaueren durch neu isolierte Arten Sulfat-reduzierter Bakterien . Thesis , Univ. Göttingen (F.R.G.) .

Widdel , F. and Pfennig , N. , 1984 . Dissimilatory sulfate -or sulfur- reducing bacteria . In : " Bergey's manual of systematic bacteriology ", 9th edn., Vol 1. pp. 663 - 679 , Krieg , N. R . and Holt , J. G. (eds.) Williams and Wilkins , (pubs.) Baltimore .

Williams , R . J. and Evans , W. S . , 1975 .The metabolism of benzoate by *Moraxella* species through anaerobic nitrate respiration . Biochem . J. 148 : 1 - 10 .

SYNTROPHIC PROPIONATE OXIDATION

F.P. Houwen, J. Plokker, C. Dijkema[a] and A.J.M. Stams

Department of Microbiology and [a]Department of Molecular Physics
Agricultural University, Wageningen, The Netherlands

ABSTRACT

In vivo high-resolution NMR with the Gram-negative Syntrophobacter wolinii and a Gram-positive syntrophic propionate oxidizing organism, indicated the involvement of an oxaloacetate:propionyl-CoA transcarboxylase in propionate oxidation. This finding was confirmed by enzyme measurement in cell-free extracts. Two sulphidogenic cocultures grew at a similar specific growth rate, whereas substitution of the hydrogenotrophic sulphate reducer by methane bacteria resulted in slower growth. The S. wolinii coculture had a lower cell yield than Desulfobulbus propionicus. This difference is explained in terms of energy conservation mechanisms.

INTRODUCTION

Propionate is an important intermediate in anaerobic breakdown of organic matter. Fermentation of carbohydrates or lactate with propionate as major reduced end product is carried by both Gram-positive and Gram-negative bacteria. Further, propionate can be formed from β-oxidation of odd numbered fatty acids, the fermentation of glycerol, some amino acids and the reduction of C_1 and C_2 compounds (see Schink, 1986; Skrabanja and Stams, 1989).

In anaerobic environments in which sulphate is present, propionate is oxidized either to acetate and carbon dioxide by Desulfobulbus species (Laanbroek and Pfennig, 1981; Widdel and Pfennig, 1982; Samain et al., 1984) or completely to carbon dioxide by other types of sulphate reducing bacteria (Widdel, 1988).

Under methanogenic conditions propionate is degraded by syntrophic consortia of bacteria. Propionate oxidation, coupled to proton reduction, is thermodynamically unfavourable and has to be linked to hydrogen uptake (Zehnder, 1978; Bryant, 1979). The syntrophic partner in such an Interspecies Hydrogen Transfer-linked process may be methane bacteria or sulphate reducing bacteria. So far, two syntrophic propionate oxidizing cocultures have been described (Boone and Bryant, 1980; Koch et al., 1983).

Desulfobulbus propionicus oxidizes propionate via the methylmalonyl-CoA pathway as shown in Figure 1 (Stams et al., 1984; Kremer

Microbiology and Biochemistry of Strict Anaerobes Involved in Interspecies Transfer
Edited by J.-P. Bélaich et al.
Plenum Press, New York, 1990

281

and Hansen, 1988). For syntrophic propionate oxidation evidence was provided for the operation of the same pathway (Buswell et al., 1951; Koch et al., 1983; Schink, 1985; Houwen et al., 1987; Robbins, 1988). Because succinate and fumarate are symmetric molecules, this pathway leads to randomization of carbon atoms in position 2 and 3, and 1 and 4.

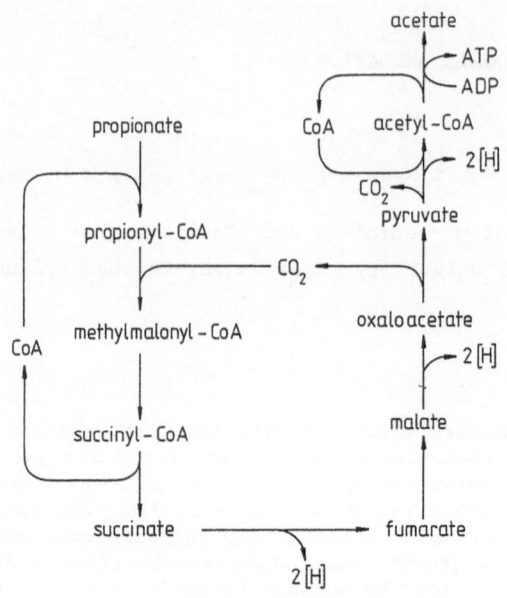

Figure 1. The succinate pathway for propionate oxidation. ΔG°'-values are calculated after Thauer et al. (1977). A transcarboxylase and CoA-transferases are not necessarily present.

It was shown for a methanogenic coulture, that propionate equilibrates with succinate (Houwen et al., 1987; Robbins, 1988). An important reaction in the interconversion of propionate and succinate is the decarboxylation of methylmalonyl-CoA to propionyl-CoA. The energy released in this step may be conserved either directly in the endergonic carboxylation of pyruvate to oxaloacetate via a transcarboxylase, an enzyme which is present in D. propionicus (Stams et al, 1984; Kremer and Hansen, 1988), or indirectly via a Na^+-gradient. Evidence for a Na^+-linked decarboxylation of methylmalonyl-CoA was provided for Veillonela alcalescens and Propionigenium modestum (Hilpert and Dimroth, 1982; Hilpert et al., 1984).

With high-resolution NMR we investigated the interconversion of propionate and succinate in D. propionicus and two syntrophic propionate oxidizing cocultures. The use of in vivo NMR with both [3-^{13}C]-propionate and $H^{13}CO_3^-$ allowed us to show the involvement of a transcarboxylase in syntrophic propionate oxidation. The NMR results were confirmed by the demonstration of transcarboxylase activities in cell-free extracts. The energy metabolism of syntrophic cultures is compared with that of D. propionicus. Growth rates and growth yields of two sulphidogenic propionate oxidizing cocultures, a methanogenic coculture and D. propionicus were determined.

MATERIALS AND METHODS

Organisms and Cultivation

In this study, three syntrophic propionate oxidizing cocultures were used. The defined sulphidogenic biculture of Syntrophobacter wolinii and Desulfovibrio G11 (Boone and Bryant, 1980). This culture (DSM 2805) was obtained from the German Collection of Microorganisms (Braunschweig, FRG) and was cultivated in a medium containing (in g/l unless otherwise stated): sodium propionate 1.9; Na_2SO_4, 2.9; $NaHPO_4$-.$2H_2O$, 0.53; KH_2PO_4, 0.41; NH_4Cl, 0.3; NaCl, 0.3; $CaCl_2$, 0.11; $MgCl_2.6H_2O$, 0.1;$NaHCO_3$; $Na_2S.9H_2O$, 0.24; yeast extract, 0.2; 1 ml of a tenfold concentrated trace element solution described by Pfennig and Lippert (1966); 1 ml of the vitamin solution described by Stams et al. (1983); 1 ml of a mixture of Na_2MoO_4 and SeO_2 (each 0.1 mM) in 10 mM NaOH. Sodium lactate (1 mM) was added to stimulate the sulphate reducing bacterium.

A methanogenic coculture, originally enriched by Koch et al. (1983) was cultivated as described before (Houwen et al., 1988). This culture is referred to as (culture) "Z". In contrast with S. wolinii, the propionate oxidizer in this culture stains Gram-positive. Because in this methanogenic culture a sulphate reducing bacterium appeared to be present (Houwen et al., 1988), the culture was also grown on propionate in the presence of sulphate (2.9 g/l). After repeated transfers, this culture was free of methanogens as evidenced by lack of methane production during incubation with hydrogen and without sulphate. This sulphidogenic coculture is referred to as (culture) "ZPS".

Desulfobulbus propionicus (DSM 2032) was a gift of D.R. Kremer, University of Groningen, The Netherlands. The organism was cultivated in a medium containing (in g/l unless otherwise stated): sodium propionate 1.9; Na_2SO_4, 2.9; $Na_2HPO_4.2H_2O$, 0.53; KH_2PO_4, 0.41; NH_4Cl, 0.54; $CaCl_2.2H_2O$, 0.15; $MgCl_2.6H_2O$, 0.4; KCl, 0.3; $NaHCO_3$, 2.4; $Na_2S.9H_2O$, 0.48; yeast extract, 0.2. Trace elements and vitamins were the same as described for the Syntrophobacter-Desulfovibrio coculture.

^{13}C-NMR Experiment

Preparation for the NMR was essentially done as described before (Houwen et al., 1987) except that the centrifuged cells were resuspended in their respective media with $H^{13}CO_3^-$,and 50 mM [3-^{13}C]-propionate was added. High-resolution NMR was carried out as described before (Houwen et al., 1987).

Enzyme Measurement

Cells were washed anaerobically in a 50 mM phosphate buffer with 2 mM $MgCl_2$, pH 7.1. Cell-free extracts were prepared anaerobically by French Press. Cells were broken at 1360 bar and the cell-debris was removed by centrifugation at 4000 rpm for 20 minutes. The supernatants were stored oxygen-free at 0 °C in glass tubes sealed with butyl rubber stoppers.

Oxaloacetate: propionyl-CoA transcarboxylase was measured in a coupled assay with malate dehydrogenase, according to Stams et al. (1984).

Cell Counts

The relative numbers of the different organisms in the sulphidogenic cocultures were determined with a Leitz Diaplan D microscope equiped with a Philips LDK 12 camera and a video recorder (Sony VO

5630, U-matic). The relative number of methanogens in the methanogenic coculture was determined using a UV-microscope (Leitz, Dialux 20 EB).

Growth Rates and Growth Yields

The growth rates of the syntrophic propionate degrading cocultures were determined using the same media as described above, except that lactate was omitted from the Syntrophobacter-Desulfovibrio coculture and yeast extract (0.2 g/l) was also added to the methanogenic coculture. 15 ml of medium with 5% inoculum was incubated in 25-ml glass tubes. The optical density at 660 nm was measured directly with a Hitachi U-1100 spectrophotometer equiped with a test tube holder. Growth rates were based on five independent experiments.
Growth yields were determined for the Syntrophobacter-Desulfovibrio coculture and for D. propionicus. The presence of 1 mM lactate in the medium of the coculture was corrected for by subtraction of the yield on lactate (1mM) only. For comparison D. propionicus was cultivated in the same medium as the coculture, without lactate. Growth yields were determined with a 5% inoculum and were averages of six and four experiments, respectively.

Analytical Methods

Propionate and acetate were measured gaschromatographically or by HPLC (LKB 2150). A CP9000 gaschromatograph (Chrompack, Middelburg) was used with a glass column (180 cm x 2 mm ID) filled with Chromosorb 101 (80 - 100 mesh). The carrier gas was nitrogen saturated with formic acid according to Ackman (1972). The temperatures of the injector, column and detector were 250 °C, 160 °C - 180 °C and 300 °C, respectively. For HPLC a Chrompack organic acids column (30 cm x 6.5 mm ID) was used. The eluence was 0.01 N H_2SO_4 with a flow rate of 0.6 or 0.8 ml/min. The column temperature was 60 °C. 20 µl sample was injected using a Spectra Physics autosampler (SP 8775). Detection was done with a differential refractometer (LKB 2142).

Hydrogen and methane were measured gaschromatographically as described before (Houwen et al., 1988).

Protein in cell-free extract was determined according to Bradford (1976). In the growth-yield experiment the microbiuret method (Kuenen and Veldkamp, 1972) was used.

Sulphide was measured according to Trüper and Schlegel (1964).

RESULTS AND DISCUSSION

^{13}C-NMR Experiment

In vivo high-resolution ^{13}C-NMR was used to study the fate of label in propionate degrading cultures. Table 1 summarizes the incorporation of label at various positions during incubation of D. propionicus, the Syntrophobacter-Desulfovibrio coculture and the methanogenic coculture in the presence of [3-^{13}C]-propionate and $H^{13}CO_3^-$. In the three cultures, 50 % of the acetate was labelled at the C-1 position (column 1). Moreover, randomization of label over the C-3 and C-2 of propionate occurred to a high degree (column 2); apparently the conversion of propionate to succinate is highly reversible in these organisms. These results are in agreement with the involvement of the succinate pathway in propionate oxidation (Figure 1), and confirm earlier findings with D. propionicus and the methanogenic coculture, (Koch et al., 1983; Stams et al, 1984; Houwen et al., 1987). Although in D. propionicus the carboxylation of propionyl-CoA occurs

Table 1. Incorporation of label during incubation of propionate oxidzing cultures with [3-¹³C]-propionate and H¹³CO₃⁻.

ORGANISM(S)	% C-1 acetate[a]	% scrambling[b]	% C-1 propionate[c]
Desulfobulbus	50	96	19
Syntrophobacter + Desulfovibrio	50	80	8.7
methanogenic coculture (Z)	50	100	17

[a]percentage of the acetate produced.
[b]percentage [2-¹³C]-propionate relative to half the amount of [2-¹³C] + [3-¹³C]- propionate.
[c]percentage [1-¹³C]-propionate relative to [2-¹³C]-propionate formed by scrambling.

via a transcarboxylase (Stams et al, 1984), some label became incorporated at the C-1 position of propionate (column 3). Therefore, either the transcarboxylase exchanges ¹³CO₂ with the environment or other carboxylation reaction are involved. Because of the similarity with D. propionicus it must be assumed that also the syntrophic organisms degrade propionate via the succinate pathway using a transcarboxylase.

Enzyme Measurement

The involvement of an oxaloacetate:propionyl-CoA transcarboxylase in syntrophic propionate oxidation, as shown by ¹³C-NMR, was confirmed by enzyme measurements. In cell-free extracts of the defined biculture with S. wolinii a transcarboxylase activity of 0.52 μmol.min⁻¹.mg⁻¹ protein was measured (Table 2). The finding that in the cell-free extract of the pure culture of Desulfovibrio G11 (the hydrogen consuming sulphate reducer), grown on hydrogen with sulphate in the presence of propionate (12.5 mM) this enzyme was not found, strongly suggests that the transcarboxylase is present in S. wolinii.
In the other syntrophic cultures (ZPS and Z), much lower transcarboxylase activities were found (Table 2). The difference in activity beween these cultures, may be due to differences in the protein contribution of the propionate oxidizer in the cell-free extracts.

Growth Rates and Inhibition

Figure 2 shows growth curves of the Syntrophobacter-Desulfovibrio coculture, the sulphidogenic coculture (ZPS) and the methanogenic coculture (Z). The growth rates determined from Figure 2, are given in Table 2. The Syntrophobacter-Desulfovibrio coculture had a higher growth rate than mentioned by Boone and Bryant (1980; Table 2). This difference can be explained by differences in growth media and the slightly different temperature. Sulphide had a strong inhibiting effect on the growth of the Syntrophobacter-Desulfovibrio coculture and on ZPS (Figure 2). Their growth rates decreased to 0.133 day⁻¹ and 0.187 day⁻¹, respectively. The propionate oxidizers grown together with

Table 2. Microbiological and biochemical characteristics of propionate oxidizing cultures.

ORGANISM(S)	% PROPIONATE OXIDIZER	TRANSCARBOXYLASE (μmol.min^{-1}.mg^{-1})	GROWTH RATE (day^{-1})	GROWTH YIELD (g protein.mol^{-1})
Desulfobulbus	100			1.69
		0.42[1]	0.89[1]	1.95a[1]
			1.66[2]	2.17a[2]
Syntrophobacter -				
Desulfovibrio G11	55	0.52	0.32	0.92
			0.19[3]	
Methanospirillum	NDb	ND	0.10[3]	ND
Sulphidogenic coculture (ZPS)	30	0.006	0.28	ND
Methanogenic coculture (Z)	10	0.036	0.23	ND

aWith the assumption that 1 g cell carbon corresponds to 1 g protein.
bNot determined.
[1]After Stams et al. (1984).
[2]After Widdel and Pfennig (1982).
[3]After Boone and Bryant (1980).

hydrogenotrophic methanogens have a lower growth rate than with sulphate reducers as hydrogenotrophs (Table 2). This may be explained by the more negative Gibbs free energy change under sulphidogenic conditions compared to methanogenic conditions (equations (i) and (ii); Thauer et al., 1977; Dolfing, 1988). Slower growth in coculture with methanogens was described earlier for S. wolinii (Boone and Bryant, 1980), the butyrate oxidizing Syntrophomonas wolfei (McInerney et al., 1979) and the benzoate degrading Syntrophus buswellii (Mountfort and Bryant, 1982).

$$\text{(i) } 4\ Pr^- + 3\ H_2O \longrightarrow 4\ Ac^- + 3\ CH_4 + HCO_3^- + H^+$$
$$\Delta G^{\circ\prime} = -102.4 \text{ kJ/mol}$$

$$\text{(ii) } 4\ Pr^- + 3\ SO_4^- \longrightarrow 4\ Ac^- + 3\ HS^- + HCO_3^- + H^+$$
$$\Delta G^{\circ\prime} = -151.3 \text{ kJ/mol}$$

Growth Yields

The growth yield of D. propionicus was higher compared with that of the Syntrophobacter-Desulfovibrio coculture (Table 2). In the coculture the relative number of propionate oxidizers was very constant (Table 2). This made it possible to determine the contribution of S. wolinii to the total protein in the coculture. Based on mean protein content per cell in the pure culture of the sulphate reducer, the total protein content in the coculture, and the relative number (55%), it was calculated that about 97% of the protein in the coculture was from the propionate oxidizing organism. Therefore, the yield of S. wolinii is about 0.89 g/mol.

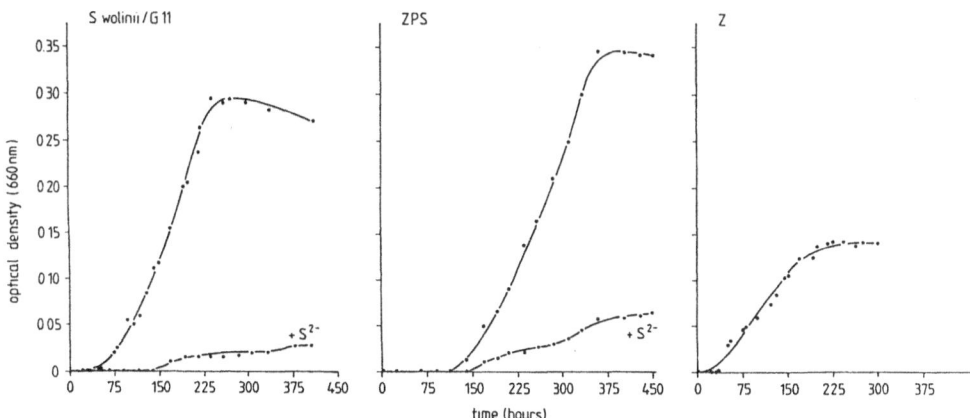

Figure 2. Growth curves of the <u>Syntrophobacter-Desulfovibrio</u>
        coculture, ZPS and Z. For the sulphidogenic
        cocultures also growth curves were determined in
        the (initial) presence of 3 mM extra sulphide.

During growth on propionate plus sulphate, <u>D. propionicus</u> conserves
energy both at substrate level and by membrane-linked electron
transport (Kremer and Hansen, 1988). The mechanism(s) by which energy
conservation takes place in <u>S. wolinii</u> is unknown. One ATP will be
formed in the conversion of acetyl-CoA to acetate. However, reoxidation
of reduced electron carries is energetically difficult. Assuming that
H^+ serves as electron acceptor, the partial pressure of hydrogen (pH$_2$)
is of great importance. At a pH$_2$ of 10^{-5} atm., (E°' $2H^+/H_2$ = -272 mV),
electrons derived from the conversion of pyruvate to acetyl-CoA (E°' =
-490 mV) and of malate to oxaloacetate (E°' = -177 mV) can be disposed
without loss of energy. At this pH$_2$, however, the oxidation of
succinate to fumarate (E°' = +33 mV) still costs 53 kJ/mol (corresponding to more than one ATP; Thauer et al., 1977). At an internal
succinate/fumarate ratio of 10^5 and a membrane potential of 150 mV to
drive the presumably membrane-linked dehydrogenase, this reaction would
just be feasable. The amount of energy required to generate the
membrane potential is not known; it can be speculated that it should be
less than 1 ATP. Alternatively, the excretion of HCO$_3^-$ in symport with
protons could contribute to the generation of the membrane potential.
Generation of metabolic energy by end-product efflux was reviewed by
Konings (1985).

ACKNOWLEDGEMENTS

        The valuable suggestions and comments of Prof. Dr. A.J.B. Zehnder
and M.S.M. Jetten are highly acknowledged. We thank W. Roelofsen for
his technical assistance. This study was supported by the Foundation
for Fundamental Biological Research (BION) and the Royal Netherlands
Academy of Arts and Sciences.

REFERENCES

Ackman, R.G., 1972, Porous polymer bead packings and formic acid vapour
    in the GLC of volatile free fatty acids, <u>J. Chromatogr. Sci.</u>,
    10:560.
Boone, D.R. and Bryant, M.P., 1980, Propionate-degrading bacterium,
    <u>Syntrophobacter</u> <u>wolinii</u> sp. nov. gen. nov., from methanogenic
    ecosystems, <u>Appl. Environ. Microbiol.</u>, 40:626.

Bradford, M.M., 1976, A rapid and sensitive method for the quantitation of microgram quantities of protein utilizing the principle of protein-dye binding, Analytical. Biochem., 72:248.

Bryant, M.P., 1979, Microbial methane production - theoretical aspects, J. Anim. Sci., 48:193.

Buswell, A.M., Fina, L., Müller, H. and Yahiro, A., 1951, Use of ^{14}C in mechanism studies of methane fermentation. II. propionic acid, J. Am. Chem. Soc., 73:1809.

Dolfing, J., 1988, Acetogenesis, in: Biology of Anaerobic Microorganisms, Zehnder, A.J.B., ed., Wiley, New York.

Hilpert, W. and Dimroth, P., 1982, Conversion of the chemical energy of methylmalonyl-CoA decarboxylation into a Na$^+$ gradient, Nature, 296:584.

Hilpert, W., Schink, B. and Dimroth, P., 1984, Life by a new decarboxylation-dependent energy conservation mechanism with Na$^+$ as coupling ion, EMBO J., 3:1665.

Houwen, F.P., Dijkema, C., Schoenmakers, C.H.H., Stams, A.J.M. and Zehnder, A.J.B., 1987, ^{13}C-NMR study of propionate degradation by a methanogenic coculture, FEMS Microbiol. Letters, 41:269.

Houwen, F.P., Cheng Guangsheng, Folkers, G.E., Heuvel v.d., W.M.J.G. and Dijkema, C., 1988, Pyruvate and fumarate conversion by a methanogenic propionate-oxidizing coculture, in: Granular Anaerobic Sludge, Microbiology and Technology, Lettinga, G., Zehnder, A.J.B., Grotenhuis, J.T.C. and Hulshoff Pol, L.W., eds.,Pudoc, Wageningen.

Koch, M., Dolfing, J., Wuhrmann, K. and Zehnder, A.J.B., 1983, Pathways of propionate degradation by enriched methanogenic cultures, Appl. Environ. Microbiol., 45:1411.

Konings, W.N., 1985, Generation of metabolic energy by end-product efflux, TIBS - August 1985:317.

Kremer, D.R. and Hansen, T.A., 1988, Pathway of propionate degradation in Desulfobulbus propionicus, FEMS Microbiol. Lett., 49:273.

Kuenen, J.G. and Veldkamp, H., 1972, Thiomicrospira pelophila gen. r. sp. n., a new obligately chemolithotroph colourless sulfur bacterium, Antonie van Leeuwenhoek, 38:241.

Laanbroek, H.J. and Pfennig, N., 1981, Oxidation of short-chain fatty acids by sulfate-reducing bacteria in freshwater and in marine sediments, Arch. Microbiol., 128:330.

McInerney, M.J., Bryant, M.P. and Pfennig, N., 1979, Anaerobic bacterium that degrades fatty acids in syntrophic association with methanogens, Arch. Microbiol., 122:129.

Mounfort, D.O. and Bryant, M.P., 1982, Isolation and characterization of an anaerobic syntrophic benzoate-degrading bacterium from sewage sludge, Arch. Microbiol., 133:249.

Pfennig, N. and Lippert, K.D., 1966, Ueber das Vitamin B12-Bedürfnis phototropher Schwefelbakterien, Arch. Mikrobiol., 55:245.

Robbins, J.E., 1988, A proposed pathway for catabolism of propionate in methanogenic cocultures, Appl. Environ. Microbiol., 54:1300.

Samain, E., Dubourguier, H.C. and Albagnac, G., 1984, Isolation and characterization of Desulfobulbus elongatus sp. nov. from a mesophilic industrial digester, System. Appl. Microbiol., 5:391.

Schink, B., 1985, Mechanisms and kinetics of succinate and propionate degradation in anoxic freshwater sediments and sewage sludge, J. Gen. Microbiol., 131:643.

Schink, B., 1986, New aspects of fatty acid metabolism in anaerobic digestion, in: Proc. IV Int. Symp. Microbiol. Ecol. Ljubljana.

Skrabanja, A.T.P. and Stams, A.J.M., 1989, Oxidative propionate formation by anaerobic bacteria, in: This book.

Stams, A.J.M., Veenhuis, M., Weenk, G.H. and Hansen, T.A., 1983,

Occurrence of polyglucose as a storage polymer in <u>Desulfovibrio</u> species and <u>Desulfobulbus propionicus</u>, <u>Arch. Microbiol.</u>, 136:54.

Stams, A.J.M., Kremer, D.R., Nicolay, K., Weenk, G.H. and Hansen, T.A., 1984, Pathway of propionate formation in <u>Desulfobulbus propionicus</u>, <u>Arch. Microbiol.</u>, 139:167.

Thauer, R.K., Jungermann, K. and Decker, K., 1977, Energy conservation in chemotrophic anaerobic bacteria, <u>Bacteriol. Rev.</u>, 41:100.

Trüper, H.G. and Schlegel, H.G., 1964, Sulphur metabolism in Thiorodaceae. I. Quantitative measurements of growing cells of <u>Chromatium okenii</u>, <u>Antonie van Leeuwenhoek J. Microbiol. Serol.</u>, 30:225.

Widdel, F. and Pfennig, N., 1982, Studies on dissimilatory sulfate-reducing bacteria that decompose fatty acids II. Incomplete oxidation of propionate by <u>Desulfobulbus propionicus</u> gen. nov., sp. nov., <u>Arch. Microbiol.</u>, 131:360.

Widdel, F., 1988, Microbiology and ecology of sulfate- and sulfur-reducing bacteria, <u>in</u>: Biology of Anaerobic Microorganisms, Zehnder, A.J.B., ed., Wiley, New York.

Zehner, A.J.B., 1978, Ecology of methane formation, <u>in</u>: Water Pollution Microbiology, vol. 2, Mitchel, R., ed., Wiley, New York.

STRUCTURAL AND FUNCTIONAL PROPERTIES OF THE CHROMOSOMAL PROTEIN MC1

ISOLATED FROM VARIOUS STRAINS OF METHANOSARCINACEAE

Bernard Laine, François Chartier, Marlène Imbert and
Pierre Sautière

URA 409 CNRS - Université de Lille II, Institut de Recherches
sur le Cancer de Lille, Place de Verdun   59045 LILLE, France

INTRODUCTION

In eukaryotes, packaging of DNA inside the nucleus is mainly ensured
by small basic proteins called histones and the chromatin is organized in
a repetitive structure of which the nucleosome represents the elementary
subunit. In contrast, the structural organization of the prokaryotic chromatin remains to be clearly elucidated. In eubacteria and archaebacteria,
investigation of chromatin structure and isolation of chromosomal proteins
have been hampered by the difficulty to prepare native chromatin because
of its great instability. Knowledge of eubacterial chromatin has been mainly
obtained on Escherichia coli and has been reviewed by Pettijohn (1). The
properties of the DNA-binding protein II (also called protein HU) which is
involved in DNA packaging in eubacteria,are described in (2). Archaebacteria comprise three groups : thermophilic sulfur-dependent bacteria, methanogens and halophiles. The organization of the chromosomal DNA in Thermoplasma acidophilum and Sulfolobus acidocaldarius has been investigated by
Searcy (3,4). Only one chromosomal protein is encountered in Thermoplasma
acidophilum (5) whereas in Sulfolobus acidocaldarius, several groups of
proteins differing by their molecular sizes have been described independently by differents authors (4,6,7). Study of the organization of chromatin and isolation of chromosomal proteins of methanogens have been performed in our laboratory. Nothing is known about the DNA associated proteins
in halophilic bacteria.

This minireview deals with the properties of the deoxyribonucleoprotein complex of Methanosarcina barkeri and with the characteristics of the
major chromosomal protein in Methanosarcinaceae, the protein MC1, with the
aim to investigate its function in the bacterial chromatin.

STUDY OF THE DEOXYRIBONUCLEOPROTEIN COMPLEXES FROM METHANOSARCINACEAE AND

OF FUNCTIONAL PROPERTIES OF THE PROTEIN MC1

The deoxyribonucleoprotein complexes have been prepared from Methanosarcina barkeri and Methanosarcina sp.CHTI 55 by a method combining ultracentrifugation and gel filtration chromatography (8,9) or alternatively by
isopycnic centrifugation (10). In these complexes protein MC1,a basic polypeptide of 93 amino acid residues, represents the major protein component.

Microbiology and Biochemistry of Strict Anaerobes Involved in Interspecies Transfer
Edited by J.-P. Bélaich *et al.*
Plenum Press, New York, 1990

291

It accounts for 90 % and 80 % of the DNA-associated proteins in Methanosarcina barkeri and Methanosarcina sp.CHTI 55 respectively (8,9). In the later strain,a second protein called MC2 has an apparent molecular size of 17 kDa and represents 14 % of the chromosomal proteins. The protein MC1-to-DNA ratio is equal to 0.1 (by weight)(8). The association of the protein MC1 with DNA in vivo has been confirmed by the localization of protein MC1 in the DNA-rich areas of M.barkeri cryosections using an immunolabelling method with protein A-colloidal gold technique (8). In order to determine the chromatin organization in Methanosarcinaceae and to elucidate the role of protein MC1, we have undertaken a study of the deoxyribonucleoprotein complexes isolated from the bacteria and of complexes reconstituted in vitro with purified protein MC1 and DNA. In electron microscopy, the M.barkeri deoxyribonucleoprotein complex appears as fibres without repetitive globular structures reminiscent of beads on a string observed with eukaryotic chromatin (8). Moreover, methods used to produce nucleoprotein subunit structures from chromatin of thermophilic archaebacteria and eukaryotes (3,4) failed to yield DNA fragments protected from staphylococcal nuclease digestion in M.barkeri chromatin (8).

The characteristics of the chromatin in eukaryotes, eubacteria and archaebacteria are compared in Table 1. By the low amount of chromosomal protein and the lack of stable repetitive globular structure, the chromatin organization in Methanosarcinaceae resembles that of eubacteria and of Sulfolobus acidocaldarius and is very different from that of eukaryotes. On the other hand, the chromatin in Thermoplasma acidophilum appears intermediate between that of other microorganisms and that of eukaryotes.

In order to elucidate the role of protein MC1 in the chromatin, we have undertaken the study of its interaction with DNA. Firstly, we have investigated whether protein MC1 can protect DNA against thermal denaturation as do eukaryotic histones, some HMG proteins and the eubacterial DNA-binding protein II.

Thermal denaturation experiments performed on the deoxyribonucleoprotein complex gave only a slight difference by comparison with data obtained with free DNA (9). This result can be explained by taking into account the low protein-to-DNA ratio in the complex (equal to 0.1). On the other hand experiments performed on complexes reconstituted with higher protein-to-DNA ratios clearly show that protein MC1 can efficiently protect DNA against thermal denaturation. Indeed when complexed with protein MC1 from Methanosarcina sp.CHTI 55, DNA from the same bacterium presents biphasic derivative melting profiles : the first melting band at 58°C is that of free DNA whereas the second melting band at 70°C is that of DNA regions protected by the protein MC1 (Fig.1). Increasing amounts of protein MC1 cause a reduction of the melting band of free DNA and a concomitant increase of the melting band of protected DNA. From these data, it has been calculated that DNA segments of 8 base pairs are protected by one molecule of protein. Identical results were obtained with DNA isolated from chicken erythrocyte DNA. Moreover, from these results it can be inferred that protein MC1 binds cooperatively with DNA.

Protein MC1 causes also a substantial increase of the extent of DNA renaturation upon cooling the protein-DNA complexes (9). This process is also dependent of the amount of protein MC1 complexed with DNA from 56°C to lower temperatures.

In addition to the effect of protein MC1 on the thermal stability of DNA we found that DNA transcription in vitro by E.coli RNA polymerase is strongly stimulated in the presence of protein MC1 at physiological protein-to-DNA ratios. On the contrary, DNA transcription is strongly inhibited when higher amounts of protein are complexed to DNA (14).

292

Table 1. Characteristics of the chromatin in eukaryotes, eubacteria and archaebacteria

	EUKARYOTES	EUBACTERIA	ARCHAEBACTERIA		
		E.coli (2)	T.acidophilum (3)	S.acidocaldarius (4) (6)	M.barkeri (8)
Protein composition	Histones	DNA-binding Protein II Protein H1	HTa	HSa 7 kDa NHSa 8 kDa 10 kDa	MC1
Protein/DNA ratio	1.0	0.2	0.6	0.4	0.1
Globular structures	++	-	+	?	-
DNA remaining resistant to staphylococcal nuclease	40-50%	0%	20%	5%	0%

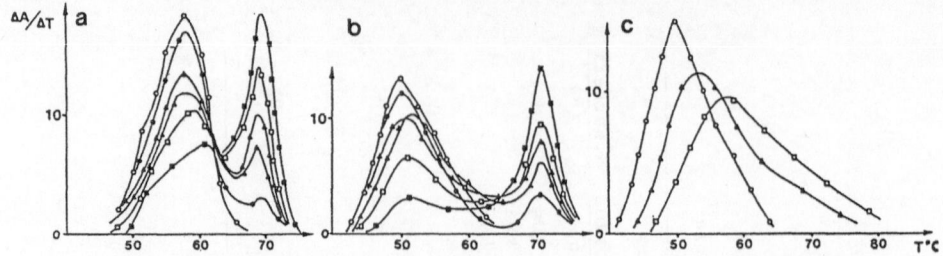

Fig. 1. Derivative melting profiles obtained with reconstituted complexes.
(A) protein MC1 and DNA from <u>Methanosarcina</u> sp.CHTI 55; (B) protein
MC1 and chicken erythrocyte DNA; (C) <u>E.coli</u> DNA-binding protein II
and chicken erythrocyte DNA. The ratios (w/w) of input protein-to-
DNA were : 0 (o——o); 0.22 (●——●); 0.55 (△——△); 0.77 (▲——▲);
1.12 (□——□); 1.60 (■——■).

    The effects of protein MC1 on thermal stability of DNA and on trans-
cription likely reflect conformational changes of DNA when complexed with
protein MC1. This hypothesis is in agreement with recent data obtained in
a study of protein MC1-DNA interaction using circular dichroism spectroscopy. In fact, as discussed in (9), the protective effect on thermal stability of DNA does not correspond to a physiological role of protein MC1
but is merely a property shared with other chromosomal proteins that
are involved in DNA packaging such as histones and the eubacterial DNA-binding
protein II. Therefore, a possible function of protein MC1 would be to
structure the chromosomal DNA of methanogenic bacteria. However DNA organization proceeds through different mechanisms since only histones form
globular and stable subunit structures with DNA. On the other hand, the
eubacterial DNA-binding protein II does not cause a biphasic derivative
melting profile of DNA but causes a shift of the DNA melting point ; this
shift is proportional to the amount of protein (Fig.1-c).

STRUCTURAL FEATURES OF PROTEIN MC1

    The protein MC1 has been isolated from four strains of the genus <u>Me-
thanosarcina</u> and one strain of the genus <u>Methanothrix</u> which is the second
genus of the family Methanosarcinaceae. The strain <u>Methanosarcina barkeri</u>
MS (DSM 800) is mesophilic whereas the strains <u>Methanosarcina</u> sp. MST-A1
(DSM 2905), sp.CHTI 55 (DSM 2906) and <u>Methanosarcina thermophila</u> TM1 (DSM
1825) are thermophilic. The genus <u>Methanothrix</u> differs strongly from the
genus <u>Methanosarcina</u> : <u>Methanothrix soehngenii</u> strain FE (DSM 3013) is mesophilic, has a high G + C content (52.6 % versus about 39 % in <u>Methano-
sarcina</u> strains) and contrary to the other strains cited above, is an
obligate acetotrophic strain and is a filamentous bacterium (11).

    The <u>Methanosarcina</u> strains contain a single protein MC1 whereas <u>Metha-
nothrix soehngenii</u> contains three variants of the protein MC1 which have
been separated by reverse-phase HPLC (12). The sequences of protein MC1
isolated from these strains (except <u>M.thermophila</u> TM1) have been determined (13-16).

    Protein MC1 is a polypeptide of 93 amino acid residues except the <u>M.
soehngenii</u> protein MC1 variants a, b and c which have 89, 87 and 90 residues respectively. Protein MC1 contains a high number of charged residues.

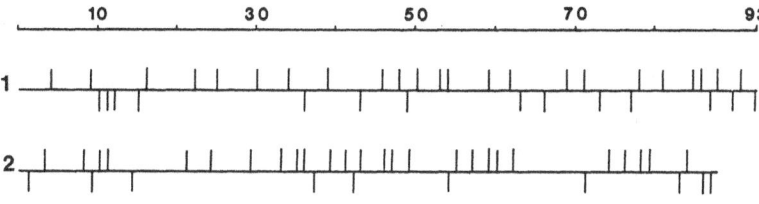

Fig. 2. Distribution of basic (⊥) and acidic (⊤) residues in protein
MC1 from <u>Methanosarcina barkeri</u> (1) and protein MC1a from <u>Methano-</u>
<u>thrix soehngenii</u> (2). The distribution of charged residues in pro-
tein MC1 from other <u>Methanosarcina</u> strains is very similar to that of
<u>M.barkeri</u> protein MC1.

With 14 acidic residues and 24 basic residues, protein MC1 from <u>M.barkeri</u>
has a net charge equal to + 10. The variant a of <u>M.soehngenii</u> protein MC1
is more basic since it contains 10 acidic residues and 26 basic residues.
The main characteristic of protein MC1 is that its charged residues are
distributed all along the polypeptide chain (Fig.2). However two sequences
appear enriched in charged residues : a basic sequence (residues 46-56)
with a net charge equal to + 5 and the carboxy-terminal sequence (residues
85-93) where positive charges are balanced by negative charges brought by
glutamic acid residues and the free α-carboxyl group. A slightly different
distribution of charged residues is encountered in protein MC1a from <u>Metha-</u>
<u>nothrix soehngenii</u> ; in this molecule, the basic character is more marked
in three sequences : the sequences 34-48 (net charge : + 7), 56-63 (net
charge : + 5) and 75-80 (net charge : + 4). Protein MC1 is also characterized by the accumulation of four proline residues in a short sequence located between residues 68 and 82.

The amino acid sequence comparison of proteins MC1 (Fig.3) indicates
that two regions are well conserved : these are the region 17-35, rich in
glycine and alanine and the region 45-58, the most basic sequence in the
protein MC1. These two conserved regions together represent 40 % of the
whole molecule. They are separated by a highly variable amino acid residue
at position 36 and a segment located between residues 37 and 44 where six
amino acid residues are deleted in the variants of <u>Methanothrix soehngenii</u>
protein MC1.

The carboxy-terminal third of protein MC1, particularly the sequence
59-73, is highly variable. Numerous non conservative changes occur in this
region. However the four proline residues at positions 68, 72, 76 and 82
and the bulky hydrophobic residues at positions 65, 74, 75 and 79 are strictly conserved.

Up to now all our attempts to crystallize the protein MC1 have failed.
Therefore the secondary structure of protein MC1 from <u>Methanosarcina barkeri</u>
has been investigated by means of predictive methods (Hydrophobic Cluster
Analysis (17), Chou and Fasman (18), Garnier et <u>al.</u>(19) and Sette <u>et al.</u>
(20)), and by circular dichroism and infrared spectroscopy. Later on, predictive methods have been applied to protein MC1 isolated from the other
species (16). Figure 4 shows the results obtained on <u>M.barkeri</u> protein MC1
with the Hydrophobic Cluster Analysis (HCA) which allows a rapid perception
of the structured regions. Protein MC1 contains only small hydrophobic clusters numbered from C1 to C7. Except clusters C4 and C7, all these clusters
display the typical shape and orientation of segments in β-sheet structure.

Fig. 3 — Comparison of amino acid sequences of proteins MC1 (protein sequence alignment)

```
                             10                            20
1  (H)Ser-Asn-Thr-Arg-Asn-Phe-Val-Leu-Arg-Asp-Glu-Glu-Gly-Asn-Glu-His-Gly-Val-Phe-Thr-Gly-Lys-Gln-Pro-Arg-
2
3
4  (H)Ala-Glu-                          Asp          Ala                          Lys-Lys           Ile
5  (H)Ala-Glu-Met                       Asp          Ala                Ser       Ala-Gln           Ile
6  (H)Met-Ile-Glu-Lys                                Ala                Ser       Lys               Ile


                   30                            40                           50
1  Gln-Ala-Ala-Leu-Lys-Ala-Ala-Asn-Arg-Gly-Asp-Gly-Thr-Lys-Ser-Asn-Pro-Asp-Val-Ile-Arg-Leu-Arg-Glu-Arg-
2  Ser                                         Ala                               Glu-Ile
3  Ser                                         Ala                               Ile
4  His---  ----------------------------
5  Tyr---  ------------Glu---  -----------                                                         Lys
6  Phe---  ----------------------------


                       60                            70
                                                                     ▶
1  Gly-Thr-Lys-Lys-Val-His-Val-Phe-Lys-Glu-Met-Val-Glu-Ala-Trp-Lys-Asn-Arg-Ala-Pro-Lys-Asn-Arg-Pro-Asp-Trp-Met-
2                        Ile              Asp     Ala-Gly-Glu-Arg-Val-Lys   Lys-Lys                  Gly-Ala       Ala
3                        Ile              Asp     Ser-Gly-Glu-Arg-Val-Gln   Asp-Lys              Ala-Gly-Ala       Ala
4                        Ile                                                                                       Ala
5                        Ile                                                                                       Ala
6                        Ile              Gln-Gly-Glu-Arg-Ile-Gln           Pro-Lys    Ser                          Lys


        ▶                             80                            90
1  Pro-Glu-Lys-Ile-Ser-Lys-Pro-Phe-Val-Lys-Lys-Glu-Lys-Lys-Glu-Lys-Ile-Glu-Lys-Ile-Glu-Glu(OH)
2                                                                         Leu        Leu-Asp-Glu-Ile(OH)
3                               Arg                                       Leu        Leu-Asp(OH)
4  Asn-Glu              Trp     Ile-Gly-Val                               Lys        Leu-Asp-Glu-Ile(OH)
5  Asp                  Trp     Gly                                       Lys        Leu-Asp(OH)
6  Ala-Asn              Trp     Leu-Gly-Val                               Asn        Leu  Asp-Ile(OH)
```

Fig. 3. Comparison of amino acid sequences of proteins MC1 from different methanogenic strains. Protein MC1 from Methanosarcina barkeri (1) is taken as reference and only the changes are indicated for the other proteins (2) Methanosarcina MST-A1, (3) Methanosarcina sp.CHTI 55,(4), (5) and (6) variants a, b and c from Methano-thrix soehngenii. The amino acid deletions are represented by ------. Strictly conserved proline residues are indicated by arrows heads.

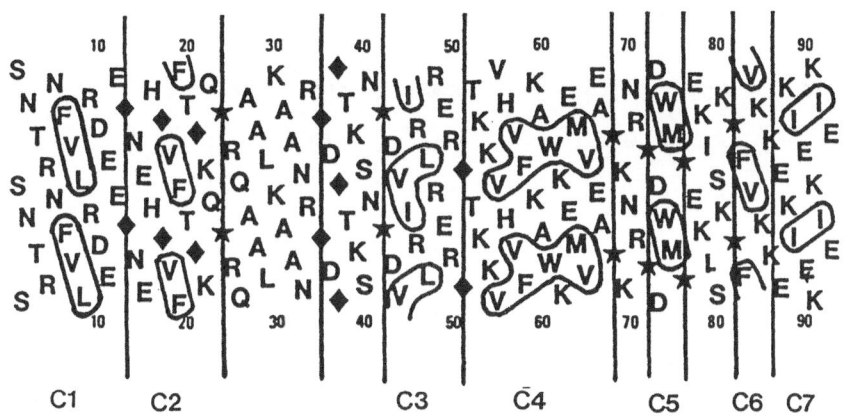

Fig. 4. Plots of M.barkeri protein MC1 obtained by Hydrophobic Cluster
Analysis (17). ★ and ◆ represent proline and glycine residues
respectively. The hydrophobic clusters are circled in bold lines.

In addition, cluster C4 probably corresponds to two β-sheet strands (Val$_{55}$-
Phe$_{58}$) and (Glu$_{63}$-Val$_{65}$) separated by a loop including the tryptophan residue at position 61. The three other methods gave similar results except for
the region covering residues 54 through 67 which would be in α-helix structure. However, since a very low α-helix content has been determined by circular dichroism spectroscopy and taking into account that the three last
methods usually overestimate the α-helix content, it is probable that the
sequence 54-67 is constituted of two β-sheet strands as presented in the
joint prediction structure of protein MC1 (Fig.5). In this structure,only
the sequence 26-32 could form an α-helix and the protein MC1 appears mainly constituted of numerous short β-sheet strands. Infrared spectroscopy
allowed us to confirm the presence of antiparallel β-sheet strands at low
ionic strength (0-50 mM NaCl). Increasing salt concentrations (200-500 mM
NaCl) produce the unfolding of these β-sheet structures without a concomitant formation of α-helix structure.

Several substitutions encountered in the variants of Methanothrix
soehngenii protein MC1 seem to be important regarding the predictions of
their secondary structure. With the replacements of alanine by glycine at
position 60 and of glutamic acid residue by a hydrophobic residue at position 63, all the methods are in agreement to predict two β-sheet strands
(residues 55-58 and 63-66) separated by a β-turn in variants a and b of
Methanothrix soehngenii protein MC1 (Fig.5). The β-turn including residue
at position 60 is not predicted in M.barkeri protein MC1. In the variant c
of M.soehngenii protein MC1, the presence of a proline residue at position
66 hinders the formation of the second β-sheet strand.

We must emphasize that the two regions of conserved amino acid sequences correspond to the structured segments of the protein : they consist of
three β-sheet strands and one α-helix separated each other by a β-turn.
One can note also that according to the predictive methods, the secondary
structure of the region comprised between residues 59 and 76 of protein
MC1 appears well conserved despite the numerous changes of amino acid sequence encountered in this region. These regions of conserved secondary
structure likely play a crucial role in the function of protein MC1 in
methanogenic bacteria.

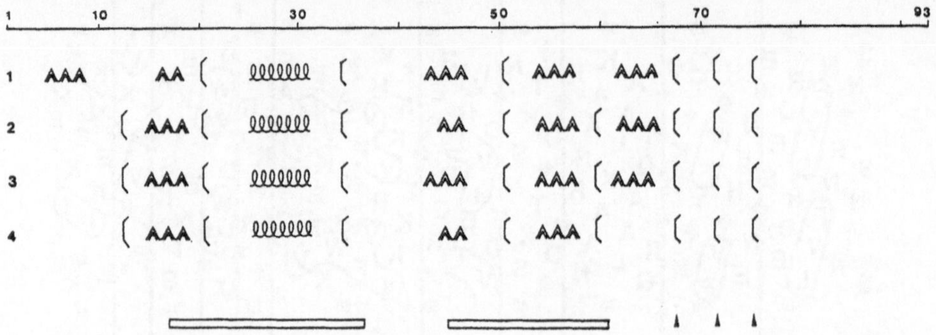

```
 1      10                30              50              70          93
 ├──────┼────────────────┼───────────────┼───────────────┼───────────┤

1   AAA      AA (   ՠՠՠՠՠՠՠ  (      AAA  ( AAA  AAA (  (  (

2        ( AAA (   ՠՠՠՠՠՠՠ  (       AA  ( AAA ( AAA (  (  (

3        ( AAA (   ՠՠՠՠՠՠՠ  (      AAA  ( AAA (AAA (  (  (

4        ( AAA (   ՠՠՠՠՠՠՠ  (       AA  ( AAA (       (  (  (

         ▭▭▭▭▭▭▭▭▭▭      ▭▭▭▭▭▭▭▭       ↓  ↓  ↓
```

Fig. 5. Joint prediction of secondary structures of protein MC1 : <u>Methano-</u>
<u>sarcina barkeri</u> (1), <u>Methanothrix soehngenii</u> variant a (2),
variant b (3), variant c (4). Four predictive methods are used
(see the text). The symbols represent residues in α-helix (ՠՠՠ),
β-sheet (ΛΛ) and β-turn (( ). Frames indicate the regions
of conserved amino acid sequence. Arrows indicate the highly conserved proline residues.

Protein MC1 mainly differs from eukaryotic histones and from the eubacterial DNA-binding protein II by the following structural features :
(i) the number and the distribution of charged residues, (ii) the lack of
large hydrophobic domains (Fig.6), (iii) the nature and the localization
of the variable domains. In protein MC1, the variable domain is rich in
basic and acidic residues and is located in the carboxy-terminal third of
the molecule. In the DNA-binding protein II, the amino-terminal hydrophobic domain is the most variable whereas in histones H2A and H2B, the less
conserved histones of the nucleosomal core particle, variations occur primarily in their basic amino-terminal sequence. Differences are also encountered in the secondary structure of these proteins. The secondary structure
of protein MC1 consists primarily of β-sheet strands whereas the DNA-
binding protein II has a large amount of α-helices and the histones, taken
individually, have no β-sheet strand. Furthermore, upon increasing salt
concentration, protein MC1 becomes unfolded whereas the other proteins
quoted above become more structured (15).

Since little is known about the exact role of protein MC1 in methanogenic bacteria, it is rather difficult to establish a relationship between
these structural features and the function of the protein. Thermal denaturation experiments suggest that protein MC1 modifies the conformation of
DNA. This property is reminiscent of that of the eubacterial DNA-binding
protein II for which several functions have been proposed : packaging of
chromosomal DNA, inhibition of DNA replication and inhibition or stimulation of transcription (2, 22).

Similarly, the function of most of the chromosomal proteins isolated
from other archaebacteria remains to be established. At least, it has been
suggested by Searcy that the physiological function of <u>Thermoplasma acido-</u>
<u>philum</u> protein HTa is to prevent complete separation of the DNA strands
during brief exposures of the organism to denaturing conditions (23). Numerous proteins have been detected in <u>Sulfolobus acidocaldarius</u> by various
authors. Searcy detected two proteins HSa and NHSa of molecular size 14 kDa
and 36 kDa (4). Three classes of proteins, called 7 kDa, 8 kDa and 10 kDa
according to their molecular masses were also isolated from <u>Sulfolobus</u>

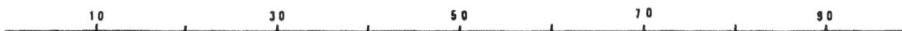

Fig. 6. Comparison of the structural characteristics of E.coli DNA-binding
protein II (1), Methanosarcina barkeri MCl (2) and calf thymus his-
tone H4 (3). Closed and open circles represent basic and acidic
residues respectively. Frames represent hydrophobic clusters,their
height is proportional to the hydrophobicity of sequences determined according to Kyte and Doolittle (21).

acidocaldarius by Dijk and Reinhardt (6). The class 7 kDa is subdivided
into 5 proteins called a, b, c, d and e in order of their increasing basi-
city and of their extent of DNA-binding. The classes 8 kDa and 10 kDa are
subdivided each into two proteins termed 8a, 8b and 10a, 10b respectively.
More recently, Reddy and Suryanarayana have isolated four proteins of molecular mass ranging between 9 kDa and 12 kDa (7). Most of these proteins can
protect DNA from thermal denaturation (4,7).

Amoung these proteins, the protein HTa from T.acidophilum and the proteins 7a, b, d and e were completely sequenced (24-26). Protein HTa is indeed a mixture of two polypeptide chains which only differ by the presence
of an additional methionine residue at the amino-terminal end of the second
chain. The S.acidocaldarius proteins of the class 7 correspond to variants
of the same protein. The proteins 7a and 7d only differ from protein 7b
(58 residues) by 3 and 6 additional residues respectively at their carboxy-
terminus. Moreover, each protein exhibits a different degree of monomethy-
lation of lysine residues at positions 4 and 6. By comparison with variant
7d, the variant 7e is identical in length but differs by three internal
substitutions and the change of arginyl and glutamic acid residues by mono-
methylated lysine residues in the carboxy-terminal sequence.

M.barkeri protein MCl is not structurally related to the chromosomal
proteins cited above. The general characteristics of these proteins are
presented in Table 2. These proteins are rich in basic residues but differ
in their amounts of acidic residues and their net charges. Furthermore they
do not exhibit the same distribution of basic residues : the basic domain
is located in the carboxy-terminal half of protein HTa whereas in S.acido-
caldarius protein of the class 7, basic residues are predominant in the
amino-terminal region. In addition these proteins do not show any sequence
similarity. It appears therefore that the various groups of archaebacteria
exhibit a wide diversity in their chromosomal proteins.

Table 2. Comparison of the general characteristics of archaebacterial
chromosomal proteins

	M.barkeri MC1	T.acidophilum HTa	S.acidocaldarius HSa	S.acidocaldarius 7 kDa variant a
Mol. size	10757	9930	14500	6967
Basic residues %	27	22	16	25
Acidic residues %	15	8	?	18
Net charge	+10	+13	?	+4
Localization of basic character	uniform	C-terminal half	?	N-terminal third
Presence of large hydrophobic sequences	-	+	?	-

REFERENCES

1. D.E. Pettijohn, Cell 30:667 (1982)
2. K. Drlica and J. Rouvière-Yaniv, Microbiol. Reviews 51:301 (1987)
3. D.G. Searcy and D.B. Stein, Biochim. Biophys. Acta 609:180 (1980)
4. G.R. Green, D.G. Searcy and R.G. Delange, Biochim. Biophys. Acta 741: 251 (1983)
5. D.G. Searcy, Biochim. Biophys. Acta 395:535 (1975)
6. J. Dijk and R. Reinhardt, in "Bacterial chromatin" C.O. Gualerzi and C.L. Pon, eds, Springer-Verlag p 186 (1986)
7. T.R. Reddy and T. Suryanarayana, Biochim. Biophys. Acta 949:87 (1988)
8. M. Imbert, B. Laine, G. Prensier, J.P. Touzel and P. Sautière, Can.J. Microbiol. 34:931 (1988)
9. F. Chartier, B. Laine and P. Sautière, Biochim. Biophys. Acta 951:149 (1988)
10. F. Chartier, B. Laine, P. Sautière, J.P. Touzel and G. Albagnac, FEBS Letters 183:119 (1985)
11. J.P. Touzel, G. Prensier, J.L. Roustan, I. Thomas, H.C. Dubourguier and G. Albagnac, Int. J. Syst. Bacteriol. 38:30 (1988)
12. F. Chartier, G. Crevel, B. Laine and P. Sautière, J. Chromatogr. 466: 331 (1989)
13. B. Laine, F. Chartier, M. Imbert, R. Lewis and P. Sautière, Eur. J. Biochem. 161:681 (1986)
14. F. Chartier, B. Laine, D. Belaïche, J.P. Touzel and P. Sautière, Biochim. Biophys. Acta in press (1989)
15. M. Imbert, Thesis, University of Lille I, France (1987)
16. F. Chartier, B. Laine, D. Belaïche and P. Sautière, J. Biol. Chem. in press (1989)
17. C. Gaboriaud, V. Bissery, T. Benchetrit and J.P. Mornon, FEBS Letters 224:149 (1987)
18. P.Y. Chou and G.D. Fasman, Ann. Rev. Biochem. 47:251 (1978)
19. J. Garnier, D.J. Osguthorpe and B. Robson, J. Mol. Biol.120:97 (1978)
20. A. Sette, G. Doria and L. Adorini, Mol. Immunol. 23:807 (1986)
21. J. Kyte and R. Doolittle, J. Mol. Biol. 157:105 (1982)
22. K. Mensa-Wilmot, K. Carroll and R. Mc Macken, Embo J. 8:2393 (1989)
23. D.B. Stein and D.G. Searcy, Science 202:219 (1978)
24. R.J. Delange, L.C.Williams and D.G. Searcy, J.Biol.Chem.256:905 (1981)
25. M. Kimura, J. Kimura, P. Davie, R. Reinhardt and J. Dijk, FEBS Letters 176:176 (1984)
26. T. Choli, B.Wittman-Liebold and R.Reinhardt, J.Biol.Chem. 263:7087 (1988)

ACKNOWLEDGEMENTS

The authors are grateful to Drs J.P. Touzel and G. Albagnac who provided the bacterial strains and for valuable discussions. They also thank Dr G. Prensier for electron microscopy studies and the service of microsequences of the Unité 409 CNRS for sequencer runs. They are indebted to A. Lemaire, M.J. Dupire and J. Herno for skilful technical assistance and to T. Ernout for editorial assistance. This work was supported by grants from the Centre National de la Recherche Scientifique, from the Pôle Régional des Anaérobies de la Région Nord-Pas-de-Calais, from the Université de Lille II and from the Fondation pour la Recherche Médicale.

CONFERENCES - G : GENETICS

AN ARCHAEBACTERIAL IN VITRO TRANSCRIPTION SYSTEM

Michael Thomm, Gerhard Frey, Winfried Hausner
and Berit Brüdigam

Lehrstuhl für Mikrobiologie, Universität
Regensburg, D-8400 Regensburg, FRG

ABSTRACT

An RNA polymerase fraction of Methanococcus vannielii
purified by gradient centrifugation synthesizes an RNA product
of 110 nucleotides in the presence of a template harbouring a
homologous tRNA^{Val} gene. The length of this transcript corresponds exactly to that of the tRNA precursor molecule synthesized in vivo. After internal deletions of 5 and 11 basepairs
from the DNA region encoding the tRNA gene, in vitro transcripts of 105 and 99 nucleotides were observed. This finding
confirms our conclusion that the RNA products originate from
the tRNA gene. Optimal transcription of the tRNA gene occurs at
a incubation temperature of 50 °C in the presence of 10 mM
MgCl$_2$ and 20 mM KCl. Synthesis of the 110 nucleotide RNA
product is maximal at a DNA-concentration of 100 µg/ml and is
inhibited at higher DNA-concentrations. By mutagenesis of the
DNA region upstream of the tRNA gene, the DNA sequence promoting in vitro transcription was located between -58 and -22.
Therefore, the TATA-box at -25 which has been proposed as an
archaebacterial consensus promoter sequence (Thomm and Wich,
1988), appears to be indispensable for initiation of transcription.

INTRODUCTION

In contrast to the RNA polymerases of eubacteria, the RNA
polymerases purified from archaebacterial cells are unable to
initiate transcription accurately in vitro (Zillig et al.,
1988). However, recently specific binding of the purified enzyme from the methanogen Methanococcus vannielii to the promoter region of both protein-encoding (Thomm et al., 1988a; Brown
et al., 1988) and tRNA/rRNA genes (Thomm and Wich, 1988) has
been demonstrated. From these footprinting experiments a TATA-
box at -25 has been inferred as an archaebacterial consensus
promoter sequence (Thomm et al., 1989). However, also the
purified RNA polymerase of M. vannielii is unable to accurately transcribe purified genes. To investigate the requirements
for cell-free transcription of archaebacterial genes, the

Microbiology and Biochemistry of Strict Anaerobes Involved in Interspecies Transfer
Edited by J.-P. Bélaich et al.
Plenum Press. New York. 1990

expression of a tRNA^{Val} gene of <u>M. vannielii</u> by less purified RNA polymerase fractions was studied. We describe here the purification and some properties of an RNA polymerase fraction, directing correct <u>in vitro</u> transcription of this template.

MATERIAL AND METHODS

Purification of the endogenous RNA polymerase from the crude extract

A soluble extract of <u>M. vannielii</u> cells (S-100) was prepared as described previously (Thomm et al., 1988b). The endogenous RNA polymerase was separated from the bulk of cellular proteins by glycerol-gradient centrifugation (Wingender et al., 1984).

DNA isolation and construction of mutated plasmids

The plasmids for the <u>in vitro</u> transcription reactions were purified by repeated centrifugation in CsCl density gradients as described previously (Thomm and Wich, 1988). Plasmid pIC31/1 contains the <u>ClaI</u> fragment of plasmid pMT31 (Wich et al., 1986b) inserted into the <u>ClaI</u> site of the cloning vector pIC-19H (Marsh et al., 1984). The different 5' deletion clones of the tRNA^{Val} gene were constructed by unidirectional digestion with exonuclease III using the protocol of Henikoff (1984). The clones pIC31/4 and pIC31/6 which contain deletions of internal sequences of the tRNA gene were constructed by the ligation of

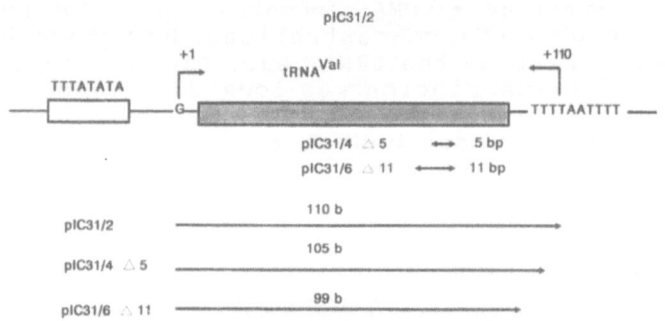

Fig.1. Genetic map of the tRNA^{Val} gene used as a template for the <u>in vitro</u> transcription reactions. Plasmid pIC31/2 contains the 5' flanking DNA sequence of the tRNA gene including the nucleotide at position -58. The plasmids pIC31/4 and pIC31/6 are subclones with internal deletions of 5 and 11 bp, respectively. The TATA-element upstream from the transcription start site is boxed, the region encoding the tRNA indicated by thick dark bars. The arrows indicate the <u>in vivo</u> initiation and termination sites of transcription determined by S1 mapping (Wich et al., 1986) and primer extension experiments (data not shown). The length of the <u>in vitro</u> transcripts from the different templates is indicated in the lower part of the figure.

appropriate DNA restriction fragments. The DNA sequences of all mutated templates were verified by dideoxy sequencing (Sanger et al., 1977).

In vitro transcription assays

The reaction mixture for the synthesis of the tRNA precursor contained 40 mM Tris-HCl, pH 8.0, 10 mM KCl, 8 mM MgCl$_2$, 0.1 mM EDTA, 0.05 mM ZnSO$_4$ and plasmid pIC31/2 (see Fig. 1) at a final DNA concentration of 50 µg/ml. Aliquots of 20 µl from the glycerol gradient fractions were added to the reaction mixtures to give a final volume of 100 µl. After 5 min preincubation at 50 °C, the transcription was started by the addition of 0.33 mM each of ATP, GTP, CTP and 0.0165 mM and 10 µCi α-^{32}P UTP (600 Ci/mmol, NEN). The transcription reaction was allowed to proceed for 30 min at 50°C. The reaction was stopped and the RNA products purified and separated by electrophoresis on 6% polyacrylamide/urea gels as described by Jahn et al. (1987).
To determine unspecific RNA synthesis the same conditions were employed exept that the plasmid DNA was replaced by poly d(A-T) (0.1 mg/ml) in the transcription reactions. The incorporation of radioactivity into acid-insoluble RNA was measured as described previously (Thomm and Stetter, 1985).

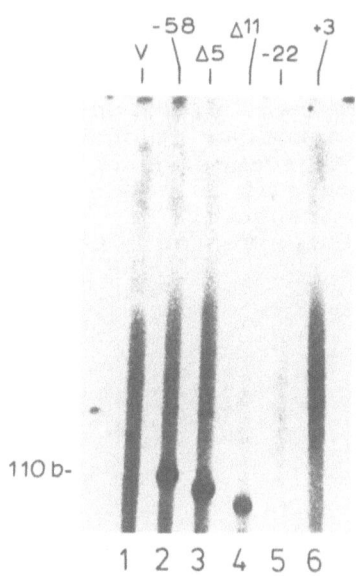

Fig. 2. Analysis of the in vitro RNA products transcribed from the templates shown in Fig. 1 and 3. The position of the RNA product with 110 nucleotides is indicated on the left side of the figure. The various templates used for the in vitro transcription reactions were: Lane 1, vector DNA (indicated by V on top); lanes 2, 5 and 6, deletion clones of the upstream DNA region generated by exonuclease III mutagenesis. The 5' boundaries of the corresponding upstream deletions are indicated on top. Lanes 3 and 4, internal deletion clones pIC31/4 and pIC31/6 (see Fig. 1); the number of nucleotides which have been deleted from the tRNA encoding region are marked by Δ5 and Δ11 on top.

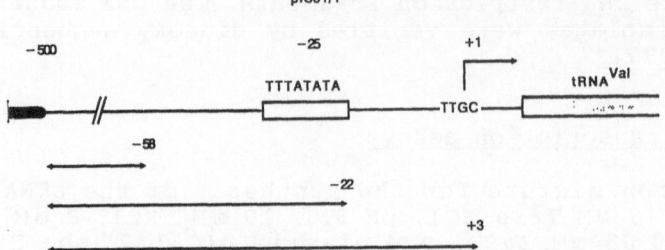

Fig. 3. Map of the upstream region of the tRNA^{Val} gene of M. vannielii. The arrows indicate the extent of the various 5' deletions. The TATA-element at -25 is boxed, the region encoding the mature tRNA marked by thick dark bars. The nucleotides at the transcription start site (+1; labelled in addition by an arrow) are shown. Plasmid pIC31/1 contains 500 bp of the wildtype upstream DNA sequence.

RESULTS AND DISCUSSION

The endogenous RNA polymerase of M. vannielii purified by gradient centrifugation of a crude extract was incubated with a supercoiled plasmid harbouring an homologous tRNA^{Val} gene and 58 nucleotides of the 5' flanking DNA region (pIC31/2; Fig. 1). When initiation and termination of transcription occurs in vitro at the same sites as in Methanococcus cells (Wich et al., 1986a), a transcript of 110 nucleotides should be expected as major RNA product. Analysis of the labelled in vitro RNA by electrophoresis in calibrated polyacrylamide/urea gels revealed that a transcript of this size was synthesized (Fig. 2, lane 2). When the vector DNA without a tRNA gene was

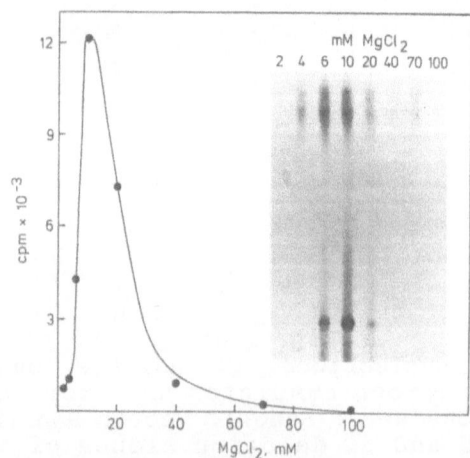

Fig. 4. MgCl$_2$-dependence of tRNA transcription. The amounts of pre-tRNA synthesized in the presence of varying amounts of MgCl$_2$ was determined after electrophoresis of the reaction products. The labelled RNA bands were excised from 6% polyacrylamide/urea gels and quantified by Cerenkov counting.

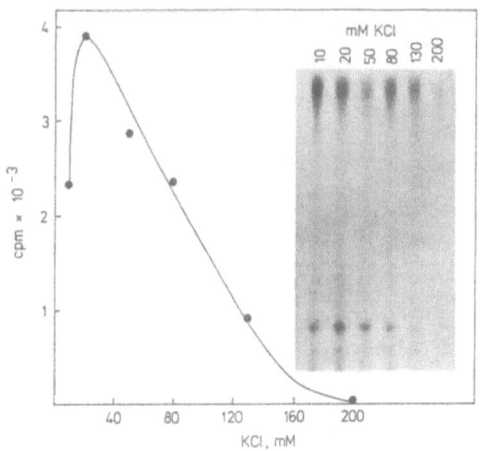

Fig. 5. Influence of KCl concentration on the in vitro expression of the tRNA^{Val} gene. Pre-tRNA synthesis was measured at different concentrations of KCl. The MgCl$_2$ concentration was 10 mM. Product analysis and quantitation was as described in the legend of Fig. 4.

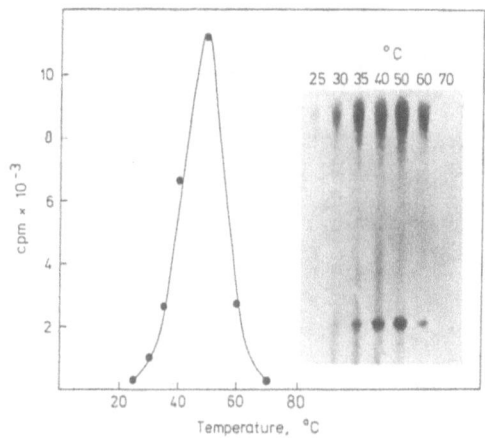

Fig. 6. Effect of the incubation temperature on transcription of the tRNA^{Val} gene. The radioactivity incorporated into the RNA product of 110 nucleotides in response to different incubation temperatures of the transcription reactions was measured as described in the legend of Fig. 4. The MgCl$_2$- and KCl concentrations were 10 and 20 mM, respectively.

employed as a template no distinct RNA product could be detected (Fig. 2, lane 1). To provide additional evidence that the in vitro transcripts originate indeed from the tRNA^{Val} gene, internal deletions of 5 and 11 basepairs were introduced into the tRNA template (Fig. 1). The transcripts from the deletion clones pIC31/4 and pIC31/6 should therefore be reduced in their size by 5 and 11 nucleotides, respectively. Analysis of the corresponding in vitro transcripts revealed that RNA products of 105 and 99 nucleotides were synthesized (Fig. 2, lanes 3 and

4). These results support the conclusion, that this RNA polymerase fraction of M. vannielii is able to faithfully transcribe
homologous tRNA genes.
When pIC31/2 was replaced in the in vitro transcription reactions by plasmid pIC31/1 which contains 500 basepairs of the 5'
flanking region instead of 58, the same rate of tRNA expression
was observed (data not shown). This finding indicates that the
DNA region upstream of -58 is not essential for in vitro transcription of the tRNA^{Val} gene. To define the DNA sequences
promoting in vitro transcription of the tRNA^{Val} gene, two
additional plasmids with deletions extending to the DNA region
downstream of -58 were constructed (Fig. 3). After deletion of
the nucleotides of the TATA-box including position -22, the
efficiency of transcription was dramatically reduced (Fig. 2,
lane 5). When the deletion extends to position +3 of the tRNA
gene (Fig. 3) no distinct in vitro transcripts from this
template could be detected (Fig. 2, lane 6). Thus, the DNA
sequence required for specific transcription of this tRNA gene
is located in the DNA region between -58 and -22. These data
strongly suggest that the TATA-box represents the main signal
promoting the expression of this tRNA gene. Since this sequence
is conserved at the same location in most archaebacterial genes
(Thomm and Wich, 1988) the TATA-box might be regarded as a
major promoter signal directing the transcription of constitutive genes in archaebacteria.
      To facilitate a further characterization of the RNA
products and the factors involved in expression of the tRNA
gene, some properties of the extract directing the cell-free
transcription were determined. Mg^{2+} is absolutely required for
the expression of the tRNA gene. Synthesis of the tRNA precursor occurs between 6 and 20 mM MgCl$_2$, with an optimum at 10
mM (Fig. 4). The rate of transcription of the tRNA^{Val} gene is
optimal at 20 mM KCl. A significant expression of this template

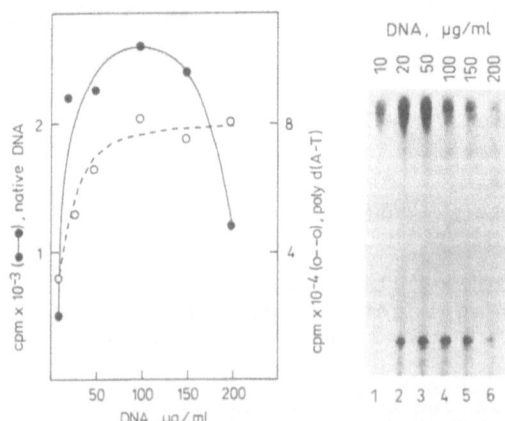

Fig. 7. Effect of DNA concentration on the synthesis of unspecific RNA (O--O) and pre-tRNA (●-●). The templates employed
for the reactions were poly d(A-T) and plasmid pIC31/2 (Fig.
1), respectively. The pre-tRNA product was quantified as described in Fig. 4, the transcripts from poly d(A-T) by liquid
scintillation counting of acid insoluble labelled RNA.

was observed up to a KCl concentration of 130 mM (Fig 5). Optimal transcription of the tRNA gene occured at a temperature of 50 ^{O}C (Fig. 6) although M. vannielii is a mesophilic strain which shows its temperature optimum for growth at 37 ^{O}C. In general, the activation profiles for the specific synthesis of the tRNA precursor resemble those obtained when the synthetic template poly d(A-T) was transcribed with the purified RNA polymerase (Frey, 1987). When the DNA-dependence of transcription was determined a striking difference between specific and unspecific RNA synthesis was observed. With polyd(A-T) as template, the rate of RNA synthesis is higher at increased DNA-concentrations in the transcription reactions until a plateau is reached (Fig. 7). However, the rate of pre-tRNA synthesis is decreased when the DNA concentration in the transcription reactions is higher than 100 μg/ml (Fig. 7). This inhibition of specific RNA synthesis suggests that a cooperative interaction of both a DNA-binding factor and the RNA polymerase with the promoter is a prerequisite for correct initiation of transcription. Assuming that a DNA-binding factor exists, at high DNA-concentration the probability is lower that the transcription factor and the RNA polymerase can form a pre-initiation complex at the same promoter. The inhibition of pre-tRNA synthesis at high DNA concentrations thus might be explained by a distribution of this factor and the RNA polymerase onto different DNA molecules.

ACKNOWLEDGEMENTS

We appreciate the excellent technical assistance of S. Hommer. This work has been supported by the Deutsche Forschungsgemeinschaft and The Fonds der Chemischen Industrie. We thank Dr. Stetter for the generous financial support of this work by funds of the Leibnitz Preis.

REFERENCES

Brown, J.D., Thomm, M., Beckler, G., Frey, G., Stetter, K.O., and Reeve , J., 1988, An archaebacterial RNA polymerase binding site and transcription initiation of the hisA gene of Methanococcus vannielii, Nucl. Acids Res., 10:135

Frey, G., 1987, Diplomarbeit, Universität Regensburg

Henikoff, S., 1984, Unidirectional digestion with exonuclease III creates targeted breakpoints for DNA sequencing, Gene, 28:351

Jahn, D., Wingender, E. , and Seifart, K. H., 1987, Transcription complexes for various class III genes differ in parameters of formation and stability towards salt, J. molec. Biol., 193:303

Sanger, F., Nicklen, S., and Coulson, A. R., 1977, DNA sequencing with chain-terminating inhibitors, Proc. Natl. Acad. Sci. USA, 74:5463

Thomm, M. and Stetter, K.O., 1985, Transcription in methanogens. Evidence for specific in vitro transcription of the purified DNA-dependent RNA polymerase of Methanococcus thermolithotrophicus, Eur. J. Biochem., 149:345

Thomm, M., Sherf, B.A. and Reeve, J.N., 1988a, RNA polymerase-binding and transcription initiation sites upstream of

the methyl reductase operon of <u>Methanococcus</u> <u>vannielii</u>,
<u>J</u>. <u>Bacteriol</u>., 170:1958

Thomm, M., Frey, G., Bolton, B.J., Laue, F., Kessler, C. and
Stetter, K.O., 1988b, MvnI: a restriction enzyme in the
archaebacterium <u>Methanococcus</u> <u>vannielii</u>, <u>FEMS</u> <u>Microbiol</u>.
<u>Letters</u>, 52:229

Thomm, M. and Wich, G., 1988, An archaebacterial promoter
element for stable RNA genes with homology to the TATA
box of higher eukaryotes, <u>Nucl</u>. <u>Acids</u> <u>Res</u>., 16:151

Thomm, M., Wich, G., Brown, J. W., Frey, G., Sherf, B. A., and
Beckler, G. S.,1989, An archaebacterial promoter sequence assigned by RNA polymerase binding experiments,
<u>Canad</u>. <u>J</u>. <u>Microbiol</u>., 35:30

Wich, G., Hummel, H., Jarsch, M., Bär, U., and Böck, A.,
1986a, Transcription signals for stable RNA genes in
Methanococcus, <u>Nucl</u>. <u>Acids</u> <u>Res</u>., 14:2459

Wich, G., Sibold, L., and Böck, A., 1986b, Genes for tRNA and
their putative expression signals in Methanococcus,
<u>System</u>. <u>Appl</u>. <u>Microbiol</u>., 7:18

Wingender, E., Jahn, D., and Seifart, K. H., 1986, Association
of RNA polymerase III with transcription factors in the
absence of DNA, <u>J</u>. <u>biol</u>. <u>Chem</u>., 261:1409

Zillig, W., Palm, P., Reiter, W.D., Gropp, F., Pühler, G., and
Klenk, H.P., 1988, Comparative evaluation of gene expression in archaebacteria, <u>Eur</u>. <u>J</u>. <u>Biochem</u>., 173:473

A SURVEY OF RECENT ADVANCES IN GENETIC ENGINEERING IN *BACTEROIDES*

M. Béchet, P. Pheulpin, J.-C. Joncquiert, Y. Tierny, and
J.-B. Guillaume

Laboratoire de Microbiologie, Université des Sciences et
Techniques de Lille Flandres-Artois, F-59655 Villeneuve
d'Ascq-Cédex, France

INTRODUCTION

Anaerobic Gram-negative bacteria belonging to the genus *Bacteroides*
are the predominant inhabitants of the gastro-intestinal tracts of man
and mammals. Some species are common members of the human oral flora, and
others are living in the rumen ecosystem. According to the Bergey's
manual of determinative bacteriology and recent reports, 50 *Bacteroides*
species have been identified, some of which being sufficiently divergent
to be subject to reclassification proposals (Holdeman et al., 1984; Okuda
ct al., 1985; Jensen and Canale-Parola, 1986; Shah and Collins, 1988;
Montgomery et al., 1988). Several species such as *Bacteroides fragilis*
are known as opportunistic pathogens able to colonize lesions of their
natural host, leading to formation of abscesses or blood stream infections
(Salyers, 1984). On the other hand, black-pigmented *Bacteroides* species
such as *Bacteroides gingivalis* are true pathogens of the oral cavity
(Mayrand and Holt, 1988), whereas *Bacteroides nodosus* is the causative
agent of ovine footrot (Elleman and Hoyne, 1984).

As chemoorganotrophs, *Bacteroides* species can be subdivided into
two main groups according to the presence of saccharoclastic properties.
Strong saccharolytic species such as *Bacteroides thetaiotaomicron* are
capable of utilizing a variety of simple sugars, and contribute actively
to the breakdown of complex carbohydrates digested or not by the host
(plant cell wall polysaccharides, starches, pectins,...) or secreted by
it (mucopolysaccharides, mucines). Enzymes involved in the latter properties are mainly located in the periplasmic space or cytoplasmic (Salyers
and O'Brien, 1980; McCarthy et al., 1985; Anderson and Salyers, 1989).
Fermentation products consist in mixtures of succinate, acetate, lactate,
formate or propionate, sometimes with short-chained alcohols. In addition,
some species such as *Bacteroides fragilis* produce hydrogen (up to 3% in
the headspace of broth cultures)(Macy and Probst, 1979; Holdeman et al.,
1984; Salyers, 1984; Lin et al., 1985). Nonsaccharoclastic species possess
important proteolytic activities which, in somes cases, are implicated in
pathogenicity, e.g., collagenases of *B. gingivalis*. Peptones catabolism
generates combinations of succinate, acetate, formate or lactate, often
with volatile fatty acids and alcohols. Hydrogen is sometimes produced
(Holdeman et al., 1984; Mayrand and Holt, 1988).

Microbiology and Biochemistry of Strict Anaerobes Involved in Interspecies Transfer
Edited by J.-P. Bélaich *et al.*
Plenum Press, New York, 1990

Ten years ago, intra- and interspecific conjugal transfers of resistance to clindamycin in *Bacteroides* were reported (Privitera et al., 1979; Tally et al., 1979; Welch et al., 1979). Characterization of the genetic elements involved in these events has allowed to devise cloning systems for these important microorganisms. Indeed, although genes of *Bacteroides* and other anaerobes such as *Clostridium* can be cloned and expressed in *Escherichia coli*, *Bacteroides* strains may represent interesting alternative and/or complementary hosts to study expression of anaerobic genes.

The purpose of the present paper is to survey: (i) the cloning systems available for introducing genes in *Bacteroides* and studying their expression; (ii) *Bacteroides* determinants which have been cloned and expressed in *E.coli*; and (iii) expression of various foreign and indigenous genes in *Bacteroides* . Potential applications in both fundamental and applied areas will be discussed.

CLONING SYSTEMS FOR TRANSFER OF GENES OF INTEREST IN *BACTEROIDES*

Before describing the different cloning systems, it appears necessary to remember basal aspects of essential genetic traits of *Bacteroides* genus, i.e., antibiotic resistance and plasmid content.

Antibiotic Resistance Patterns and Plasmid Content

Bacteroides strains are uniformly resistant to aminoglycosides. Most of gastro-intestinal isolates are resistant to β-lactam antibiotics such as ampicillin, and to nalidixic acid. About half of them are resistant to erythromycin and two thirds to tetracycline. On the other hand, the majority of strains are susceptible to chloramphenicol, clindamycin, fusidic acid, metronidazole and rifampin. However, the emergence of strains resistant to clindamycin and metronidazole is constantly increasing (Tally and Malamy, 1984; Hill and Ayers, 1985; Mary et al., 1986; Scher, 1988; Breuil et al., 1989). *Bacteroides* species of the oral cavity are generally resistant to chloramphenicol, and sensitive to clindamycin, β-lactams, erythromycin, metronidazole and tetracycline (Baker et al., 1985; Mayrand and Holt, 1988).

Many *Bacteroides* strains belonging to different species have been found to harbour one to five plasmid(s) with sizes ranging from 2.7 to more than 80 kilobases-pairs (kb). Most of them are smaller than 10 kb, and no phenotypic character has been assigned to them. Several classes of homology among these cryptic plasmids have been distinguished according to size, restriction endonucleases patterns, and ability of replicating/coexisting within different clinical isolates (Callihan et al., 1983; Beul et al., 1985; Mary et al., 1986). However, demonstration of transferable constitutive resistance to clindamycin in *Bacteroides* was associated with the presence of a conjugative (Tra+) plasmid in donor strain: pBF4 (also called pIP410; 41 kb; from *B. fragilis*; Welch and Macrina, 1981), pBFTM10 (14.6 kb; from *B. fragilis*; Tally et al., 1982) and closely related pCP1 (from *B. thetaiotaomicron*; Guiney et al., 1984b), and pBI136 (82 kb; from *Bacteroides ovatus* ; Smith and Macrina, 1984). Two other transferable plasmid-linked resistance determinants have been recently reported: pRRI4 (19.5 kb; Tra+) confers tetracycline resistance in *Bacteroides ruminicola* (Flint et al., 1988), and pIP417 (7.7 kb; mobilizable by the coresident plasmid pIP418) mediates resistance to 5-nitroimidazoles in *Bacteroides vulgatus* (Breuil et al., 1989).

Extensive studies of the Cc^r-carrying plasmids first revealed homologies in regions involved in that resistance (Guiney et al., 1984b; Smith and Gonda, 1985). These latter were then identified as transposons: Tn*4351* for pBF4 (Shoemaker et al., 1985), Tn*4400* for pBFTM10 (Robillard et al.,

1985), and Tn4551 for pBI136 (Smith and Spiegel, 1987). All three transposons are flanked by two 1155 bp insertion sequences (IS) with the same
orientation and designated IS4351, IS4400 and IS4551, respectively. Tn4351
and Tn4400 are highly homologous, while Tn4551 displays homology with them
only at the level of Cc^r/Em^r (ermF) gene and IS (Odelson et al., 1987).
Nucleotide sequences of IS4351, IS4551 and adjacent ermF genes have shown
that the -35 and -10 promoter sequences of ermF genes were located within
the end of their respective IS (Rasmussen et al., 1986; Rasmussen et al.,
1987; Smith, 1987). In addition, Tn4351 and Tn4400 harbour a tetracycline
resistance determinant in the vicinity of the Cc^r/Em^r gene, but not Tn4551
(Guiney et al., 1984a; Robillard et al., 1985; Smith and Gonda, 1985).
This $*Tc^r$ element, cryptic in Bacteroides, confers tetracycline resistance
only in aerobically grown E. coli. Here, phenotypic expression involves
both detoxification and efflux of tetracycline in presence of oxygen (Park
and Levy, 1988; Speer and Salyers, 1989). In other respects, these trans-
posons-borne Cc^r/Em^r and $*Tc^r$ genes have proved essential for construction
of various plasmid vectors for Bacteroides genus.

Plasmid Vectors for Gene Transmision in Bacteroides

None of Tra⁺ plasmids belonging to various incompatibility groups
(Inc) in E. coli was found to establish in Bacteroides, suggesting narrow-
host-range of this genus (Guiney et al., 1984c). Thus, it appeared necessary to construct composite plasmids carrying (i) a Bacteroides replicon;
(ii) an antibiotic resistance marker allowing selection in Bacteroides;
and (iii) similar characters for maintenance in E. coli. Such a chimeric
'shuttle' plasmid, pDP1, was first described by Guiney et al. (1984c). It
consisted of pDG5, a Tc^s derivative of E. coli vector pBR322 carrying the
oriT region of the broad-host-range IncPα plasmid RK2, coupled to a large
part of the B. thetaiotaomicron plasmid pCP1. This hybrid plasmid (19 kb)
replicated in E. coli and, upon mobilization by the Ap^s RK2-derivative
pRK231, in Bacteroides. Resistances to Ap and Tc were used as markers in
E. coli, while the Cc^r/Em^r gene of pCP1 permitted detection of Bacteroides
transconjugants at a frequency of 3.0×10^{-6} per recipient. As expected,
the latter did not contain pRK231, indicating narrow-host-range of the
recipient. However, the main disadvantages of pDP1 were its relatively
large size and the lack of convenient restriction sites to serve as cloning
vector. Since this report, various vehicles have been achieved, most of
them having sizes around 10 kb or less. According to their specificity,
they can be classified into the three main groups shown in Table 1.

E. coli-Bacteroides shuttle vectors. These vectors contain a pBR322-
or RSF1010-based replicon for maintenance in E. coli, and a Bacteroides
replicon. To select Bacteroides transconjugants/transformants, the sole
marker available is the transposons-originating Cc^r/Em^r gene. The majority
of the plasmids are mobilizable by helper IncP Tra⁺ plasmids such as R751
or pRK231, which cannot replicate in Bacteroides. For our part, we have
developed a series of shuttle vectors by combining the pBR322 derivative
pKC7 (5.9 kb) with (i) pBF367, a 4.6 kb cryptic plasmid from B. fragilis
367 (Mary et al., 1986), and (ii) the EcoRI-B fragment of pBFTM10 which
bears one IS4400 and adjacent Cc^r/Em^r and $*Tc^r$ genes (Robillard et al.,
1985). The resulting composite plasmid, pKBF367-1 (14.7 kb) was mobilized
from E. coli by either the IncPβ plasmid R751 (pBF367 replicon involved in
transfer) or the ColE1 derivative pRK2013 carrying tra genes of RK2 (pKC7
component involved in transfer), into a strain of Bacteroides distasonis
at frequencies of 10^{-5}. Successive deletions of non-essential regions of
pKBF367-1 led to derivatives of 12.8, 10.5 and 9.3 kb (Pheulpin et al.,
1988), and then 8.2 kb (this paper). This latter retained properties of
the parental plasmid (Fig. 1a). Using them, we introduced transposon Tn501
(mediating mercury resistance) into the B. distasonis strain, but no
expression of the Hg^r phenotype was observed in the transconjugants.

Table 1. Characteristics of different plasmid vectors
constructed for *Bacteroides*

Name	Replicon E. coli	Bacter	Markers[a] E. coli	Bacter	Mob[b]	Size (kb)	References
1. Shuttle vectors							
pDP1	pBR322	pCP1	Ap, Tc	Cc/Em	+	19.0	Guiney et al. (1984c)
pE5-2	RSF1010	pB8-51	Su, Tc	Cc/Em	+	17.1	Shoemaker et al. (1985)
pFD176	pUC19	pBI143	Ap, Lac	Cc/Em	-	7.3	Smith (1985a)
pEG920	pUC19	pB8-51	Ap, Tc	Cc/Em	±	11.0	Shoemaker et al. (1986b)
pFD214	pUC19	pBI143	Ap, Lac	-	-	6.3	Smith (1987)
pVAL-1	pBR328	pB8-51	Ap, Tc	Cc/Em	+	11.0	Valentine et al. (1988)
pKBF367 series	pKC7	pBF367	Ap, Tc and/or Km	Cc/Em	+	14.7 -8.2	Pheulpin et al. (1988) and here
pDP/pDK series	pBR322	pCP1	Ap, Tc +/- Lac	Cc/Em or -	+	15.0 -7.4	Guiney et al. (1988)
pOA10 (cosmid)	pBR322	pCP1	Ap, Tc	-	+	10.0	Guiney et al. (1988)
pNJR1 (cosmid)	RSF1010	pB8-51	Km, Sm	-	+	14.4	Shoemaker et al. (1989)
2. *Bacteroides*-specific vector							
pBI191	-	pBI143	-	Cc/Em	-	5.3	Smith (1985a)
3. Suicide vectors							
pSS-2	RSF1010	-	Su, Tc	Cc/Em	+	46.0	Shoemaker et al. (1985)
pE3-1	pBR328	pB8-51 (Rep⁻)	Ap, Tc	Cc/Em	+	13.0	Guthrie and Salyers (1986)
pFD197 (unstable)	pUC19	pBI143	Ap, Km Lac	Cc/Em	-	16.7	Smith and Spiegel (1987)
pVAL-7	pBR328	-	Ap	Cc/Em	+	9.5	Smith and Salyers (1989)

[a] Resistances to: Ap, ampicillin; Cc, clindamycin; Em, erythromycin; Km, kanamycin; Sm, streptomycin; Su, sulphonamide; and Tc, tetracycline. Lac, β-galactosidase (*lacZ* gene). [b] Mobilizable by a helper plasmid.

Attempts to transfer the pKBF367 series into *Bacteroides ruminicola* 23 (type strain) were unsuccessful. This strain contains one single plasmid of 9.5 kb (pRRI7) and displays an antibiotic resistance pattern similar to those of intestinal *Bacteroides* spp. (Flint and Stewart, 1987). Therefore, we constructed chimeric plasmids consisting of *Cla*I-digested pRRI7 ligated to pKC71 (*Eco*RI-B fragment of pBFTM10 inserted within the *Eco*RI site of pKC7; Pheulpin et al., 1988) partially cleaved by *Cla*I. The resulting hybrid plasmids, pKBR23-1 and pKBR23-2 (19.6 kb; Fig. 1b), were mobilized by pRK2013 into *B. distasonis* at a frequency of 10^{-5}, but not by R751. Mobilization by pRK2013 into *B. ruminicola* 23 yielded transconjugants at low frequencies ($2.0-7.0 \times 10^{-7}$ per recipient). Comparison of total DNA restriction profiles showed that these transconjugants were identical to parent strain 23, and analysis of their plasmid content revealed only the shuttle plasmids (data not shown).

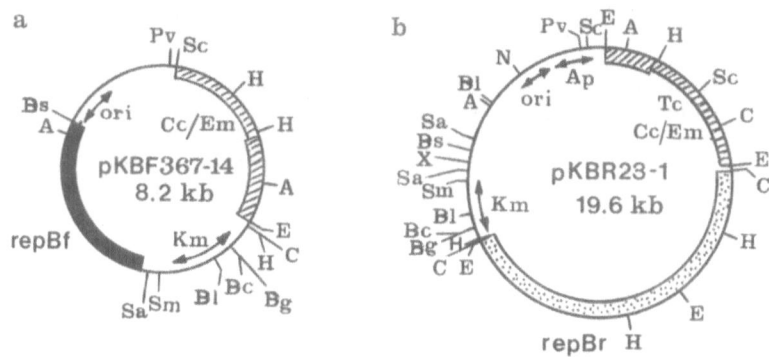

Fig. 1. Restriction maps of the *E. coli–Bacteroides* shuttle
plasmids (a) pKBF367-14; (b) pKBR23-1. The hatched zone
represents the remaining of Th*4400*.
The pRRI7 replicon is inverted in pKBR23-2.

Smith has achieved a series of non-mobilizable shuttle vehicles which
can be introduced in *Bacteroides* strain 638 and some others, using poly-
ethylene-glycol-mediated transformation (Smith, 1985b). pFD176 (7.3 kb) is
the smaller shuttle vector conferring clindamycin resistance, whereas
pFD214, which carries a promoterless Cc^r/Em^r gene from Tn*4551*, can serve
to detect promoter activity in cloned *Bacteroides* DNA fragments. Recently,
two shuttle cosmid vectors, which are devoid of antibiotic resistance
marker for *Bacteroides*, have been described: pOA10 was used to test
expression of a foreign Tc^r determinant in *B. fragilis* (Guiney et al.,
1988), while pNJR1 was employed to isolate and characterize a chromosomal
Tra^+ Tc^r-Em^r element from *B. thetaiotaomicron* (Shoemaker et al., 1989).

   Bacteroides-specific cloning vector. To date, there is only one
cloning vector harbouring a *Bacteroides* replicon and Cc^r/Em^r gene from
Tn*4551*: pBI191 (5.3 kb) contains also a small fragment from the pUC19
region bearing a multiple cloning sequence, and is introduced into
B. fragilis 638 by PEG-mediated transformation (Smith, 1985a).

   Suicide plasmids for transposon or insertional mutagenesis in
Bacteroides. pSS-2 was the first suicide plasmid carrier used for delivery
of Tn*4351* in *Bacteroides* . Upon mobilization from *E. coli* to *Bacteroides*
uniformis by R751, all Em^r transconjugants contained single insertions of
Tn*4351*. However, introduction of Tn*4351* with R751 as suicide vehicle
(since it cannot replicate in *Bacteroides*) led to cointegration of R751::
Tn*4351* into the chromosome of half of the transconjugants. Auxotrophic
mutants arose in 13% of Em^r transconjugants and were mainly methionine-
requiring. These mutations were observed to revert at relatively high frequency (Shoemaker et al., 1986a). Three other suicide plasmids have been
reported: pFD197, highly unstable in absence of selective pressure for
clindamycin resistance, generated Tn*4551*-induced auxotrophs in *B. fragilis*
transformants (Smith and Spiegel, 1987), whereas pE3-1 and pVAL-7 were
used for insertional inactivation of genes involved in complex carbohydrates in *B. thetaiotaomicron*(see below).

   An important feature of several vectors replicating in *Bacteroides*
is their ability to be mobilized from their *Bacteroides* host strain to
other *Bacteroides* , e.g., rifampin-resistant mutants, owing to presence in
donor strain of either a conjugal Tc^r/Em^r-Tc^r chromosomal element (Shoemaker et al., 1986b; Valentine et al., 1988), or a Tra^+ mobilizing

transposon such as Tn4399 (Hecht and Malamy, 1989). Since few *Bacteroides* strains receive different shuttle vectors from *E. coli* at satisfactory frequencies, secondary *Bacteroides*-to-*Bacteroides* matings should allow to disseminate cloned genes of interest into suitable recipients.

In other respects, none of the antibiotic resistance genes carried by the *E. coli* part of the shuttle vehicles was found to confer an increment of resistance to the *Bacteroides* transconjugants (Guiney et al., 1984c; Shoemaker et al., 1985). Except for the cryptic *Tc^r gene borne by *Bacteroides* transposons, neither the Cc^r/Em^r genes from the latter nor *Bacteroides* chromosomal Em^r and Tc^r determinants are expressed in *E. coli* (Guiney et al., 1984c; Shoemaker et al., 1985; Shoemaker et al., 1989). However, *Bacteroides* transposons and flanking IS can insert in the *E. coli* genome, and activate a plasmid-borne promoterless chloramphenicol resistance gene in *E. coli* (Robillard et al., 1985; Rasmussen et al., 1987; Hwa et al., 1988). In addition, cloned *Bacteroides* DNA fragments were found to contain promoter sequences functional in *E. coli*, using fusions with a galactokinase (*galK*) gene (Roberts et al., 1988). Thus, non- expression in *E. coli* appears limited to antibiotic resistance and host-range of *Bacteroides* plasmids, since various genes encoding hydrolytic activities or synthesis of surface components have been expressed in *E. coli* too.

CLONING AND EXPRESSION OF ·*BACTEROIDES* GENES IN *E. COLI*

Since five years, molecular cloning of *Bacteroides* genes in *E. coli* has focused mainly on two characteristics of *Bacteroides* strains: (i) pilins synthesized by the ovine pathogen *B. nodosus*; (ii) various hydrolytic activities of several *Bacteroides* spp., involved in complex polymers degradation. Besides these determinants, some genes of interest, such as those implicated in amino acids synthesis or recombination events, were found to function in *E. coli* (Table 2).

Pilin genes from *B. nodosus* of different serogroups

This asaccharolytic species induces ovine footrot, an economically worrying disease. Fimbriae (pili) carried by the bacterium mediate attachment to the feet of sheep, and bring on immunological response of the host. Nine serogroups (designated A to I) have been identified on the basis of immunological cross-reactivity (Finney et al., 1988). Vaccines achieved from either killed bacteria or purified pili induce effective immunity against strains of the same serogroup. However, *B. nodosus* displays fastidious growth requirements, and fimbriation is an unstable feature. To develop multivalent vaccines, cloning of pilin subunits in other bacteria has been undertaken. Using a general-purpose or an expression cloning vector, genomic banks from 6 *B. nodosus* strains belonging to different serogroups have been established in *E. coli*. Pilin-producing transformants were detected by either colony immunoassays with antisera induced by purified pili, or hybridization with pilin determinants already cloned. Nucleotide sequences of cloned pilin genes showed homologies ranging from 77 to 98%. Putative *E. coli* RNA-polymerase recognition (-35) and binding (-10) sites, Shine-Dalgarno sequence (ribosome-binding site) and termination signals (regions of hyphenated dyad symmetry) were identified on both sides of open reading frames (ORF) corresponding to pilin subunits. Sizes of the latter were similar to that of *B. nodosus* (17 kD). Pilin was located within the *E. coli* inner membrane, but no mature fimbriae were produced at the surface of the cells, because a 80 kD polypeptide thought to be the basal protein linking the fimbriae to the cells was missing. By subcloning a 734 bp fragment into a vector bearing a strong *trp* promoter, serogroup A pilin synthesis was elevated at least 1000-fold in *E. coli*, but no efficient vaccine could be prepared, presumably because of lack of protein

Table 2. *Bacteroides* genes cloned and expressed in *E. coli*

Species	Genes involved in	References
B. nodosus	Pilin subunits of serogroups A	Elleman et al. (1984), Anderson et al. (1984)
	B	Boulos and Rood (1986)
	C	Finney et al. (1988)
	G	Elleman and Von Ahlefeldt (1987)
	H1	Elleman et al., (1986b)
	H2	Hoyne et al. (1989)
	Protease	Moses et al. (1989)
B. gingivalis	Fimbrial subunit	Dickinson et al. (1988)
B. thetaiotaomicron	Chondroitin-lyase II	Guthrie et al. (1985)
	Pullulanase	Smith and Salyers (1989)
B. succinogenes	Cellulases (Endoglucanases)	Crosby et al. (1984), Taylor et al. (1987)
	Xylanase	Sipat et al. (1987)
	Lichenase	Irvin and Teather (1988)
	Cellodextrinase	Gong et al. (1989)
B. ruminicola	Xylanase	Whitehead and Hespell (1989)
B. fragilis	Glutamine-synthetase	Southern et al. (1986)
	Recombination event	Goodman et al. (1987)

processing in *E. coli*. Thus, a 'prepilin' was made in *E. coli* (Elleman et al., 1986a). However, further subcloning of pilin genes into a broad-host-range thermoregulated vector led to expression in *Pseudomonas aeruginosa*, and subsequent realization of effective vaccines. Owing to strong homology between pilins of both species, *B. nodosus* pilins were processed and generated pili in this host (Elleman et al., 1986c; Elleman and Stewart, 1988).

Protease gene from *B. nodosus*. Chymotrypsin-like proteases produced by *B. nodosus* contribute to pathogenesis of ovine footrot. Using colony imunoassays with sera against the purified 38 kD serine protease, one *E. coli* clone expressing a 50 kD immunoreactive polypeptide was isolated from a genomic library of *B. nodosus*. The vector carried a 2.8 kb insert, but no protease activity was detected. Upon subcloning in pUC8, a 1.4 kb fragment still conferred immunological response. Sequencing revealed an ORF with a Shine-Dalgarno sequence and a potential signal peptidase cleavage site. Putative -10 and -35 consensus promoter regions were identified upstream the 1.4 kb fragment. These results indicated partial cloning of a *prvA* (protease-virulent) gene with synthesis of an immunologically reactive portion of the enzyme. Hybridization of this cloned fragment to *B. nodosus* total DNA digests suggested a multiplicity of protease-encoding genes in the strain (Moses et al., 1989).

Fimbrilin gene from *B. gingivalis*. Experiments similar to cloning of pilin genes from *B. nodosus* have been reported with *B. gingivalis*. This species is implicated in periodontal disease and harbours fimbriae whose function is unknown. Determination of N-terminal sequence of the fimbrial subunit (fimbrilin; 43 kD) allowed to design oligonucleotide probes. Using the latter for screening total DNA digests, and then a genomic library consisting of pUC13 with fragments of selected sizes, led to detection of one strongly positive clone carrying a 2.5 kb insert. Sequencing revealed an ORF with putative Shine-Dalgarno sequence, leader sequence and translation initiation codon. Size of the mature fimbrilin polypeptide was estimated at 36 kD, and no homology was observed with fimbrial subunit of *B. nodosus* (Dickinson et al., 1988).

Hydrolytic activities involved in complex carbohydrate degradation

Sacharolytic species of the colon and rumen ecosystems have evolved elaborated enzyme systems for the breakdown of complex polysaccharides such as cellulose and hemicellulose, dextrans, galactanes, pectins, mucopolysaccharides, starches, xylans,... Expression of some of these activities with potential interest for industry has been done in *E. coli*.

Chondroitin- lyase II and pullulanase genes from B. thetaiotaomicron. This species synthesize two inducible chondroitin-lyases (Cases I and II), which cleave chondroitin-sulphate and related mucopolysaccharides into disaccharides, and are located in the periplasmic space (Linn, et al., 1983). After enzymatic screening of a cosmid pHC79—mediated genomic bank in *E. coli*, two positive clones were found. Properties of the cloned activity were similar to those of Case II of the original strain, but not identical (size of 95 kD vs. 104 kD). Upon subcloning into pBR328 and mutagenesis by Tn*1000*, a 3.3 kb fragment was sufficient to confer activity. Use of suicide plasmid pE3-1 for insertional inactivation of Case II gene in *B. thetaiotaomicron* showed that Case II activity was not essential for utilization of chondroitin sulphate (Guthrie and Salyers, 1986). Additional studies revealed the presence of an adjacent gene encoding chondro-4— sulphatase activity, next step of the breakdown. Promoter governing expression in *E. coli* did not function in the original strain, as judged by no increase in specific activity of Case II, suggesting a more complex regulation in *B. thetaiotaomicron* (Guthrie and Salyers, 1987).

The cosmid bank also revealed two clones able to degrade pullulan, a starch-like polymer. Minimal size of the fragment conferring activity was less than 3.0 kb and, as in *B. thetaiotaomicron*, the enzyme was mainly cytoplasmic. It was expressed at high levels in *E. coli*, and migrated as a doublet (71.6 + 73.2 kD vs. 77 kD) in maxicells experiments followed by SDS-PAGE, suggesting either protein processing in *E. coli* or occurrence of two translational start sites. Unlike in the original strain, the pullulanase was constitutive. Insertional inactivation using suicide plasmid pVAL-7 carrying a 0.5 kb segment of the cloned gene did not abolish activity upon transfer in *B. thetaiotaomicron*, showing existence of a second pullulanase-encoding gene (Smith and Salyers, 1989).

Genes mediating ß-glucanases activities in Bacteroides species of the rumen. Bacteroides (Fibrobacter) succinogenes is one of the major cellulolytic organisms in the bovine rumen. This activity is due to a multiplicity of endoglucanases (Schellorn and Forsberg, 1985). The corresponding genes were cloned in *E. coli* using expression vector pUC8 (Crosby et al., 1984). Subcloning of a 1.9 kb fragment increased the activity, which was located in the periplasmic space. The Cel enzyme was repressed by glucose, cleaved carboxy-methyl-cellulose more randomly than the complex of *B. succinogenes*, and was associated with cellobiosidase and lichenase activities, suggesting a lower specificity in *E. coli* (Taylor et al.,1987).

Xylanolytic activity greatly contribute to hemicellulose degradation since xylan is a major polymeric component of the latter. Screening of a *B. succinogenes* genomic bank in *E. coli* on Remazol brilliant blue-xylan agar led to isolation of two stable identical clones with a 9.5 kb insert conferring high xylanase activity. Upon subcloning into pUC19, a 3.0 kb fragment was enough to encode a reduced activity. Most of the enzyme was located in the periplasmic space, indicating that the gene coded for a signal peptide allowing the protein to traverse the cytoplasmic membrane.

The activity was neither subject to catabolite repression nor inducible by xylan or xylose (Sipat et al., 1987). Such an activity has been cloned from another ruminal species, *B. ruminicola*, type strain 23. Identical screening detected several clones with pUC18 carrying the same 5.7 kb insert, which was shortened to 2.7 kb. Equal activities were obtained for both orientations on pUC18 in presence or absence of IPTG, showing that the gene was translated from its own promoter. Xylanase specific activity was similar to that of the original strain (Whitehead and Hespell, 1989).

A cellodextrinase gene of *B. succinogenes*, involved in conversion of cellooligosacharides into cellobiose and glucose, has been recently cloned in *E. coli* . The positive clones carried a 7.7 kb insert in pBR322, which was reduced to 2.5 kb without affecting the activity. Although not inducible by cellobiose, the cloned enzyme was subject to catabolite repression by glucose, and had properties similar to those of the original enzyme. Unlike in the *B. succinogenes* strain, the enzyme was cytoplasmic, suggesting that sequences coding for signal peptides allowing periplasmic localization were missing (Gong et al., 1989).

Further analyses of the pUC8-mediated *B. succinogenes* library seen above gave rise to 6 clones displaying a high mixed-linkage glucanase activity on plates containing lichenan, a 1,3-1,4-β-D-glucan, after Congo Red staining. The common 5.2 kb insert was deleted to a final size of 1.35 kb, and the specific activity of the cytoplasmic enzyme was increased 5-fold in glycerol-grown cells, suggesting alteration of gene regulation. After subcloning of the 1.35 kb segment into pUC18/19, and though IPTG had no effect on expression of the β-glucanase, different levels of activity were observed, a catabolite repression-like phenomenon arising with fortuitous juxtaposition of *lac* promoter region of pUC18 and the β-glucanase regulatory region (Irvin and Teather, 1988).

Dextranases genes from *B. thetaiotaomicron*. Some *Bacteroides* species of the oral cavity (*Bacteroides oralis*) and the colon (*B. ovatus* ; *B. thetaiotaomicron*) are capable of utilizing dextrans (1,6-linked α-D-glucans) as sole carbon source. In our laboratory, a *B. thetaiotaomicron* strain was found to grow satisfactorily in minimal medium containing 0.5% dextran 10,000. A genomic bank was achieved in *E.coli* HB101, which consisted of partially *Sau*3AI-digested total DNA inserted within the Tc[r] gene of pBR322. Screening of about 14,000 Tc[s] clones on complex medium with blue dextran led to detection of 10 clones producing clear halos typical of dextran hydrolysis (Mencier, 1972). Restriction endonucleases analyses of several inserts did not reveal homology between them. HPLC analyses of the supernatants from cultures of some clones showed significant alteration of the peak corresponding to blue dextran. Characterization and possible amplification of the enzymatic activities are under way.

Other *Bacteroides* determinants expressed in *E. coli*.

Genes involved in amino acids synthesis. The suicide plasmid pSS-2 seen above contains a 4.0 kb chromosomal fragment from a *B. fragilis* strain and was found to complement *trpE* mutants of *E. coli*. However, no further characterization of the fragment has been reported (Shoemaker et al.,1985). Using a direct selection vector, a glutamine synthetase (GS) gene from *B. fragilis* by pooling all clones of the genomic bank, extracting recombinant plasmid DNAs, and transforming a *glnA* deletion mutant of *E. coli* for complementation. The GS gene was located within a 4.2 kb fragment of the 8.0 kb insert, and the conferred activity was fully repressed by glutamate and glutamine, indicating that the gene was expressed from its own promoter. No homology with the *E. coli glnA* gene was detected. Apparent size of the polypeptide product was about 75 kD, as the GS subunit purified from *B. fragilis* cells (Southern et al, 1986; 1987).

<u>Genes governing recombination events</u>. A similar screening procedure
based on <u>en masse</u> transformation of an *E. coli* recombination-deficient
(RecA⁻) mutant with selection for resistance to methyl-methane-sulphonate,
- a DNA damaging agent lethal to RecA⁻ cells -, allowed a *recA*-like gene
to be cloned from the same *B. fragilis* strain. The vector carried a 5.2 Kb
insert showing no homology with a restriction endonuclease digest of
E. coli RecA⁺ DNA. The fragment complemented defects of its host in homologous recombination, *recA*-controlled phage Pl lysis, DNA repair and
prophage λ induction. Immunological responses in Western blots to *E. coli*
RecA protein- specific antibodies revealed involvement of two polypeptides
of 39 and 37 kD (Goodman et al., 1987).

EXPRESSION IN *BACTEROIDES* OF CLONED GENES OF FOREIGN OR *BACTEROIDES* ORIGIN

No resistance determinant of foreign origin such as those borne by
the *E. coli-Bacteroides* shuttle vectors and helper plasmids has been found
to express in *Bacteroides* (see above). An exception to this rule has been
recently reported with the cloning in shuttle vehicle pDK3 of a Tcʳ (*tetM*)
gene originally described in streptococci, and found in both Gram⁺ and
Gram⁻ bacteria too. Upon mobilization in *B. fragilis*, a small but reprodu-
cible increment in tetracycline resistance was found (Guiney et al., 1988).

A chromosomal *B. thetaiotaomicron* determinant involved in conjugal
transfer and Tc + Em resistances, inserted in the cosmid vector pNJR1, was
expressed in *B. uniformis* upon <u>en masse</u> transfer of the genomic library
pool from *E. coli*. The latter served as transitory host for cloning and
did not express the resistance genes (Shoemaker et al., 1989). At the
moment, however, there is no report dealing with expression of other
foreign or *Bacteroides*-originating genes, e.g., involved in metabolic
properties, in a *Bacteroides* strain devoid of them.

CONCLUSIONS

Recent development of cloning systems for introducing various genes
in *Bacteroides* has allowed (i) to thoroughly characterize genes involved
in plasmid-borne resistance to Cc/Em, only expressed in their natural host;
(ii) to isolate and analyse chromosomal elements encoding resistances to
Em + Tc. This was not feasible by conventional genetic techniques, since
no Hfr-like or transduction-based genetic mapping system is available for
Bacteroides; (iii) to demonstrate occurrence of at least two genes governing the same degradative property in colonic *Bacteroides* (chondroitinases
and pullulanases of *B. thetaiotaomicron*), and begin to study their regulation; and (iv) to introduce foreign antibiotic resistance determinants
for searching new markers. Use of *Bacteroides* insertion sequences carrying
several putative consensus promoter sequences such as IS*4351* (Rasmussen
et al., 1987) might activate expression of simple antibiotic resistance
genes, as it has been shown in *E. coli*. In addition, plasmid-linked resistances to tetracycline in *B. ruminicola* (Flint et al., 1988) and to
metronidazole in *B. vulgatus* (Breuil et al., 1989) should prove useful for
achievement of other vectors.

Several determinants implicated in complex polymers breakdown have
been expressed in *E. coli*, in most cases from their own promoter. Comparative studies of their expression in *E. coli* and suitable *Bacteroides*
recipients, e.g., heterologous *Bacteroides* strains, should contribute to
deepen the knowledge of their organization. In another way, molecular
cloning in *E. coli* of a *recA*-like gene from *B. fragilis* (Goodman et al.,
1987) may facilitate isolation of RecA⁻ mutants by insertional inactivation of the corresponding gene upon transfer in a convenient strain.

Concerning usefulness of *Bacteroides* cloning systems in applied area, the activity conferred by some genes cloned in *E. coli* was significantly altered: it became constitutive (pullulanase) or was higher than that of the original strain (xylanase). This opens the possibility of producing *Bacteroides* enzymes of interest by industrial microorganisms using secretion vectors. In other respects, potentialities and limitations of genetic engineering on the rumen microflora have been discussed by several authors (Teather, 1985; Forsberg et al., 1986; Russell and Wilson, 1988). We have constructed composite plasmids able to replicate in both *E. coli* and *B. ruminicola*. However, transfer frequency from *E. coli* was low, and their size was relatively high. Improvement of the latter might lead to introduce foreign genes in these important ruminal bacteria.

Lastly, expression of cloned foreign genes in *Bacteroides* remains exceptional. Thus, studies on regulation of genes, particularly using metabolic mutants, are required.

Acknowledgments. We are grateful to Dr. H.-C. Dubourguier for his help in communicating this review at the Symposium.

REFERENCES

Anderson, B.J., Bills, M.M., Egerton, J.R., and Mattick, J.S., 1984, Cloning and expression in *Escherichia coli* of the gene encoding the structural subunit of *Bacteroides nodosus* fimbriae, J.Bacteriol.,160:748

Anderson, K.L. and Salyers, A.A., 1989, Biochemical evidence that starch breakdown by *Bacteroides thetaiotaomicron* involves outer membrane starch binding sites and periplasmic starch-degrading enzymes, J. Bacteriol., 171:3192.

Baker, P.J., Evans, R.T., Slots, J., and Genco, R.J., 1985, Antibiotic susceptibility of anaerobic bacteria from the human oral cavity, J. Dent. Res.,64:1233.

Beul, H.A., Von Eichel-Streiber, C., Schreiner, M., Schwindling, F.P., Weinblum, D., Zollner, E.J., and Dierich, M, 1985, Characterization of cryptic plasmid in clinical isolates of *Bacteroides fragilis* , J. Med. Microbiol., 20:39.

Boulos, S., and Rood, J.I., 1986, Molecular cloning of the fimbrial subunit gene from a benign type B isolate of *Bacteroides nodosus* , FEMS Microbiol. Lett., 33:73.

Breuil, J., Dublanchet, A., Truffaut, N., and Sebald, M., 1989, Transferable 5- nitroimidazole resistance in the *Bacteroides fragilis* group, Plasmid, 21:151.

Callihan, D., Young, F., and Clark, V., 1983, Identification of three homology classes of small, cryptic plasmids in intestinal *Bacteroides* species, Plasmid, 9:17.

Crosby, B., Collier, B., Thomas, D.Y., Teather, R.M., and Erfle, J.D., 1984, Cloning and expression in *Escherichia coli* of cellulase genes from *Bacteroides succinogenes*, in: Proc. 5th Can. Bioenergy R. and D. Semin, S. Hasnain, ed., Elsevier Applied Sciences Publishers, Barking, England, p. 573.

Dickinson, D.P., Kubiniec, M.A., Yoshimura, F., and Genco, R.J., 1988, Molecular cloning and sequencing of the gene encoding the fimbrial subunit protein of *Bacteroides gingivalis*, J. Bacteriol., 170:1658.

Elleman, T.C., Hoyne, P.A., Emery, D.L., Stewart, D.J., and Clark, B.L., 1984, Isolation of the gene encoding pilin of *Bacteroides nodosus* (strain 198), the causal organism of ovine footrot, FEBS Lett., 173:103.

Elleman,T.C., and Hoyne, P.A., 1984, Nucleotide sequence of the gene encoding pilin of *Bacteroides nodosus*, the causal organism of ovine footrot, J. Bacteriol., 160:1184.

Elleman, T.C., Hoyne, P.A., Emery, D.L., Stewart, D.J., and Clark, B.L.,1986a

Expression of the pilin gene from *Bacteroides nodosus* in *Escherichia coli*, Infect. Immun., 51:187.

Elleman, T.C., Hoyne, P.A., McKern, N.M., and Stewart, D.J., 1986b, Nucleotide sequence of the gene encoding the two-subunit pilin of *Bacteroides nodosus* 265, J. Bacteriol., 167: 243.

Elleman, T.C., Hoyne, P.A., Stewart, D.J., McKern, N.M., and Peterson, J.E., 1986c, Expression of pili from *Bacteroides nodosus* in *Pseudomonas aeruginosa*, J. Bacteriol., 168:574.

Elleman, T.C., and Von Ahlefeldt, D.A., 1987, Nucleotide sequence of the pilin gene from *Bacteroides nodosus* strain 238 (serogroup G). Nucl. Acids Res., 15:7189.

Elleman, T.C., and Stewart, D.J., 1988, Efficacy against footrot of a *Bacteroides nodosus* 265 (serogroup H) pilus vaccine expressed in *Pseudomonas aeruginosa*, Infect. Immun., 56:595.

Finney, K.G., Elleman, T.C., and Stewart, D.J., 1988, Nucleotide sequence of the pilin gene of *Bacteroides nodosus* 340 (serogroup D) and implications for the relatedness of serogroups, J. Gen. Microbiol., 134:575.

Flint, H.J., and Stewart, C.S., 1987, Antibiotic resistance patterns and plasmid contents of ruminal strains of *Bacteroides rumunicola* and *Bacteroides multiacidus*, Appl. Microbiol. Biotechnol., 26:450.

Flint, H.J., Thomson, A.M., and Bisset, J., 1988, Plasmid-associated transfer of tetracycline resistance in *Bacteroides ruminicola*, Appl. Environ. Microbiol., 54:855.

Forsberg, C.W., Crosby, B., and Thomas, D.Y., 1986, Potential for manipulation of the rumen fermentation through the use of recombinant DNA techniques, J. Anim. Sci., 63:310.

Gong, J., Lo, R.Y.C., and Forsberg, C.W., 1989, Molecular cloning and expression in *Escherichia coli* of a cellodextrinase gene from *Bacteroides succinogenes* S85, Appl. Environ. Microbiol., 55:132.

Goodman, H.J.K., Parker, J.R., Southern, J.A., and Woods, D.R., 1987, Cloning and expression in *Escherichia coli* of a *recA*-like gene from *Bacteroides fragilis*, Gene, 58:265.

Guiney, D.G., Hasegawa, P., and Davis, C., 1984a, Expression in *Escherichia coli* of cryptic tetracycline resistance genes from *Bacteroides* R plasmids, Plasmid, 11:248.

Guiney, D.G., Hasegawa, P., and Davies, C., 1984b, Homology between clindamycin resistance plasmids in *Bacteroides*, Plasmid, 11:268.

Guiney, D.G., Hasegawa, P., and Davis, C., 1984c, Plasmid transfer from *Escherichia coli* to *Bacteroides fragilis* : differential expression of antibiotic resistance genes, Proc. Nat. Acad. Sci. USA, 83:7203.

Guiney, D.G., Bouic, K., Hasegawa, P., and Matthews, B., 1988, Construction of shuttle cloning vectors for *Bacteroides fragilis* and use in assaying foreign tetracycline resistance gene expression, Plasmid, 20:17.

Guthrie, E.P., Shoemaker, N.B., and Salyers, A.A., 1985, Cloning and expression in *Escherichia coli* of a gene coding for a chondroitin lyase from *Bacteroides thetaiotaomicron*, J. Bacteriol., 164:510.

Guthrie, E.P., and Salyers, A.A., 1986, Use of targeted insertional mutagenesis to determine whether chondroitin lyase II is essential for chondroitin sulfate utilization by *Bacteroides thetaiotaomicron*, J. Bacteriol., 166:966.

Guthrie, E.P., and Salyers, A.A., 1987, Evidence that the *Bacteroides thetaiotaomicron* chondroitin lyase II gene is adjacent to the chondro-4-sulfatase gene and may be part of the same operon, J. Bacteriol., 169:1192.

Hecht, D.W., and Malamy, M.H., 1989, Tn4399, a conjugal transposon of *Bacteroides fragilis*, J. Bacteriol., 171: 3603.

Hill, G.B., and Ayers, O.M., 1985, antimicrobial susceptibilities of bacteria isolated from female genital tract infections, Antimicrob. Agents Chemother., 27:324.

Holdeman, L.V., Kelly, R.W., and Moore, W.E.C., 1984, Family I, *Bacteroi-*

daceae in: Bergey's manual of determinative bacteriology, N.R. Krieg et al., eds., Williams and Wilkins, Baltimore/London, p.602.

Hoyne, P.A., Elleman, T.C., McKern, N.M., and Stewart, D.J., 1989, Sequence of pilin from *Bacteroides nodosus* 351 (serogroup H) and implications for serogroup classification, J. Gen. Microbiol., 135:1113.

Hwa, V., Shoemaker, N.B., and Salyers, A.A., 1988, Direct repeats flanking the *Bacteroides* transposon Tn*4351* are insertion sequence elements, J. Bacteriol., 170:449.

Irvin, J.E., and Teather, R.M., 1988, Cloning and expression of a *Bacteroides succinogenes* mixed-linkage β-glucanase (1,3-1,4-β-D-glucan 4-glucanohydrolase) gene in *Escherichia coli*, Appl. Environ. Microbiol., 54:2672.

Jensen, N.S., and Canale-Parola, E., 1986, *Bacteroides pectinophilus* sp. nov. and *Bacteroides galacturonicus* sp. nov.: two pectinolytic bacteria from the human intestinal tract, Appl. Environ. Microbiol., 52:880.

Lin, K.W., Patterson, J.A., and Ladisch, M.R., 1985, Anaerobic fermentations: microbes from ruminants, Enzyme Microb. Technol., 7:98.

Linn, S., Chan, T., Lipeski, L., and Salyers, A.A., 1983, Isolation and characterization of two chondroitin lyases from *Bacteroides thetaiotaomicron*, J. Bacteriol., 156:859.

Macy, J.M., and Probst, I., 1979, The biology of gastrointestinal *Bacteroides*, Annu. Rev. Microbiol., 33:561.

Mary, P, Pheulpin, P., Béchet, M. and Guillaume, J.B., 1986, Plasmids and antibiotic resistances in clinical isolates of *Bacteroides* spp., in: Biology of anaerobic bacteria, H.-C. Dubourguier et al., eds., Elsevier Science Publishers, Amsterdam, p. 165.

Mayrand D., and Holt S.C., 1988, Biology of asaccharolytic black-pigmented *Bacteroides* species, Microbiol. Rev., 52:134.

McCarthy, R.E., Kotarski, S.F., and Salyers, A.A., 1985, Location and characteristics of enzymes involved in the breakdown of polygalacturonic acid by *Bacteroides thetaiotaomicron*, J. Bacteriol., 161:493.

Mencier, F., 1972, Méthode simple et rapide de mise en évidence des microorganismes producteurs de dextranase, Ann. Inst. Pasteur, 122:153.

Montgomery, L., Fleshner, B., and Stahl, D., 1988, Transfer of *Bacteroides succinogenes* (Hungate) to *Fibrobacter* gen. nov. as *Fibrobacter succinogenes* comb. nov. and description of *Fibrobacter intestinalis* sp. nov., Int. J. Syst. Bacteriol., 38:430.

Moses, E.K., Rood, J.I., Yong, W.K., and Riffkin, G.G., 1989, Molecular analysis of one of multiple protease-encoding genes from the prototype virulent strain of *Bacteroides nodosus*, Gene, 77: 219.

Odelson, D.A., Rasmussen, J.L., Smith, C.J., and Macrina, F.L., 1987, Extrachromosomal systems and gene transmission in anaerobic bacteria, Plasmid, 17:87.

Park, B.H., and Levy, S.B., 1988, The cryptic tetracycline resistance determinant on Tn*4400* mediates tetracycline degradation as well as tetracycline efflux, Antimicrob. Agents Chemother., 32: 1797.

Pheulpin, P., Tierny, Y., Béchet, M., and Guillaume, J.-B., 1988, Construction of new shuttle plasmid vectors for *Escherichia coli-Bacteroides* transgeneric cloning, FEMS Microbiol. Lett., 55:15.

Privitera, G., Dublanchet, A., and Sebald, M., 1979, Transfer of multiple antibiotic resistance between subspecies of *Bacteroides fragilis*, J. Infect. Dis., 139:97.

Rasmussen, J.L., Odelson, D.A., and Macrina, F.L., 1986, Complete nucleotide sequence and transcription of *ermF*, a macrolide-lincosamide-streptogramin B resistance determinant from *Bacteroides fragilis*, J. Bacteriol., 168:523.

Rasmussen, J.L., Odelson, D.A., and Macrina, F.L., 1987, Complete nucleotide sequence of insertion element IS*4351* from *Bacteroides fragilis*. J. Bacteriol., 169:3573.

Roberts, I., Hylemon, P.B., and Holmes, W.M., 1988, Isolation of promoters from two anaerobic bacteria, Microbios, 54:87.

Robillard, N.J., Tally, F.P., and Malamy, M.H., 1985, Tn*4400*, a compound isolated from *Bacteroides fragilis*, functions in *Escherichia coli* , J. Bacteriol., 164:1248.

Russell, J.B., and Wilson, D.B., 1988, Potential opportunities and problems for genetically altered rumen microorganisms, J. Nutr., 118:271.

Salyers, A.A., and O'Brien, M., 1980, Cellular location of enzymes involved in chondroitin sulfate breakdown by *Bacteroides thetaiotaomicron*, J. Bacteriol., 143:772.

Salyers, A.A., 1984, *Bacteroides* of the human lower intestinal tract, Annu. Rev. Microbiol., 38:293.

Schellorn, H.E., and Forsberg, C.W., 1984, Multiplicity of extracellular β-(1,4)-endoglucanases of *Bacteroides succinogenes* S85, Can J. Microbiol., 30:930.

Scher, K.S., 1988, Emergence of antibiotic resistant strains of *Bacteroides fragilis*, Surg. Gynec. Obst., 167:175.

Shah, H.N., and Collins, M.D., 1988, Proposal for reclassification of *Bacteroides gingivalis* , and *Bacteroides endodontalis* in a new genus, *Porphyromonas*, Int. J. Syst.Bacteriol., 38:128.

Shoemaker, N.B., Guthrie, E.P., Salyers, A.A., and Gardner, J.F., 1985, Evidence that the clindamycin-erythromycin resistance gene of *Bacteroides* plasmid pBF4 is on a transposable element, J. Bacteriol., 162:626.

Shoemaker, N.B., Getty, C., Gardner, J.F., and Salyers, A.A., 1986a, Tn*4351* transposes in *Bacteroides* spp. and mediated the integration of plasmid R751 into the *Bacteroides* chromosome, J. Bacteriol., 165:929.

Shoemaker, N.B., Getty, C., Guthrie, E.P., and Salyers, A.A., 1986b, Regions in *Bacteroides* plasmids pBFTM10 and pB8-51 that allow *Escherichia coli-Bacteroides* shuttle vectors to be mobilized by IncP plasmids and by a conjugative *Bacteroides* tetracycline resistance element, J. Bacteriol., 166:959.

Shoemaker, N.B., Barber, R.D., and Salyers, A.A., 1989, Cloning and characterization of a *Bacteroides* conjugal tetracycline-erythromycin resistance element by using a shuttle cosmid vector, J. Bacteriol.,171:1294.

Sipat, A., Taylor, K.A., Lo, R.Y.C., Forsberg, C.W., and Krell, P.J., Molecular cloning of a xylanase gene from *Bacteroides succinogenes* and its expression in *Escherichia coli*, Appl. Environ. Microbiol., 53:477.

Smith, C.J., 1985a, Development and use of cloning systems for *Bacteroides fragilis*: cloning of a plasmid-encoded clindamycin resistance determinant, J. Bacteriol., 164:294.

Smith, C.J., 1985b, Polyethyleneglycol-facilitated transformation of *Bacteroides fragilis* with plasmid DNA, J. Bacteriol., 164:466.

Smith, C.J., and Gonda, M.A., 1985, Comparison of the transposon-like structures encoding clindamycin resistance in *Bacteroides* R-plasmids, Plasmid, 13:182.

Smith, C.J., and Spiegel, H., 1987, Transposition of Tn*4551* in *Bacteroides fragilis*: identification and properties of a new transposon from *Bacteroides* spp., J. Bacteriol., 169: 3450.

Smith, C.J., 1987, Nucleotide sequence analysis of Tn*4551*: use of *ermFS* operon fusions to detect promoter activity in *Bacteroides fragilis*, J. Bacteriol., 169:4589.

Smith, K.A., and Salyers, A.A., 1989, A cell-associated pululanase from *Bacteroides thetaiotaomicron*: cloning, characterization, and insertional mutagenesis to determine its role in pullulan utilization, J. Bacteriol., 171: 2116.

Southern, J.A., Parker, J.R., and Woods, D.R., 1986, Expression and purification of glutamine synthtase cloned from *Bacteroides fragilis*, J. Gen. Microbiol., 132:2827.

Southern, J.A., Parker, J.R., and Woods, D.R., 1987, Novel structure, properties and inactivation of glutamine synthetase cloned from *Bacteroides fragilis*, J. Gen. Microbiol., 133:2437.

Speer, B.S., and Salyers, A.A., 1989, Novel aerobic tetracycline resistance that chemically modifies tetracycline, J. Bacteriol., 171:148.

Tally, F.P., Snydman, D.R., Gorbach, S.L., and Malamy, M.H., 1979, Plasmid mediated, transferable resistance to clindamycin and erythromycin in *Bacteroides fragilis*, J. Infect. Dis., 139:83.

Tally, F.P., Snydman, D.R., Shimell, M.J., and Malamy, M.H., 1982, Characterization of pBFTM10, a clindamycin-erythromycin resistance transfer factor from *Bacteroides fragilis*, J. Bacteriol., 151:686.

Tally, F.P., and Malamy, M.H., 1984, Antimicrobial resistance and resistance transfer in anaerobic bacteria, Scand. J. Gastroenterol., 19:21.

Taylor, K.A., Crosby, B., McGavin, M., Forsberg, C.W., and Thomas, D.Y., 1987, Characteristics of the endoglucanase encoded by a *cel* gene from *Bacteroides succinogenes* expressed in *Escherichia coli*, Appl. Environ. Microbiol., 1987, 53:41.

Teather, R.M., 1985, Application of gene manipulation to rumen microflora, Can. J. Anim. Sci., 65:563.

Valentine, P.J., Shoemaker, N.B., and Salyers, A.A., 1988, Mobilization of *Bacteroides* plasmids by *Bacteroides* conjugal elements, J. Bacteriol., 170:1319.

Welch, R.A., Jones, K.R., and Macrina, F.L., 1979, Transferable lincosamide-macrolide resistance in *Bacteroides*, Plasmid, 2:261.

Welch, R.A., and Macrina, F.L., 1981, Physical characterization of *Bacteroides fragilis* R plasmid pBF4, J. Bacteriol., 145:867.

Whitehead, T.R., and Hespell, R.B., 1989, Cloning and expression of a xylanase gene from *Bacteroides ruminicola* 23, Appl. Environ. Microbiol., 55:893.

Okuda, K., Kato, T., Shiozu, J., Takazoe, I., and Nakamura, T., 1985, *Bacteroides heparinolyticus* sp. nov. isolated from humans with periodontis, Int. J. Syst. Bacteriol., 35:438.

Smith, C.J., and Macrina, F.L., 1984, Large transmissible clindamycin resistance plasmid in *Bacteroides ovatus*, J. Bacteriol., 158:739.

THE MEMBRANE-BOUND HYDROGENASE OF THE PHOTOSYNTHETIC BACTERIUM *RHODOBACTER CAPSULATUS*

Paulette M. Vignais, Annette Colbeau, Béatrice Cauvin
Pierre Richaud

Biochimie Microbienne (CNRS UA 1130), Département de Recherche
Fondamentale, Centre d'Etudes Nucléaires de Grenoble, 85 X
38041 Grenoble cedex, France

SUMMARY

The photosynthetic bacterium *Rhodobacter capsulatus* contains a respiratory
chain-linked membrane-bound hydrogenase which functions as a H_2-uptake enzyme
and enables the cell to use H_2 as electron donor. The enzyme consists of two
non identical subunits and contains both Ni and Fe at the active site. The
structural genes and other genes encoding hydrogenase activity were isolated by
two methods: a) by complementation of hydrogenase-deficient (Hup$^-$) mutants
using a gene bank of *R. capsulatus* constructed in *Escherichia coli* and b) by
colony hybridization using the hydrogenase structural genes of *Bradyrhizobium
japonicum*. All identified *hup* DNA was reisolated from a second gene bank
constructed from 40-kb *R. capsulatus* DNA fragments inserted in the cosmid
vector pHC79. The *hup* genes were shown to be linked to chromosomal markers;
they were physically mapped by the use of ^{32}P-labeled probes and by restriction
analyses. *R. capsulatus hup* genes were identified over a 15-kb region of
chromosomal DNA. The two structural genes comprised the gene encoding the
small subunit (*hupS*) preceding the gene encoding the large one (*hupL*). The two
genes were separated by only three nucleotides and transcribed in the same
direction; they probably belong to the same operon and are capable of encoding
a polypeptide of 34256 Da and 65839 Da, respectively. The deduced amino acid
sequences of the two subunits showed strong homologies with the membrane-
bound hydrogenase from *Bradyrhizobium japonicum*, *Rhodocyclus gelatinosus* (80%
identical amino acids between the small subunits) and from *Azotobacter
chroococcum*. Upstream of the small subunit, a putative signal peptide was
identified which shared also a great degree of identity with the signal peptide
preceding other membrane-bound [NiFe] hydrogenases. Nine nucleotides
downstream of the *R. capsulatus* gene encoding the large subunit (*hupL*) another
open reading frame was identified which contained 786 nucleotides capable of
encoding a largely hydrophobic polypeptide of 262 amino acids (30195 Da).

INTRODUCTION

Hydrogenases are found in widely distributed procaryotes, either anaerobic,
aerobic or photosynthetic, and also in some lower eucaryotes (algae, protozoa).
By catalyzing the oxidation of molecular hydrogen they enable the cell to use
hydrogen gas as an electron and an energy source.

Microbiology and Biochemistry of Strict Anaerobes Involved in Interspecies Transfer
Edited by J.-P. Bélaich *et al.*
Plenum Press, New York, 1990

329

Although the reaction catalyzed by hydrogenases: $H_2 \rightleftharpoons 2H^+ + 2e^-$ is simple, several different proteins with hydrogenase activity have been identified, even in the same cell. The reason for the diversity of hydrogenase molecules may be found in their physiological function and may be related to their cellular localization. During autotrophic growth, hydrogenase catalyzes hydrogen consumption and channels hydrogen electrons to be used in carbon dioxide reduction, or in respiration in the presence of oxygen. In fermenting cells hydrogenase fulfills another physiological purpose, namely the reoxidation of coenzymes under anaerobic conditions (cf. Gest, 1954; Gray and Gest, 1965 for early reviews). By the production and consumption of molecular hydrogen, either within the same cell or in bacterial consortia, hydrogenases contribute to hydrogen cycling; hydrogen cycling has been proposed as a bioenergetic mechanism for energy coupling in sulfate-reducing bacteria (cf. Fauque et al., 1988 for a review).

Photosynthetic bacteria are endowed with a great metabolic versatility. They can grow aerobically in the dark, anaerobically in the light, or can have a fermentative type of metabolism (in the presence of accessory electron acceptors). Under each type of conditions they can synthesize hydrogenase for oxidizing hydrogen via a membrane-bound respiratory chain under aerobic conditions, for recycling hydrogen produced by nitrogenase under phototrophic conditions or for evolving hydrogen under fermentative conditions (cf. Vignais et al., 1985). Early studies concerning the occurrence of those enzymes and their physiological role in photosynthetic bacteria (Vignais et al., 1985) and in cyanobacteria (Lambert and Smith, 1981; Houchins; 1984) were reviewed recently.

HYDROGENASES OF PHOTOSYNTHETIC BACTERIA

In the photosynthetic bacteria, hydrogenases have been mostly studied in purple bacteria. Although preponderant soluble hydrogenase activity was found in *Rhodospirillum rubrum*, *Ectothiorhodospira shaposhnikovii* and *Thiocapsa roseopersicina* (reviewed by Gogotov, 1986), hydrogenases isolated from the photosynthetic bacteria and purified to homogeneity were membrane-bound enzymes; they were isolated from the species *Chromatium*, *Rs. rubrum*, *T. roseopersicina* and *Rhodobacter capsulatus* (cf. Vignais et al., 1985; Gogotov, 1986 for reviews). These membrane-bound hydrogenases serve in the cell for hydrogen consumption; they are αβ heterodimers and contain Ni besides Fe at their active site.

Two types of approaches were applied to the study of these [NiFe] enzymes: the physical-chemical approach to elucidate the mechanism of electron transfer at the molecular level in the active site and the genetic approach to characterize the genetic organization and mechanisms of regulation of gene expression. We will not report on the catalytic role of Ni; studies related to this topic were recently reviewed (Cammack et al., 1986; Hausinger, 1987; Cammack, 1988). We will rather discuss recent results obtained for the *R. capsulatus* hydrogenase by the genetic and molecular biology approach.

GENES ENCODING HYDROGENASE ACTIVITY IN *R. CAPSULATUS*

One of the first questions raised about the hydrogenase (*hup*) genes of *R. capsulatus* concerned their location in the cell; were those genes located on the chromosome or did they reside on an endogenous megaplasmid as is the case in *Alcaligenes eutrophus* (Kortlücke et al., 1987) for example?

We came to the conclusion that, in *R. capsulatus*, *hup* genes are located on the chromosome. That conclusion was based on two types of evidence: a) strains cured of the endogenous plasmid of 86 MDa found in the wild type strain B10, did not systematically lose hydrogenase activity (Willison et al., 1987) and b) genetic mapping experiments demonstrated that *hup* genes were linked to chromosomal markers (Magnin, 1987; Colbeau et al., 1989).

We then endeavoured to isolate genes involved in the synthesis of hydrogenase. To that end, hydrogenase-deficient (Hup⁻) mutants were isolated after chemical mutagenesis. By complementation of those mutants, using a gene bank of *R. capsulatus* constructed in pLAFR1 (Colbeau et al., 1986), three different cosmids were isolated which could complement three distinct groups of Hup⁻ mutants (Fig. 1).

COSMIDS COMPLEMENTING HUP⁻ MUTANTS

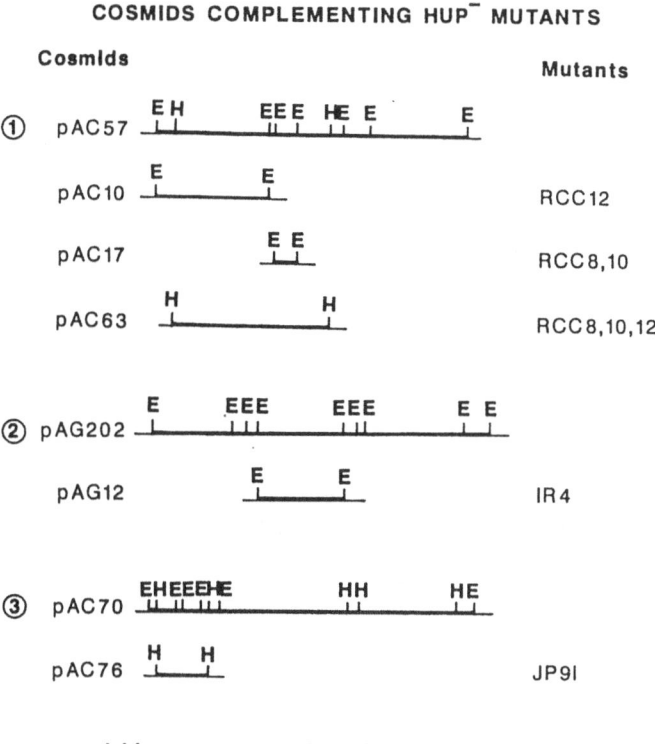

Fig. 1. List of cosmids complementing Hup⁻ mutants. Cosmids pAC57, pAC202 and pAC70 were isolated by mating the first cosmid library of *R. capsulatus* DNA with the indicated Hup⁻ mutants. Endonuclease digested fragments were separated and eluted from agarose gels. *Eco*RI and *Hind*III fragments were subcloned into plasmids pRK290 and pRK292, respectively, and used in further complementation experiments (Colbeau et al., 1986; Leclerc et al., 1988).

The *Hind*III-*Hind*III fragment of 3.5 kb complementing mutant JP91 (Fig. 1) was completely sequenced. It was shown to contain the structural genes (Leclerc et al., 1988) encoding the two subunits of *R. capsulatus* hydrogenase (Seefeldt et al., 1987), namely *hupS* capable of encoding a protein of 34256 Da which presents 80% identical amino acids with the small subunit of *Bradyrhizobium japonicum* hydrogenase (Sayavedra-Soto et al., 1988) and *hupL* capable of encoding a protein of 65839 Da which shares 70% identity with the large subunit of *B. japonicum* hydrogenase. The 3.5 kb *Hind*III insert of pAC76 (Fig. 1) also contained immediately downstream of *hupSL*, separated by only 9 nucleotides, a third open reading frame called ORFX by Richaud et al. (1990). ORFX with 786 nucleotides was capable of encoding a polypeptide of 30195 Da having 63% hydrophobic amino acids; its function is still unknown. Preliminary Northern blot analyses of total RNA from the wild type strain B10 indicated the presence of transcripts the size of which would be compatible with a co-expression of ORFX and *hupSL* (A. Colbeau, unpublished data).

The ^{32}P-labeled 3.5 kb *Hin*dIII fragment carrying the hydrogenase structural genes was used in hybridization experiments to probe total RNA from the two Hup⁻ mutants RCC8 and RCC12 and from the wild type strain B10 as a control. The mRNA of the two mutant strains did not hybridize with the *hup* structural genes; on the other hand, good hybridization was observed when the structural genes of nitrogenase (*nifHDK*) were used as a probe in control experiments (Leclerc, 1988). These results suggest that RCC8 and RCC12, unable to synthesize transcripts of hydrogenase structural genes, are regulatory mutants.

Genetic mapping experiments had shown that the mutations in the Hup⁻ mutants RCC8, RCC12 and JP91 all mapped in the same region of the chromosome (Magnin, 1987; Colbeau et al., 1989). Since the DNA fragments complementing the above-mentioned mutants happened to reside at the edge of the insert in pAC57 and in pAC76 (Fig. 1), a second gene bank of the *R. capsulatus* genome was constructed in cosmid pHC79 with *Bam*HI fragments of 40 kb in size. Using hybridization probes constructed from *hup*-specific DNA, two clones from the new gene bank, BC1 and BC2, were isolated which could hybridize with both the insert of pAC63 (9.6 kb *Hin*dIII fragment)ʹ and of pAC76 (3.5 kb *Hin*dIII fragment). It is shown on Fig. 2 that these two fragments are present in their entirety in the insert of cosmid pBC2. Figure 2 also shows the position of an 8 kb *Bam*HI fragment which was found by Colbeau et al., (1989) to fully complement the *R. capsulatus* Hup⁻ mutant, RS20.

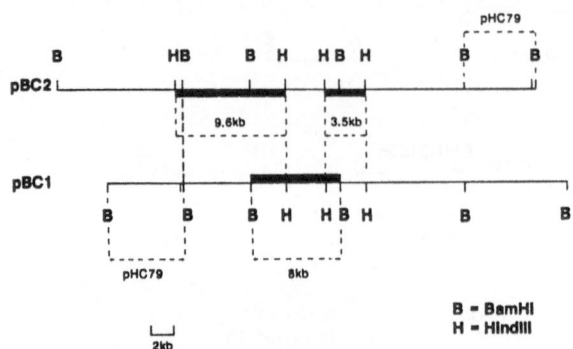

Fig. 2. Simplified restriction map of cosmids pBC1 and pBC2 isolated from the second gene bank of *R. capsulatus* DNA constructed with 40 kb insert DNA into pHC79 (Leclerc et al., 1988; Colbeau et al., 1989). The structural genes *hupS* and *hupL*, and ORFX (Richaud et al., 1990) are located on the 3.5 kb *Hin*dIII fragment. The 9.6 kb *Hin*dIII fragment is the same as the insert of pAC63 shown on Fig. 1.

In conclusion, cosmid pBC2 contained *hup* genes on a stretch of 15 kb DNA capable of complementing the Hup⁻ mutants, JP91, RS20, RCC8, RCC10 and RCC12; indeed the sites of mutation carried by those mutants were shown to map in the same chromosomal DNA region (Magnin, 1987; Colbeau et al., 1989). Mutation *aut-4*, carried by strain IR4 which has a reduced hydrogenase activity (5-10% that of the wild type strain B10) (Willison et al., 1984), mapped close to the *nif* cluster *nifHDKR4AB* (cf. Willison et al., 1985; Masepohl et al., 1988) and was fully complemented by plasmids pAG202 and pAG12 (Fig. 1). If one includes in the list of DNA fragments carrying *hup* genes, the insert of plasmid pRHP20 isolated by Xu et al. (1989) which is not present among those shown on figures 1 and 2, it appears that several genes are necessary for the synthesis of *R. capsulatus* hydrogenase and that the organization of these *hup* genes is quite complex.

Earlier studies by Colbeau and Vignais (1981) and Colbeau et al. (1983) had shown that the hydrogenase of *R. capsulatus* was an intrinsic membrane protein which required Triton X-100 to be extracted from the membrane. In the latter studies, it was shown that antibodies raised against the 65 kDa protein (large subunit) were adsorbed by chromatophores, which are inside-out particles, and not by spheroplasts indicating that the large subunit is accessible from and protuding into the cytoplasmic compartment. However, Kovacs et al. (1983) using viologen dyes as electron acceptors from H_2 concluded that in *R. capsulatus*, as well as in *T. roseopersicina* or *Chromatium minutissimum*, the hydrogenase active center is oriented towards the outer aspect of the cytoplasmic membrane. Further work is required to obtain a clear picture of the hydrogenase orientation in the cytoplasmic membrane of *R. capsulatus*. It is possible that the large subunit extends into the cytoplasmic compartment but that the active site hypothetically comprised between the small and the large subunit and buried inside the membrane is more accessible from the periplasm. Indeed observations by electron microscopy of the *A. eutrophus* membrane-bound hydrogenase indicated that the large subunit surrounds the small one (Gerberding et Mayer, 1989). A greater accessibility to antibodies of the large subunit may explain why Kovacs et al. (1989) who studied the immunological relationship of several hydrogenases concluded that the large subunits of [NiFe] hydrogenases are more conserved than the small subunits, a conclusion not entirely supported by sequence data (cf. for ex. Richaud et al., 1990).

Another matter of reflection is the presence of a signal peptide at the N-terminus of the small hydrogenase subunit (Fig. 3). Although the signal peptide itself could not be isolated, the N-terminal amino acids of the mature small subunits were determined; the striking similarities between the putative leader sequences found in the hydrogenase from *R. capsulatus*, *R. gelatinosus*, *B. japonicum* and *A. chroococcum* (Fig. 3) lead one to conclude that these sequences do belong to a signal peptide with a specific function.

```
                                             hydrophobic
                  - -     -   ++      ++   +<--------------> +   -  ↓
     Rc     ------LSDIETFYDVMRRQGITRRSFMKFCSLTAAALGLGPSFVPKIGEA M
     Rg     ---------METFYEVMRRQGISRRSFLKYCSLTATSLGLAPSFVPQIAHA M
     Bj     -----MGAATETFYSVIRRQGITRRSFHKFCCLTATSLGLGPLAASRIANA L
     Ac     ---------------MRRQGITRRSFLKYCSLTGRP-CLGPTFAPQIAHA M
                        RRQGI RRSF K CSLT        L P     I A
                        ** *  ** * * *                       *
     Dg     MKCYIGRGKDQVEERLERR-GVSRRDFMKFCTAVAVAMGMGPAFAPKVAEA L
     Db     M----S---------------LSRREFVKLCSAGVAGLGISQIYHPGIVHA M
     DvM    M--QI------------VNLTRRGFLKAACVVTAAALIS-IRMTGKAVA A
     DvH    M--QI------------ASITRRGFLKVACVTTGAALIG-IRMTGKAVA A
```

Fig. 3. Amino acid sequence and distribution of charged polar residues of the putative leader peptide of *R. capsulatus* (Rc) hydrogenase (Leclerc et al., 1988) aligned with those of *R. gelatinosus* (Rg) (Uffen et al., 1990), of *B. japonicum* (Bj) (Sayavedra-Soto et al., 1988), of *A. chroococcum* (Ac) (Ford et al., 1989) and of the periplasmic hydrogenase from *D. gigas* (Dg) and *D. baculatus* (Db) (Voordouw et al., 1989a), from *D. vulgaris* (Monticello) (DvM) and *D. vulgaris* (Hildenborough) (DvH) (Voordouw et al., 1989b). Conserved amino acids in the first four leader peptides are written in bold type below the Ac sequence and those also conserved in the *Desulfovibrio* hydrogenases are marked by asterisks. The arrow indicates the signal peptidase cleavage site.

The hydrogenases from the *Desulfovibrio* species listed in Fig. 3 are periplasmic enzymes and, classically contain a signal peptide which contribute to the translocation of the proteins accross the membrane. In the case of *R. capsulatus*, no hydrogenase activity was detected in the periplasmic space; the hydrogenase was always found to be a membrane-associated enzyme. However the signal peptide of *R. capsulatus* has an amino acid composition which is consistent with the "residue distribution rules" deduced by von Heijne (1983, 1985) from a statistical study of about 150 leader sequences. It contains at its N-terminus a stretch of 25 amino acids with charged amino acids resulting in a positive charge of +2, followed by a hydrophobic region of 13-15 amino acids and the carboxyl end of the signal peptide, which is 5 amino acid long and includes the consensus site (Gly-X-Ala) of signal peptidase I (cf. Perlman and Halvorson, 1983).

Until new experimental data are available one may speculate on the possible mechanisms which stop transfer of *R. capsulatus* hydrogenase to the periplasmic compartment. Hydrophobic interactions inside the membrane may contribute to anchor the protein in the membrane. These hydrophobic interactions could be provided by a third protein, e.g. the gene product of ORFX (Richaud et al., 1990) which seems to belong to the same transcription unit as *hupSL* and is highly hydrophobic, or by the distribution of hydrophobic and hydrophilic domains on the hydrogenase subunits themselves (Leclerc et al., 1988). Indeed, it was shown by Davis and Model (1985) for pIII, the coliphage f1 gene III protein in *Escherichia coli* or by Moore and Miura (1987) for the leader peptidase of *E. coli*, that small hydrophobic domains may stop the transfer of protein across the membrane. On the other hand, the polar, cytoplasmic domain of *E. coli* leader peptidase was suggested by von Heijne et al. (1988) to serve as a "translocation poison" sequence. Hydrophobic α helices predicted by calculation (Garnier et al., 1978; Taylor and Thornton, 1984) to occur in the small subunit between amino acids 89 and 111 and in the large subunit between amino acids 196 and 218 (Leclerc et al., 1988) could contribute to anchor hydrogenase to the *R. capsulatus* cytoplasmic membrane.

In conclusion, the *R. capsulatus* hydrogenase represents a demanding, challenging but fascinating research subject: both at the genetic level for the study of *hup* gene organization and regulation; and at the biochemical level in studies of the enzyme protein, its insertion in the membrane, its tridimentional structure and the mode of electron transfer at its active site.

REFERENCES

Cammack, R. (1988) Nickel in metalloproteins. In Advances in Inorganic Chemistry, vol. 32, pp. 297-333.

Cammack, R., Fernandez, V.M. and Schneider, K. (1986) Nickel in hydrogenases from sulfate-reducing photosynthetic, and hydrogen-oxidizing bacteria. In Bioinorganic Chemistry of Nickel (Lancaster, J.R., Jr., ed.), pp. 167-190, VCH Publishers, Deerfield Beach, FL.

Colbeau, A. and Vignais, P.M. (1981) The membrane-bound hydrogenase of *Rhodopseudomonas capsulata*. Stability and catalytic properties. Biochim. Biophys. Acta 662, 271-284.

Colbeau A., Chabert J. and Vignais, P.M. (1983) Purification, molecular properties and localization in the membrane of the hydrogenase of *Rhodopseudomonas capsulata*. Biochim. Biophys. Acta 784, 116-127.

Colbeau, A., Godfroy, A. and Vignais, P.M. (1986) Cloning of DNA fragments carrying hydrogenase genes of *Rhodopseudomonas capsulata*. Biochimie 68, 147-155.

Colbeau, A., Magnin, J.P., Cauvin, B., Champion, T. and Vignais, P.M. (1989) Genetic-physical mapping of a hydrogenase gene cluster from *Rhodobacter capsulatus*. Mol. Gen. Genet (in press).

Davis N.G. and Model, P. (1985) An artificial anchor domain: hydrophobicity suffices to stop transfer. Cell 41, 607-614.

Fauque, G., Peck Jr., H.D., Moura, J.J.G., Huynh, B.H., Berlier, Y., Der Vartanian, D.V., Teixeira, M. Przybyla, A.E., Lespinat, P.A., Moura, I. and LeGal, J. (1988) The three classes of hydrogenases from sulfate-reducing bacteria of the genes *Desulfovibrio*. FEMS Microbiol. Reviews 54, 299-344.

Ford, C.M., Garg, N., Garg, R.P., Tibelius, K.H., Yates, M.G., Arp, D.J. and Seefeldt, L.C. (1989) The identification, characterization and sequencing of the genes (*hupSL*) encoding the small and large subunits of the H_2-uptake hydrogenase of *Azotobacter chroococcum*. Mol. Microbiol. (submitted).

Garnier, J., Osguthorpe, D.J. and Robson, B. (1978) Analysis of the accuracy and implications of simple methods for predicting the secondary structure of globular proteins. J. Mol. Biol. 120, 97-125.

Gerberdin, H. and Mayer, F. (1989) Structural organization of the membrane-bound hydrogenase isolated from *Alcaligenes eutrophus* as revealed by electron microscopy. FEMS Microbiol. Lett. 60, 159-164.

Gest, H. (1954) Oxidation and evolution of molecular hydrogen by microorganisms. Bacteriol. Rev. 18, 43-73.

Gogotov, I.N. (1986) Hydrogenases of phototrophic microorganisms. Biochimie 68, 181-187.

Gray, C.T. and Gest, H. (1965) Biological formation of molecular hydrogen. Science 148, 186-191.

Hausinger, R.P. (1987) Nickel utilization by microorganisms. Microbiol. Rev. 51, 22-42.

Houchins, J.P. (1984) The physiology and biochemistry of hydrogen metabolism in cyanobacteria. Biochim. Biophys. Acta 768, 227-255.

Kortlücke, C., Hogrefe, C., Eberz, G., Pühler, A. and Friedrich, B. (1987) Genes of lithoautotrophic metabolism are clustered on the megaplasmid pHG1 in *Alcaligenes eutrophus*. Mol. Gen. Genet. 210, 122-128.

Kovacs, K.L., Bagyinka, C. and Serebriakova, T. (1983) Distribution and orientation of hydrogenase in various photosynthetic bacteria. Curr. Microbiol. 9, 215-218.

Kovacs, K.L., Seefeldt, L.C., Tigyi, G., Doyle, C.M., Mortenson, L.E. and Arp, D.J. (1989) Immunological relationship among hydrogenases. J. Bacteriol. 171, 430-435.

Lambert, G.R. and Smith, D.G. (1981) Hydrogen metabolism of cyanobacteria. Biol. Rev. 56, 589-660.

Leclerc, M. (1988) Génétique de l'hydrogénase chez *Rhodobacter capsulatus*: séquençage de gènes et identification des ARNm. Ph.D. Thesis, Grenoble University.

Leclerc, M., Colbeau, A., Cauvin, B. and Vignais, P.M. (1988) Cloning and sequencing of the genes encoding the large and the small subunits of the H_2-uptake hydrogenase (*hup*) of *Rhodobacter capsulatus*. Mol. Gen. Genet. 214, 97-107. Erratum (1989) 215, 368.

Magnin, J.P. (1987) Isolement d'une souche Hfr de la bactérie photosynthétique *Rhodobacter capsulatus* et cartographie du chromosome. Ph.D. Thesis, Grenoble University.

Masepohl, B., Klipp, W. and Pühler, A. (1988) Genetic characterization and sequence analysis of the duplicated *nifA/nifB* gene region of *Rhodobacter capsulatus*. Mol. Gen. Genet. 212, 27-37.

Moore, K.E. and Miura, S. (1987) A small hydrophobic domain anchors leader peptidase to the cytoplasmic membrane of *Escherichia coli*. J. Biol. Chem. 262, 8806-8813.

Perlman D. and Halvorson, H.A. (1983) A putative signal peptidase recognition site and sequence in eukaryotic and prokaryotic signal peptides. J. Mol. Biol. 167, 391-409.

Richaud, P., Vignais, P.M., Colbeau, A., Uffen, R.L. and Cauvin, B. (1990) Molecular biology studies of the uptake hydrogenase of *Rhodobacter capsulatus* and *Rhodocyclus gelatinosus*. FEMS Microbiology Reviews (in press).

Sayavedra-Soto, L.A., Powell, G.K., Evans, H.J. and Morris, R.O. (1988) Nucleotide sequence of the genetic loci encoding subunits of *Bradyrhizobium japonicum* uptake hydrogenase. Proc. Natl. Acad. Sci. USA 85, 8395-8399.

Seefeldt, L.C., McCollumn, L.C., Doyle, C.M. and Arp, D.J. (1987) Immunological and molecular evidence for a membrane-bound, dimeric hydrogenase in *Rhodopseudomonas capsulata*. Biochim. Biophys. Acta 914, 299-303.

Taylor, W.R. and Thornton, J.M. (1984) Recognition of super secondary structure in proteins. J. Mol. Biol. 173, 487-514.

Vignais, P.M., Colbeau, A., Willison, J.C. and Jouanneau, Y. (1985) Hydrogenase, nitrogenase and hydrogen metabolism in the photosynthetic bacteria, in Advances in Microbial Physiology (Rose, A.H. and Tempest, D.A., eds), vol. 26, pp. 155-234, Academic Press Inc., London.

von Heijne, G. (1983) Patterns of amino acids near signal sequence cleavage sites. Eur. J. Biochem. 133, 17-21

von Heijne, G. (1985) Signal sequences: the limits of variation. J. Mol. Biol., 184, 99-105.

von Heijne, G., Wickner, W. and Dalbey, R.E. (1988) The cytoplasmic domain of *Escherichia coli* leader peptidase is a "translocation poison" sequence. Proc. Natl. Acad. Sci. USA 85, 3363-3366.

Uffen, R.L., Colbeau, A., Richaud, P. and Vignais, P.M. (1990) Cloning and sequencing the genes encoding hydrogenase subunits of *Rhodocyclus gelatinosus*. Mol. Gen. Genet. (submitted).

Voordouw, G., Menon, N.K., LeGall, J., Choi, E.S., Peck, H.D., Jr. and Przybyla, A.E. (1989a) Analysis and comparison of nucleotide sequences encoding the genes for [NiFeSe] hydrogenases from *Desulfovibrio gigas* and *Desulfovibrio baculatus*. J. Bacteriol. 171, 2894-2899.

Voordouw, G., Strang, J.D. and Wilson, F.R. (1989b) Organization of the genes encoding [Fe] hydrogenase in *Desulfovibrio vulgaris* subsp. *oxamicus* Monticello. J. Bacteriol. 171, 3881-3889.

Willison, J.C., Madern, D. and Vignais, P.M. (1984) Increased photoproduction of hydrogen by non-autotrophic mutants of *Rhodopseudomonas capsulata*. Biochem. J. 219, 593-600.

Willison, J.C., Ahombo, G., Chabert, J., Magnin, J.P. and Vignais, P.M. (1985) Genetic mapping of the *Rhodopseudomonas capsulata* chromosome shows non clustering of genes involved in nitrogen fixation. J. Gen. Microbiol. 131, 3001-3015.

Willison, J.C., Magnin, J.P. and Vignais, P.M. (1987) Isolation and characterization of *Rhodobacter capsulatus* strains lacking endogenous plasmids. Arch. Microbiol. 147, 1134-1142.

Xu, H.W., Love, J., Borghese, R. and Wall, J.C. (1989) Identification and isolation of genes essential for H_2 oxidation in *Rhodobacter capsulatus*. J. Bacteriol. 171(2), 714-721.

HYDROGENASE MUTANTS OF Escherichia coli DEFECTIVE IN NICKEL UPTAKE

Marie-Andrée Mandrand[1], Long-Fei Wu[1] and David Boxer[2]

1- Laboratoire de Microbiologie, Laboratoire de Biologie et
Technologie des Membranes du CNRS
I.N.S.A.
69621 Villeurbanne Cedex, France
2- Department of Biochemistry
Medical Sciences Institute
University of Dundee
Dundee DD1 4HN, U.K.

INTRODUCTION

Two pathways for H_2 metabolism have been identified in Escherichia coli and other members of the family Enterobacteriaceae under anaerobic growth conditions. The first pathway, known as the formate hydrogenlyase system, occurs during fermentative growth on carbohydrates in the absence of an external electron acceptor (Peck et Gest, 1957). It consists at least of two enzymes, a benzyl viologen-linked formate dehydrogenase (FDH-BV) and a hydrogenase, which catalyze the oxidation of formate produced by glycolysis to carbon dioxide and molecular H_2 (Gray and Gest, 1965). The overall reaction is scalar and functions to remove reducing equivalents exchangeable with formate and to help offset acidification of the growth medium. In the second pathway the bacteria are able to utilize H_2 as an energy source in the presence of a nonfermentable carbon source, such as fumarate, which is acting as (or generating) a terminal electron acceptor (Macy et al., 1976). In this case, a respiratory hydrogenase catalyzes the oxidation (uptake) of H_2 in an energy-conserving manner by proton translocation across the cytoplasmic membrane (Jones, 1980).

Three immunologically distinct hydrogenase isoenzymes have been recently demonstrated in anaerobically grown E. coli (Sawers et al., 1985) and S. typhimurium (Sawers et al., 1986). In E. coli, two of these membrane-bound isoenzymes, named hydrogenases 1 and 2, have been purified and shown to contain nickel (Ballantine and Boxer, 1986; Sawers and Boxer, 1986). They are thought to function as uptake hydrogenases, but under different growth conditions : hydrogenase 2 catalyzes respiration-linked H_2 uptake, whereas hydrogenase 1 is proposed to recycle H_2 produced during fermentative growth. Hydrogenase 3 must participate to the formate hydrogenlyase pathway (Sawers et al., 1985; Sawers et al., 1986).

Genetic analysis has revealed that the majority of genes involved in H_2 metabolism are pleiotropic, affecting both hydrogenase and FDH-BV activities (Graham et al., 1980; Lee et al., 1985; Chaudhuri and Krasna, 1987). Most of the isolated mutants carried a mutation located in the

Microbiology and Biochemistry of Strict Anaerobes Involved in Interspecies Transfer
Edited by J.-P. Bélaich et al.
Plenum Press, New York, 1990

hydABE cluster near minute 58 of the E. coli genome. Some of these genes have been cloned, but their precise role in the cell has not been elucidated at present (Karube et al., 1984; Sankar et al., 1985; Waugh and Boxer, 1986; Chaudhuri and Krasna, 1987). In contrast, the hyd-17 mutation which specifically affects hydrogenase 3 isoenzyme could define the structural gene for hydrogenase 3 (Birkmann et al., 1987). Recently, a new hydrogenase locus mapping at 65 min, hydFL also called hup, was reported to be essential for H_2 uptake activity (Lee et al., 1985; Stoker et al., 1989). Finally, another gene required for growth of E. coli on fumarate and H_2 and located near 17 min was also described (Chaudhuri and Krasna, 1988).

Among the various hydrogenase mutants isolated so far by ourselves, two distinct classes, hydE and hydC, deserve special attention as they were found to be restored to the wild-type hydrogenase phenotype by the presence of excess nickel salts in the growth medium (Waugh and Boxer, 1986; Wu and Mandrand-Berthelot, 1986). We report here the genetic and biochemical characterization of hydC mutants and we explore the relationship between hydC and fnr, the regulatory gene required for the expression of several anaerobically expressed genes. We present evidence suggesting that hydC encodes a specific transport system for nickel. We also describe initial cloning of the hydC locus by complementation of two independent deletion mutants and identification of the encoded polypeptides.

MATERIALS AND METHODS

Organisms

Bacterial strains and plasmids are listed in Table 1.

Media and growth conditions

The organisms were grown routinely at 37°C in LB medium (Miller, 1972). Anaerobic growth was achieved in LB supplemented with 2 µM sodium selenite, 2 µM ammonium molybdate in 250 ml or 50 ml bottles filled almost to the top. 30 mM sodium formate was added to the medium to insure maximal hydrogenase and formate hydrogenlyase activities (Ruiz-Herrera and Alvarez, 1972). Nickel chloride was added to give the indicated concentrations. Minimal media were M63 (Miller, 1972) and CR-Hyd medium (Wu et al., 1989). The carbon source used was glucose (0.4 % w/v). Where required, L-amino acids were used at 20 µg/ml, and antibiotics were added at the following final concentrations : ampicillin 50 µg/ml; chloramphenicol 25 µg/ml; kanamycin 20 µg/ml. MacConkey formate-fumarate medium has been previously described (Wu and Mandrand-Berthelot, 1986). Gas Pak anaerobic jars (BBL Microbiology Systems) were used for the anaerobic incubation of plates.

Preparation of cells and enzyme assays

The cells were harvested by sedimentation at 5000 g for 10 min at 4°C, washed twice with 50 mM potassium buffer pH 6.8 and resuspended in 1 to 4 ml of the same buffer. They were made permeable by addition of toluene (2 %). For the preparation of membrane fractions the cells were ruptured in a French press and treated as described previously (Sawers et al., 1985).

Spectrophotometric enzyme assays were performed as already described (Wu and Mandrand-Berthelot, 1986). One unit of hydrogenase (H_2:benzyl viologen oxidoreductase) activity was 1 µmole benzyl viologen reduced per min. The proton-deuterium exchange reaction was performed as previously described (Berlier et al., 1985) using a mass spectrometer. One unit of β-galactosidase activity was 1 nanomole of o-nitrophenyl-β-D-galactopyranoside hydrolysed per min. Protein was estimated by the method of Lowry et

Table 1. Strains and plasmids used

Strain or plasmid	Genotype	Source
Strains		
MC4100	F⁻ araD139 Δ(argF−lac)U169 ptsF25 deoC1 relA1 flb5301 rpsL150 λ⁻	Casadaban and Cohen, 1979
MC4100NI1	MC4100 fnr zcj261::Tn10	Wu and Mandrand-Berthelot, 1986
HYD71, 74,75	MC4100 hydX::Mu cts dI (AmpR lac)	This work
HYD72,79	MC4100 hydC::Mu cts dI (AmpR lac)	"
HYD76	MC4100 hydB::Mu cts dI (AmpR lac)	"
HYD77,78	MC4100 hydA::Mu cts dI (AmpR lac)	"
HYD720	MC4100 ΔhydC72	HYD72:ts⁺ ApS selection
HYD790	MC4100 ΔhydC79	HYD79: "
HYD723	MC4100 hydC72::Mu dI (AmpR lac)	HYD72:ts⁺ ApR selection
HYD72NI1	HYD723 fnr zcj261::Tn10	Wu and Mandrand-Berthelot, 1986
P4X	Hfr metB	Wollman
HPX1	P4X hydC72::Mu dI (AmpR lac)	Wu et al., 1989
P4XN	P4X fnr zcj261::Tn10	"
FD12	P4X hydB12	Graham et al., 1980
Plasmids		
pBR322	AmpR TetR	Maniatis et al., 1982
pBS8⁺	KanR lacZ'	Spratt et al., 1986
pPH126	KanR TetR	"
pGP1-2	KanR T7 gene 1 (RNA polymerase)	Tabor and Richardson, 1985
pT7-4	AmpR T7 φ10	"
pT7-6	AmpR T7 φ10	"

al. (1951) using bovine serum albumin as the standard.

Analytical procedures

The nickel content of cells was estimated by liquid scintillation counting after growth in the presence of added 0.1–0.3 μM [^{63}Ni] NiCl$_2$ (specific radioactivity 0.97 Ci mgatom^{-1}) and following collection of the bacteria on fibre glass filters by a filtration technique (Wu et al., 1989).

Genetic techniques and DNA manipulation

Random lac operon fusions in the E. coli chromosome were isolated by the method developed by Casadaban and Cohen (1979) using the MuctsdI (AmpR lac) bacteriophage. P1cml-mediated transductions were carried out as described by Miller (1972). Extraction of plasmid DNA, restriction enzyme digestion, ligation, agarose gel electrophoresis and transformation experiments were performed according to the methods described by Maniatis et al.

(1982). Tn5 insertion mutagenesis was achieved following the procedure of O'Hoy and Krishnapillai (1985).

In vivo expression of the plasmid pT7 encoded polypeptides

Various DNA fragments containing the totality or portions of the hydC locus were subcloned into the multiple cloning site of plasmids pT7-4 and pT7-6 (Tabor, unpublished data) downstream the T7 ϕ10 promoter in either orientation. Recombinant plasmids were transformed into strain K38 harboring the compatible plasmid pGP1-2 which contains the gene coding for T7 RNA polymerase under the control of the heat-inducible λpL promoter (Tabor and Richardson, 1985). The synthesis of plasmid-encoded gene products was followed by the incorporation of L-[^{35}S] methionine according to the procedure of Tabor and Richardson (1985).

RESULTS

Isolation and characterization of mutants deficient in hydrogen metabolism

Strain MC4100 mutagenized by infection with phage Mu dI (Amp^R lac) was used to generate mutants deficient in hydrogenase activity, on the basis of red colony color (acidification) on MacConkey-formate fumarate medium (Wu and Mandrand-Berthelot, 1986). Eight acid-producing strains could be first assigned to two separate groups after test for transductional linkage to the srlC-cysC area at 58 min on the E. coli chromosome. Five mutants, designated HYD71, HYD74 through HYD77 (Table 1) possessed lesions mapping in this region (Wu, 1988). They were further differentiated into three different classes following transformation with either plasmid pLW19 (Wu, 1988), carrying the same 5 kb EcoRI-SalI insert as plasmid pEH3 (Karube et al., 1984) and plasmid pSE-201 (Sankar et al., 1985) which complement mutations at the hydA locus, or plasmid pRW1 (Waugh and Boxer, 1986) which is able to complement both hydB and the nickel-restorable hydE locus. Mutant HYD77 was shown to belong to the hydA class like mutant HYD78 which had previously been falsely attributed to the hydD locus at min 77 (Wu and Mandrand-Berthelot, 1986), due to artefact in transduction analysis. Mutant HYD76 was assigned to the non nickel-restorable hydB class and mutants HYD71, 74 and 75 which are not restored by either of these plasmids represented at least a third category of genes present in this complex region. In contrast, lesions of mutants HYD72 and HYD79 did not show any linkage to srlC or cysC. They were subsequently found to cotransduce with dnaM at 77 min on the E. coli chromosome, thus defining a new locus which was called hydC (Wu and Mandrand-Berthelot, 1986).

Anaerobic enzyme activities of formate hydrogenlyase, H$_2$ uptake and formate-nitrate reductase (Ruiz-Herrera and DeMoss, 1969) pathways were determined. All mutants totally lacked hydrogenase activity as measured by H$_2$-dependent reduction of benzyl viologen. FDH-BV activity was reduced to around 30-50 % of the wild type value in hydA and hydC mutants and it was totally absent in the last three hydX (min 58) mutants. As expected, enzyme activities of the formate-nitrate reductase complex as well as fumarate reductase activity were not affected (Wu and Mandrand-Berthelot, 1986; Wu, 1988). As the hydC mutations represented a new category of hyd genes, we decided to concentrate on those.

Although hydC mutants failed to reduce benzyl viologen with H$_2$ as electron donor, they produced both hydrogenase-dependent activities, formate hydrogenlyase and fumarate-linked H$_2$ uptake, but to a reduced level (20-30 %) compared to the wild-type (Wu and Mandrand-Berthelot, 1986). Closer examination of hydrogenase activity, employing a D$_2$/H$^+$ exchange method (Berlier et al., 1985) which permits the direct measurement of the

Table 2. Hydrogenase activities in the hydC mutant measured by proton-deuterium exchange reaction with a mass-spectrometer

Strain[a]	Genotype	Hydrogenase[b] (μmoles HD+H$_2$/min/ mg dry weight)		Formate hydrogenlyase[b] (μmoles H$_2$/min/mg dry weight)	
		$-Ni^{2+}$	$+Ni^{2+c}$	$-Ni^{2+}$	$+Ni^{2+c}$
MC4100	wild-type	1.19	0.86	0.25	0.19
HYD723	hydC	0.12	0.61	0.13	0.26

[a]Cells were grown anaerobically in minimal medium supplemented with 0.4 % glucose, 30 mM formate, 2 μM selenite and 2 μM molybdate.
[b]Hydrogenase activity was measured by the D_2/H^+ exchange reaction (Berlier et al., 1985).
[c]Nickel was added at the concentration of 500 μM.

activation of H$_2$, provided evidence that the hydC mutants still retained about 10 % of activity of the wild-type which could account for the persistence of H$_2$-related functions (Table 2).

Nickel can restore hydrogenase activity to the hydC mutants

Our interest in the hydC mutants was further enhanced once we discovered that they could be restored to wild-type hydrogenase phenotype by the presence of excess NiCl$_2$ (500 μM) in the growth medium (Tables 2 and 3).

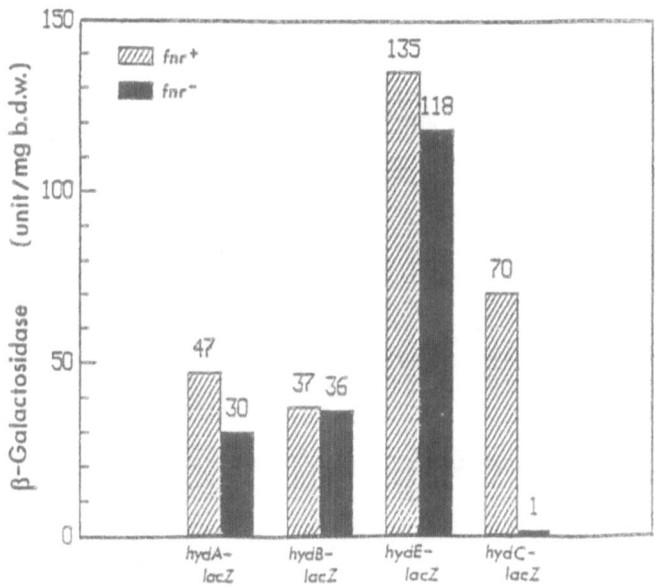

Fig. 1. Effect of the fnr mutation on the expression of hyd-lac fusions. β-galactosidase activities of mutants HYD78, HYD76, HYD74 and HYD723 and their fnr derivatives were assayed after growth and preparation of cultures as described in Table 3.

341

This effect was further strengthened by the simultaneous recovery of both H_2 evolution and H_2 uptake activities. Restoration was highly specific for nickel salts, since other divalent cations, such as Ca^{2+}, Zn^{2+}, Mn^{2+}, Mg^{2+}, Cu^{2+}, Fe^{2+} and Co^{2+} cannot substitute for nickel (Wu et al., 1989).

Furthermore, immunological studies using antiserum directed to either hydrogenase 1 or 2 revealed that in the absence of nickel the hydC mutants produced negligible amounts of hydrogenase 1 or 2 antigens, which were restored to wild-type levels after growth with high concentration of nickel in the medium. Hydrogenase 3 also exhibited the same behaviour (Wu et al., 1989). These observations strongly suggested that the lack of an active hydC gene product abolishes the expression of hydrogenases 1 and 2 and that nickel-restoration was not the consequence of the direct reactivation of a preformed inactive protein. Confirmation of this point was assessed by the fact that nickel did not stimulate hydrogenase activity when an inhibitor of protein synthesis, chloramphenicol, preceded the introduction of nickel (Wu et al., 1989). This result shows that the restoration of hydrogenase activity is dependent on protein synthesis.

The regulatory fnr gene controls hydrogenase synthesis via the hydC locus

Expression of the hydC locus could be monitored following β-galactosidase activity measured from the two Mu dI (Amp^R lac) operon fusion mutants HYD723 and HYD79. It was induced by anaerobiosis (Wu and Mandrand-Berthelot, 1986). Since the product of the pleiotropic fnr gene is required for the anaerobic induction of several oxidoreduction activities including hydrogenase (Shaw et al., 1983) it was worthwhile to investigate its effect on the hydC locus. Introduction of an fnr mutated allele in the hydC-lac mutants led to the total suppression of β-galactosidase activity, while it did not influence the expression of the three other hyd genes tested (Fig. 1). Since transcription of hydC is positively controlled by fnr and that fnr mutants are also devoid of hydrogenase activity, we suggested that fnr affects hydrogenase via its effect on hydC (Wu and Mandrand-Berthelot, 1986). If this is the case, then the hydrogenase phenotype of an fnr mutant should be equivalent to that of hydC. Table 3 provides further evidence for

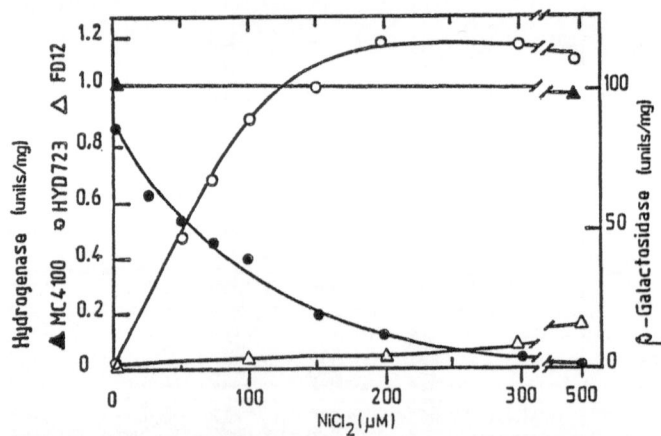

Fig. 2. Influence of nickel on hydrogenase restoration and hydC expression in a hydC mutant. Strains HYD723 (hydC-lac)(O-O), MC4100 (wild-type)(▲-▲) and FD12 (hydB12)(△-△) were grown in LB medium containing 30 mM formate with $NiCl_2$ added to the concentration indicated, harvested and assayed for hydrogenase activity. β-galactosidase (●-●) was also determined in HYD723.

Table 3. Reversion of the hydrogenase phenotype in the hydC and fnr mutants

Strain[a]	Genotype	Hydrogenase[b]		Formate[b] dehydrogenase		Formate[b] hydrogenlyase		Nickel[c]
		$-Ni^{2+}$	$+Ni^{2+}$	$-Ni^{2+}$	$+Ni^{2+}$	$-Ni^{2+}$	$+Ni^{2+}$	
MC4100	wild-type	0.99	0.98	0.20	0.21	0.37	0.35	
MC4100NI1	fnr	0.02	0.55	0.03	0.11	0.02	0.14	
HYD72	hydC72	0.01	1.40	0.07	0.36	0.10	0.32	
HYD72NI1	hydC72 fnr	0.01	0.66	0.02	0.14	0.02	0.13	
P4X	wild-type	1.34		0.41		0.53		43.20
HPX1	hydC72	0.01		0.14		0.18		0.27
HPX1/pLW20	hydC/hydC+	1.22		0.45		0.51		62.70
P4XN	fnr	0.01		0.10		0.14		0.50
P4XN/pLW20	fnr/hydC+	0.64		0.14		0.28		58.00

[a]All strains were grown in LB medium supplemented with 30 mM Na formate.
[b]Expressed as units of activity per mg bacterial dry weight, as defined in Materials and Methods.
[c]Expressed as pgatom/mg of protein.

such an hypothesis. Addition of nickel in the fnr strain resulted in the concomitant restoration of hydrogenase, FDH-BV and formate hydrogenlyase activities to 50 % of the values found in the related wild-type strain.

Recovery pattern of hydrogenase activity was further explored by varying the concentration of $NiCl_2$ in the growth medium. Hydrogenase activity in the hydC mutant increased progressively with increasing amounts of $NiCl_2$ (Fig. 2). The $NiCl_2$ dependence of the activation curve for the fnr mutant was closely similar, while restoration of the hydE class requi- red much higher levels of added $NiCl_2$ (Wu et al., 1989). Conversely, expression of the hydC-lac operon fusion decreased with increasing nickel concentrations (Fig. 2). The striking correlation between these two events suggested the hypothesis that hydC could encode a specific transport system for nickel. This system would supply the needs of the cells for nickel, when grown in media containing very low nickel concentrations. In the presence of high external nickel concentrations, the synthesis of the hydC product is greatly reduced, as alternate processes are sufficient to supply nickel. The large capacity magnesium transport system has indeed been shown to efficiently take up nickel (Jasper and Silver, 1977), thus being able to circumvent the requirement for the hydC product at high nickel concentra- tions. Additional evidence supporting this interpretation is given : first, by the very low nickel content of hydC mutants (1 % of that of the parental strain) which can be relieved, along with hydrogenase and related activi- ties, after introduction of plasmid pLW20 (hydC+) (see below) (Table 3); second, by the ability to suppress the hydC hydrogenase phenotype by growth in media with very low (0.01 mM) $MgCl_2$, so facilitating nickel uptake via the magnesium transport system by reducing magnesium competition for nickel entry (Wu et al., 1989). The behaviour of the fnr mutant was closely similar to that of hydC in all respects examined.

Initial cloning and expression of the hydC locus

One hybrid cosmid complementing the hydC72 mutation of mutant HYD723 was selected from an E. coli genomic cosmid library (Touati, 1983). Its DNA was purified and subcloned into plasmid pBS8⁺ (Spratt et al., 1986) resulting in plasmid pLW20 which contains a 9.5 kb chromosomal DNA insert. After establishing a physical map of pLW20, DNA fragments containing parts of this hydC region were further subcloned into different vectors and tested for their ability to complement two hydC deletion mutants HYD720 and HYD790. For example, pLW22 complemented both mutants, plasmid pLW23 was found to exclusively complement HYD790 and plasmid pLW28 was unable to restore the hydrogenase positive phenotype to either mutant (Fig. 3). Results indicated that the mutants were distributed into two complementation groups, hydC72 and hydC79. Determination of extent of chromosomal deletions revealed that mutant HYD720 possesses a large deletion encompassing that of HYD790 and thus covering at least two genes (Wu, 1988). The region of the E. coli DNA necessary for the hydC locus was determined using Tn5 transposon mutagenesis of plasmids pLW23 and pLW28 which in association carry the totality of hydC, and subsequent integration by homologous recombination in the E. coli chromosome (Fig. 3). Interestingly, the Tn5 insertion site 4115 did not eliminate hydrogenase activity in contrast to insertions 4001, 4014 and 4017 that confer an hydrogenase negative phenotype which could be restored by nickel. As a consequence, the hydC locus did not comprise the DNA segment located on the left side of EcoRV3.

Characterization of the polypeptide products of the hydC72 and hydC79 genes was achieved after cloning of a series of chromosomal inserts downstream of the strong phage T7 φ10 promoter. Figure 4 demonstrates that there were five polypeptides encoded by the totality of the 7.1 kb BamHI-EcoRI fragment of plasmid pLW22 (Fig. 3), provided that it is cloned with

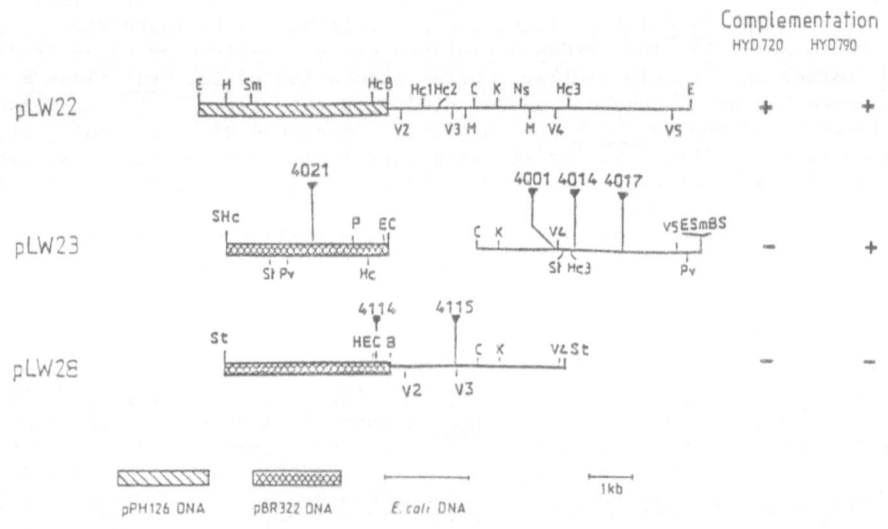

Fig. 3. Physical and genetic map of the E. coli hydC locus. At the top is shown plasmid pLW22 which is derived from an hybrid cosmid carrying hydC⁺. Vertical lines above the map of pLW23 and pLW28 indicate the position of Tn5 insertion mutations.
+ : restoration of hydrogenase activity; - : absence of hydrogenase activity. Symbols : B, BamHI; C, ClaI; E, EcoRI; V, EcoRV; H, HindIII; Hc, HincII; K, KpnI; M, MluI; Ns, NsiI; P, PstI; Pv, PvuII; S, SalI; Sm, SmaI;, St, StyI.

the BamHI to EcoRI orientation (lanes B and D). Deletion of the 1.6 kb MluI fragment resulted in the absence of complementation of both mutants HYD720 and HYD790, and in the simultaneous lack of two protein bands with apparent molecular weights of 59 and 27.5 kDa (lanes A and E). In addition, the 5.1 kb EcoRI fragment of plasmid pLW23 which complemented mutant HYD790 direc- ted the synthesis of the 27.5 kDa polypeptide, but not that of the 59 kDa protein (lane C). This result strongly suggests that the 27.5 kDa protein is encoded by hydC79 whereas the 59 kDa protein is encoded by hydC72.

DISCUSSION

According to its biochemical and genetic properties, we have identi- fied a new class of mutants which is pleiotropically defective in hydroge- nase activity, to which all three hydrogenase isoenzymes are restored by growth in media containing excess NiCl$_2$. The phenotype of the hydC mutants is unlikely to originate from a lesion affecting a structural gene for any of the hydrogenases, since all of them are simultaneously impaired. It is also difficult to envisage a regulatory role for this locus as the regula- tion of each of the hydrogenases is distinct (Sawers et al., 1985). Several lines of evidence are consistent with the phenotype of hydC arising from nickel limitation : its very low nickel content; inability of hydC strains to regulate their cellular nickel content which apparently reflects the external availability of nickel ions (data not shown); repression of the hydC expression with increasing external nickel concentration. It is most likely that the primary defect is in a specific nickel transport system of high affinity and low capacity. However, our attempts to measure this transport directly have so far been unsuccessful due to the inability to distinguish a specific rate in the presence of the high background uptake of the magnesium system (Waugh, Holt and Boxer, manuscript in preparation).

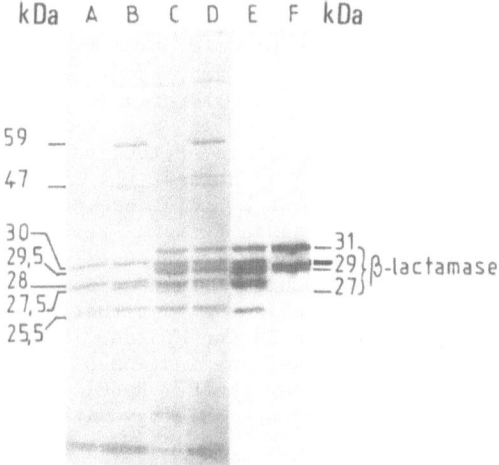

Fig. 4. Expression of the hydC locus in the E. coli T7 RNA polymerase/ promoter system. An autoradiograph of a 12.5 % polyacrylamide gel is loaded with SDS-solubilized cells of Escherichia coli K38/pGP1-2 transformed with plasmid pT7-6 carrying the BamHI-EcoRI insert of pLW22 deleted from the 1.6 kb MluI fragment (lane A), or the BamHI-EcoRI insert of pLW22 (B); pT7-4 carrying the EcoRI insert of pLW23 (C), or the BamHI-EcoRI insert of pLW22 (D) and the same insert deleted from the 1.6 kb MluI frag- ment (E), pT7-4 as a standard (F). Molecular weight of ß-lacta- mase protein bands are indicated.

Mutants defective in the magnesium transport system would provide a suitable background for kinetic analysis. Additional evidence in favor of this idea is the ability to restore hydrogenase along with internal nickel content by growth in media with very low magnesium salt concentration, thus affecting nickel uptake by the magnesium uptake system.

In E. coli, mutations in the fnr gene prevent the anaerobic expression of several anaerobically expressed genes. We have shown that fnr is necessary for the anaerobic induction of hydC. Behaviour of fnr mutants markedly mimics that of hydC mutants with regard to hydrogenase phenotype, nickel content and restoration by added nickel, reduced magnesium salts or presence of multicopies of the cloned hydC gene. This leads to the important conclusion that fnr affects H_2 metabolism indirectly via the hydC locus. This distinguishes the role of the Fnr protein on H_2 metabolism from its proposed direct interaction with the structural genes of other anaerobic respiratory enzymes such as nitrite reductase (Jayaraman et al., 1987).

Besides the well-known involvement of nickel as a component of the hydrogenase isoenzymes, we report evidence that this metal can act as a regulator, possibly a corepressor, of the hydC gene whose expression is repressed when high nickel levels are available. Whether nickel can regulate the expression of hydrogenases at the molecular (transcriptional) level is suggested by the lack of immunoprecipitable material in the absence of nickel. Such a conclusion has been reached for hydrogenase in Bradyrhizobium japonicum (Stults et al., 1986).

Preliminary cloning of the hydC locus indicates that the region consists of at least two genes which are transcribed in the same direction (from BamHI to EcoRI : see Fig. 3, pLW22) and which are similarly regulated by anaerobiosis, fnr and nickel (Wu and Mandrand-Berthelot, 1986). This suggests that genes hydC72 and hydC79, which are shown to encode 59 and 27.5 kDa polypeptides respectively, belong to the same transcriptional unit, with the gene order hydC72-hydC79 (Fig. 3). However the complementation pattern of mutant HYD790 by plasmid pLW23 (hydC79$^+$) deleted from an hypothetical promoter site in front of gene hydC72 rises the possibility of the existence of a secondary promoter located just upstream gene hydC79. Measurements of hydrogenase activities in mutant HYD790 harboring plasmid pLW23 which contained the various Tn5 insertions outlined in Fig. 3 allowed us to map the end of the hydC79 gene near the HincII-3 site (data not shown). Taking into account the 1.6 kb distance separating insertion 4014 and insertion 4017 which still confers a nickel restorable hydrogenase phenotype after chromosomal integration, we propose that the hydC locus is a complex region including a minimum of three genes. Recent cloning and sequencing of the chlD locus which shows a situation analogous to hydC in relation to molybdate uptake (Johann and Hinton, 1987) have prompted us to investigate the hydC locus. Further examination of the nucleotide sequence of hydC will allow us to determine whether hydC, like chlD, encodes proteins which share amino acid sequence homology with binding protein-dependent transport systems.

ACKNOWLEDGEMENTS

This work was supported by grants from the CNRS (ATP Microorganismes d'Intérêt Industriel, Agricole ou Médical, Fermentations) and from the Science and Engineering Research Council. We thank people listed in Table 1 for donation of strains or plasmids, C. Navarro, Y. Hussain, N. Razafitsara, R. Waugh, C. J. Edmonds and S. E. Holt for participating to some experiments cited, Y. Berlier and P. A. Lespinat for generous help with D_2/H^+ exchange reaction. We also want to thank J. Pellissier, G. Robert and J. M. Prost for skillful technical assistance and G. Luthaud for typing the manuscript.

REFERENCES

Ballantine, S. P., and Boxer, D. H., 1986, Isolation and characterization of a soluble active fragment of hydrogenase isoenzyme 2 from membranes of anaerobically grown Escherichia coli, Eur. J. Biochem. 156:277.

Berlier, Y. M., Dimon, B., Fauque, G., and Lespinat, P. A., 1985, Direct mass-spectrometric monitoring of the metabolism and isotope exchange in enzymatic and microbiological investigation, in: "Gas Enzymology", H. Degn, R. P. Cox, and H. Toftlund, eds., Reidel Publishing Company, p. 17.

Birkmann, A., Zinoni, F., Sawers, G., and Böck, A., 1987, Factors affecting transcriptional regulation of the formate-hydrogen-lyase pathway of Escherichia coli, Arch. Microbiol., 148:44.

Casadaban, M. J., and Cohen, S. N., 1979, Lactose genes fused to exogenous promoters in one step using a Mu-lac bacteriophage : in vivo probe for transcriptional control sequences, Proc. Natl. Acad. Sci. USA, 76:4530.

Chaudhuri, A., and Krasna, A. I., 1987, Isolation of genes required for hydrogenase synthesis in Escherichia coli, J. Gen. Microbiol., 133:3289.

Chaudhuri, A., and Krasna, A. I., 1988, Isolation of a gene required for growth of Escherichia coli on fumarate and H_2, J. Gen. Microbiol., 134:2155.

Graham, A., Boxer, D. H., Haddock, B. A., Mandrand-Berthelot, M. A., and Jones, R. W., 1980, Immunological analysis of the membrane-bound hydrogenase of Escherichia coli, FEBS Lett., 113:167.

Gray, C. T., and Gest, H., 1965, Biological formation of molecular hydrogen, Science, 148:186.

Jasper, P., and Silver, S., 1977, Magnesium transport in microorganisms, in: "Microorganisms and Minerals", E. D. Weinberg, ed., Marcel Dekker, New-York, p. 7.

Jayaraman, P. S., Peakman, T. C., Busby, S. J. W., Quincey, R.V., and Cole, J.A., 1987, Location and sequence of the promoter of the gene for the NADH-dependent nitrite reductase of Escherichia coli and its regulation by oxygen, the Fnr protein and nitrite, J. Mol. Biol., 196:781.

Johann, S., and Hinton, S. M., 1987, Cloning and nucleotide sequence of the chlD locus, J. Bacteriol., 169:1911.

Jones, R. W., 1980, The role of the membrane-bound hydrogenase in the energy-conserving oxidation of molecular hydrogen by Escherichia coli, Biochem. J., 188:345.

Karube, I., Tomiyama, M., and Kikuchi, A., 1984, Molecular cloning and physical mapping of the hyd gene of Escherichia coli K-12, FEMS Microbiol. Lett., 25:165.

Lee, J. H., Patel, P., Sankar, P., and Shanmugam, K. T., 1985, Isolation and characterization of mutant strains of Escherichia coli altered in H_2 metabolism, J. Bacteriol., 162:344.

Lowry, O. H., Rosebrough, J., Farr, A. L., and Randall, R. J., 1951, Protein measurement with Folin phenol reagent, J. Biol. Chem., 193:265.

Macy, J., Kulla, H., and Gottschalk, G., 1971, H_2-dependent anaerobic growth of Escherichia coli on L-malate : succinate formation, J. Bacteriol., 125:423.

Maniatis, T., Fritsch, E. F., and Sambrook, J., 1982, Molecular cloning : a laboratory manual. Cold Spring Harbor Laboratory, Cold Spring Harbor, New York.

Miller, J. H., 1972, Experiments in molecular genetics. Cold Spring Harbor Laboratory, Cold Spring Harbor, New York.

Miller, J. H., 1972, Experiments in molecular genetics. Cold Spring Harbor Laboratory, Cold Spring Harbor, New York.

O'Hoy, K., and Krishnapillai, V., 1985, Transposon mutagenesis of the Pseudomonas aeruginosa PAO chromosome and the isolation of high frequency of recombination donors, FEMS Microbiol. Lett., 29:299.

Pecher, A., Zinoni, F., Jatisatienr, C., Wirth, R., Hennecke, H., and Böck, A., 1983, On the redox control of synthesis of anaerobically induced enzymes in enterobacteriaceae, Arch. Microbiol., 136:131.

Ruiz-Herrera, J., and DeMoss, J. A., 1969, Nitrate reductase complex of Escherichia coli K-12 : participation of specific formate dehydrogenase and cytochrome b1 components in nitrate reduction, J. Bacteriol., 99:720.

Ruiz-Herrera, J., and Alvarez, A., 1972, A physiological study of formate dehydrogenase, formate oxidase and hydrogenlyase from Escherichia coli K-12, Antonie van Leeuwenhoek J. Microbiol Serol., 38:479.

Sankar, P., Lee, J. H., and Shanmugam, K. T., 1985, Cloning of hydrogenase genes and fine structure analysis of an operon essential for H2 metabolism in Escherichia coli, J. Bacteriol., 162:353.

Sawers, R. G., and Boxer, D. H., 1986, Purification and properties of membrane-bound hydrogenase isoenzyme 1 from anaerobically grown Escherichia coli K-12, Eur. J. Biochem., 156:265.

Sawers, R. G., Ballantine, S. P., and Boxer, D.H., 1985, Differential expression of hydrogenase isoenzymes in Escherichia coli K-12 : evidence for a third isoenzyme, J. Bacteriol., 164:1324.

Sawers, R. G., Jamieson, D. J., Higgins, C. F., and Boxer, D. H., 1986, Characterization and physiological roles of membrane-bound hydrogenase isoenzymes from Salmonella typhimurium, J. Bacteriol., 168:398.

Shaw, D. J., Rice, D. W., and Guest, J. R., 1983, Homology between CAP and Fnr, a regulator of anaerobic respiration in Escherichia coli, J. Mol. Biol., 166:241.

Spratt, B. G., Hedge, P. G., te Heesen, S., Edelman, A., and Broome-Smith, J.K., 1986, Kanamycin-resistant vectors that are analogues of plasmids pUC8, pUC9, pEMBL8 and pEMBL9, Gene, 41:337.

Stoker, K., Oltmann, L. F., and Stouthamer, A. H., 1989, Randomly induced Escherichia coli K-12 Tn5 insertion mutants defective in hydrogenase activity, J. Bacteriol., 171:831.

Stults, L. W., Sray, W. A., and Maier, R. J., 1986, Regulation of hydrogenase biosynthesis by nickel in Bradyrhizobium japonicum, Arch. Microbiol., 146:280.

Tabor, S., and Richardson, C. C., 1985, A bacteriophage T7 RNA polymerase/promoter system for controlled exclusive expression of specific genes, Proc. Natl. Acad. Sci. USA, 82:1074.

Touati, D., 1983, Cloning and mapping of the manganese superoxide dismutase gene (sodA) of Escherichia coli K-12, J. Bacteriol., 155:1078.

Waugh, R., and Boxer, D. H., 1986, Pleiotropic hydrogenase mutants of Escherichia coli K-12 : growth in the presence of nickel can restore hydrogenase activity, Biochimie, 68:157.

Wu, L. F., 1988, Caractérisation génétique et physiologique du système formiate-hydrogènelyase chez Escherichia coli : clonage et régulation de l'expression des gènes fdhF et hydC, Ph. D. Thesis, Institut National des Sciences Appliquées de Lyon.

Wu, L. F., and Mandrand-Berthelot, M. A., 1986, Genetic and physiological characterization of new Escherichia coli mutants impaired in hydrogenase activity, Biochimie, 68:167.

Wu, L. F., Mandrand-Berthelot, M. A., Waugh, R., Edmonds, C. J., Holt, S. E., and Boxer, D. H., 1989, Nickel deficiency gives rise to the defective hydrogenase phenotype of hydC and fnr mutants in Escherichia coli, Mol. Microbiol., in press.

POSTERS - 1 - MICROBIOLOGY

THERMOPHILIC ANAEROBIC OXIDATION OF BUTYRATE IN AXENIC CULTURE

Birgitte Ahring [a] , Peter Westermann [b] and R.A. Mah [c]

   a) Department of Biotechnology
      The Technical University of Denmark
      DK 2800 Lyngby, Denmark
   b) Department of General Microbiology
      University of Copenhagen
      DK 2800 Lyngby, Denmark
   c) School of Public Health
      University of California
      Los Angeles, CA 90024, USA

Introduction

Normally, propionate and butyrate account for approximately 20% of the methane produced in an anaerobic digestion. The degradation of these volatile fatty acids involves at least two groups of bacteria, the obligately hydrogen-producing acetogenic bacteria oxidizing the acids and the methane-producing bacteria utilizing the hydrogen produced (for a review see reference 1). Owing to the unfavorable thermodynamics of fatty acid oxidation under standard conditions, the metabolism of the acetogenic bacteria demands a low partial pressure of hydrogen normally maintained by the hydrogen - utilizing methanogenic bacteria.

Detailed physiological and biochemical studies require pure culture of the bacterium investigated. Recently a mechanism for non-biological hydrogen-removal has been described (2). We here report on growth of a thermophilic butyrate degrading organism in axenic culture using a palladium catalyst as a hydrogen scavenger.

Materials and Methods

The isolation procedure for the thermophilic butyrate - degrading bacterium has been reported previously (3,4,5). The medium used was basically as previously described (3). The butyrate concentration was 10 mM and the medium was supplemented with clarified rumen fluid (5%, vol/vol). Experiments were performed in serum tubes (27 ml) with 5 ml medium and $N_2 : CO_2$ (80 : 20%) in the gas phase. The hydrogenation catalyst was Pd - $BaSO_4$ (5% Pd on $BaSO_4$ matrix , SIGMA).

For experiments with the hydrogenation catalyst, Pd - $BaSO_4$ was added to the medium in a concentration of 14 mg/ml and the headspace was supplemented with ethylene giving a final concentration of 0.3 atm. During the action of the catalyst, ethylene was reduced to ethane. Gases were analysed by gas chromatography using a Porapak Q column and a thermal conductivity detector.

Results

The presence of ethylene (30 KPa) inhibited methanogenesis in the coculture containing the thermophilic butyrate degrader together with *Methanobacterium thermoautotrophicum* . By repeated transfer of the coculture in the presence of ethylene and catalyst and by addition of methanogenic inhibitor (2 -BES) to the medium, the H_2 - utilizing methanogen was eliminated from the culture.

Table 1 shows the mass balance for the oxidation of butyrate by the thermophilic butyrate degrader in the presence of Pd - $BaSO_4$ and C_2H_4[a] .

Time (days)	Butyrate (μmol/vial)	Acetate (μmol/vial)	C_2H_4 (μmol/vial)	C_2H_6 (μmol/vial)
0	251.2	10.2	252.0	0
7	156.3	209.1	85.7	149.2

a The data are means of 3 replicates. No methane was produced. A hydrogen partial pressure of 1.10^{-3} atm. was found at day 7. For each mol butyrate oxidized, 2.1 mol acetate was produced under simultaneous reduction of 1.75 mol ethylene to nearly the same amount of ethane (1.6 mol). This results in a carbon and hydrogen recovery of 105 and 87.5%, respectively.

Table 2 shows the maximum specific growth rate of the thermophilic butyrate degrader during growth in coculture and in axenic culture at different temperature of incubations.

Incubation temperature	μ days^{-1} methane [a]	μ days^{-1} ethane[b]	μ days^{-1} ethane · 100% μ days^{-1} methane
37°C	0	0	-
45°C	0.150	0	-
52°C	0.387	0.115	29.7
56°C	0.497	0.195	39.2
60°	0.509	0.220	43.2
65°C	0.075	0.053	70.7
70°C	0	0	-

a) The specific methane production rate during growth of the thermophilic butyrate degrading coculture.
b) The specific ethane production rate during growth of the thermophilic butyrate degrader in the presence of catalyst and ethylene.

An optimal and maximal growth rate at 60 °C and 65 °C, respectively, was found for both experiments with the coculture and with axenic culture of the thermophilic butyrate degrader.

The data clearly demonstrate that the function of the catalyst is improving with temperature indicating that this mechanism of non-biological hydrogen-removal is especially suitable for growth-experiments with thermophilic acetogenic bacteria.

Conclusion

The study presented demonstrate the feasibility of using catalytic reduction instead of hydrogen consumption by a methanogenic bacterium during oxidation of volatile fatty acids.

The carbon balance confirms the results previously found for the stoichiometry of butyrate oxidation by the thermophilic butyrate degrader (3). Furthermore, ethylene was quantitatively reduced to ethane. From the hydrogen balance it can be seen that not all of the hydrogen produced was used for reduction of ethylene probably due to absorption by the catalyst. However, the amount retained by the catalyst was low in our experiments (approximately 12.5%) compared to a previous study at mesophilic temperatures (half of the hydrogen produced) (2).

References

1. R.A. Mah. Methanogenesis and methanogenic partnerships. Philos. Trans. R. Soc. London Ser . B. 297 : 599 - 616. (1982).
2. H.F. Kaspar , A.J. Holland and D.O. Mountfort. Simultaneous butyrate oxidation by *Syntrophomonas wolfei* and catalytic olefin reduction in absence of interspecies hydrogen transfer. Arch. Microbiol. 147 : 334 - 339. (1987).
3. B.K. Ahring and P. Westermann. Thermophilic anaerobic degradation of butyrate by a butyrate - utilizing bacterium in coculture and triculture with methanogenic bacteria. Appl. Environ. Microbiol. 53 : 429 - 433. (1987).
4. B.K. Ahring and P. Westermann. Kinetics of butyrate, acetate, and hydrogen metabolism in a thermophilic, anaerobic, butyrate-degrading coculture. Appl. Environ. Microbiol. 53 : 434 - 439. (1987).
5. B.K. Ahring and P. Westermann. Product inhibition of butyrate metabolism by acetate and hydrogen in a thermophilic coculture. Appl. Environ. Microbiol. 54 : 2393 - 2397. (1988).

EUBACTERIUM ACIDAMINOPHILUM, AN ORGANISM ABLE TO INTERACT IN INTERSPECIES H-TRANSFER REACTIONS OR TO TRANSFER ELECTRONS TO DIFFERENT TERMINAL REDUCTASE SYSTEMS

J.R. Andreesen, K. Hormann, K. Granderath, M. Meyer and D. Dietrichs

Institut für Mikrobiologie der Universität
Grisebachstr. 8
D-3400 Göttingen, FRG

Eubacterium acidaminophilum is an acetate forming anaerobe utilizing a variety of amino acids whose utilization mostly requires the addition of an oxidant such as sarcosine or betaine or of H_2-scavenging organisms as methanogenic, sulfidogenic, or acetogenic bacteria. Although all enzymes required for acetate synthesis from CO_2 via glycine (Dürre and Andreesen 1982) are present in crude extracts and CO_2 can be reduced to formate, the organism does not grow autotrophically nor on formate as C_1-compound, but it grows excellent on glycine and has to synthesize glycine from CO_2 in most cases due to the lack of serine hydroxymethyltransferase (Zindel et al. 1988).

Formate can act as electron donor for the reduction of glycine, sarcosine, and betaine allowing good growth and high molar growth yields indicating a yield of about 1 ATP per mol acetate formed. Control experiments using formate plus [2-^{14}C]-glycine or formate plus [^{14}CH$_3$]-betaine definitely show that acetate is only formed by reductive deamination reactions. Thus, energy has to be conserved by these reactions (Hormann and Andreesen 1989).

Growth experiments (substrate utilization, product formation) indicate the existence of a hierarchial order between glycine reduction and that of sarcosine and betaine. The differences are also obvious for the individual optimized enzyme assays. Only glycine reductase activity requires the presence of ADP + AMP in the assay (an indication of a direct role of ADP + P_i in the energy conservation mechanism?). Both sarcosine reductase and betaine reductase share many common properties, but they are separate enzymes for they are only inductively formed by their substrate, whereas glycine reductase is also present in sarcosine- and betaine-, but not in serine-grown cells (Hormann and Andreesen 1989). The proper induction of the glycine reductase might constitute a problem for *E. acidaminophilum* , thus, it does not form acetate from CO_2, although glycine has to be synthesized from CO_2 (except for growth on glycine). *E. acidaminophilum* requires 1 μM selenite only when grown on glycine, sarcosine, or betaine. [^{75}Se]-selenite labeled about the same proteins in glycine-, sarcosine-, or betaine-grown cells of Mr´s of 13, 20, and 50 kDa (Hormann and Andreesen 1989).

Electron flow to glycine reductase

During our study of the glycine decarboxylase proteins (Freudenberg and Andreesen 1989; Freudenberg et al. 1989a) a peculiarity of *E. acidaminophilum* became evident: the complex contains an atypically small dihydrolipoamide dehydrogenase (P_3) of Mr 68 kDa preferring NADP instead of NAD. By immunochemical methods it is shown that this enzyme is not directly part of the glycine decarboxylase, but is more closely connected with the selenoprotein P_A of glycine reductase, which both are associated with the cytoplasmic membrane in contrast to the other proteins of glycine decarboxylase. Thus, it is assumed that P_3 is a component of glycine reductase being reduced by protein P_2 of glycine decarboxylase, for the latter is present in huge molar surplus and could act as a hydrogen carrier between both complexes (Freudenberg et al. 1989b).

The actual electron flow seems to be more complicated after studying an analogous system in bacterium W6 (DSM 5388), which is physiologically related to *C.sporogenes* and is able to grow on glycine. This organism contains a conventional dihydrolipoamide dehydrogenase involved in glycine oxidation and, additionally, a flavoprotein which cross reacts with the dihydrolipoamide dehydrogenase of *E. acidaminophilum* and can exhibit dihydrolipoamide dehydrogenase activity, if at least one other protein is added which shares many similarities with thioredoxin. A second protein also stimulating some of the electron transfer reactions seems to be related to the selenoprotein P_A of glycine reductase. In *E. acidaminophilum* dihydrolipoamide dehydrogenase is also stimulated by a thioredoxin protein and the selenoprotein P_A. The analogy to thioredoxin reductase and thioredoxin became clearly evident by comparison of their N-terminal sequences. Except for the common FAD-binding site close to the N-terminal, the enzyme of *E. acidaminophilum* exhibits no relations even to other dihydrolipoamide dehydrogenases of other glycine utilizing anaerobes isolated by us, but again to the cross reacting electron transferring flavoprotein of *C.sporogenes* . Antibodies directed against the dihydrolipoamide dehydrogenase inhibited the NADPH-dependent reduction of glycine, which proves its involvement. Since the whole glycine metabolizing complexes are soluble or cytoplasmatically orientated, no proton gradient can be established. Thus, a mechanism postulated by Barnard and Akhtar (1979) might be responsible for energy conservation by glycine reduction. The same proteins (P_3, thioredoxin, and P_A) might be involved in sarcosine and betaine reduction, too. Thus, only the terminal reductases should be different.

Energetical considerations using amino acids as reductants

Amino acids are commonly degraded in the cytoplasma. Except for glycine and serine, the reducing equivalents have to be transferred in case of *E. acidaminophilum* to a H_2-scavenging organism or to sarcosine as an oxidant in order to allow growth. If sarcosine acts only as electron sink, no energetical advantage should be observed. After elucidation of the enzymatic outfit involved in amino acid degradation and of the coenzyme specificity, three different stoichiometries were observed for the conversion of substrate per mol sarcosine

1.: 2 mol substrate/1 mol sarcosine (substrate: serine)
2.: 1 mol substrate/1 mol sarcosine (substrate: malate)
3.: 0.64 mol substrate/1 mol sarcosine (substrate: alanine, aspartate, valine, and leucine)

Except for malate, these stoichiometries can be explained if all the reduced pyridine nucleotides should get reoxidized by sarcosine reduction and about half of the "reduced ferredoxin"-equivalents. Serine, a substrate not requiring the addition of an oxidant for growth, shows the lowest stoichiometry for it provides only electrons at the ferredoxin level, which can form hydrogen (Zindel et al. 1988). The molar growth yield increases significantly if sarcosine is added. In addition no hydrogen and ethanol are formed any longer. Data obtained for the other amino acids also reveal that the cells can draw a similar energetical advantage only from those electrons which are generated at the ferredoxin level.

Hydrogen cycling would be an attractive explanation by which a proton gradient could be established via the membrane. By that, part of the electrons are reshuffled inwards creating similar conditions as during growth on formate/sarcosine. As a consequence, hydrogen should be evolved by a different hydrogenase separate from that enzyme which is responsible for an unidirectional uptake. The latter should be located at the outer face of the cytoplasmic membrane. Using special conditions, hydrogenase can be recovered as membrane bound enzyme. If H_2-scavenging organisms compete with sarcosine reductase as electron sink, both systems coexist. Especially with serine and alanine, most of the electrons still reduce sarcosine. This demonstrates (i) that part of the electrons are efficiently coupled to sarcosine reductase as a regenerating and energy conserving system, (ii) that another part might get cycled outside the cytoplasmic membrane, thus, becoming available to H_2-scavenging organisms such as *Acetobacterium woodii, Desulfovibrio vulgaris* , and *Methanospirillum hungatei* (in an increasing order). An interspecies formate tranfer can be established with *D. baarsii* (Zindel et al. 1988). During growth on glycine, *E.acidaminophilum* does not require an oxidant and exhibits fast growth, thus, outcompeting the H_2-scavenging organisms. Electrons generated by the oxidation of glycine do not form H_2 and are not transferred to sarcosine reductase, again indicating a tight coupling of the electron flow.

Summary

So far no acetogenic organism is known which grows autotrophically by using the glycine pathway of CO_2 reduction to acetate, although *Eubacterium acidaminophilum* contains all the enzymes to be necessary and conserves energy by reducing glycine, sarcosine, or betaine. Glycine reductase is an enzyme system different from sarcosine reductase and betaine reductase, which might correlate with a different mechanism for energy conservation. The electron flow towards the different reductases from reduced pyridine nucleotides involves an atypically small dihydrolipoamide dehydrogenase as a membrane associated "thioredoxin reductase" reacting with "thioredoxin" and selenoprotein P_A. Energy might be conserved via hydrogen cycling if serine, alanine, aspartate, valine, and leucine are substrates. Some of this hydrogen can be used to allow an interspecies transfer of hydrogen or formate.

Literature

Barnard GF, Akhtar M (1979) Mechanistic and stereochemical studies on the glycine reductase of *Clostridium sticklandii* . Eur J Biochem 99: 593-603

Dürre P, Andreesen JR (1982) Pathway of carbon dioxide reduction to acetate without net energy requirement in *Clostridium purinolyticum* . FEMS Microbiol Lett 15: 51-56

Freudenberg W, Andreesen JR (1989) Purification and partial characterization of the glycine decarboxylase multienzyme complex from Eubacterium acidaminophilum . J Bacteriol 171: 2209-2215

Freudenberg W, Dietrichs D, Lebertz H, Andreesen JR (1989a) Isolation of an atypically small lipoamide dehydrogenase involved in the glycine decarboxylase complex from *Eubacterium acidaminophilum* . J Bacteriol 171: 1346-1354

Freudenberg W, Mayer F, Andreesen JR (1989b) Immunocytochemical localization of proteins P1, P2, P3 of glycine decarboxylase, and of the selenoprotein P_A of glycine reductase,all involved in anaerobic glycine metabolism of *Eubacterium acidaminophilum* . Arch Microbiol 152: 182-188

Hormann K, Andreesen JR (1989) Reductive cleavage of sarcosine and betaine by *Eubacterium acidaminophilum* via enzyme systems different from glycine reductase. Arch Microbiol, in press

Zindel U, Freudenberg W, Rieth M, Andreesen JR, Schnell J, Widdel F (1988) *Eubacterium acidaminophilum* sp. nov., a versatile amino acid-degrading anaerobe producing or utilizing H_2 or formate. Description and enzymatic studies. Arch Microbiol 150: 254-266

IMMUNOLOGICAL PROPERTIES OF *DESULFOBACTER*

Janiche Beeder, Torleiv Lien and Terje Torsvik

Departement of Microbiology and Plant Physiology
University of Bergen
Jahnebakken 5, 5007 Bergen, Norway

Abstract

Sulfate reducing bacteria of the genera *Desulfobacter, Desulfococcus* and *Desulfobulbus* were shown to possess common antigens. Using monoclonal antibodies and polyclonal antiserum, it was possible to identify bacteria, within these genera, at genus, species and strain level.

Introduction. Acetate utilizing sulfate reducing bacteria (SRB) may play an important role in corrosion of oil installations, leading to a demand for their rapid detection. For this purpose immunological methods may be useful, especially if antibodies can be made towards antigens common to a group of SRB. In this study, antibodies were produced against strain B 54, a sulfate reducing bacterium isolated from an oilfield installation in the North Sea. B 54 is a curved, polary flagellated marine bacterium utilizing acetate as the only substrate with sulfate as the terminal electron acceptor. According to these characteristics, B 54 was included in the genus *Desulfobacter*. The antibodies were used in a search for common antigenes in the *Desulfobacter* group.

Methods

Polyclonal antiserum (PAb) was produced (using rabbits) against whole cells of B 54. Monoclonal antibodies (MAb's), were produced (using mice) according to the hybridoma technique described by Galfre and Milstein (1981). Antigenic properties were compared using SDS-PAGE gel electroforesis and immunoblotting with nitrocellulose sheet.

Results and discussion

Using PAb it was found that B 54 and four other *Desulfobacter* species (*D. postgatei, D. hydrogenophilus, D. curvatus* and *D. latus*) contained several common protein antigens. Many differences in the antigenic properties were also evident. The differences between strain B 54 and the other *Desulfobacter* type species were as great as the differences between the individual type species. Thus the immunological properties indicated that strain B 54 should be considered as a new species belonging to the *Desulfobacter* group.

Using MAb, it was shown that a 54 kD protein was present in all the *Desulfobacter* type species. The same protein was also found in *Desulfococcus multivorans*, a bacterium known to oxidize fatty acids of a longer chain length, up to C-18 (Widdel 1988). In *Desulfobulbus*, the MAb reacted with a 56 kD protein. Other fatty-acid-oxidizing SRB's, such as *Desulfovibrio sapovorans* and *Desulfotomaculum acetoxidans*, and the sulfur reducing bacterium *Desulfuromonas acetoxidans*, did not contain the 54 kD protein.

The function of the 54 kD protein is not known. This protein is present in *Desulfobacter* and in the related genera (by 16S rRNA homology, Fowler et al.,1986) *Desulfococcus* and *Desulfobulbus*. In *Desulfobulbus* the protein had a slightly larger molecular weight. The protein could not be detected in more distantly related sulfate reducing bacteria.

MAb's made against the LPS of strain B 54 did not react with any of the other sulfate or sulfur reducing bactera tested. These Mab's appeared to react monospecifically with strain B 54, and could be used to recognize this strain using immunofluorescence, ELISA or in immunoblotting procedures.

In conclusion, one MAb reacting with a 54 kD protein could be used to detect SRB belonging to *Desulfococcus, Desulfobulbus* and *Desulfobacter*. The antigenic pattern obtained by using PAb's, allowed the identification of the different genera. Also, within *Desulfobacter*, the different species could be identified using PAb's. Thus the combination of MAb's and PAb's could be used for identification of *Desulfobacter* at genus, species and strain level.

References

Fowler, V.J. Widdel, F., Woese, C.R. and Stackebrant, E.:
  Phylogenetic relationships of sulfate- and sulfur-reducing bacteria. Syst. Appl. Microbiol. 8, 32-42 (1986).

Galfre, G., Milstein, C.: Preparation of monoclonal
  antibodies: strategies and procedures. Meth. in Enzymol. 73, 3-47 (1981)

Widdel, F.: Microbiology and ecology of sulfate and sulfur-reducing bacteria. In: Biology of anaerobic microorganism spp. 469-586,(J.B. Zender, Eds.) John Wiley & Sons Inc., New York, 1988.

FERMENTATION PROPERTIES OF FOUR STRICTLY ANAEROBIC RUMEN FUNGAL SPECIES:

H2-PRODUCING MICROORGANISMS

Annick BERNALIER, G. FONTY and Ph. GOUET

Laboratoire de Microbiologie - INRA - CR de Clermont-Theix
63122 - ST GENES-CHAMPANELLE - FRANCE

INTRODUCTION

In ruminants, the main part of the digestion takes place in the rumen thanks to a complex and diversified strictly anaerobic microbial population, composed of bacteria, fungi and protozoa which interact to
form a trophic chain (nutritional and metabolic interactions). As in all
anaerobic ecosystems, the interspecies transfert of hydrogen between H2-
producing and H2-utilizing microorganisms plays an important role in the
orientation and regulation of fermentations. Among H2-producing microorganisms, there are several species of bacteria, protozoa and
fungi. The objective of this paper is to describe the fermentation properties of four H2-producing anaerobic fungi, isolated from the rumen in
our Laboratory : Neocallimastix frontalis, Piromonas communis, Sphaeromonas communis and Neocallimastix joyonii.

MATERIALS AND METHODS

The ability of the four species to use different carbon sources :
xylan, arabinogalactan, glucomanan, galactan, polygalacturonate, pectin,
soluble starch, D-glucose, D+cellobiose, D+maltose, L+arabinose, D+-
xylose, L-fructose, D-mannose, fucose, D+raffinose, galactose, lactose,
sucrose, gentobiose, glycerol and mannitol was studied, using the medium
described by LOWE et al (1985). Monosaccharides and disaccharides were
added to 10 ml of medium at a concentration of 0,2%. The polysaccharides
were added at a concentration of 0,1%. The culture were made under 100%
CO2 according to the method of HUNGATE (1969). The ability to use these
substrates was considered as positive when the fungi maintained their
growth after 3 or 4 transferts on the respective substrates.

We have also studied the ability of these fungi to degrade various
types of pure cellulose (Whatman n°1 filter paper, Avicel, cellulose
fibre, Sigmacell 50 and 100, Solka floc and Cotton) and three natural
cellulosic substrates (milled Rye-grass hay, milled ammonia-treated
wheat straw and fragments of wheat straw). Each substrate was added to
10 ml of medium at the concentration of 0,1%. The dry matter loss was
measured, in triplicates, after 5 and 8 days of incubation for the pure
celluloses and the natural substrates, respectively.

We have determined the end-products of cellulose fermentation (filter paper) after 8 days of incubation. Volatile fatty acids, ethanol and gas were analysed by gas chromatography (JOUANY, 1982), formate, L- and D-lactate by enzymatic assay according to the method of Boerhinger. The amount of reducing sugars remaining was determined by the method of MILLER (1959).

RESULTS AND DISCUSSION

The four fungal species used a wide range of sugars and polysaccharides : glucose, L-fructose, D+xylose, cellobiose, D+maltose, gentobiose, soluble starch, cellulose, xylan, glucomanan and arabinogalactan but were unable to grow on L+arabinose, galactose, fucose, polygalacturonate and mannitol. Nevertheless we observed a poor culture of Sphaeromonas communis on D+raffinose. The four fungal strains utilised lactose except Piromonas communis and galactan except Neocallimastix joyonii. Sucrose could not support the growth of Sphaeromonas communis and Neocallimastix frontalis. Sphaeromonas communis was unable to use glycerol like Neocallimastix joyonii. Mannose was only used by Piromonas communis.

All the strains were cellulolytic and were able to degrade different types of cellulose (Tab.1). The more efficient species were those which possess rhizoïds (Neocallimastix frontalis, Neocallimastix joyonii and Piromonas communis). Sphaeromonas communis which don't have these structures, was less efficient whatever the type of cellulose tested. The high ordered celluloses (Whatman n°1 filter paper, Avicel, cotton, Solka floc and Sigmacell 50) were more degraded by fungi than amorphous celluloses (Sigmacell 100).

Tab. 1 Dry Matter disappearance (%) of various types of cellulose

	Neocallimastix frontalis	Piromonas communis	Sphaeromonas communis	Neocallimastix joyonii
Whatman n°1 filter paper	63,0	77,0	12,0	45,0
Avicel	34,5	39,1	2,9	13,6
cellulose fibre	46,0	20,1	4,7	22,4
Sigmacell 50	42,4	26,1	3,4	21,7
Sigmacell 100	21,4	16,1	5,6	18,3
Solka floc	37,7	35,8	4,1	33,3
Cotton	35,7	26,8	1,0	N.D.

N.D. : no determined

At the end of the culture, the percent (%) of dry matter disappearance of the three natural substrates was comprised between 25% and 35% for the 3 rhizoïdal species and less than 10% for Sphaeromonas communis. Milled Rye-grass hay and ammonia treated wheat straw were more degraded than fragments of Wheat straw.

The end-products of cellulose fermentation were qualitatively the same for the four species but the relative proportions of each product varied according to the species (Tab.2). The major products were formate, acetate and lactate except for Neocallimastix joyonii which was characterized by a low production of D-lactate. These fungi were also high H2-producing microorganisms. Previous studies (BAUCHOP and MOUNTFORT, 1981, MOUNTFORT et al, 1982, FONTY et al 1988) have shown that these fungal species were involved in interspecies transfert with H2-utilizing rumen methanogenic bacteria. When cocultured with Methanobrevibacter ruminantium, the cellulolytic activity of the rumen fungi increased and their metabolisms were shifted towards a high production of acetate with a concomittant decrease in the production of reduced compounds (lactate and ethanol).

Tab. 2 End-products of cellulose fermentation by the four strains of rumen anaerobic fungi

	Neocallimastix frontalis	Piromonas communis	Sphaeromonas communis	Neocallimastix joyonii
Formate ([1])	48,6	85,5	151,4	71,1
Acetate	39,8	32,1	121	50,7
L-Lactate	0,5	1,0	3,9	0,0
D-Lactate	54,1	63,5	24,8	8,9
Succinate	trace	trace	0,0	0,0
Ethanol	6,5	7,9	0,0	15,2
H_2 (%) (2)	28,5	25,7	22,2	15,0

(1) Moles/100 Moles hexose fermented
(2) % of total gas phase

REFERENCES

BAUCHOP , T. and MOUNTFORT, D.O., 1981 . Cellulose fermentation by a rumen anaerobic fungus in both the absence and the presence of methanogens, Appl. Environ. Microbiol. , 42 : 1103 .

FONTY , G., GOUET , Ph. and SANTE , V., 1988 . Influence d'une bactérie méthanogène sur l'activité cellulolytique de deux espèces de champignons du rumen . Résultats préliminaires . Reprod. Nutr. Develop. , 28 : 133 .

HUNGATE , R.E., 1969 . A roll tube method for cultivation of strict anaerobes . In : " Methods and Microbiology " , 38 : 117 J.R. Norris & D.W. Gibsons (eds.) , Acad . Press , New York and London (pub.) .

JOUANY, J.P., 1982 .Volatile fatty acids and alcohol determination in digestive contents, silage juices , bacterial cultures and anaerobic fermentor contents . Sci. Al. , 2 : 131 .

LOWE , S.E., THEODOROU , M.K., TRINCI , A.P.J. and HESPELL , R.B., 1985 . Growth of anaerobic rumen fungi on defined and semi - defined media lacking rumen fluid . J. Gen. Microbiol . 131 : 2225 .

MILLER , G.L., 1959 . Use of dinitrosalicylic reagent for determination of reducing
   sugars. Anal. Chem. , 31 : 426 .
MOUNTFORT , D., ASHER , R. and BAUCHOP , T., 1982 . Fermentation of cellulose
   to methane and CO_2 by a rumen anaerobic fungus in a triculture with
   Methanobrevibacter sp. strain RA1 and *Methanosarcina barkeri* . Appl.
   Environ. Microbiol . 44 : 128 .

GROWTH OF BACTEROIDES XYLANOLYTICUS IN THE PRESENCE AND ABSENCE OF A
METHANOGEN

S. Biesterveld and A.J.M.Stams

Agricultural University
Department of Microbiology
H. v. Suchtelenweg 4
6703 CT Wageningen
The Netherlands

INTRODUCTION

Bacteroides xylanolyticus X5-1 is a Gram-negative motile xylanolytic bacterium, isolated from cattle manure (I.Scholten-Koerselman et
al., 1986). It ferments xylan, but not cellulose to ethanol, acetate,
hydrogen and carbon dioxide as the main end products. In addition, both
C5 and C6 sugars are used for growth.
In coculture of Bacteroides xylanolyticus X5-1 with a hydrogen consuming methanogen, acetate and hydrogen are the only fermentation
products.
The aim of this research was to investigate a) which pathway is used
for xylose degradation in the hemicellulolytic organism Bacteroides
xylanolyticus X5-1, and b) how Interspecies Hydrogen Transfer influences product formation on enzyme level in B. xylanolyticus X5-1.

MATERIALS AND METHODS

Bacteroides xylanolyticus X5-1 and Methanospirillum hungatei JF1
were cultivated in a medium which contained per liter: KH_2PO_4, 0.41 g;
Na_2HPO_4, 0.53 g; NH_4Cl, 0.3 g; NaCl, 0.3 g; $CaCl_2.2H_2O$, 0.11 g;
$MgCl_2.6H_2O$, 0.1 g; $NaHCO_3$, 4 g; $Na_2S.9H_2O$, 0.24 g; Bio-trypticase, 0.5
g; Yeast extract, 0.2 g; resazurine, 0.5 mg; vitamin solution according
to Wolin et al. (1963), 1 ml; trace element solution according to
Zehnder et al.(1980), 1 ml.
B. xylanolyticus X5-1 was grown on 0.5% xylose either in pure culture
or in a dense H_2/CO_2 (80%/20%) grown culture of M. hungatei JF1.
All fermentation products were determined with a Varian aerograph
gaschromatograph with a Chromosorb 101 column or by HPLC (LKB, Bromma,
Sweden) with an Chrompack Organic Acids Column (Chrompack, Middelburg,
the Netherlands), using a 2142 Refractive Index Detector (LKB, Bromma,
Sweden) and 5 mM H_2SO_4 as eluent at a temperature of 60°C.
All gases were analyzed on a Packard gaschromatograph with a molecular
sieve column.
For the preparation of cell extracts cells were washed twice with 50 mM

Tris-HCl buffer (pH 7.5) containing 1 mM dithiothreitol and 1 mM MgCl₂ and disrupted by sonification (10 times of 30 s, with intermittent cooling) using a Sonics & Materials sonifier (Sonics & Materials, CT, USA). The sonified cells were centrifuged for 20 min. at 13000 rpm and the supernatant was used for enzyme assays.

Standard procedures were used for the determination of enzyme activities. Enzyme activities of B. xylanolyticus X5-1 in extracts of cocultures were corrected for enzyme activities of Methanospirillum hungatei by relating the protein content of the methanogen to the amount of cofactor F420.

Protein was assayed according to Bradford (1976).

RESULTS AND DISCUSSION

Pathway of xylose degradation

Enzyme activity measurements in cell extracts of Bacteroides xylanolyticus X5-1 indicate that the pentose phosphate pathway in combination with the glycolysis is involved in the degradation of xylose. Key enzymes of the 2-keto-3-desoxy-6-phosphogluconate pathway were not detected (table 1, figure 1).

Low activities of the enzyme phosphoketolase were present. Whether a direct cleavage of xylulose-5-phosphate to glyceraldehyde-3-phosphate and acetyl-phosphate contributes significantly to xylose degradation remains to be investigated in detail. The ratio (acetate + ethanol) : CO_2 which was determined to be 0.96, indicates that one CO_2 is formed per C2 molecule. This value would have been much higher if the phosphoketalase reaction is important.

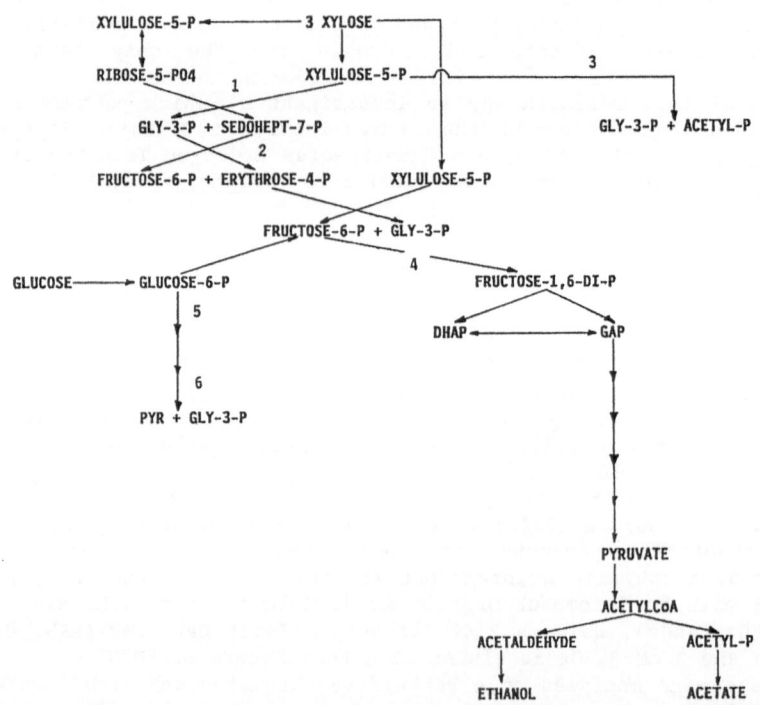

Figure 1. Pathways involved in xylose fermentation in B. xylanolyticus.

Table 1. specific enzyme activities measured in cell extract of Bacteroides xylanolyticus X5-1.

enzyme		spec. Act (nmol/min. mg)
1) transketolase	PPP	40
2) transaldolase	PPP	120
3) phosphoketolase		10
4) phosphofructokinase	GLY	790
5) glucose-6-phosphate dehydrogenase	KDPG	0
6) glucose dehydrase/ 6-fosfo-2-keto-3-desoxy aldolase	KDPG	0

PPP = pentose phosphate pathway
GLY = glycolysis
KDPG = 2-keto-3-deoxy-6-phospho-gluconate pathway

Influence of Interspecies Hydrogen transfer

In the coculture of Bacteroides xylanolyticus X5-1 with Methano-spirillum hungatei no ethanol was produced. Acetate and presumably hydrogen were the only fermentation products. Analysis of enzyme levels in cell extract of B. xylanolyticus X5-1 in pure and in mixed cocultures showed an increase in acetate kinase and a decrease in alcohol dehydrogenase (table 2). These findings indicate that the presence of a methanogen does not only regulate product formation by differences in concentrations of metabolites and kinetic properties of the enzymes involved, but that the synthesis of enzymes is regulated as well.

Table 2. Specific activities of acetate kinase and alcohol dehydrogenase in cell extracts of Bacteroides xylanolyticus X5-1 in pure culture and in a mixed culture with Methanospirillum hungatei JF1.

	B. X5-1 in pure culture	B. X5-1 in coculture
acetate kinase	800	1600
alcohol dehydrogenase	1100	0

CONCLUSIONS

- Bacteroides xylanolyticus X5-1 degrades xylose mainly through the pentose phosphate pathway.

- Product formation by Bacteroides xylanolyticus X5-1 is affected by the presence of a methanogen.

- The synthesis of the enzyme alcohol dehydrogenase of Bacteroides xylanolyticus X5-1 is regulated by the hydrogen partial pressure.

ACKNOWLEDGMENT

This research was supported by grants from the innovation oriented program on agricultural biotechnology (PcLB) and the Royal Netherlands Academy of Arts and Sciences.

REFERENCES

Bradford, M.M., 1976, A rapid and sensitive method for the quantification of microgram quantities of protein utilizing protein-dye binding, Anal. Biochem., 72:248.

Scholten-Koerselman, I., Houwaard, F., Janssen, P., and Zehnder, A.J.B., 1986, Bacteroides xylanolyticus sp. nov., a xylanolytic bacterium from methane producing cattle manure. Antonie van Leeuwenhoek, 52:543.

Wolin, E.A., Wolin, M.J., and Wolfe, R.S., 1963, Formation of methane by bacterial extracts, Journ. Biol. Chem., 137:420.

Zehnder, A.J.B., Huser, B.A., Brock, T.D., and Wuhrmann, K., 1980, Characterization of an acetate-carboxylating, non-hydrogenoxidizing methane bacterium, Arch. Microbiol., 124:1-11.

HYDROGEN METABOLISM BY TERMITE GUT MICROBES

A. Brauman[1], M. D. Kane[2], M. Labat[1], J. A. Breznak[2]

[1]Laboratoire de Microbiologie, ORSTOM, Universite de Provence
Marseille, France, and [2]Department of Microbiology and Public
Health, Michigan State University, East Lansing, MI, USA

INTRODUCTION

In *Reticulitermes flavipes* termites, H_2/CO_2 acetogenesis outcompeted
methanogenesis as the main H_2-consuming (i.e., electron sink) reaction of
wood fermentation by the hindgut microbiota. Moreover, acetate produced from
$H_2 + CO_2$ supported up to 1/3 of this termite's respiratory requirement[1,2].
To increase our understanding of the nature and nutritional significance of
H_2 metabolism by termite gut microbes in general, we have begun extending
our studies to taxonomically diverse termites including soil-feeding
species. In this paper, we report rates of H_2/CO_2 acetogenesis and
methanogenesis in various termite species, as well as the enumeration and
isolation of some of the gut bacteria involved in H_2 metabolism.

RESULTS

Relationship between termite diet and H2 metabolism by gut microbes

Rates of H_2-dependent reduction of $^{14}CO_2$ to ^{14}C-acetate by gut
homogenates from xylophagous termite species were generally significantly
greater than rates of CH_4 emission (Table 1). By contrast, two soil-feeding
termite species emitted CH_4 at rates usually three to ten times greater than
those of xylophagous species, and no H_2-dependent fixation of $^{14}CO_2$ to ^{14}C-
acetate was observed in the one soil-feeder tested.

Relationship between termite diet and populations of hydrogenotrophic bacteria in guts

H_2/CO_2 acetogenic bacteria were present in greater numbers in the guts
of two xylophagous termite species than were methanogens. By contrast, the
reverse was true for soil-feeding termites (Table 2). These data were
consistent with the idea that H_2/CO_2 acetogenesis is the main H_2-consuming
reaction in guts of xylophagous termites, but not in guts of soil-feeding
termites where methanogenesis appears to be more important in H_2
utilization.

Table 1. H_2-Dependent Reduction of $^{14}CO_2$ to ^{14}C-acetate by Termite Gut Microbiota and CH_4 Emission by Live Termites[a]

	μmol product x g termite^{-1} x h^{-1}	
	^{14}C-Acetate	CH_4
Xylophagous termites[b]:		
Coptotermes formosanus	1.7	0.0
Prorhinotermes simplex	0.6	0.3
Pterotermes occidentis	1.6	0.0
Reticulitermes flavipes	1.1	0.0-0.1
Zootermopsis angusticollis	0.5	0.0-1.3
Amitermes sp.	4.1	0.1
Gnathamitermes perplexus	1.7	0.2
Nasutitermes costalis	5.0	0.1
Nasutitermes lujae	1.7	0.0
Nasutitermes nigriceps	2.8	0.2
Microcerotermes parvus	4.0	0.0
Tenuirostritermes tenuirostris	0.9	0.1
Soil-feeding termites:		
Cubitermes speciosus	0.0	1.0
Thoracotermes macrothorax	n.d.[c]	0.9

[a]Rates were determined as described previously.[2]
[b]The first 5 species listed are "lower" termites and have a hindgut flora of bacteria and cellulolytic, flagellate protozoa. The remaining species in this table are "higher" termites and have a hindgut flora consisting only of bacteria.
[c]n.d., not determined

Table 2. Enumeration of Hydrogenotrophic Bacteria in Termite Gut Fluid[a]

	10^5 bacteria/ml	
	H_2/CO_2 acetogenic bacteria[b]	H_2/CO_2 methanogenic bacteria
Xylophagous termites:		
Nasutitermes lujae	1150.0	15.0
Microcerotermes parvus	815.0	156.0
Soil-feeding termites:		
Cubitermes speciosus	0.8	304.0
Thoracotermes macrothorax	0.4	236.0

[a]Enumeration was by the most probable number method using selective media with H_2/CO_2 (80/20) as substrates.[3]
[b]Bromoethane sulfonate (45 mM) was included in the medium to inhibit methanogenesis.

H_2/CO_2 acetogens isolated from termite guts

H_2/CO_2 acetogenic bacteria were isolated from two wood-feeding and one soil-feeding termite species. The general characteristics of these strains are given in Table 3. Recently isolated[4] strains APO-1 and SFC-5 are different from the previously isolated *Sporomusa termitida*[5] and may represent new species of bacteria.

Table 3. Characteristics of H_2/CO_2 Acetogenic Bacteria Isolated From Termite Guts

Origin	*Sporomusa termitida* *Nasutitermes* *nigriceps* (xylophagous)	Strain APO-1 *Pterotermes* *occidentis* (xylophagous)	Strain SFC-5 *Cubitermes* *speciosus* (soil-feeding)
Cell dimension (μm)	0.5-0.8 x 2-8	0.3 x 6-60	1 x 2-6
Cell wall type	Gram -	Gram -	Gram +
Motility	+	+	+
Endospore location	terminal/subterminal	terminal	subterminal
pH optimum/range	7.2/6.2-8.1	7.8/6.4-8.6	7.3/5.4-9.0
Temp.optimum/range	30/19-37	33/19-40	33/19-37
Catalase	+	+	-
Oxidase	-	-	-
Mol% G+C in DNA	48.6	51.5	not determined

Possible role of methanogenesis in guts of soil-feeding termites

The nutrition of soil-feeding termites is presumably derived from utilization of the organic-humic fraction of soil which is rich in aromatic compounds derived from lignin. To assess the potential role of microbes in degradation of aromatic compounds in the guts of such termites, benzoate-degrading bacteria were enumerated in the gut of the soil-feeding termite *Cubitermes speciosus*. Results indicated that a minimum of 3.2×10^5 benzoate-degrading bacteria were present in guts of *C. speciousus*. Moreover, benzoate degradation always coincided with methane production in enrichment tubes. No benzoate-degrading bacteria could be detected when bromoethane sulfonate (1 mM) and MoO_4(20 mM) (inhibitors of methanogenesis and sulfate reduction respectively) were included in the medium. These observations suggested that benzoate degradation in the gut of *C. speciousus* termites may occur by a syntrophic relationship involving interspecies transfer of H_2 from benzoate-degrading bacteria to H_2-consuming methanogens.

CONCLUSIONS

1. The ability of H_2/CO_2 acetogenic bacteria to outcompete methanogenic bacteria for H_2 is a widespread phenomenon in the gut of xylophagous termites. However, preliminary results suggest that the reverse is true in the gut of soil-feeding termites.

2. The process of H_2/CO_2 acetogenesis in termite guts is not restricted to a single species of bacteria.

3. Methanogenic bacterial consortia may be important to the nutrition of soil-feeding termites by degrading aromatic compounds ingested with soil.

LITERATURE CITED

1. D. A. Odelson and J. A. Breznak, Appl. Environ. Microbiol. 45:1602-1613 (1983).
2. J. A. Breznak and J. M. Switzer, Appl.Environ.Microbiol. 52:623-630 (1986).
3. A. Brauman, Etude du metabolisme bacterien de termites superiurs a regimes alimentaires differencies. Ph.D. Thesis, Universite de Provence, Marseille (1989).
4. M. D. Kane and J. A. Breznak, Abstr. Ann. Mtg. Amer. Soc. Microbiol. p. 234 (1989).
5. J. A. Breznak, J. M. Switzer, and H.-J. Seitz, Arch. Microbiol. 150:282-288 (1988).

HYDROGEN AND METHANOGENESIS IN RUMEN LIQUOR AND IN RUMEN CILIATE/METHANOGEN COCULTURES

Jayne Ellis, Kevin Hillman, Alan G. Williams* and David Lloyd

Microbiology Group, PABIO
University of Wales College of Cardiff
Museum Avenue
Cardiff CF1 3TL

and

*Hannah Research Institute
Ayr KA6 5HL
Scotland

INTRODUCTION

Interspecific hydrogen transfer is necessary for methanogenesis by rumen microorganisms, and as the production of methane represents a loss of carbon from the host ruminant, is a process of great economic significance.[1] Ruminal hydrogenogens include bacteria,[2] chitridomycete fungi[3] and holotrich ciliate protozoa.[4] Close physical associations between rumen ciliates and methanogenic bacteria have been demonstrated[5] as well as metabolic interactions.[6] In this report we show how direct measurements of dissolved hydrogen and methane can provide information on the kinetics and stoichiometries of species interactions both in crude rumen liquor and in a defined methanogenic coculture.

METHODS

Rumen fluid was obtained from fistulated cattle receiving twice daily equal portions of ryegrass hay (4.1kg) and concentrates (2.8kg). Samples obtained before the morning feed were transported to the laboratory in sealed insulated containers and strained under N_2 through muslin. Holotrich protozoa were separated as previously described.[7] *Methanosarcina barkeri* type strain MS was from the Deutsche Sammlung von Mikroorganismen and grown as directed by the suppliers. Mass spectrometric measurement of dissolved gases employed the system described previously[8] with either a 25ml reaction mixture stirred at 450 rpm (for crude rumen liquor) or a 6ml working volume stirred at 600 rpm (for cocultures). Rumen liquor was diluted two-fold with Simplex buffer[9] and the same buffer was used in experiments with cocultures.

RESULTS AND DISCUSSION

Figure 1 shows the effects of sequential additions of glucose to rumen fluid stirred under a mobile gas phase of N_2. No detectable increases in either hydrogen or methane were observed until more than 1.4mM glucose had been added. Then hydrogen evolution commenced (16μM min^{-1}), diminished slightly as the rate of methanogenesis increased to 16μM min^{-1}. Exhaustion of substrate led to decreased hydrogen production before the decline in methanogenesis was observed. Further additions of glucose gave identical sequences of events. Accumulating hydrogen attained 70μM (e.g. after adding 2.8mM glucose). This level depends on the reaction conditions (surface area/volume ratio and stirring rate) but was much higher than the dissolved H_2 measured in situ in the rumen (usually < 3μM).[10] Approximately 13.5% of the glucose-carbon was converted to methane in these experiments: this compares with a figure of 6-10% obtained for cattle[11]. We have previously shown that when oxygen inhibits glucose supported methanogenesis in rumen liquor, levels of hydrogen rise.[12,13] The kinetics of gas production from formate was different from that observed for glucose. Thus additions of formate above a threshold of 1.2mM gave an immediate stimulation of methanogenesis before any increase in hydrogen was detectable. The most likely explanation for this is that at the very high methanogenic rates attained with formate (90μM min^{-1}), hydrogen remains undetectable because its provision is initially rate-limiting. Alternatively, the production of methane from formate may not require interspecies interaction and may be carried out by a single bacterium (e.g. Methanobacterium formicicum).

Artificial cocultures of holotrich ciliate protozoa and a methanogenic bacterium mimic some of the characteristics of the rumen liquor system.[6] Thus in the presence of Dasytricha ruminantium and a mixed Isotrich

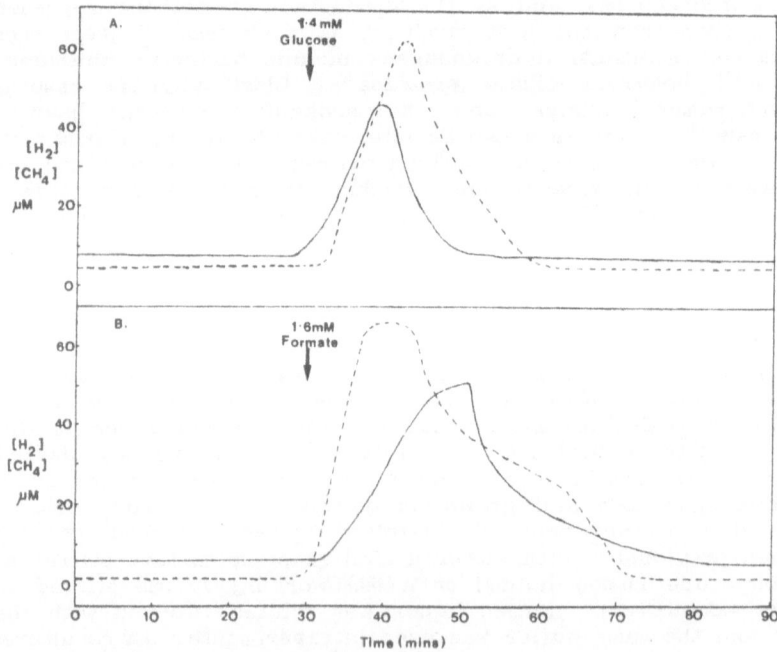

Fig. 1.   Mass spectrometric monitoring of hydrogen (---) and methane (-) on addition of (a) 1.4mM glucose and (b) 1.6mM formate to rumen liquor.

374

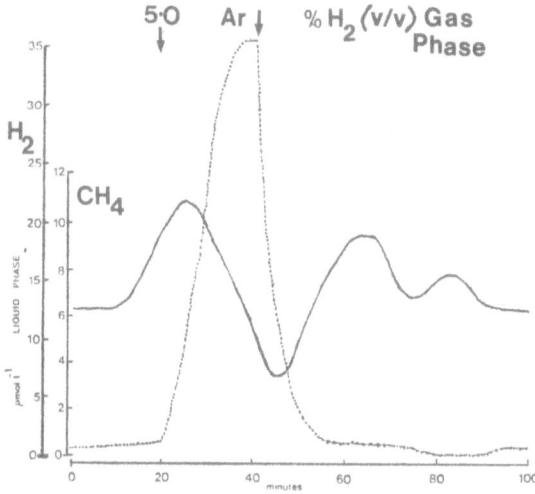

Fig. 2.   The effect of high levels of hydrogen on methane production by a coculture of *Methanosarcina barkeri* (4 x 10⁶ cells ml⁻¹) *Dasytricha ruminantium* (25 cells ml⁻¹) and *Isotricha* spp. (100 cells ml⁻¹).

population, *Methanosarcina barkeri* produced methane, and inhibition by O₂ led to an accumulation of hydrogen produced by the ciliate protozoa. Figure 2 shows that high levels of dissolved hydrogen (> 10µM) reversibly inhibit methanogenesis by the coculture.

References

1.   J. W. Czerkawski, Methane production in ruminants and its significance, *W. Rev. Nut. Diet.* 11: 240 (1969).

2.   M. Chen and M. J. Wolin, Influence of methane production by *Methanobacterium ruminantium* on the fermentation of glucose and lactate by *Selenomonas ruminantium*, *Appl. Environ. Microbiol.* 34: 756 (1977).

3.   T. Bauchop and D. O. Mountfort, Cellulose fermentation by a rumen anaerobic fungus in both the absence and presence of a rumen methanogen, *Appl. Environ. Microbiol.* 42: 1103 (1981).

4.   A. G. Williams, Rumen holotrich ciliate protozoa, *Microbiol. Rev.* 50: 25 (1986).

5.   C. K. Stumm, H. J. Gijzen and G. D. Vogels, Association of methanogenic bacteria with ovine rumen ciliates, *Brit. J. Nutrit.* 47: 95 (1982).

6.   K. Hillman, D. Lloyd and A. G. Williams, Interactions between the methanogen *Methanosarcina barkeri* and rumen holotrich ciliate protozoa, *Lett. Appl. Microbiol.* 7: 49 (1988).

7.   A. G. Williams and N. Yarlett, An improved technique for the isolation of holotrich protozoa from rumen contents by differential filtration with defined aperture textiles, *J. Appl. Bact.* 52: 267 (1982).

8.   D. Lloyd and R. I. Scott, Direct measurement of dissolved gases in microbiological systems using membrane-inlet mass spectrometry, *J. Microbiol. Meth.* 1: 313 (1983).

9. G. S. Coleman, Rumen entodiniomorphid protozoa, *in*: "Methods of Cultivation of Parasites In Vitro", pp. 39-53, A. E. R. Taylor and J. R. Baker, eds., Academic Press, London (1978).

10. K. Hillman, D. Lloyd and A. G. Williams, Use of a portable mass spectrometer for the measurement of dissolved gas concentrations in ovine rumen liquor *in situ*, *Curr. Microbiol.* 12: 335 (1985).

11. J. W. Czerkawski, "An Introduction to Rumen Studies", Pergamon Press, Oxford (1986).

12. R. I. Scott, N. Yarlett, K. Hillman, T. N. Williams, A. G. Williams and D. Lloyd, The presence of oxygen in rumen liquor and its effects on methanogenesis, *J. Appl. Bacteriol.* 55: 143 (1983).

13. R. I. Scott, T. N. Williams and D. Lloyd, Oxygen sensitivity of methanogenesis in rumen and anaerobic digester populations using mass spectrometry, *Biotech. Lett.* 5: 375 (1983).

HYDROGEN PRODUCTION BY RUMEN CILIATE PROTOZOA

Jayne Ellis, Kevin Hillman, Alan G. Williams* and David Lloyd

Microbiology Group, PABIO
University of Wales College of Cardiff
Museum Avenue
Cardiff CF1 3TL

and

*Hannah Research Institute
Ayr KA6 5HL
Scotland

INTRODUCTION

The exact role of rumen protozoa in ruminant nutrition is not well defined, but it is implicit that a group of organisms which is present in all wild and domesticated ruminants and contributes as much as half the biomass of the microbial population must make a significant contribution to the economy of the system. Rumen ciliates are classified into holotrichs, which ferment a wide range of soluble carbohydrates[1] and entodiniomorphs, which are principally particle feeders (i.e. cellulolytic and amylolytic).[2] Some species of the latter have a limited ability to utilize soluble carbohydrates. The production of hydrogen by rumen ciliates[3,4] occurs in a specialized organelle, the hydrogenosome.[5,6] At some times oxygen is present in the rumen at low concentrations[7] and rumen ciliates show high affinity oxygen consumption. Here we show that four different species of ciliates have oxygen-sensitive hydrogenases, so that the availability of hydrogen for interspecies hydrogen transfer will fluctuate depending on ambient oxgyen concentrations.

METHODS

Separation of individual protozoal species from rumen liquor and removal of contaminating bacteria was as previously described[8]: for the entodiniomorphs, *Eudiplodinium maggii* and *Polyplastron multivesiculatum*, defined populations had been established in the ruminal contents of ciliate-free (defaunated) sheep.[9] Mass spectrometric measurements of hydrogen and oxygen in a stirred vessel open for gas flow were as previously described.[10]

RESULTS AND DISCUSSION

Figure 1 shows the effects of increasing oxygen on the rates of oxygen consumption and hydrogen production by a suspension of rumen ciliates. At the low concentrations of oxygen sometimes present in the rumen (< 1.5µM[7]), inhibition of hydrogen production was slight and reversible. Higher oxygen levels gave extensive and irreversible inactivation. Oxygen consumption is itself decreased above an inhibition threshold.[11] In one species of entodiniomorph, *E.maggii*, oxygen at undetectable levels (< 0.25µM) gave stimulation of hydrogen production;[9] this effect has not been observed in *P.multivesiculatum* or in the holotrichs. Table 1 compares these features for the four ciliates.

Table 1. Hydrogen Production and its Sensitivity to Oxygen in Rumen Ciliates

	V_{MAX} (μMH_2/min/10^5orgs)	K_iO_2 (µM)
Dasytricha ruminantium	1.1	1.11
Isotricha spp.	20.3	2.33
Polyplastron multivesiculatum	12.0	N.D.
Eudiplodinium maggii	3.7	*< 2

* stimulated at O_2 < 0.25µM

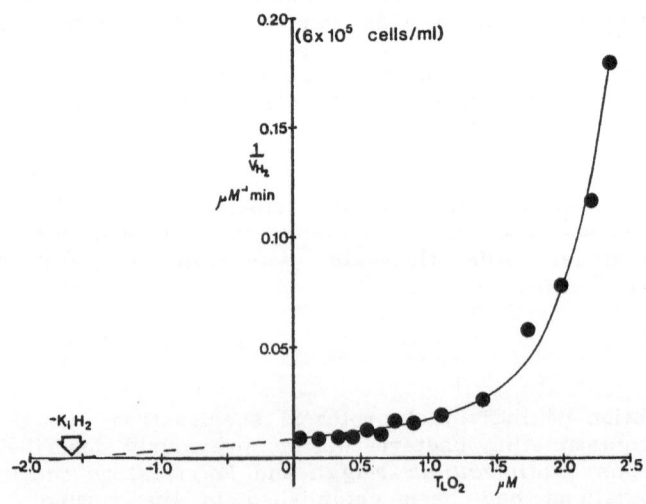

Fig. 1.   The inhibitory effect of oxygen on hydrogen production in *Isotricha* spp determined by mass spectrometry.

We conclude that the ciliates may play two important roles in the rumen, *i.e.* (i) the production of hydrogen for interspecies transfer (*e.g.* to methanogens), and (ii) the maintenance of near anaerobiosis by their efficient oxygen scavenging systems.

References

1. A. G. Williams and C. G. Harfoot, Factors affecting the uptake and metabolism of soluble carbohydrates by the rumen ciliate *Dasytricha ruminantium* isolated from ovine rumen contents by filtration, *J. Gen. Microbiol.* 96: 125 (1976).

2. G. S. Coleman, Rumen ciliate protozoa, *in:* "Biochemistry and Physiology of the Protozoa", 2nd ed., pp. 381–408, M. Levandowsky and S. H. Hutner, eds., Academic Press, New York (1979).

3. R. A. Prins and W. Van Hoven, Carbohydrate fermentation by the rumen ciliate *Isotricha prostoma*, *Protistologica* 13: 549 (1977).

4. V. Van Hoven and R. A. Prins, Carbohydrate fermentation by the rumen ciliate *Dasytricha ruminantium*, *Protistologica* 13: 599 (1977).

5. N. Yarlett, A. C. Hann, D. Lloyd and A. G. Williams, Hydrogenosomes in the rumen protozoon *Dasytricha ruminantium* Schuberg, *Biochem. J.* 200: 365 (1981).

6. N. Yarlett, G. S. Coleman, A. G. Williams and D. Lloyd, Hydrogenosomes in known species of rumen entodiniomorphid protozoa, *FEMS Microbiol. Lett.* 21: 15 (1984).

7. K. Hillman, D. Lloyd and A. G. Williams, Use of a portable quadrupole mass spectrometer for the measurement of dissolved gas concentrations in ovine rumen liquor, *Curr. Microbiol.* 12: 335 (1985).

8. A. G. Williams and N. Yarlett, An improved technique for the isolation of holotrich protozoa from rumen contents by differential filtration with defined aperture textiles, *J. Appl. Bact.* 52: 267 (1982).

9. J. E. Ellis, A. G. Williams and D. Lloyd, Oxygen consumption by rumen microorganisms: protozoal and bacterial contributions, *J. Appl. Environ. Microbiol.* in press (1989).

10. D. Lloyd and R. I. Scott, Direct measurement of dissolved gases in microbial systems using membrane-inlet mass spectrometry, *J. Microbiol. Meth.* 1: 313 (1983).

11. J. E. Ellis, D. Lloyd and A. G. Williams, Protozoal contribution to ruminal oxygen utilization, *in:* "Biochemistry and Molecular Biology of "Anaerobic" Protozoa", pp. 32–41, D. Lloyd, G. H. Coombs and T. A. Paget, eds., Harwood Academic, Chur (1989).

METHANOGENIC BACTERIA AND THEIR ACTIVITY IN A SUBSURFACE RESERVOIR OF TOWN GAS

M. Greksák[a], P. Šmigáň[a], J. Kozánková[a], F. Buzek[b]
V. Onderka[c] and I. Wolf[c]

[a]Institute of Animal Physiology, Slovak Academy of Sciences
900 28 Ivanka pri Dunaji; [b]Geological Survey, 150 00 Prague
[c]Institute of Geological Engineering, 601 88 Brno, Czecho-
slovakia

INTRODUCTION

Identification of different microorganisms in deep terrestrial subsurface environments and modern sampling methods have opened the way to the
understanding of microbial life in this kind of allobiosphere. Methanogenic
bacteria also have been found to be a member of bacterial communities in
some underground environments (Ward and Olson, 1980; Brassel et al., 1981;
Ward ct al., 1985). In a water-saturated structure of an artificially
created subsurface reservoir of town gas, some constituents of which are
hydrogen and carbon dioxide, a diminution of stored town gas and its enrichment with methane was observed. Since methanogenic bacteria exhibit an
extreme diversity, and most of them can grow with hydrogen and carbon dioxide as their sole carbon and energy source (Balch et al., 1979), we regard them as one of the chief agents for the changes observed.

Here we report a search for methanogenic bacteria in this extreme
environment, their enrichment, partial characterization and their potential
contribution to the observed town gas volume diminution and to an enrichment
of the stored gas with methane.

MATERIALS AND METHODS

The subsurface reservoir of town gas of an aquifer type has been
artificially created in a water-saturated strata of an anticlinal structure.
The main sorage strata of the reservoir is heterogenous, formed by Myocene
rock-sand, gravel and sandstone. The reservoir serves for more than twenty
years as a seasonal stock of town gas (a mixture of coal gas and syngas)
at working pressure around 4.0 MPa and the inside temperature from 25 to
45 ^{o}C.

Samples of underground stratal water were collected via wells located
near the water-gas contact phase from the depth of 400 to 500 meters.
A special sampler "Subsurface Sampler Model 60" of Leutert Co. F.R.G. for
this purpose was used. The sampling procedure was done in a manner that
prevented contamination of samples from surface microorganisms and oxygen.

The common procedure (Bryant, 1972) for enrichment of methanogenic
bacteria was used. The samples were anaerobically transferred to the anaerobic cultivating medium No. 2 (Balch et al., 1979), pressurized to 150 KPa

with hydrogen and carbon dioxide (4:1) and incubated on a gyratory shaker at 37 and 60 ^{O}C. After several transfers of bacterial suspensions yielding methane to the new cultivating medium, specific antibiotics (penicilin G 500, cycloserine 10 and kanamycin 100 /ug/ml) were included into the growth media to minimize growth of nonmethanogenic microorganisms. This procedure was repeated twice. After subcultivation of these bacteria in antibiotic--free medium only one type of methane producing cells was obtained.

Methane formation by bacterial suspensions was measured by gas chromatography. The $^{13}C/^{12}C$ ratios of CO_2 samples expressed as $\delta^{13}C$ were analyzed with Finnningan MAT 250 mass spectrometer in which PDP carbonate was used as a standard. Prior to measurement, town gas samples containing CO_2, CO and methane were passed through liquid nitrogen traps to remove CO_2 which was subsequently used for isotopic analysis. Methane and CO were separated from gas mixtures by preparative gas chromatography and after separation converted to CO_2 by combustion at 950 ^{O}C on columns filled with CuO.

RESULTS AND DISCUSSION

Samples of underground stratal water can be characterized as a slightly opalescent, yellowish and not very turbid liquid, pH 6.5 to 7.0, exhibiting an oxidoreduction potential around -330 mV. Collected samples exerted a population density of methanogen-like microorganisms about 10^3 to 10^4 cells per ml as obseved by direct epifluorescence microscopy.

Since there was no evidence that changes in the volume and composition of town gas stored in the underground reservoir (Table 1) could be coupled with microbial activity, the possibility of a whole microbial population of the stratal water withdrawn from the reservoir to perform changes observed was examined. This microbial population cultivated at 37 ^{O}C on hydrogen and carbon dioxide transformed them to methane in media prepared from the natural environment of the following composition: (i) original stratal water, (ii) original stratal water containing 10% of gently powdered rocks obtained from the underlying paleosoic structure of the reservoir, (iii) 10% of the rock powder in distilled water. The result indicated that metahnogenic bacteria being a member of the bacterial community of the reservoir could be responsible for the diminution of the stored gas volume and for its enrichment in methane content.

Using the enrichment procedure the samples having been incubated at 37 ^{O}C yielded methane-forming cultures containing rods fluorescent at 420 nm. No methanogenic bacteria were enriched in the samples incubated at 60 ^{O}C.

The enriched bacteria were Gram+ nonmotile rods without filaments with a tendecy to form aggregates. They grew and produced methane with hydrogen and carbon dioxide but not with acetate, formate, methanol or methylated amines. The temperature optimum for growth and methane production was between 37 to 40 ^{O}C, the optimal pH value at 6.5 to 7.0.

To verify our suggestion that methanogenic bacteria can play an important role in the transformation of the stored town gas, isotopic analyisi of the gas was performed. The values of $\delta^{13}C$ of methane which we have obtained for the town gas stored in the reservoir were around -80°/oo. This result indicates that the methane formed during storage of town gas in the subsurface reservoir is of biological origin.

It should be noticed that town gas taken from the reservoir was able to support the growth of the enriched methanogenic bateria, though the growth was very slow comparing to that when H_2 + CO_2 mixture was used as a substrate. However, it could be assumed that at the working pressure of the reservoir (around 4.0 MPa), the stored town gas might serve as a sufficient substrate for the growth of and methane formation by methanogenic bacteria present there.

The results obtained suggest that methanogenic bateria present in the subsurface town gas reservoir can take part in the consumption of some constituents of the town gas in that reservoir. This biological transforma-

Table 1. Chemical composition of town
gas before and after storage
in the subsurface reservoir

COMPONENT	INPUT vol%	OUTPUT vol%
CH_4	21.90	40.00
C_2H_4	0.05	0.01
C_2H_6	0.36	0.52
C_3-hydrocarbons	0.08	0.16
C_4-hydrocarbons	0.01	0.02
CO	9.00	3.30
CO_2	11.67	8.78
N_2	2.50	8.60
H_2	54.00	37.00

tion is associated with gas volume diminution and with an enrichment of the
starting gas with methane, and represents a serious economic and technolo-
gical problem.

REFERENCES

Balch, W. E., Fux, G. E., Magrum, R. G., Woese, C. R. and Wolfe, R. S.,
1979, Methanogens: Reevaluation of a unique biological group,
Microbiol. Rev., 43:260.

Brassell, S. C., Wardroper, A. M. K., Thomson, I. D., Maxwell, J. R. and
Eglinton, G., 1981, Specific acyclic isoprenoids as biological
markers of methanogenic bacteria in marine sediments, Nature, 290:
:693.

Bryant, M. P., 1972, Commentary on Hungate technique for culture of
anaerobic bacteria, Am. J. Clin. Nutr., 25:1324.

Ward, M. D., Brassell, S. C. and Eglinton, G., 1985, Archaebacterial lipids
in hot-spring mats, Nature, 318:656.

Ward, M. D. and Olson, G. J., 1980, Terminal processes in the anaerobic
degradation of an algal-bacterial mat in a high-sulfate hot spring,
Appl. Environ. Microbiol., 40:67.

STUDY OF THE TRANSITION OF ACTIVATED SLUDGES TO AN ADAPTED ANAEROBIC INOCULUM FOR ANAEROBIC DIGESTION

J.P. Guyot, C. Fajardo, A. Noyola and C. Barrena

Universitad Autonoma Metropolitana - Iztapalapa / ORSTOM
Departemento de Biotechnologica , Apartado Postal 55 - 535
Iztapalapa 09340 , Mexico D.F. , Mexico

INTRODUCTION

Countries which do not possess a well established network of anaerobic digesters, such as many developing countries, may have great difficulty in purchasing large amounts of anaerobic inocula, for use as sludge which is well adapted to seed anaerobic reactors for wastewater treatment.

We successfully used an aerobically adapted activated sludge to inoculate UASB reactors operating a wide range of effluents at lab level. Previous studies (WU *et al.*, 1987) have demonstrated the suitability of activated sludge for use as seed for anaerobic digesters. However, no data have been published characterizing the microbial and physical changes which occur during the transition from the " activated sludge " state to the " anaerobically adapted activated sludge " state.

Here we describe the results obtained with activated sludges, left in batch at two different temperatures under anoxic conditions, with no feeding for two months. After this starvation period, the same sludges were fed with acetate for one further month.

MATERIALS AND METHODS

Freshly activated sludges were collected in an aerated tank at a conventional aerobic processing plant, travelling urban wastewaters at Mexico City. The sludges were incubated at 25°C and 35°C. They were kept with no mixing or feeding for two months . At day 61, the sludges were fed 4 times with acetate. Each time the acetate depletion was monitored, and the initial acetate concentration was ajusted to between 10 and 15 mM, depending on the remaining substrate in the batches.

The sludge volumetric index (SVI) and total and volatile solids (TSS, VSS) were determined using standard methods (1985). All cultivation media were prepared anaerobically and bacterial counts were made by the MPN method with 5 tubes per level. Acetate was analyzed by gas chromatography.

RESULTS AND DISCUSSION

In the freshly sampled activated sludge, it was possible with the MPN method to detect only hydrogenophilic methanogens (6.10^2 bact./g VSS). The fact that no acetoclastic methanogens and obligate hydrogen producing acetogens (OHPA) were detected does not therefore mean that they were not present. In any case, the presence of hydrogenophilic methanogens means that strict anaerobes can survive in the aeration tank of aerobic processing plants. This may be attributable to the existence of micro-aggregates which provide anaerobic micro-environments.

It was not possible to determine the SVI of the freshly activated sludge due to their poor settling characteristics. After one month of incubation at 25°C and 35°C the sludges had acquired better settling capacities. The percentage of VSS decreased during this period, which is consistent with sludge digestion.

Furthermore 30 days after the beginning of incubation, bacterial counts showed an increase in hydrogenophilic methanogens (1.10^6 bact. /g VSS); acetoclastic methanogens and OHPA reached detectable levels.

This may be due, not only to bacterial growth, but probably also to an increase in anaerobe concentrations in the VSS, due to hydrolysis of micro-organisms unable to survive in anoxic and starving conditions. During an incubation period of 48 h, activity tests in serum-bottles with fresh and one month old sludges, performed with acetate, did not show any acetate consumption to have occured. On the contrary, acetate, propionate and butyrate productions were detected, which corresponded to the hydrolysis of the sludge.

After 2 months, the same sludges were fed at day 61 with acetate, and the liquid phase was sampled to determine the course of acetate degradation. With the sludges incubated at 35 °C, only 33% of acetate was degraded after 6 days. A similar pattern was observed at 25 °C.

At the third feeding (day 81), it was observed that both sludges showed identical acetate degradation time-courses, and after a period of 12 days, acetate concentrations remained relatively high.

At the 4th feeding (day 93), in the case of the sludge incubated at 35 °C, the acetate degradation evolved faster than during the previous feeding, which may mean that the sludge was adapting. Unexpectedly, the rate of acetate degradation at 25 °C was lower than at 3rd feeding. This is difficult to explain, since at day 93, sludges incubated at both temperatures, had nearly the same methanogenic bacteria concentrations. The only difference between the two sludges were their respective OHPA content : a higher concentration was measured at 35 °C.

At both temperatures, the VSS content decreased over the period of feeding. At day 81, the SVI decreased to 104 ml/g at 25 °C and increased to 107 ml/g, at which stage the sludge presented fair settling characteristics.

Lastly, we observed that similar adapted sludges used for other experiments to inoculate UASB reactors showed better characteristics after 135 days of reactor feeding with a mixture of volatile fatty acids as substrate : nowdays, activity tests performed with these sludges can be carried out in 6 hours instead of 12 days. Furthermore, the granular sludges from these UASB showed higher concentrations of methanogens. The adapted sludge used as inoculum therefore, still needs to evolve in order to reach compatible with bacteria concentrations normal reactor operation : this is possible during the period of reactor start-up.

CONCLUSION

Preparation of inoculum for anaerobic reactors is possible using freshly activated sludges. Leaving activated sludges under anoxic conditions in batch will allow methanogens and OHPA to grow, and the sludges to acquire good settling characteristics During the start-up period of anaerobic reactors, the final adaptation step can take place. During this period, the acetoclastic activity of the sludges and the concentration of anaerobes increase, reaching levels compatible with normal reactor operation.

REFERENCES

APHA . AWWA . WPCF., 1985 . Standard methods for the evaluation of wastewaters, 16th edition.

WU , W. , HU , J . , GU , X. , KHAO , Y. , ZANG , H. and GU , G. , 1987. Cultivation of anaerobic granular sludge in UASB reactors with aerobic activated sludge as seed. Wat. Res. 21, 789 - 799.

EFFECTS OF ELEVATED HYDROGEN PARTIAL PRESSURES

ON ANAEROBIC TREATMENT OF CARBOHYDRATE

Stephen R. Harper[1] and Frederick G. Pohland[2]

[1]Georgia Tech Research Institute, Atlanta, GA, USA

[2]University of Pittsburgh, Pittsburgh, PA, USA

BACKGROUND

The theory of interspecies hydrogen transfer (Wolin and Miller, 1982) is now nearly two decades old, and is considered the biochemical cornerstone of methanogenesis in all natural and man-made habitats (soils, sediments, intestines; anaerobic digesters and treatment systems). Therefore, it is surprising that only a handful of engineering studies (Table 1) have focused on the production and effects of hydrogen during continuous treatment of various waste substrates. A review of this information reveals: 1) a lack of consensus on the inhibitory effects of hydrogen, and 2) an insufficiency of information to allow generalization of interspecies hydrogen effects under all possible treatment scenarios (i.e., all combinations of reactor configuration and wastewater type).

Recent evidence (Harper and Pohland, 1988; Denac, et al., 1988; Thiele and Zeikus, 1988) has demonstrated that situations exist where substrate turnover and methanogenesis are possible at hydrogen concentration orders of magnitude higher than prescribed by the theory of interspecies hydrogen transfer. In some cases, successful treatment has been demonstrated to be essentially independent of hydrogen concentration.

TREATMENT OF SOFT DRINK WASTEWATER

During treatment of soft drink wastewater in packed bed reactors, Harper and Pohland (1988) found that all gaseous hydrogen originated from glucose (none from volatile acids). External hydrogen additions (up to 35 % in the gas) had no effect on propionic or butyric acid conversion rates (Figure 1), but glucose fermentation was slightly disturbed (Figure 2).

Biological removal of gaseous hydrogen was very slow (Harper, 1989), and removal rates were linear with respect to gaseous hydrogen concentration (the cultures exhibited a very low substrate affinity and/or mass transfer rates). While some serum bottle studies with pure cultures of hydrogen-consuming methanogens have suggested very high substrate affinities, other studies with more realistic reactors have revealed Ks values of 0.2 atm H_2 or higher (Shea, et al. 1968; Kaspar and Wuhrmann, 1978). Compared with the hydrogen consumption rates of the enrichment culture of Shea, et al. (1968), the biomass specific rates observed for the packed-bed cultures studied here (Harper, 1989) suggest that less than 5 % of the total biomass was hydrogenotrophic.

TABLE 1. RESEARCH ON HYDROGEN EFFECTS IN ANAEROBIC TREATMENT

REFERENCE	SUBSTRATE	REACTOR TYPE	ORIGINAL CULTURE	HYDROGEN OBSERVED	OPERATIONAL OBSERVATIONS
Sykes (1970) (A)	Glucose	CSTR	Sludge digester	0.001 atm	Response to 4 g/l glucose pulse
(B)		CSTR		0.03 atm	Response to 8 g/l glucose pulse
(C)		CSTR		0.5 atm	Response to 13 g/l glucose pulse
Kasper and Wuhrmann (1978) (A)	Propionic acid	CSTR	Sludge digester	0.005 atm	Propionic acid degradation inhibited
(B)	Ethanol	CSTR		0.07 atm	Ethanol degradation inhibited
Smith, (1980) (A)	Sewage sludge	CSTR	Sludge digester	0.09 atm	No inhibition
(B)		CSTR		0.18 atm	Propionic acid degradation inhibited
Barnes, et al., (1983) (A)	Molasses and yeast	Fluidized bed	Sludge digester	0.0002 atm	During normal operation
(B)		Fluidized bed		0.00015 atm	Immediately following organic shock load
Boone, (1982)	Cattle manure	Batch CSTR	Cattle manure	0.7 atm	Propionic acid accumulated when 70% H2 added
Heyes and Hall, (1983) (A)	Propionic acid	CSTR	Sludge digester	0.00006 atm	Normal operation at 8-day HRT
(B)		CSTR		0.00004 atm	Normal operation at 14-day HRT
Poels, et al., (1985)	Cattle manure	CSTR	Cattle manure	0.005 atm	Normal operation at 2-day HRT
Whitmore and Lloyd, (1986)	Glucose	CSTR	Sludge digester	(40-60 uM)	Average H2 for English municipal digesters
Collins and Paskins, (1987)	Sewage sludge	CSTRs	Sludge digesters	0.0002 atm	Normal operation
Denac, et al., (1988) (A)	Molasses	Packed bed reactor	Sludge digester	>0.5 atm	No inhibition in reactor or with removed biomass
(B)	Propionic acid	Packed bed reactor		>0.5 atm	No inhibition in reactor or with removed biomass
(C)	Butyric acid	Packed bed reactor		>0.5 atm	No inhibition in reactor or with removed biomass
Harper and Pohland, (1988) (A)	Glucose	Packed bed reactor	Sludge digester	>0.3 atm	Slight inhibition of glucose fermentation
(B)	Propionic acid	Packed bed reactor		>0.3 atm	No decrease in continuous turnover rates
(C)	Butyric acid	Packed bed reactor		>0.5 atm	No decrease in continuous turnover rates
(D)	Soft-drink waste	Packed bed reactor		>0.1 atm	Greater than 90% COD removal on continuous basis
de Santis and Friedman (1988) (A)	Glucose	CSTR	Sludge digester	0.0015 atm	During normal feeding with vacuum on gas phase
(B)		CSTR		0.2 atm	When pulsed with excess glucose
Mosey and Fernandes (1988) (A)	Skimmed milk waste	CSTR	Sludge digester	<0.0002 atm	Immediately after daily feeding (HRT=20d)
(B)		CSTR		<0.0002 atm	Several hours after feeding
Thiele and Zeikus (1988)	Ethanol	Serum vials	Whey digester	<0.21 atm	Ethanol degradation not inhibited
Smith and McCarty, (1988) (A)	Ethanol/propionic	CSTR	Sludge digester	0.0001 atm	Normal conversion
(B)				0.01 atm	Immediately following pulse load of EtOH and HPr
Wilkie, et al., (1986)	Napier grass	Two-stage reactor	Sludge digester	0.00002 atm	Normal operation in second stage reactor

SUBSTRATE CONVERSION RATES

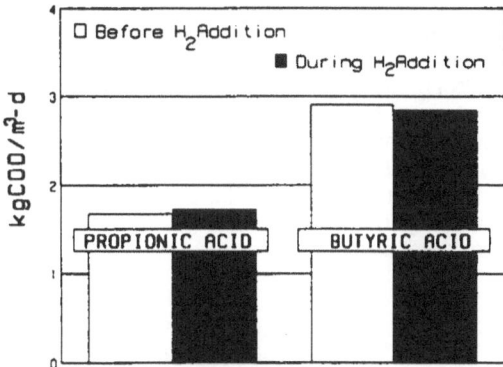

Fig. 1.
Effects of Hydrogen on
Conversion of Volatile Acids

SUBSTRATE CONVERSION RATES

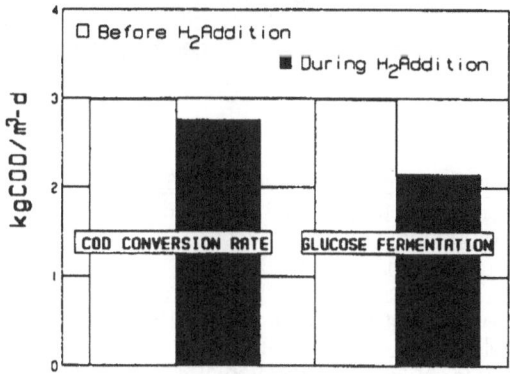

Fig. 2.
Effects of Hydrogen
on Conversion of Glucose

Based on these findings, hydrogen was concluded to be inconsequential as a process inhibitor during treatment of carbohydrates in packed bed reactors. Although hydrogen production could be managed to a large degree by reactor configuration, this was not necessary to achieve satisfactory treatment results. Greater than 90 % COD removal was possible despite continuous gaseous hydrogen concentrations exceeding 5 % and often approaching 10 %.

As a process monitoring tool, hydrogen was not particularly useful for indicating process upsets. On a daily basis, variabilities were too high to allow an operational distinction between 1 and 10 % H_2, and hydrogen fluctuations (when truly discernible) were due more to differences in carbohydrate loading than to process upsets. Accordingly, hydrogen monitoring was found to have some use in tracking the presence or absence of influent carbohydrates, and may have some utility in a daily (or on-line) response scenario, but only after a careful data base has been established correlating observed hydrogen concentrations with substrate loading rates and reactor operational conditions.

RECOMMENDATIONS

Much work remains to be done to define the practical limits of interspecies hydrogen transfer and its implications on anaerobic treatment of industrial wastewaters. Studies involving the external addition of hydrogen per se to reactors treating different wastewaters are strongly encouraged, to observe if and where inhibitory effects exist. The use of labelled substrates under different treatment scenarios (different hydrogen partial pressures, reactors, and wastewater types) should be pursued to help determine the exact biochemical pathways used in each case; these are most often generalized but may in fact be different. Special attention should be paid to the mass transfer characteristics of different reactors (e.g., mixing intensity, depth from gas phase, mass transfer rates) and the resulting effects on equilibria between gas and liquid phases, such that a complete representation of all influences on hydrogen production and consumption can be accurately presented for different treatment circumstances.

REFERENCES

Ahring, B.K. and Westermann, P. Appl Env. Microbiol., 53, 2, 434, (1987).
Archer, D.B. et al., Biotechnology Letters, 8, 3, 197, (1986).
Barnes, et al., Proc 37th Purdue Industrial Waste Conf., 715, (1983).
Boone, D.R. Appl Env. Microbiol., 43, 1, 57, (1983).
de Santis, J. and Friedman, A.A. Proc. 14th IAWPRC Conf. (Brighton), (1988).
Collins, L.J., and Paskins, A.R. Water Research, 21, 12, 1567, (1987).
Denac, M. et al., Biotechnol. Bioeng., 31, 1, (1988).
Harper, S.R. and Pohland, F.G. Anaerobic Digestion 1988, 79, (1988).
Harper, S.R. Ph.D. Dissertation, Georgia Institute of Technology (1989).
Kaspar, H.F. and Wuhrmann, K. Appl Env. Microbiol., 36, 1, 1, (1978).
Heyes, R.H. and Hall, R.J. Appl Env. Microbiol., 46, 3, 710, (1983).
Mosey, F.E. and Fernandes, X. A. 14th IAWPRC Conf. (Brighton) (1988).
Poels, J. et al., Biotechnology and Bioengineering, 27, 1692, (1985).
Schonheit, et al., Arch. Microbiol., 131, 285, (1982).
Shea, T.G. et al., Water Research, 2, 833, (1968).
Smith, P.H. USEPA 600/2-80-093 (1980).
Smith, D.P. and McCarty, P.L. Anaerobic Digestion 1988, Poster, 75, (1988).
Sykes, R.M. Ph.D. Dissertation, Purdue University, June (1970).
Thiele, J.H. and Zeikus. J.G. Appl Env. Microbiol., 54, 1, 20, (1988).
Whitmore,T.N. and Lloyd, D. Biotechnology Letters, 8, 3, 203, (1986).
Wilkie, A.et al., In:Methane from Biomass: A Systems Approach, Elsevier (1988).
Wolin, M.J. and Miller, T.L. ASM News, 48, 12, 561 (1982).

HYDROGEN PRODUCTION IN ANAEROBIC BIOFILMS

Elizabeth Henry, Timothy Ford, and Ralph Mitchell

Laboratory of Microbial Ecology
Harvard University
Cambridge, MA 02138 USA

Anaerobic syntrophic associations and microbially-induced hydrogen damage to metals depend on production and consumption of hydrogen. In both instances, hydrogen-producing bacteria such as homoacetogens and other fermentors are present. In syntrophy, hydrogen is directly transferred to bacteria such as methanogens which, by the consumption of hydrogen, permit the otherwise thermodynamically unfavorable hydrogen production reaction to occur. Hydrogen damage to metal results from entry of atomic hydrogen into the metal lattice. The atomic hydrogen recombines at irregularities in the metal to form gaseous molecular hydrogen. Cracks then propagate, leading to loss of ductility and premature failure of metal structures. In both cases, gaseous and dissolved hydrogen concentrations are found to be lower than predicted by rates of methanogenesis[1] and metal corrosion. To our knowledge, no techniques have yet been developed to measure hydrogen production within the tightly coupled community of anaerobic biofilms.

The method we describe in this paper is a modification of an electrochemical technique. The method is nondestructive and hydrogen permeation can be measured over time. We have compared hydrogen production by attached cells of Clostridium butyricum, C. acetobutylicum and Desulfovibrio desulfuricans.

The apparatus is a variation of the cell developed by Devanathan and Stachurski to measure diffusion of electrolytic hydrogen in palladium[2]. The cell consists of two compartments, one containing culture medium and the other, 0.1N NaOH solution. The compartments are separated by a thin membrane of palladium. Hydrogen generated by the bacterial film on the input surface establishes a uniform gradient across the palladium membrane. At the output surface, the electrochemical conditions are set so that hydrogen atoms are oxidized. The hydrogen current is directly related to hydrogen production at the input surface.

Prior to running each experiment, the palladium was sanded with 600 grit sandpaper and then degreased by soaking in hexane followed by ethanol. The apparatus was then assembled and autoclaved with the exception of heat sensitive components which were sterilized in ethanol and added after autoclaving. The apparatus was purged with sterile-filtered, oxygen-free gas: 20% CO_2/80% N_2 on the input side and N_2 on the output side. 700ml of culture medium (ATCC Media Handbook) was added anaerobically and aseptically to the input cell while 700ml 0.1N NaOH was added to the output cell.

The open circuit potential was monitored on the input side using a Keithley 616 digital electrometer. On the output side an AIS potentiostat (Model V-2LR-D) was used to polarize the palladium at +0.05V (SCE) and to monitor the hydrogen permeation current. When both a stable open circuit potential and a stable baseline current were achieved, the input cell was inoculated with 2ml stationary phase Clostridium sp. or 35ml D. desulfuricans. Values for the open circuit potential and the hydrogen permeation current were recorded on a chart recorder for later calculations.

C. butyricum, C. acetobutylicum and D. desulfuricans were obtained from ATCC. D. desulfuricans was grown fermentatively on pyruvate to avoid interference with sulfide. Cultures were maintained and experiments run at 37°C. During the fermentation of glucose to butyrate, carbon dioxide and hydrogen[3], C. butyricum produces 235 mol H_2 for each 100 mol glucose fermented. C. acetobutylicum also ferments glucose according to this reaction but reutilizes some of the hydrogen to synthesize butanol, acetone and 2-propanol. Thus 135 mol H_2 are formed for each 100 mol glucose fermented. D. desulfuricans ferments pyruvate yielding acetate, carbon dioxide and hydrogen.

The hydrogen permeation current for all species remained at background for several hours after inoculation. The current then rose to a peak before dropping back to initial values. The hydrogen permeation current for C. acetobutylicum peaked at 422 uA/cm^2 and then dropped off sharply, indicating the shift from hydrogen production to net hydrogen consumption. The peak in current for C. butyricum was 158uA/cm^2. C. butyricum does not shift to net hydrogen consumption and the hydrogen permeation current dropped off more gradually.

The peak in hydrogen permeation current for D. desulfuricans was 8.9uA/cm^2. D. desulfuricans produced hydrogen over a much longer period: 25 hours as compared to 13 hours and 10 hours for C. butyricum and C. acetobutylicum, respectively. The low hydrogen permeation current and relatively long time period for hydrogen production reflect the slower rate of metabolism and lower yield of hydrogen per mole substrate by D. desulfuricans.

The maximum flux of hydrogen and the total amount of hydrogen produced were calculated from the hydrogen permeation current. The hydrogen permeation current is related to the amount of hydrogen oxidized per second at the output surface by the following equation:

$$\text{moles } H^o \text{ oxidized} = \frac{A \times T}{Z \times F}$$

where A = hydrogen permeation current (amp/cm^2), T = time (sec), Z = valency of hydrogen = 1, and F = Faraday's constant = 96,487 coulombs/mol.

The amount of hydrogen (H^O) oxidized at the output surface is 93% of the hydrogen (H^O) produced at the input surface[4]. This, in turn, is twice the amount of hydrogen gas (H$_2$) produced at the input surface that dissociates to atomic hydrogen (H^O).

The total amount of hydrogen produced during the course of the experiment was derived by integrating under the curve of the hydrogen permeation current and following the above calculation. The total amounts of hydrogen produced by C. acetobutylicum and C. butyricum were 145 umol/cm^2 and 103 umol/cm^2, respectively. D. desulfuricans produced much less hydrogen, 2.3umol/cm^2.

The modified Devanathan cell is a novel method for studying hydrogen metabolism in anaerobic biofilms. By its maximum value, shape and duration, the hydrogen permeation current measured by the cell reflects the different metabolic processes of the bacteria studied. The modified Devanathan cell provides an opportunity to study hydrogen transfer in adherent or "juxtapositioned" anaerobic bacteria.

ACKNOWLEDGEMENTS

This research was financed in part by the Massachusetts Division of Fisheries and Wildlife as part of the Cooperative Aquatic Research Program and in part by the United States Department of Interior, Geological Survey, both through the Massachusetts Water Resources Research Center. Contents of the publication do not necessarily reflect the views of the United States Department of Interior or Commonwealth of Massachusetts, nor does mention of trade names or commercial products constitute their endorsement by the U.S. Government or Commonwealth of Massachusetts. This work was also supported by the Office of Naval Research Contract Number N00014-87-K-0121.

REFERENCES

1. R. Conrad, T.J. Phelps, and J.G. Zeikus, Gas metabolism evidence in support of the juxtaposition of hydrogen-producing and methanogenic bacteria in sewage sludge and lake sediments, Appl. Env. Micro. 50:595 (1985).
2. M.A.V. Devanathan and Z. Stachurski, The adsorption and diffusion of electrolytic hydrogen in palladium, Proc. Roy. Soc. A270:90 (1962).
3. G. Gottschalk, Bacterial Metabolism, 2nd Edition, Springer-Verlag, New York (1986).
4. T.E. Ford and R. Mitchell, Hydrogen embrittlement: a microbiological perspective, CORROSION/89, Paper No. 189, National Assoc. Corr. Eng. Houston, TX (1989).

EFFECTS OF VARIOUS HEADSPACE GASES ON THE PRODUCTION OF VOLATILE FATTY ACIDS BY RUMEN CILIATE PROTOZOA

Kevin Hillman[+], Alan G. Williams and David Lloyd[*]

Hannah Research Institute
Ayr KA6 5HL
Scotland

and

[*]Microbiology Group PABIO
University of Wales College of Cardiff
Cardiff CF1 3TL

[+]Present address: Rowett Research Institute
Bucksburn
Aberdeen AB2 9SB
Scotland

INTRODUCTION

Because the rumen is usually regarded as an anaerobic environment, almost all studies on the activities of rumen protozoa *in vitro* have been performed under an atmosphere of oxygen-free nitrogen or carbon dioxide. However, the rumen headspace gas is typically composed of 25-67% CO_2, 18-33% CH_4, 6-36% N_2, 0.01-4% H_2 and 0.5-6.5% O_2.[1] In this study we report the effects of CO_2, CH_4, H_2 and O_2 on the production of acetate, butyrate and propionate by the total protozoal population and by partially separated subpopulations. The presence of CO_2 decreased acetate production, whereas H_2 increased butyrate production by holotrichs. Mixed results were obtained in incubations under CH_4. The inhibitory effects of O_2 were most apparent on entodiniomorphid protozoa.

METHODS

Rumen liquor samples were obtained from fistulated sheep fed a diet of sugar-beet pulp (0.6 kg at 07:00) and hay (0.3 kg at 16:00), 5h after the morning feed, as sampling at this time provided adequate quantities of protozoa, with low endogenous activities. Separation into four fractions (total population, mainly entodiniomorphid protozoa, mainly *Isotricha* spp., and mainly *Dasytricha ruminantium*) was by differential filtration.[2] Volatile fatty acids (VFA) were measured by gas chromatography[3] after incubation under appropriate gas mixtures in N_2, obtained using a gas mixer.[4] Headspace gas mixtures employed were 0.05% O_2, 1% O_2, 20% CO_2, 70% CO_2, 15% CH_4, 2.5% H_2 and 5% H_2 in N_2.

RESULTS AND DISCUSSION

Total VFA production (both endogenous and glucose-supported) was higher in all cases under 100% N_2 than under any of the other atmospheres tested (some data are presented in Table 1). Therefore it seems that many of the *in vitro* experiments with rumen ciliates previously reported may have given artificially high results. H_2 and CO_2, as products of rumen metabolism, may inhibit by feedback on protozoal metabolism. The inhibitory effects of O_2 are probably mediated by its effects on the oxygen-sensitive enzymes of the hydrogenosomes, especially pyruvate-ferredoxin oxidoreductase which converts pyruvate to acetyl-CoA,[5] a precursor of both acetate and butyrate.[6] The entodiniomorphid-enriched population was the most sensitive to O_2 inhibition.

Both in the mixed protozoal fraction and in the three subpopulations, increasing H_2 in the headspace increased the proportion of butyrate formation relative to that of acetate. The production of butyrate provides a route for the disposal of excess reducing equivalents (*i.e.* butyrate is a "hydrogen-sink" product). The inclusion of H_2 in the headspace increased the proportion of propionate produced by the fraction enriched in entodiniomorphid protozoa and by that containing *D. ruminantium*.

Table 1. Proportions of VFA produced by population fractions of rumen protozoa under various headspace gases

Fraction	Headspace Gas Concentrations				
	100% N_2	0.05% O_2	15% CH_4	70% CO_2	2.5% H_2
1E	1.5:0:1	0:0:1	5.4:1:0	0.5:1:0	1.8:0.2:1
1G	1.8:0:1	0.7:0:1	1.9:0.4:1	1.1:0.7:1	1.1:0:1
2E	13.6:0:1	0:0:0	6.7:1.2:1	1:0:0	0.5:0.1:1
2G	2.0:0.7:1	8.6:4.1:1	3.8:1:0	0.3:0.7:1	0.5:1:3:1
3E	9.8:0:1	1.7:0:1	1.8:0.5:1	0.4:0.1	4.3:0.6:1
3G	1.4:0.3:1	0.8:0:1	1.2:0.2:1	0.8:0.7:1	1.1:0.5:1
4E	5.6:0:1	0:0:0	8.3:0:1	0.6:2:1	4.9:0.9:1
4G	2.3:0.5:1	1:0:1	0.4:0.1:1	0.9:1.5:1	0.5:1.2:1

(Ratios are Acetate:Propionate:Butyrate; E = endogenous; G = with 5mg ml^{-1} glucose. Fraction numbers refer to population fractions containing principally; 1 = Whole population; 2 = Entodiniomorphid protozoa; 3 = *Isotricha* spp; 4 = *Dasytricha ruminantium*).

The presence of CH$_4$ in the headspace reduced the proportion of acetate produced endogenously by those fractions enriched with entodiniomorphid protozoa or *Isotricha* spp., although the converse was true of the whole population and of *D.ruminantium*. In incubations with glucose, the proportions of acetate:butyrate:propionate were similar to those obtained under N$_2$ in all fractions except that enriched with *D.ruminantium,* in which the relative production of acetate was reduced. The significance of the effects of CH$_4$ are unclear, but may involve the attachment of bacteria to the protozoa, and the ingestion of bacteria by these.

REFERENCES

1. J. M. McArthur and J. E. Miltimore, Rumen gas analysis by gas-solid chromatography, *Can. J. Animal Sci.* 62: 299 (1961).
2. A. G. Williams and N. Yarlett, An improved technique for the isolation of holotrich protozoa from rumen contents by differential filtration with defined aperture textiles, *J. Appl. Bact.* 52: 267 (1982).
3. A. G. Williams and G. C. Harfoot, Factors affecting the uptake and metabolism of soluble carbohydrates by the rumen ciliate *Dasytricha ruminantium,* isolated from rumen contents by differential filtration, *J. Gen. Microbiol.* 96: 125 (1976).
4. J. S. Lundsgaard and H. Degn, Digital regulation of gas flow rates and composition of gas mixtures, *IEEE Trans. Biomed. Eng.* BME=20: 384 (1973).
5. N. Yarlett, A. C. Hann, D. Lloyd and A. G. Williams, Hydrogenosomes in the rumen protozoon *Dasytricha ruminantium* Schuberg, *Biochem. J.* 200: 365 (1981).
6. N. Yarlett, D. Lloyd and A. G. Williams, Butyrate formation from gluocse by the rumen protozoon *Dasytricha ruminantium, Biochem. J.* 228: 187 (1985).

MASS TRANSFER OF HYDROGEN IN A CULTURE OF METHANOBACTERIUM

THERMOAUTOTROPHICUM STRAIN HVERAGERDI

Gaudenz Jud, Kurt Schneider and Reinhard Bachofen

Institute for Plant Biology
Zollikerstrasse 107
CH-8008 Zürich, Switzerland

INTRODUCTION

Interspecies hydrogen transfer has been studied mostly in cultures maintained in closed vessels. In this way mass balances are obtained, but it is not possible to get information on the actual chemical environment in the vicinity of the cells. It is thus not surprising that apparent K_S values for hydrogen of 0.05 to 80 µM have been obtained for methanogens[1]. These values are extremely high compared to the actual H_2 concentration observed in natural habitats (0.03 - 6 µM). H_2 concentrations in the headspace of the vessel are not a true indication of the conditions at the cell surface, the mass transfer from the gas phase into the cell being the limiting process. This situation seems similar to the one, which occured with CO_2 over ten years ago, when algae were cultured by bubbling the unstirred cell suspension with 5 - 10% CO_2, resulting in a rather unreallistically high K_S value for CO_2[2].
Experiments in a chemostat culture with varying concentrations of molecular hydrogen suggest that there is no constant K_S value for hydrogen, but during limitation of the substrate hydrogen the K_S value decreases and the affinity of the cells to hydrogen is increased.

MATERIALS AND METHODS

The microrganisms (Methanobacterium thermoautotrophicum strain hveragerdi[3]) were grown in a minimal salt solution[3] in a 2 l bioreactor (MBR/Switzerland) containing 1.7 l medium at 58^0 C and a dilution rate of 0.176 h^{-1}. The pH was kept constant at 6.9. The bioreactor was stirred at 1200 rpm. At the beginning the gassing rate was 0.141 vvm with a gas composition of 66.7% H_2, 16.7% CO_2 and 16.6% Ar. When a steady state was reached after about 48 hours, the proportion of H_2 in the inlet gas was reduced by about 8% and replaced by argon, thus keeping the gasing rate constant until finally the proportion of H_2 at the bioreactor inlet was about 4%. Mixing of gases was made by gas mass flow controllers (MKS/Germany). The composition of the gas at the bioreactor gas inlet and outlet was monitored by a quadrupole mass spectrometer (QMG112/Balzers FL), using a capillary inlet system. The gas flow rate at the reactor outlet was calculated, using the argon as an internal standard. The gas solubilized in the medium was also monitored by mass spectrometry using a membrane probe connected with a second mass spectrometer inlet. The flow rate of the medium was computer controlled. The biomass was determined internally by a biomass probe (Aquasant/Switzerland) and externally by a spectrophotometer at 660 nm (Kontron/Zürich) and by determination of the dry weight. Na_2S supply to the culture was also computer controlled. All datas were sampled on–line with the PCS68020 process control system (MBR/Switzerland). A scheme of the experimental set up is given in figure 1.

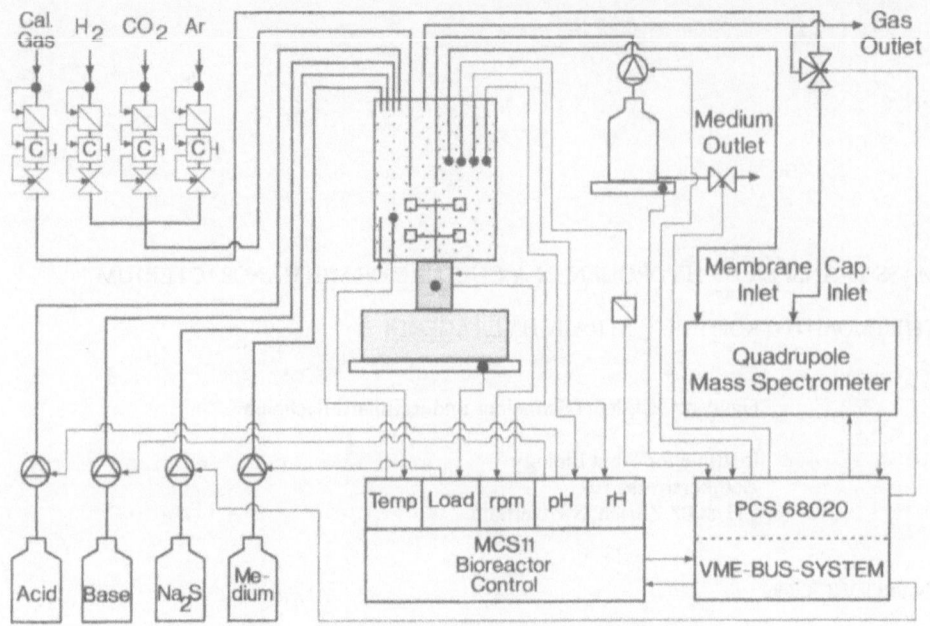

Figure 1. Scheme of the experimental set up.

RESULTS AND DISCUSSION

As seen in figure 2, hydrogen became limiting when the hydrogen at the reactor gas inlet was smaller than 60%. When hydrogen was reduced further, biomass decreased. When the hydrogen at the inlet was decreased, the hydrogen yield coefficient increased while the solubilized hydrogen dropped as shown in figur 2.

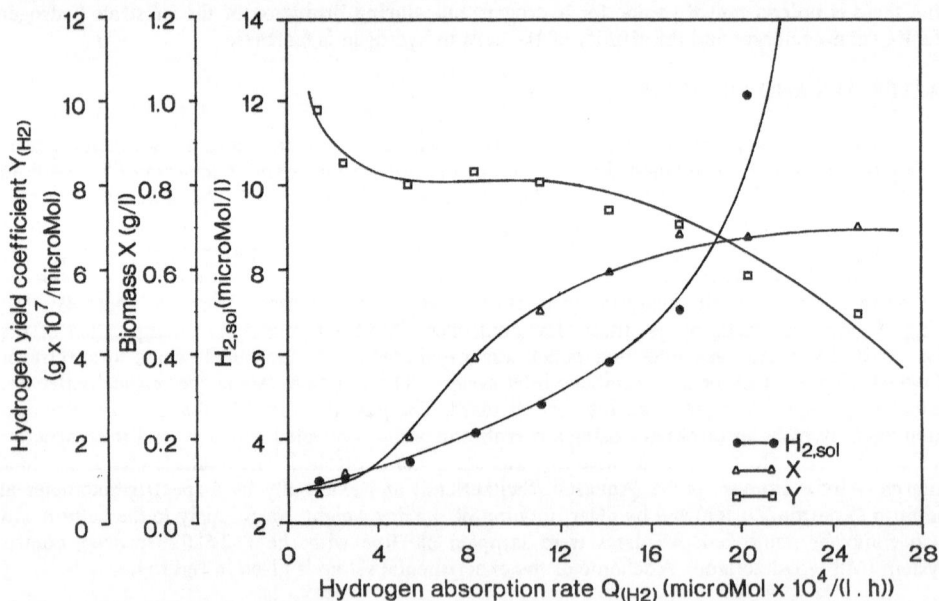

Figure 2. Solubilized hydrogen, biomass and hydrogen yield coefficient for different hydrogen uptake rates

Figure 3 shows that the K_s value for H_2 was linear decreasing when the hydrogen supply was further reduced in the limiting phase. The $K_{s(H2)}$ value was calculated from the formula:

$$Q_{(H2)} = \frac{1}{Y} \cdot \mu_{max} \cdot \frac{S}{K_S + S} \cdot X \quad {}^{4,5}$$

μ_{max} has been determined earlier and be found to be $0.31 (h^{-1})$. $Q_{(H2)}$ is the rate of hydrogen uptake.

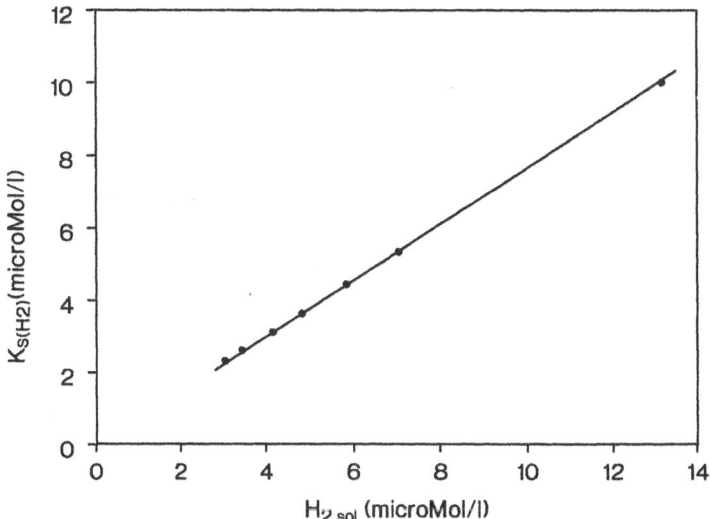

Figure 3. K_S values for different hydrogen concentrations

The results obtained demonstrate that the Monod equations cannot be used to explain the growth of a hydrogen limited culture of Mb. thermoautotrophicum. Further investigation may prove whether the phenomenon described applies also for other gaseous substrates.

REFERENCE

1. J. T. Keltjens and C. van der Drift, Elektron transfer reactions in methanogens, FEMS Microbiol. Revs., 39:259 (1986)
2. K. Schneider, Determination of the O2 and CO2 KLa values in fermenters with the dynamic method measuring th step responses in the gas phase, J. Appl. Chem. Biotechnol., 27:631 (1977)
3. B. M. Butsch and R. Bachofen, The membrane potential in whole cells of Methanobacterium thermoautotrophicum, Arch. Microbiol., 138:293 (1984)
4. S. J. Pirt, Principles of Microbe and Cell Cultivation, Blackwell Scientific Publications, London (1975)
5. J. E. Bailey and D. F Ollis, Biochemical Engineering Fundamentals, McGraw-Hill Book Company, New York (1986)

PHYSIOLOGY AND METABOLIC FEATURES OF A NOVEL METHANOGENIC ISOLATE

Shanthi Krishnan and K. Lalitha

Department of Chemistry (Biochemistry)
Indian Institute of Technology
Madras - 600 036, India

INTRODUCTION

Methanogenic species have been classified in three orders, Methanobacteriales, Methanococcales and Methanomicrobiales[1]. Methanogenic species capable of using methylated compounds as growth substrates include Methanococcoides, Methanosarcina, Methanothrix, Methanolobus, and certain species of Methanococcus all belonging to Methanomicrobiales[2]. Among the species utilising methylated compounds all except those belonging to the genera of Methanothrix utilise methanol. A recently described species belonging to Methanobacteriales, M.stadtmaniae has been shown to utilise methanol for its growth along with H_2[3]. Rod shaped methanogens so far reported utilise H_2/CO_2 and/or formate as substrates.

Characteristics of a rod shaped methanogen growing on methanol isolated in our laboratory from anaerobically degrading leaves of Leucaena leucocephala (a leguminous plant), degrading medium is reported here.

MATERIALS AND METHODS

Throughout the studies, deionised water (conductivity 10 mega Ohms) and AR grade chemicals were used. Anaerobic mud samples from fish pond and sewage sludge (collected from I.I.T., Madras) were used as seed inocula for the anaerobic digestion of leaves of L.leucocephala operated under mesophilic conditions[4]. Digested slurry from the reactor operated semi-continuously for over two years was serially diluted in a minimal basal medium containing $MgCl_2$ (1.0 mM), NH_4Cl (0.02 M) and CH_3OH (0.1 M) in potassium phosphate buffer (0.1 M) at pH 7.4, in an atmosphere of 100% N_2 gas devoid of any added nutrients such as vitamins, minerals and sulphur source. Isolation, growth and maintenance was achieved in this medium.

Estimations of activities of ATPase, aspartate and alanine amino-transferases, and protein and F_{420} were carried out. Gas analyses were performed in a Tracor Model 540 Gas Chromatograph equipped with Nelson software Chromatographic Package. After isolation of DNA, the mole% G+C of the DNA was determined from the Tm curve as described by Marmur[5] with calf thymus DNA as reference. Cells stained negatively with 1% (W/V) phosphotungstic acid were studied using Transmission Electron Microscopy, Philips Model CM12, under standard conditions.

RESULTS AND DISCUSSION

The growth of the isolated methanogen acclamatised to methanol as growth substrate was also tested for other carbon sources including CO_2, ethanol, 2-propanol, n-butanol, formate, acetate and propionate. Growth on CO_2 and 2-propanol was found to be as much as that with methanol while the organism failed to grow on other carbon sources. Utilisation of 2-propanol as H_2 donor for methanogenesis from CO_2 has been shown for spirullum (SK) and coccoid (CV) species[6]. No growth was observed in the media containing cellulose, sucrose and glucose as indvidual carbon sources supporting the purity of methanogenic culture. Added glutathione did not serve as carbon source and gaseous N_2 was not utilised as N_2 source during experimental period of 30 days. Added glutathione upto 2 mM level resulted in linear response in growth and no growth was observed with cysteine hydrochloride and sodium sulfide.

Optimal growth of the isolate was achieved at 0.1 M methanol concentration in the medium, at pH 7.4 and at temperatures between $30-37^\circ C$.

Growth was insensitive to added anitbiotics including penicillin, erythromycin, rifampicin, chloroamphenicol, and streptomycin (at levels of 60 µg/ml and 120 µg/ml).

Enhanced growth rate by 50% was recorded with supplemented vitamins in the absence of added minerals in the basal medium. For growth, required vitamins were riboflavin, biotin, thiamin, vitamin B_{12} and folic acid and non-essential vitamins being pyridoxine, calcium pantothenate, nicotinic acid and p-amino benzoicacid. Requirement of thiamin, p-amino benzoic acid, folic acid, riboflavin, vitamin B_{12} were reported for the growth of the methanogenic species[2,4].

Effect of NaCl on the growth of the isolate was evaluated by adding varying concentrations of NaCl from 0.02 to 0.1 M to the basal medium at series of K^+ concentration in the medium. The studies revealed that F_{420} levels increased as a function of K^+ in the medium from 0.02 to 0.1 M with an inverse relationship with Na^+ levels in the medium while absorbance was not altered significantly. Protein levels were optimal at 0.05 M each of K^+ and Na^+ ions in the medium. Requirement of Na^+ for growth, aminoacid transport and for methanogenesis has been reported for other methanogenic species[2]. For the growth of this isolate addition of metals Ni, Fe, Mo, Zn, Cu, Mn to the basal medium as chlorides in concentration ranging from 0.05 to 10 µM was found enhancing growth to varying extent, the order of response being Mo, Mn, Fe > Ni, Se, Co > Cu, Zn. Enzyme activities detected were as reported earlier[7].

Cells of this isolate were pink colored, gram positive, rod shaped of about 2 to 3.8 μm in length and 0.6 μm in diameter, occurring singly or in short chains of about 10 μm in length. The cell membrane of ca. 20 nm thickness was enveloped by a fuzzy cell coat material of unknown composition (electron micrographs available with the author). The G+C content of this species is 52.8 mole%. The characteristics of this pink colored, rod shaped methanogen with substrate specificity limited to methanol, H_2/CO_2 and 2-propanol appears to be novel necessitating further studies for suitable classification.

ACKNOWLEDGEMENT

The study was supported by the Department of Non-conventional Energy Sources Grant No. 5/2/65/85-BE, India.

REFERENCES

1. W.E. Balch, G.E. Fox, L.J. Magrum, C.R. Woese, and R.S. Wolfe, Methanogens reevaluation of a unique biological group, Microbiol. Rev. 43:260 (1979).
2. K.F. Jarrell, and M.L. Kalmokoff, Nutritional requirements of the methanogenic archaebacteria, Can.J.Microbiol. 34:557 (1988).
3. W.J. Jones, D.P. Nagle,J.R, and W.B. Whitman, Methanogens and diveristy of archaebacteria, Microbiol.Rev. 51:135 (1987).
4. N. Vasanthy, K. Shankar, P.M. Chandra Sekaran, and K. Lalitha, Biomethanation of L.leucocephala: a potential biomass substrate, Fuel. 65:1129 (1986).
5. M. Mandel, and J. Marmur, Use of ultraviolet absorbance temperature profile for determining the guanine plus cytosine content of DNA, Methods Enzymol. 12: (1968).
6. F. Widdel, P.E. Rouviere, and R.S. Wolfe, Classification of secondary alcohol utilising methanogens including a new thermophilic isolate, Arch.Microbiol 150:477 (1988).
7. K. Shanthi, N. Vasanthy, and K. Lalitha, Effect of Selenium on the metabolism of a novel methanogen, "Selenium in Medicine and Biology," J. Neve and A. Favier eds., Walter de Gruyter, New York (1989).

THERMOPHILIC ANAEROBIC OXIDATION OF ETHANOL

S. Larsen and B. K. Ahring

Department of Biotechnology
The Technical University of Denmark
DK - 2800 Lyngby

INTRODUCTION

A variety of facultative and obligate anaerobic bacteria
ferment carbohydrates to primary aliphatic alcohols. These
compounds are, therefore, important intermediates during the
anaerobic degradation of organic matter to methane and carbon
dioxide.

Under mesophilic anaerobic conditions primary alcohols are
oxidized to the corresponding fatty acids with additional
reduction of an external electron acceptor (Postgate and
Campbell, 1966).
If an external electron acceptor is not available, the electrons derived from the alcohol oxidation can be released as
molecular hydrogen. The hydrogen is subsequently used by
hydrogen-utilizing methanogenic bacteria, keeping the hydrogen
partial pressure at the low level necessary for the process.
Methanobacillus omelianskii is an example of such a syntrophic
association of two different species of bacteria, which
together convert ethanol to acetate and methane (Barker, 1941;
Bryant et al., 1967).
A third way of degrading primary alcohols is fermentation with
the production of fatty acids as reduced end products (Lanbroek
et al., 1982; Eichler and Schink, 1984).

MATERIALS AND METHODS

The anaerobic techniques for all culture work were those of
Hungate (1950) and Bryant (1972). All the experiments were made
in serum vials using a mineral media as previously described
(Ahring and Westermann, 1988). All cultures were incubated at
60°C.

A stable, anaerobic thermophilic enrichment culture (E26) was
established with ethanol (20 mM) as substrate. The original
inoculum was obtained from a stable thermophilic (60°C) bench-
scale reactor, operating on sewage sludge. The degradation path

of ethanol and the influence of bromoethanesulfonate (BES) on the conversion of ethanol was examined by growth experiments with the enrichment culture.

The effect of the addition of hydrogen on the specific ethanol consumption rate of the exponential growing enrichment culture was also studied. The appropriate partial pressure of hydrogen was produced by adding hydrogen gas to the headspace of the culture vessels.
Analysis of alcohols, volatile fatty acids and gasses were performed by the use of gas chromatographic systems.

RESULTS AND CONCLUSIONS

Growth experiments were performed in 50 ml triplicate vials with 25 ml of mineral media. After inoculation with 1 ml of an exponential growing enrichment culture, ethanol (20 mM) was completely degraded to acetate and methane within 23 to 26 days (Fig. 1). The stoichiometry of substrate utilization showed that 2 moles of ethanol were converted to 2 moles of acetate and 1 mole of methane. Neither propionate nor butyrate was found in the samples.

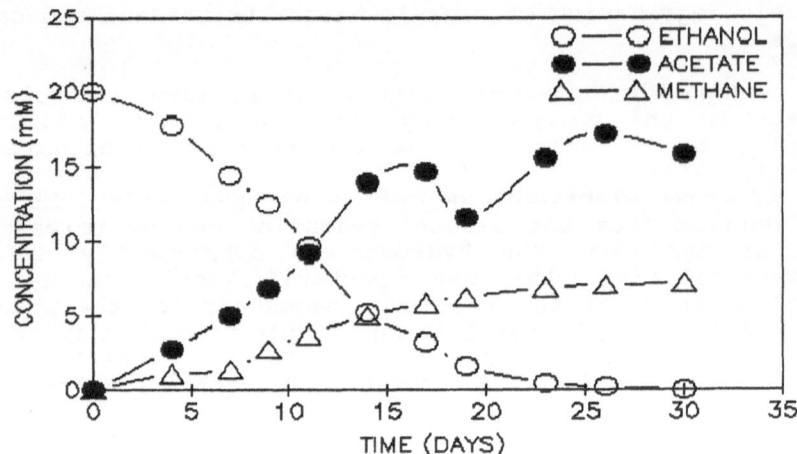

Fig. 1. Conversion of ethanol (20 mM) to acetate and methane by the thermophilic enrichment culture E26.

The addition of BES (5 mM) completely inhibited methane formation and decreased the ethanol degradation rate by 95 % (Fig. 2). At the end of the experiment, a partial pressure of approximately 0.1 atm hydrogen was found in the headspace.

The effect of hydrogen on the enrichment culture is shown in Table 1. The partial pressure of hydrogen in headspace was checked every third hour, and more hydrogen was added. Addition of hydrogen produced an immediate inhibitory effect on ethanol consumption by the culture. A partial pressure of 0.5 atm was

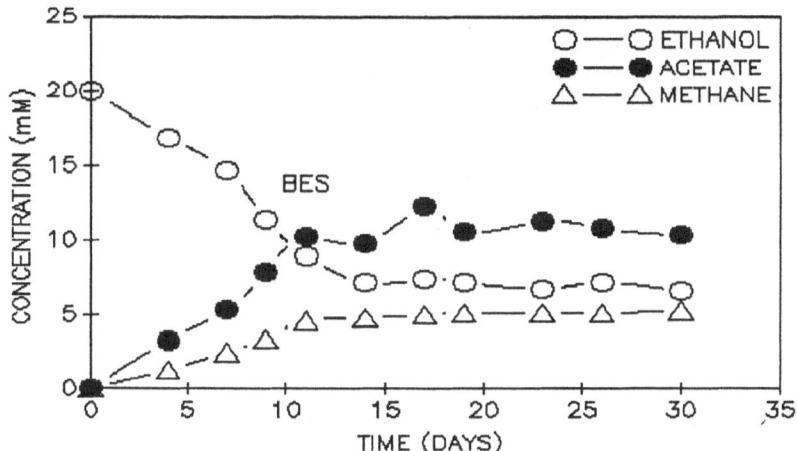

Fig. 2. Influence of BES on the enrichment culture E26 with ethanol (20 mM) as substrate. BES (5 mM) was added after 11 days.

found totally inhibitory for growth of the culture, while a partial pressure of 0.1 to 5 x 10^{-2} atm gradually inhibited the degradation. The partial pressure causing the total inhibition was much higher, than the corresponding value found for inhibition of thermophilic butyrate-degrading cocultures (Ahring and Westermann, 1988).

Table 1. Effect of hydrogen on the specific ethanol consumption rate of the thermophilic enrichment culture[a].

H_2 partial pressure (10^{-2} atm).	μ(EtOH) (d^{-1}).	% Inhibition.
0	0.587	
0.1	0.550	6.3
0.5	0.449	23.5
5.0	0.158	73.1
50.0	0	100

a) Experiments were performed in 50 ml serum vials with 25 ml of exponential growing enrichment culture (E26). μ(EtOH) is the specific ethanol consumption rate. Values are means of 3 independent experiments. The standard deviations were less than 0.1.

The ethanol-oxidizing bacterium was isolated in pure culture by using roll tubes with Gelrite as a gelling agent in the media. The cells were rod-shaped, 0.5-1.0 x 1.0-2.5 μm, with slightly pointed ends, appearing single or in pairs. Our further studies will concentrate on characterization of this bacterium.

ACKNOWLEDGEMENT

This study was supported by grant 5.17.4.6.17 from the Danish Technical Research Council.

REFERENCES

Ahring , B.K. and Westermann , P . , 1988 . Product inhibition of butyrate metabolism by acetate and hydrogen in a thermophilic coculture. Appl. Environ. Microbiol. 54 : 2393 - 2397.

Barker , H.A. , 1941. Studies on the methane fermentation .V. Biochemical activities of *Methanobacillus omelianskii* . J . Biol. Chem. 137: 153 - 167.

Bryant , M.P. , Wolin , E.A. , Wolin , M.J. and Wolfe , R. S. , 1967 . *Methanobacillus omelianskii* , a symbiotic association of two species of bacteria . Archiv. für Mikrobiol. 59 : 20 - 31.

Eichler , B. and Schink , B. , 1984 . Oxidation of primary aliphatic alcohols by *Acetobacterium carbinolicum* sp. nov. , a homoacetogenic anaerobe . Arch. Microbiol. 140 : 147 - 152 .

Lanbroek , H.J. , Abee , T. and Voogd , T. L. , 1982 . Alcohol conversions by *Desulfobulbus propionicus* Lindhorst in the presence and absence of sulphate and hydrogen . Arch. Microbiol. 133 : 178 - 184 .

Postgate , J.R. and Campbell , L. L. , 1966 . Classification of *Desulfovibrio* species , the non - sporulating sulphate - reducing bacteria . Bact . Reviews 30 : 732 - 738 .

EFFICIENCY OF BACTERIAL PROTEIN SYNTHESIS AND METHANOGENESIS DURING ANAEROBIC DEGRADATION

Roderick I. Mackie[1] and Marvin P. Bryant[1,2]

Department of Animal Sciences[1], and Microbiology[2]
University of Illinois at Urbana-Champaign
Illinois, 61801

INTRODUCTION

During methanogenesis in anaerobic environments degradable organic matter is converted to CH_4 and CO_2. During this process, most of the available energy is retained in the CH_4 produced with a relatively low yield of microbial cells (1). Little information exists on bacterial growth on complex, heterogeneous substrates where it is difficult to distinquish microbial and other proteins in substrate and effluent. An $^{15}NH_3$-N tracer technique was used to measure microbial protein synthesis (2, 3). Efficiency of methanogenesis was also measured.

MATERIALS AND METHODS

Experiments were carried out in 31, stirred benchtop fermentors with cattle waste (feces & urine without bedding) as substrate. Fermentation was carried out at 40°C and 60°C. Four different loading rates were employed with a decrease in retention time (RT) as loading rate increased (Table 1 and 2). Biogas production was measured by fluid volume displacement and gas composition analyzed using GC. The tracer, $(^{15}NH_4)_2SO_4$, was infused at a constant measured rate into the fermentors. Serial samples were removed and deproteinized. For ^{15}N analysis, NH_3 was steam-distilled and trapped in HCl. ^{15}N enrichment was analyzed by MS (Nuclide 3-60 RMS) after sample oxidation with alkaline hypobromite on the MS inlet system. Gross energy determinations were made using a bomb calorimeter (Parr Instrument Co.) on substrate and effluent dried after alkali addition.

RESULTS AND DISCUSSION

Nitrogen kinetics in the fermentors were characterized by large NH_3-N pool sizes and slow fractional turnover rates. The rate of bacterial protein synthesis was calculated assuming 85% of NH_3-N assimilated was utilized for protein synthesis (2, 3, 4) and a value of 10.6% N (66% CP) in bacterial cells (5). Despite a 4-fold increase in loading rate, bacterial cell production increased only 2-fold (Table 1). Cell yield was higher in the thermophilic digestor at the two lower loading rates with shorter RT's, VFA's (mainly

Table 1. Rate and efficiency of bacterial protein synthesis in mesophilic and thermophilic digestors.

Loading rate (gVS/ℓ reactor vol per day)	RT (days)	Rate of CH_4 production (mmol/ℓ/h)	Rate of NH_3-N incorporation (mg/ℓ/h)	Rate of Bacterial[a] cell production (mg/ℓ/h)	Cell Yield[b] (g cell/ mole CH_4)
(Mesophilic)					
3	13	1.28	0.40	3.25	2.66
6	10	1.97	0.68	5.37	2.42
9	9	2.39	0.88	7.11	3.07
12	5	2.31	0.91	7.30	3.18
(Thermophilic)					
3	13	1.50	0.54	4.34	2.95
6	10	2.65	0.84	6.60	2.47
9	9	3.78	0.97	7.74	2.08
12	5	4.10	1.01	8.13	1.98

[a](Rate of NH_3-N incorporation x 6.25 x 0.85) 100/66.
[b]Rate of bacterial cell production/rate of CH_4 production.

Table 2. Efficiency of digestion of gross energy (GE) and conversion to CH_4 in mesophilic and thermophilic reactors.

Loading rate (gVS/ℓ reactor vol per day)	GE(KJ/day) in			% Degradin of GE[b]	% Degraded GE converted to CH_4[c]
	Substrate	Effluent	CH_4[a]		
(Mesophilic)					
3	185.3	119.4	62.3	46.2	94.6
6	370.5	247.3	109.3	43.3	88.7
9	555.8	416.2	118.0	32.6	84.5
12	741.0	589.8	127.0	26.5	83.9
(Thermophilic)					
3	185.3	112.9	69.7	50.8	96.2
6	370.5	233.0	127.6	48.2	92.8
9	555.8	349.4	190.9	48.2	92.5
12	741.0	483.9	234.3	45.1	91.1

[a]Calculated assuming heat of combustion of CH_4 = 33.3 kJ/ℓ.
[b]Calculated as GE (substrate) - GE (effluent)/GE (substrate) multiplied by 100/77.7 for volatile solids content.
[c]Calculated as GE (methane)/GE (substrate) - GE (effluent).

412

propionate) accumulated in the mesophilic digestor with a resultant decrease in CH_4 production and hence higher yield. The data set for the thermophilic digestor was used to plot $1/Y_{CH_4}$ against $1/\mu$. A coefficient of 0.9 mmole CH_4/g cell/ℓ was obtained for the maintenance energy requirement and a $Y_{CH_4}^{max}$ value of 4.26 g dry cells/mole CH_4. These values are important in evaluating and predicting performance.

The thermophilic fermentor was more efficient than the mesophilic fermentor at all loading rates in respect of CH_4 production, degradation of GE, and conversion of GE to CH_4 (Table 2). Values for % conversion of degraded GE into CH_4 ranged from 84-95 and 91-96 in the mesophilic and thermophilic digestor, respectively.

REFERENCES

1.   P.L. McCarty, The methane fermentation, in: "Principles and Applications in Aquatic Microbiology," H. Heukelekian and N.C. Dondero, eds, John Wiley, New York (1964).

2.   M.F. Al-Rabbat, R.L. Baldwin and W.C. Weir, In vitro ¹⁵Nitrogen-tracer technique for some kinetic measures of ruminal ammonia, J. Dairy Sci. 54:1150 (1971).

3.   G.W. Mathison, and L.P. Milligan, Nitrogen metabolism in sheep, Br. J. Nutr. 25:351 (1971).

4.   J.V. Nolan and R.A. Leng, Dynamic aspects of ammonia and urea metabolism in sheep, Br. J. Nutr. 27:177 (1972).

5.   R.E. Speece and P.L. McCarty, Nutrient requirements and biological solids accumulation in anaerobic digestion, in: "International Conference on Water Pollution Research," Pergamon Press, London (1964).

TRACE METHANE IN SOME PROTEOLYTIC NONGLUCIDOLYTIC CLOSTRIDIA:

THE ROLE OF SOME S-METHYL AND N-METHYL COMPOUNDS

Philippe NIEL[1], Georges LELUAN[1], Henri VIRELIZIER[2]
and Alain RIMBAULT[1]

Laboratoire de Microbiologie, Faculté des Sciences Pharmaceuti-
ques et Biologiques de Paris, 4 Avenue de l'Observatoire,
F-75270 Paris Cedex 06, France[1] and Section d'Etudes et d'Analyse Isotopique et Nucléaire, Centre d'Etudes Nucléaires de
Saclay, F-91191 Gif-sur-Yvette Cedex, France[2]

INTRODUCTION

Clostridia form a very heterogeneous bacterial group involved in the
various aspects of the decomposition of organic matter[1-3]. The reduction
into methane of various substrates[4-8] formed in the prior steps by members
of microbial communities, e.g., clostridia[1,2,9], is carried out by methanogens.

Low amounts of various organic volatiles, particularly methane (at
nmol levels in the headspace gas per ml of culture) and volatile organosulfur
compounds (VOSCs) associated with growing cultures of six type strains of
mesophilic clostridia were detected by static headspace analysis with gas
chromatography (GC)-mass spectrometry (MS)[10].

To elucidate the origin of trace methane, a previous study was focused
on S-methyl compounds (L-methionine, methanethiol) with four Clostridium
strains of the proteolytic nonglucidolytic group[11]. Since an increased formation of methane was observed for Clostridium sp. DSM 1786 after addition
of 100-mM L-methionine[11], the aim of the present work was the study of the
methane formation for this strain with different concentrations of either
L-methionine or N-methylamine, a N-methyl compound described as a precursor
of methane for some methanogens[4].

MATERIALS AND METHODS

Clostridium sp. DSM 1786 was grown under reduced pressure at 37°C in
a culture tube containing 5 ml of filter-sterilized thioglycolate-Trypcase-
yeast extract (TTY) medium[10] supplemented or not with either L-methionine,
or L-(methyl-^{2}H$_3$) methionine, or N-methylamine hydrochloride, or N-(^{2}H$_3$)
methylamine hydrochloride. After a 7-day incubation period, the headspace
gas was analyzed by GC with flame ionization detection[10]. The methane isotopic species were separated by HS capillary GC[12].

RESULTS

After addition of different concentrations of L-methionine (1mM up to 100 mM), a progressive increased formation of methane was observed, the maximal amount of methane being almost reached with a 25-mM concentration (Table 1). The proportion of ($^2H_3$) methane in the methane peak increased from 46% to 91% for 1-mM and 100-mM concentrations, respectively. Whether N-methylamine was added or not, amounts of methane were similar and addition of N-($^2H_3$) methylamine did not lead to the detection of ($^2H_3$) methane (Table 1).

Table 1. Amounts of methane in the headspace gas for
Clostridium sp. DSM 1786 grown for 7 days
in 5 ml of TTY medium supplemented or not
with either L-methionine, N-methylamine,
L-(methyl-$^2H_3$) methionine*, or N-($^2H_3$)
methylamine*

Compound added to TTY medium		Methane (nmol per tube)	
		Control	Clostridium sp. DSM 1786
None		0.22[a] (0.01)[b]	61 (3)
1-mM	L-methionine	0.24 (0.04)	90 (4) (46:54)*
10-mM	L-methionine	ND	121 (1) (80:20)
25-mM	L-methionine	ND	144 (6) (80:20)
50-mM	L-methionine	ND	151 (6) (90:10)
100-mM	L-methionine	1.2 (0.03)	146 (5) (91:9)
1-mM	N-methylamine	0.15 (0.05)	62 (4) (0:100)*
10-mM	N-methylamine	ND	62 (5) (0:100)
25-mM	N-methylamine	0.22 (0.01)	64 (2.4) (0:100)

* : proportions of the two isotopic species of
methane [($^2H_3$) methane : methane] in labeling
experiments (as determined by capillary GC)
with the corresponding concentration of
L-(methyl-$^2H_3$) methionine or N-($^2H_3$) methylamine.
a : means calculated from triplicate experiments
for cultures and from duplicate experiments
for controls.
b : standard deviation.
ND : not determined.

CONCLUSION

From our results, we can conclude that for Clostridium sp. DSM 1786 methane originates from the S-methyl group of L-methionine, but not from the N-methyl group of N-methylamine. This trace methane formation may inter-fere when studying the metabolism of the S-methyl compounds by microbial

consortia containing methanogens along with clostridia. The ecological
significance of this observation needs further evaluation.

REFERENCES

1. G. Gottschalk, J. R. Andreesen, and H. Hippe, The genus *Clostridium*
   (nonmedical aspects), p. 1767-1803, in: "The prokaryotes, a
   handbook on habitats, isolation, and identification of bacteria," M. P. Starr, H. Stolp, H. G. Trüper, A. Balows, and H. G.
   Schlegel, ed., Springer-Verlag KG, Berlin (1981).
2. E. P. Cato, W. L. George, and S. M. Finegold, Genus *Clostridium*,
   p. 1141-1200, in: "Bergey's manual of systematic bacteriology,"
   9th ed., P. H. A. Sneath, N. S. Mair, M. E. Sharpe, and J. G.
   Holt, ed., The Williams and Wilkins Co., Baltimore (1986).
3. S. M. Finegold, V. L. Sutter, and G. E. Mathisen, Normal indigenous
   intestinal flora, p. 3-31, in: "Human intestinal microflora
   in health and disease," D. J. Hentges, ed., Academic Press, Inc.,
   New York (1983).
4. H. Hippe, D. Caspari, K. Fiebig, and G. Gottschalk, Utilization
   of trimethylamine and other N-methyl compounds for growth and
   methane formation by *Methanosarcina barkeri*, Proc. Natl. Acad.
   Sci. U.S.A. 76:494 (1979).
5. R. S. Wolfe, Microbial biochemistry of methane. A study in contrast. Methanogenesis, Int. Rev. Biochem. 21:270 (1979).
6. R. A. Mah, and M. R. Smith, The methanogenic bacteria, p. 948-
   977, in: "The prokaryotes, a handbook on habitats, isolation,
   and identification of bacteria," M. P. Starr, H. Stolp, H. G.
   Trüper, A. Balows, and H. G. Schlegel, ed., Springer-Verlag
   KG, Berlin (1981).
7. R. P. Kiene, R. S. Oremland, A. Catena, L. G. Miller, and D. G.
   Capone, Metabolism of reduced methylated sulfur compounds in
   anaerobic sediments and by a pure culture of an estuarine
   methanogen, Appl. Environ. Microbiol. 52:1037 (1986).
8. R. P. Kiene, and P. T. Visscher, Production and fate of methylated
   sulfur compounds from methionine and dimethylsulfoniopropionate
   in anoxic salt marsh sediments, Appl. Environ. Microbiol. 53:
   2426 (1987).
9. J. L. Pons, A. Rimbault, J. C. Darbord, and G. Leluan, Gas chro-
   matographic-mass spectrometric analysis of volatile amines
   produced by several strains of *Clostridium*, J. Chromatogr.
   337:213 (1985).
10. A. Rimbault, P. Niel, J. C. Darbord, and G. Leluan, Headspace
    gas chromatographic-mass spectrometric analysis of light hydro-
    carbons and volatile organosulphur compounds in reduced-pres-
    sure cultures of *Clostridium*, J. Chromatogr. 375:11 (1986).
11. A. Rimbault, P. Niel, H. Virelizier, J. C. Darbord, and G. Leluan,
    L-methionine, a precursor of trace methane in some proteolytic
    clostridia, Appl. Environ. Microbiol. 54:1581 (1988).
12. G. Berger, C. Prenant, J. Sastre, and D. Comar, Separation of
    isotopic methanes by capillary gas chromatography. Application
    to the improvement of $^{11}CH_4$ specific radioactivity, Int. J.
    Appl. Radiat. Isot. 34:1525 (1983).

INHIBITION KINETICS BY H_2, ACETATE AND PROPIONATE IN METHANOGENESIS FROM PROPIONATE IN A MIXED CULTURE

Naomichi Nishio, Satoshi Fukuzaki and Shiro Nagai

Dept. of Fermentation Technology, Fac. of Engineering
Hiroshima University
Saijo, Higashi-Hiroshima 724, Japan

ABSTRACT

A propionate-acclimatized sludge consisting of flocs(size, 150-300 μm) was used to analyze the inhibitory effects of H_2, acetate and propionate on the rate of propionate consumption. Inhibition by propionate could be analyzed by second order substrate inhibition model giving that K_s=15.9 μM, K_i=0.79 mM and q_m=2.15 mmol/g MLVSS/day. For the inhibition by H_2 and acetate, non-competitive product inhibition model was adopted giving that for H_2 inhibition, $K_{p(H_2)}$=0.11 atm(71.3 μM), q_m=2.4 mmol/g MLVSS/day, n=1.51, and for acetate inhibition, $K_{p(HAc)}$=48.6 μM, q_m=1.85 mmol/g MLVSS/day, n=0.96.

INTRODUCTION

The methanogenesis of organic matters occasionally accumulates H_2, acetate, propionate and butyrate as intermediates which act as potent growth-inhibitors for the bacteria in anaerobic digestion. Therefore, to achieve an efficient anaerobic digestion, the intermediates have to be further decomposed to CH_4. Generally, the degradation of propionate to acetate, CO_2 and H_2 is thermodynamically unfavorable and may often be a rate-limiting step of the whole methanogenesis. The methanogenesis of propionate requires both obligate proton-reducing acetogen and H_2- and acetate-utilizing methanogens:

$$CH_3CH_2COO^- + 3H_2O \rightarrow CH_3COO^- + HCO_3^- + H^+ + 3H_2 \quad (\Delta G°'= +76.1 \text{ KJ/reaction})$$

$$4H_2 + HCO_3^- + H^+ \rightarrow CH_4 + 3H_2O \quad (\Delta G°'= -135.6 \text{ KJ/reaction})$$

$$CH_3COO^- + H_2O \rightarrow CH_4 + HCO_3^- \quad (\Delta G°'= -31.0 \text{ KJ/reaction})$$

It is well-known that the growth of proton-reducing acetogens is remarkably inhibited by the accumulated H_2, hence, to accelerate the metabolism of propionate, the partial pressure of H_2 in the reactor must be kept at a low level of 10^{-6} to 10^{-4} atm.[1] On the other hand, in spite of the acetate is a major substrate for CH_4 production,[2] the influence of acetate on propionate degradation has not been fully investigated.

To elusidate this, we acclimatized anaerobic digestion sludge capable of producing CH_4 from propionate. Then, the inhbition kinetics by H_2, acetate, and propionate to the methanogenesis from propionate were analyzed.

MATERIALS AND METHODS

Microorganisms

The granulated sludge of anaerobic digestion of starch wastewater was supplied from Biotechnol. Research Lab. (Kobe Steel, Ltd.). Besides, a hydrogenotrophic methanogen isolated from the above sludge was used.

Acclimatization

To acclimatize the sludge to a propionate-minimal medium, the centrifuged sludge was inoculated into a 700 ml serum bottle. Then, static culture(37°C) was carried out by monitoring CH_4 evolution and when the gas evolution was stopped, e.g., after 2 weeks culture, propionate(20 mM) was resupplied to the culture. Through such repeated batch cultures for 4 months, the propionate could be completely consumed after a week without detecting H_2 and acetate. Thus, the propionate-acclimatized sludge obtained was inoculated to fresh medium every 2 weeks to maintain its activity.

RESULTS AND DISCUSSION

Propionate conversion to methane

A time course of cultivation in the dense propionate-acclimatized sludge(4.23 g MLVSS) showed that propionate consumption and CH_4 production proceeded without a lag time. It seems likely to have established a well-balanced ecological system consisting of proton reducing acetogens and H_2- and acetate-consuming methanogens, since H_2 and acetate could not be detected, and CH_4 yield of propionate was 1.74 mol CH_4/mol propionate being stoichiometrically assented. The microscopic observation showed that numerous flocs were observed together with a few dispersed microorganisms such as rod, sarcina, and thrix-like bacteria. The size of floc was mostly from 150 to 300 μm, occasionally 700 μm in diameter. These flocs were absolute majority during the culture without deformation.

Propionate inhibition to propionate consumption

The effect of propionate on the rate of propionate consumption was studied by adjusting the initial propionate concentration as a function of pH. The results showed that the maximum rates of propionate consumption were observed at pH 6.8 to 7.3 and the rates were gradually decreased with the increase of propionate. In the cases of the lower pH of 6.0 and 6.4, the rates were remarkably inhibited compared to other runs, suggesting that the lowered pH caused to increase the undissociated acid in the medium.

To analyze the relationship between propionate consumption rates and undissociated propionic acid concentration, the second order substrate inhibition kinetic model was adopted[3].

$$q_s = q_m S/(K_s + S + S^2/K_i) \tag{1}$$

where, q_s: specific rate of substrate(propionate) consumption, mol/g MLVSS /day; q_m: maximum value of q_s; S: substrate concentration, mM; K_s: substrate saturation constant, mM; K_i: substrate inhibition constant, mM.

The data fitting by nonlinear least-squares regression method based on eq. 1 gave that K_s=15.9 μM, K_i=0.79 mM, and q_m=2.15 mmol/g MLVSS/day.

Hydrogen inhibition by propionate consumption

H_2 inhibition to propionate consumption was tested during propionate conversion to CH_4. When H_2 was injected to 69 h culture, propionate degradation was sharply inhibited depending on the H_2 partial pressure. For the analysis of the propionate consumption rate against the H_2 partial pressure, a non-competitive inhibition equation was used.[4]

$$q_s = q_m/[1 + (P/K_p)^n] \tag{2}$$

where, P: product(H_2 or acetate) concentration; K_p: inhibition constant, atm; n: exponent of inhibition.

The data fitting by nonlinear regression method based on eq. 2 gave that $K_{p(H_2)}$=0.11 atm(71.5 μM), q_m=2.40 mmol/g MLVSS/day, and n=1.51.

The oxidation of propionate is thermodynamically feasible only at extremely low partial H_2 pressures of 10^{-6} to 10^{-4} atm.[1] However, the propionate oxidation against the H_2 partial pressure in this experiment indicated that 0.07 atm of H_2 did not remarkably inhibited the propionate degradation ($K_{p(H_2)}$=0.11 atm). This gap may be attributed by the H_2 removal of hydrogenotrophic methanogens which are existing together with propionate consuming bacteria in the flocs, i.e., interspecies H_2 transfer.[5,6] Therefore, it seemed likely that propionate-consuming bacteria inside the flocs might be partially protected from exogenous H_2 by adjacent hydrogenotrophic methanogens.

Acetate inhibition to propionate consumption

The inhibitory effect of acetate on propionate degradation was investigated by adding acetate(2.5 to 300 mM). From a linear decrease of propionate concentration for the biginning of each culture(for 69 h), specific rates of propionate consumption, q_s were estimated. In the case of initial pH of 6.5, much more severe inhibition was observed compared to those of pH 7.2 suggesting the inhibition by undissociated acetic acid which was calculated from pK_a=4.77 at 37°C. Kinetic pattern between q_s and undissociated acetic acid concentration seemed likely in the case of H_2 inhibition against q_s, hence the relationship was analyzed by eq. 2 giving that $K_{p(HAc)}$=48.6 μM , q_m=1.85 mmol/g MLVSS/day, and n=0.96.

It is of interest to note that $K_{p(HAc)}$ (=48.6 μM) was in the same order as $K_{p(H_2)}$ (=71.5 μM) suggesting that both products might have a high toxicity to propionate oxidation, hence, both hydrogenotrophic and acetoclastic methanogens were required to enhance the propionate oxidation.

REFERENCES

1. S. R. Harper and F. G. Pohland, Recent developments in hydrogen management during anaerobic biological wastewater treatment, Biotechnol. Bioeng. 28:585 (1986).
2. W. Gujer and A. J. B. Zehnder, Conversion processes in anaerobic digestion, Water Sci. Technol. 15:49 (1983).
3. V. H. Edwards, The influence of high substrate concentrations on microbial kinetics, Biotechnol. Bioeng. 12:679 (1970).
4. O. Levenspiel, The Monod equation: a revisit and a generalization to product inhibition situations, Biotechnol Bioeng. 22:1671 (1980).
5. M. P. Bryant, E. A. Wolin, M. J. Wolin, and R. S. Wolfe, "Methanobacillus omelianski," a symbiotic association of two species of bacteria, Arch. Microbiol. 59:20 (1967).
6. E. L. Ianotti, P. Kafkewitz, M. J. Wolin, and M. P. Bryant, Glucose fermentation products of Ruminococcus albus grown in continuous culture with Vibrio succinogenes: changes caused by interspecies transfer of H_2, J. Bacteriol. 114:1231 (1973).

HYDROGEN CONTENT IN BIOGAS AS A STATE INDICATOR OF METHANOGENESIS FROM WASTES

A.N.Nozhevnikova, I.V.Bodnar, A.I.Slobodkin
T.G.Sokolova

The Institute of Microbiology of the USSR
Academy of Sciences
Moscow, USSR

H_2 is a central metabolite and a regulator in the methanogene microbial community (1). The example of methane fermentation of cattle manure allowed us to demonstrate that hydrogen concentration in the gas phase under the balanced community activities does not exceed threshold levels for H_2-utilizing methanogenic bacteria: 10 ppm for mesophillous and 40-60 ppm for thermophillous ones (2). There was suggested the possibility of applying hydrogen concentration as a control parameter (3). The aim of the work was to examine the impact of some physico-chemical factors on the change of hydrogen concentration in the produced biogas under the thermophilic cattle manure fermentation. The experiments were conducted under the conditions of periodical cultivation. Hydrogen and methane concentrations in the gas phase were determined on the chromatograph with a solid electrolytic cell (4).

When the temperature increased from 55 to 61° after 20 hours, hydrogen concentration in the gas phase increased from 40 to 120 ppm, and when the temperature reached 90° it made up 500 ppm. When the initial values of the medium pH from 4,5 to 10 was determined by HCl and NaOH increased hydrogen content was observed in alkaline and acidic zones. After 15 hours it was $20.10^{-3}\%$ under pH 8,7 and $25.10^{-3}\%$ under pH 5,5 whereas maximum methane concentration (15%) was revealed in conditions of neutral reaction of the medium. Hydrogen and methane content in the gas phase were inversely proportional.

Methanogene community reacts in different ways to the medium acidification by various acids (table 1). Though the increase of medium acidification in all cases led to the increase of hydrogen content in the gas phase, methane concentration varied greatly, depending on the kind of the acid. Maximum decelerating impact on methanogenesis was observed in the variants with the propionic acid. The addition

Table 1. The influence of pH and of organic acids on hydrogen and methane formation, when fermenting cattle manure at 55°C

Acid, (g/l)	Initial pH	After 36 hours			
		pH	$H_2(\%)$	$CH_4(\%)$	excess pressure (atm)
control	7,5	7,0	3×10^{-3}	18	1,7
HCl	5,3	5,3	10^{-2}	1	0,1
	6,36	6,4	4×10^{-3}	16	1,6
acetate, 4	5,4	5,4	10^{-4}	10^{-2}	0
	6,0	6,0	10^{-2}	0,15	0,2
	8,5	7,5	3×10^{-3}	20	2,7
propionate, 4	5,2	5,2	10^{-2}	10^{-2}	0,1
	6,0	6,0	9×10^{-3}	10^{-2}	0,2
	6,85	6,8	8×10^{-3}	4	0,3
lactate, 4	5,35	5,3	0,45	10^{-2}	0,1
	6,25	6,4	5×10^{-2}	3,5	1,0
	7,15	7,0	2×10^{-2}	9	1,5

of acetate stimulated methane formation in conditions of neutral reaction of the medium. Maximum hydrogen quantity was produced when lactate was added, but sufficient decrease of methane formation was not observed. Hence, both the nature of the introduced substrate and the pH of the medium influence considerable increase of hydrogen content in the gas phase.

The rate changes of substrate addition, its composition changes, may cause variations in the composition of intermediate products. Table 2 demonstrates the data on the influence of exogene substrates' addition on hydrogen and methane content in biogas. Calculations were made after 23 and 46 hours. The system reacts to concentration changes of intermediate products of methanogenesis by the increase of hydrogen content. Maximum effect was observed when propionate, glucose, lactate were added. Hydrogen concentration increase does not always correlate with the rate decrease of methane formation. In some cases, for example, when glucose, starch, cellulose, acetate were added, intensification of methane formation was observed.

The dynamics study of hydrogen release and consumption from different exogenous methane precursors by the microflora of fermented cattle manure showed that the time of hydrogen "release" and observed absolute values of its concentration differ for the studied substances. Maximum hydrogen concentration was observed upon glucose decomposition; after 10 hours it made up 1100 ppm. When cellulose was decomposed, maximum hydrogen concentration was observed after 23

Table 2. The influence of exogenous substrates on hydrogen
formation under thermophillic fermentation of
cattle manure

| Substrate | g/l | 23 hours | | 46 hours | | |
		excess pressure atm	H$_2$ concentr. ppm	excess pressure atm	H$_2$ concentration ppm	pH
control	–	0,7	50	0,8	42	7,2
cellulose	10	1,0	87	2,4	23	6,9
starch	10	1,5	90	2,2	23	6,55
casein	10	0,8	83	1,1	34	7,1
lactate	5	0,6	92	0,67	40	7,05
glucose	5	1,8	116	2,5	40	6,4
butirate	5	1,0	53	1,7	35	7,15
propionate	5	0,7	200	0,6	40	7,2
pyruvate	5	1,5	48	1,2	45	7,15
acetate	5	1,5	30	0,6	36	7,25

hours. Immediate methane precursors – methanol and acetate-
increased hydrogen concentration up to 220 ppm after 19
hours. By the end of the experiment, after 72 hours, hydrogen concentration in all the variants decreased to 50-60 ppm
in the result of its consumption by methane bacteria.

Thus the results obtained show, that methanogenous
microbic system responds by the increase of hydrogen concentration in biogas to the change of physico-chemical conditions of the medium. Evidently, this indicator may be
used as a control one, while estimating the work of methane-
tanks. Qualitative characteristics of hydrogen content variations in biogas and the reasons, which cause them, call
for further investigations, in particular, in the systems
with continuous cultivation. However, it is evident, that
the excess of hydrogen content in biogas over threshold concentrations of H$_2$-utilizing methanogenes is indicative of
malfunction of the system.

REFERENCES

1. Zavarzin G.A., 1986, Izvestija USSR Acad. Sci., Ser.
   Biol.,N° 3: 341-360
2. Nozhevnikova A.N., Simankova M.V., Slobodkin A.I., et
   al., 1988, in "Archebacteria". Zavarzin G.A. ed. Puschi-
   no: 29-39
3. Whitmore T.N., Lloyd D., Jones G., et al., 1987, Appl.
   Microbiol. Biotechnol., 26, 383-388
4. Kunin L.L., Serebrennikov V.S., Fiodorov M.S., 1984,
   Geochemistry, 11: 1791-1796

INTERSPECIES TRANSPORT OF HYDROGEN IN THERMOPHILIC ANAEROBIC CELLULOSE DECOMPOSITION

A.N. Nozhevnikova and M.V. Simankova

Institute of Microbiology
Academy of Sciences of the USSR
Moscow, USSR

Hydrogen is the key metabolite in the process of anae-robic decomposition of complex organic compounds and plays the regulatory role in the methanogenic comunity. A conside-rable amount of hydrogen, which is consumed by methanogenic and homoacetogenic bacteria, is formed in the process of cellulose fermentation by anaerobic cellulolytic bacteria. The effect of interspecies hydrogen transfer on cellulose decomposition by combined cultures of Clostridium thermocellum with hydrogen utilizing methanogenic bacterium Methanobacte-rium thermoformicicum and homoacetogenic baoterium Clostri-dium thermoautotrophicum was stadied.

When pure culture C.thermocellum grown on cellulose the main fermentation products were hydrogen and ethanol: hydro-gen concentration in gas phase was 180 mM, ethanol content reached 40 mM. Organism produced acetate, lactate and car-bon dioxide.

During the development of C.thermocellum and M.thermo-formicum in co-culture on cellulose the ratio of fermenta-tion products changed in comparison with the monoculture C.thermocellum: acetate production increased by 30%, hydro-gen content in gas phase descreased to 50 ppm V. (fig. 1). This concentration corresponds to the threshold level of hydrogen consumption by pure cultures of thermophilic metha-nogenic bacteria. However, allowing for hydrogen converted into methane, its formation in co-culture increased 2-2,5 as much in comparison with production of pure culture.

In co-culture C.thermocellum and C.thermoautotrophicum only during the first two days some increase of hydrogen concentration was observed; then it decreased and didn't exceed 300 ppm V. (0,56 mM) (fig. 2). The increase of etha-nol concentration to 3,8 mM was also observed only during the first two days of cultivation; lactate was not produced. Later acetate was the only product of cellulose fermentation; its concentration reached 43,5 mM. The ratio acetate: etha-nol in monoculture was 0,16, in co-culture it increased to 12.8. H_2, CO_2 and sugars glucose and cellobiose are the

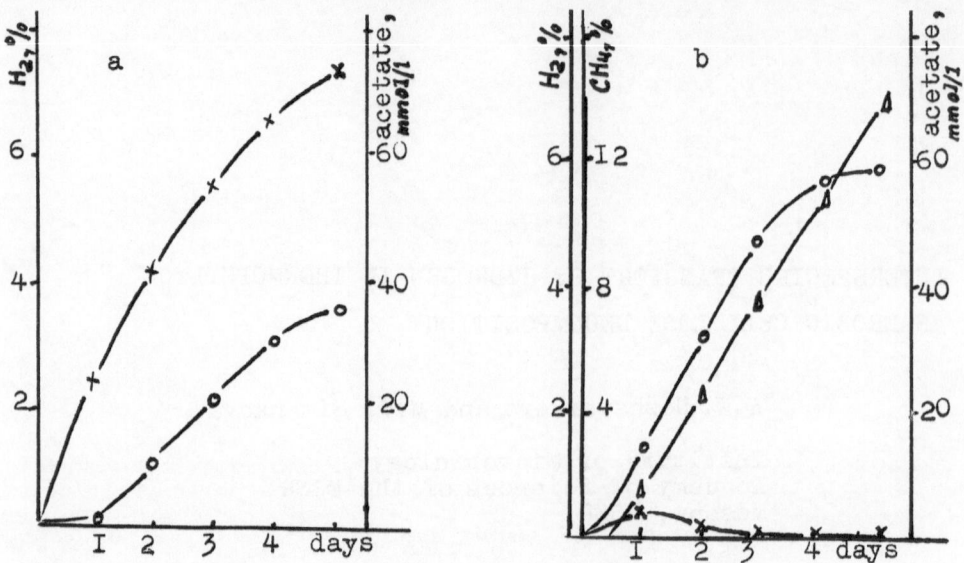

Fig. 1. Products of cellulose fermentation by C.thermocel-
lum in the absence (a) and presence (b) M.thermoformicicum:
hydrogen (—×—), acetate (—o—), methane (—▲—).

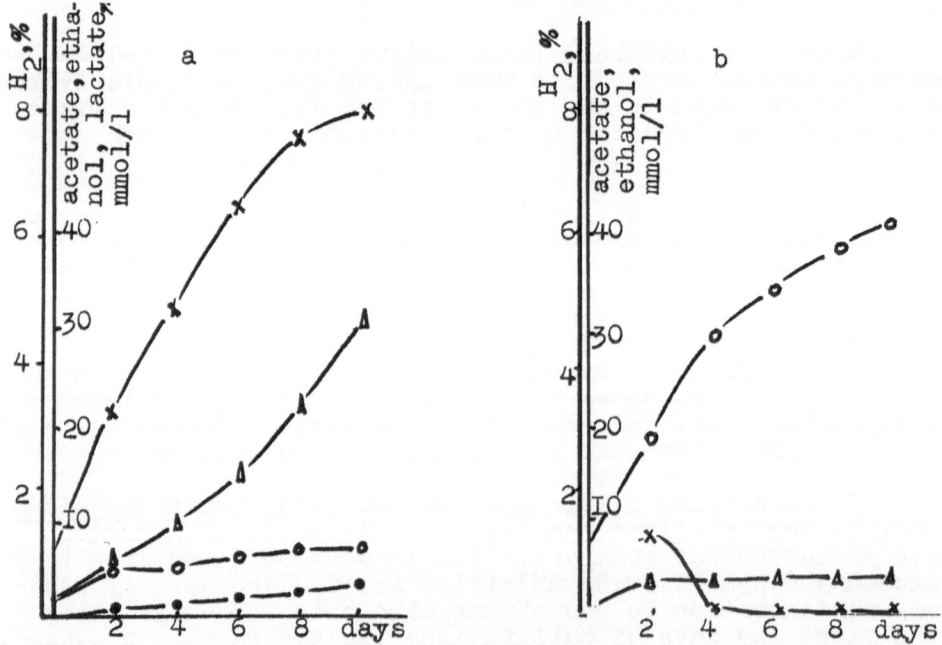

Fig. 2. Products of cellulose fermentation by C.thermocel-
lum in the absence (a) and presence (b) C.thermoautotrop-
hicum: hydrogen (—×—), acetate (—o—), ethanol (—▲—),
lactate (—●—).

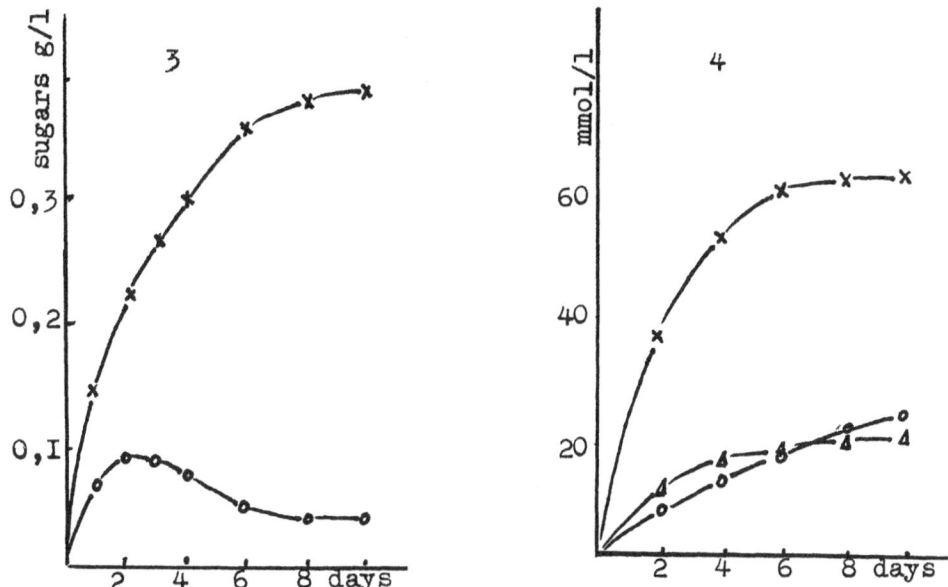

Fig. 3. Sugars production by C.thermocellum in the absence
(—x—) and presence (—o—) C.thermoautotrophicum.

Fig. 4. Acetate production by C.thermoautotrophicum: chemolithotrophic growth with H_2 (—▲—), heterotrophic growth
on glucose (—o—), simultaneous utilization of H_2 and glucose (—x—).

substrates for acetate formation by C.thermoautotrophicum.
Carbohydrates concentration in co-culture by the end of cultivation was 0,045 g/l, in monoculture it was 0,38 g/l
(fig. 3).

Pure culture C.thermoautotrophicum was shown to be capable of growing with acetate formation by simultaneous glucose and hydrogene utilization. In this case acetate production increased by 1,5-2,5 times in comparison with the
growth on each substrate (fig. 4).

The positive effect of homoacetogenic bacterium on the
cellulose decomposition rate and cellulitic activity of C.
thermocellum was not observed. In monoculture C.thermocellum
the number of degraded cellulose was 61%, cellulolytic activity was 0,777 unit/ml. 50% of cellulose was hydrolysed by
co-culture C.thermocellum and C.thermoautotrophicum, cellulolytic activity was 0,652 unit/ml.

Hence, hydrogen comsuption by methanogenic and homoacetogenic bacteria causes acetogenic shift in metabolism of
cellulose decomposition by C.thermocellum. In the case of
co-culture with C.thermoautotrophicum homoacetic cellulose
fermentation was observed.

REFERENCES

Simankova M.V., Nozhevnikova A.N. 1989, Microbiologija,
58, N° 6 (in press).

THE PARTICIPATION OF LITOTROPHIC HOMOACETOGENIC BACTERIA AND METHANOTHRIX IN THERMOPHILIC ANAEROBIC ETHANOL DEGRADATION WITH METHANE FORMATION

A.N. Nozhevnikova and A.I. Slobodkin

Institute of Microbiology
Academy of Sciences of the USSR
Moscow, USSR

Ethanol is one of the important intermediates during anaerobic biodegradation of organic substrates. Anaerobic ethanol degradation with methane formation can occur by the action of syntrophic microorganisms' associations such as consortium "Methanobacillus omelianskii" under mesophilic conditions (1) and Thermoanaerobium brokii and Methanobacterium thermoautotrophicum under thermophilic conditions (2). It is removed by methanogenic bacteria.

Process of ethanol degradation by methanogenic microorganisms' associations from cattle waste under thermophilic conditions (T=55°C) has been studied (3). The degradation of 80 mM exogenic ethanol lasted 5-6 days with hydrogen and acetate formation as intermediate products. However it was rather difficult to obtain stable thermophilic elective cultures degradating ethanol with methane formation. Some elective cultures were obtained from samples of fermented cattle waste after 10-12 passages on mineral media with ethanol as the single substrate. The ability of cultures to degrade ethanol with methane formation was not stable. The pH value often decreased up to 5,0-5,5 because of acetate accumulation and methane was not produced. Microscopy reviled 3-4 types of rods and among them H_2-utilizing methanogens shone under ultraviolet but there were no acetate-utilizing Methanosarcina and Methanothrix. The investigation of H_2-utilizing bacteria in this elective culture showed the presence of Methanobacterium sp. and homoacetogenic clostridium-type rods.

Acetogenic hydrogen-utilizing bacterium strain Z-55 was isolated from elective culture E5, which degraded 60-70 mM of ethanol during 20-30 days with 10-15 mM methane and 50-60 mM acetate formation. Morphologically the isolate represented rods which were sililar to Clostridium thermoautotrophicum capable of endospore-forming. Growth of

Table 1. Thermophilic ethanol degradation and methane formation by culture E5 and methanogenic bacteria after 30 days cultivation

| Culture | Concentration mM | | | | |
| | Ethanol | | Acetate | Methane | pH |
	Initial	Remaining			
E5	65,0	5,4	60,1	14,4	5,0
E5+Mb.thermofor-micicum	65,0	0,1	60,8	31,5	5,0
E5+Mt.thermoace-tophila	65,0	0,1	1,4	64,9	6,5

microorganism observed only under strictly anaerobic conditions. H_2:CO_2, methanol, glucose were utilized. Ethanol was not utilized. Acetic acid was the only product in the process of lithotrophic, methylotrophic and organotrophic growth. Specific rate of growth in H_2:CO_2 was 0,016 hour$^-$, specific rate of hydrogen consumption and acetate formation was 0,020 hour^{-1}. The threshold level of hydrogen consumption was 1500-2000 ppmV. The optimal temperature for acetogenesis of strain Z-55 was 55-57°C, the growth occured at 46-70°C. The optimal pH value was found to be 6,0-6,2.

The investigations of the influence of high methanogenic cell concentration on ethanol degradation and methane formation were carried out. Table 1 illustrates the results obtained after 30 days cultivation of the culture E5 together with H_2-utilizing Methanobacterium thermoformicicum and acetate-utilizing Methanothrix thermoacetophila (4,5). The culture E5 alone did not degrade all the amount of ethanol. With M.thermoformicicum all the amount of ethanol was degraded because of interspecies hydrogen transfer, but pH decreased to 5 because of acetate accumulation. The balanced ethanol degradation was obtained with M.thermoace-tophila; acetate was transformed to methane. Later on we obtained a stable culture containing Methanothrix and degradating ethanol with formation of methane and carbon dioxide as the end products.

Experiments with methanogenic microbial association obtained from fermented cattle waste by 2% inoculation of mineral media containing 65mM ethanol were carried out for more detailed investigation of Methanothrix thermoacetophila role in thermophilic ethanol degradation. The initial content of methanogenic association biomass was 0,425 g/l. After 30 days cultivation the remaining concentration of ethanol was 39 mM, acetate concentration was 24,5. pH decreased to 5,7 and 15,5 mM methane was formed. Concentration of hydrogen in the gas phase was 45 ppmV, which was in accordance with the threshold for H_2-utilizing methanogenic bacteria. These results indicated that there were not enough acetate utilizing methanogens in the association or they grew more slowly than ethanol-degradating microorganisms.

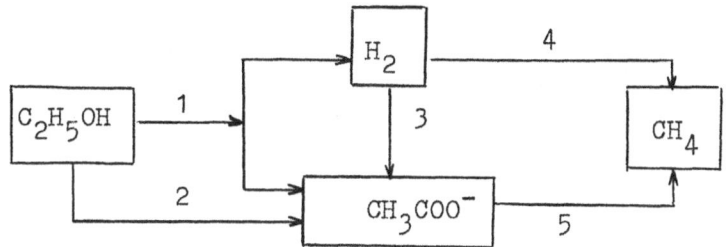

Fig. 1. Possible pathways of ethanol degradation by methanogenic microbial association: 1 - H_2-producing acetogens, 2 - ethanol-utilizing homoacetogenic bacteria, 3 - lithotrophic homoacetogens, 4 - H_2-utili-
zing methanogens, 5 - acetate-utilizing methanogens.

When 0,02 g/l of Methanothrix thermoacetophila was
inoculated together with 0,425 g/l of the association, the
remaining concentration of ethanol after of 30 days cultivation was 0,1 mM, of acetate - 0,8 mM, of hydrogen - 50
ppm V; pH did not change. The bottoms of the flasks used for
cultivation were covered by a thick bacterial film. The main
bacterial form in it was Methanothrix thermoacetophila.

The influence of pH and of the acetate-ion concentration on ethanol degradation was investigated in order to
elucidate the reason of the stimulation effect of Methanothrix on the process. The neutral reaction of the media which
was kept constant by NaOH titration was favourable for methanogenesis from ethanol by association. In 30 days all
amount of the ethanol was utilized with 35 mM methane and 40 mM
acetate formation. The addition of 50 mM acetate together
with 65 mM of ethanol decreased the ethanol consumption by
methanogenic association approximately 2-fold. The remaining
concentration of ethanol after 30 days cultivation was 55 mM,
the acetate 60 mM, only 7 mM methane was formed, the pH value did not change significantly. So both decreasing of pH
value and high concentration of acetate-ion inhibit the
anaerobic ethanol degradation. The stimulation of ethanol
degradation by Methanothrix thermoacetophila appears to be
the result of both maintaining optimal pH value and acetate-
ion removal. Possible pathways of ethanol degradation by
methanogenic association containing homoacetogenic bacteria
are shown in figure 1. Hence, it is possible that the presence of acetate-utilizing methanogenes in the system
allows the functioning of methanogenic association in which
hydrogen is removed by lithotrophic homoacetogenic bacteria.

REFERENCES

1. Bryant M.P., Wolin M.J., Wolfe, 1967, Arch.Microbiol.,
   59: 20-31.
2. Ben-Bassat A., Lamed R., Zeikus J.G. 1981. J.Bacteriol.
   146, N° 1: 192-199.
3. Slobodkin A.I., Nozhevnikova A.N. 1989. Prikladnaja
   biochim. microbiol. 25, N° 6 (in press).
4. Nozhevnikova A.N., Chudina V.I., 1985, Microbiol., 53;
   618-624.
5. Nozhevnikova A.N., Jagodina T.G., 1983, Microbiol.
   51: 534-541.

CHARACTERIZATION OF A SULFATE REDUCING BACTERIUM ISOLATED FROM A HYPERSALINE AFRICAN LAKE

B. Ollivier [1], C.E. Hatchikian [2], G. Prensier [3] J. Guézénec [4] and J.L. Garcia [1]

[1] Laboratoire de Microbiologie ORSTOM, case 87, Université de Provence
3, Place Victor Hugo, 13331 Marseille Cédex 3, France
[2] Laboratoire de Chimie Bactérienne, CNRS, BP 7
13277 Marseille Cédex 9, France
[3] Laboratoire de Microbiologie, Université Blaise Pascal
63177 Aubière, France
[4] Laboratoire de Chimie-Corrosion Marine, IFREMER, B.P. 337
29273 Brest Cédex, France

INTRODUCTION

Retba lake is a hypersaline African lake (340 g / l total salt) near Dakar (Senegal). Preliminary work has shown the existence of methanogenic, cellulolytic and sulfate reducing bacteria in sediments. Although several results have shown an active sulfate reduction to occur in hypersaline environments (Nissenbaum et al., 1976; Zeikus, 1983), no sulfate reducer has been described from these ecosystems up to now (Oren 1988).

Here we report on the isolation and partial characterization of a halophilic sulfate reducing bacterium.

RESULTS AND DISCUSSION

Strain HR is a halophilic sulfate reducing bacterium which incompletely oxidizes a limited range of substrates (Table 1).

Growth was inhibited above 250 g NaCl per l and was optimum with 100 g / l NaCl. Since the Na^+ concentration in Retba lake is 82.66 g per l (210 g / l NaCl), strain HR has to grow in this ecosystem beyond its salt optimum, which apparently diminuishes the metabolic activity. The isolate required Mg^{2+} for growth. In the presence of 100 g / l NaCl, the upper $MgCl_2$ concentration limit was 1.2 M.

Cells are straight to slightly curved rods (0.7-0.9 x 1-10 μm), which are motile by means of one or two flagella.

Slight growth was served in a mineral medium containing acetate and vitamins. Higher cell densities were reached when Biotrypcase was added to the medium.

Sulfate, thiosulfate and elemental sulfur were used as electron acceptors.

Table 1 . Substrates used as energy sources by strain HR

	$+ SO_4^{2-}$	$- SO_4^{2-}$
H_2	-	-
Formate	+	-
Ethanol	-	-
Pyruvate	+	+
Lactate	+	-

Culture medium containing 1 g / l Yeast extract and Biotrypcase.

Substrates tested and not used : acetate, propionate, butyrate, fumarate, malate, succinate, glycerol, fructose, choline, casaminoacids, yeast extract, Biotrypcase.

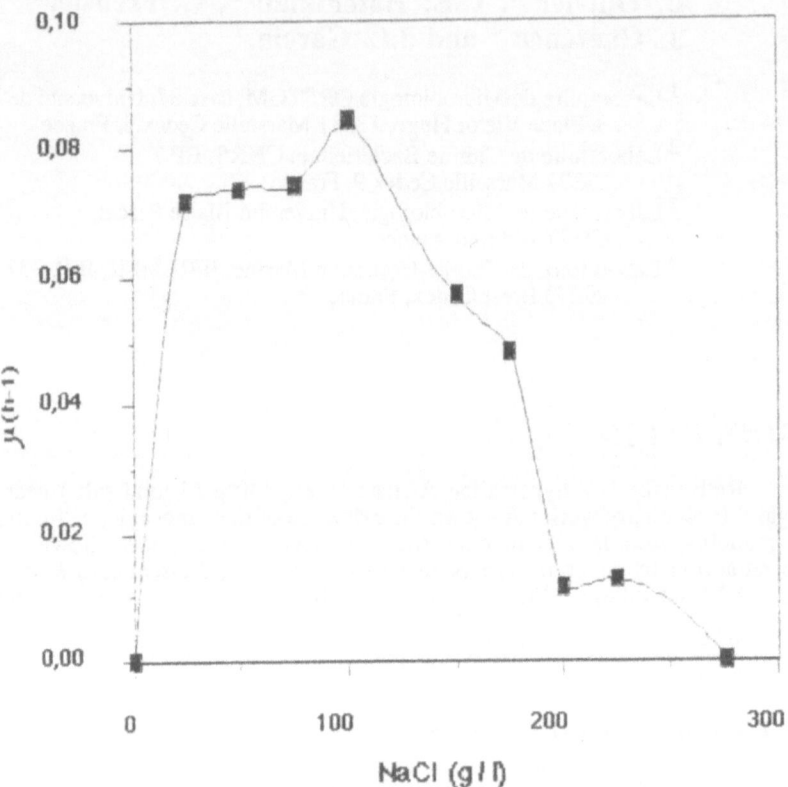

Fig 1 . Effect of NaCl concentration on growth rate of strain HR cultivated on lactate at 35°C with 5 g / l yeast extract and Biotrypcase .

Strain HR clearly belongs to the first group of sulfate reducing bacteria. It differs from species of *Desulfovibrio* and *Desulfomonas* genera in its morphology and the absence of desulfoviridin, and from the genus *Desulfotomaculum* in the absence of spore. Furthermore, contrary to *Desulfobulbus* species, strain HR does not oxidize propionate. It cannot belong to the genus *Thermodesulfobacterium*, due to its growth range temperature. Strain HR probably constitutes a new genus among the sulfate-reducers.

REFERENCES

NISSEBAUM , A. and KAPLAN , I.R. (1976). Sulfur and carbon isotopic evidence for biogeochemical processes in the Dead Sea ecosystem. In : " Environmental Biogeochemistry ", Vol. 1, pp. 309 - 325; J.O. Nriagu (ed.), Ann Arbor Science (Pub.), Ann Arbor, Michigan.

OREN , A. (1988). Anaerobic degradation of organic compounds at high salt concentrations. Antonie Van Leeuvenhoek , 54 : 267 - 277.

ZEIKUS , J.G. (1983). Metabolic communication between biodegradative populations in nature. In : " Microbes in their natural environments ". Symposium 34, pp. 423 - 462. J.H. Slater , R. Wittenbury and J.W.T. Wimpenny (ed.), Society for General Microbiology, Cambridge University Press, Cambridge (U.K.).

ISOLATION AND CHARACTERIZATION OF AN ETHANOL-DEGRADING ANAEROBE FROM METHANOGENIC GRANULAR SLUDGE

C.M.Plugge, J.T.C.Grotenhuis and A.J.M.Stams

Agricultural University
Department of Microbiology
H.v.Suchtelenweg 4
6703 CT Wageningen
The Netherlands

INTRODUCTION

Bacteria present in UASB-reactors (Upflow Anaerobic Sludge Blanket reactors) are immobilized in granules. These granules were grown in a 5 l. laboratory scale UASB-reactor with ethanol as the sole carbon and energy source. Bacteria were isolated from these granules to investigate the importance of different physiological groups of bacteria in the granule formation process. The syntrophic growth and the ability to form aggregates were studied with one of the isolated strains.

METHODS

All bacteria used were cultivated in a medium which contained per liter:
KH_2PO_4 0.41 g; Na_2HPO_4 0.53 g; NH_4Cl 0.3 g; NaCl 0.3 g; $CaCl_2.2H_2O$ 0.11 g; $MgCl_2.6H_2O$ 0.1 g; $NaHCO_3$ 4 g; $Na_2S.9H_2O$ 0.24 g; cysteine.HCl 0.5 g; resazurine 0.5 mg; vitamin solution according to Wolin et al. (1963) 1 ml; trace element solution according to Zehnder et al.(1983) 1 ml. Carbon sources were added from anaerobic, sterile stock solutions in desired concentrations.

All fermentation products were assayed with a Varian aerograph gaschromatograph with a Chromosorb 101 column.
All gases were analyzed on a Packard gaschromatograph with a molecular sieve column.

To investigate the granule formation a UASB-recycle system was used. This system was developed to prevent that bacteria growing in suspension in a UASB are washed out of the system (Grotenhuis et al. 1988). A schematic diagram is given in figure 1.

RESULTS

Characterization of the ethanol-degrading strain EE121

Ethanol degrading bacteria were isolated by the serial dilution method. The most abundant strain (EE121) was characterized further. EE121 is a Gram-positive, non-motile, spore-forming, strictly anaerobic rod and could grow on a wide range of substrates (Table 1).

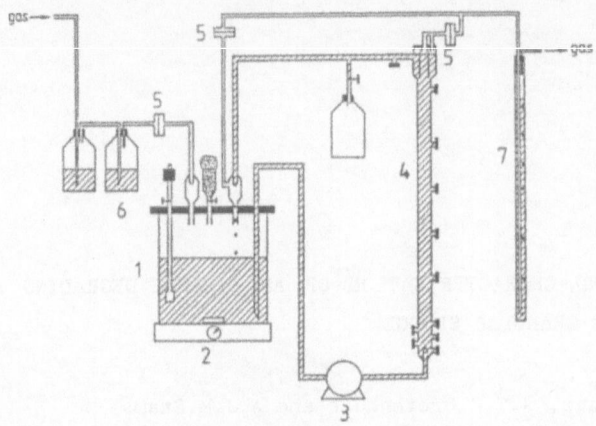

<u>Figure 1.</u> Schematic diagram of the UASB-recycling system. 1) fermentor; 2) magnetic stirrer; 3) pump; 4) UASB-reactor; 5) bacterial filter; 6) bottle with Na_2S; 7) 1 m water column.

<u>Table 1.</u> Substrates utilized by EE121.

substrate	growth
glucose	+
xylose	+
sucrose	+
galactose	+
fructose	+
ribose	+
lactate	+
methanol	+
ethanol	+
propanol	+
2,3 butanediol	+
acetoin	+
H_2/CO_2	+

Growth of strain EE121

In pure culture strain EE121 grows as a homo-acetogen: 1 mol of ethanol is degraded to 1.5 mol acetate and less than 0.05 mol of hydrogen. However, in mixed culture with the hydrogenotroph <u>Methanobrevibacter arboriphilicus</u>, ethanol was degraded stoichiometrically to acetate and H_2 (Figure 2).
The specific growth rate was the highest on the substrate glucose and decreased in the range acetoin, ethanol and H_2/CO_2.

Formation of aggregates

In a UASB-recycling system (Figure 1) the strain was tested on the ability to form aggregates in pure culture or in syntrophy with <u>M. arboriphilicus</u>. With the pure culture aggregates were observed after 14 days, and with the coculture already after 4 days (Figure 3). The aggregates consisted of densely packed cells and inorganic precipitates. In batch cultures aggregate formation could never be observed.

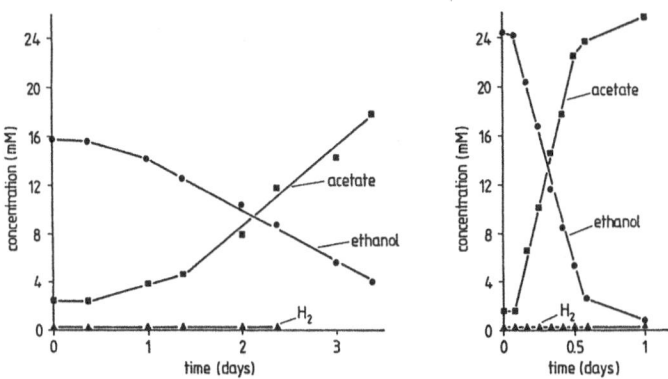

Figure 2. Degradation of ethanol by a) a pure culture of EE121 and b) coculture of EE121 and M.arboriphilicus.

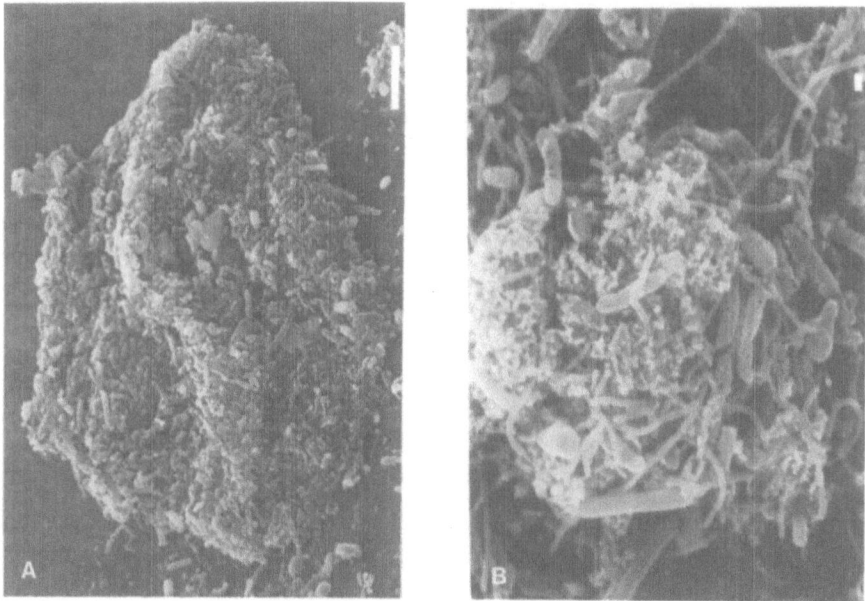

Figure 3. Scanning electron micrographs of aggregates formed in the UASB-recycle system. A) the strain EE121 and B) the coculture after 18 respectively 16 days of incubation in the recycle system. Bar indicates 1 μm.

DISCUSSION

The ethanol-degrading strain EE121 is an abundant bacterium in ethanol-adapted, methanogenic granular sludge. It is a Gram-positive, spore-forming, homo-acetogenic bacterium. In the presence of a hydrogen-consuming methanogen, reducing equivalents are disposed as hydrogen, rather than used for CO_2-reduction to acetate. In addition the specific growth rate of EE121 clearly increased in the presence of the hydro-genotroph (Figure 2). Similar findings were described for other homo acetogenic bacteria (Eichler and Schink, 1984). A remarkable property of strain EE121 is its ability to form aggregates in a UASB-recycle system. Good adherence properties are a prerequisite for bacteria present in methanogenic granular sludge, because of the high hydraulic loading rates which are applied in UASB-reactors. Bacteria, which are not able to adhere, will be washed out of the reactor. The surface-properties (charge and hydrophobicity) as well as further physiological and morphological characteristics of strain EE121 are presently investigated.

LITERATURE

Eichler, B., Schink, B., 1984, Oxidation of primary aliphatic alcohols by Acetobacterium carbinolicum sp. nov., a homoacetogenic anaerobe, Arch. Microbiol., 140:147-152.

Grotenhuis, J.T.C., Koornneef, E., Plugge, C.M., 1988, Immobilization of anaerobic bacteria in methanogenic aggregates, in: Granular anaerobic sludge; microbiology and technology, G. Lettinga, A.J.B. Zehnder, J.T.C. Grotenhuis and L.W. Hulshoff Pol, Pudoc Wageningen, 1988.

Wolin, E.A., Wolin, M.J., Wolfe, R.S., 1963, Formation of methane by bacterial extracts, Journ. biol. chem., 137:420-432.

Zehnder, A.J.B., Huser, B.A., Brock, T.D., Wuhrmann, K., 1980, Characterization of an acetate-carboxylating, non-hydrogenoxidizing methane bacterium, Arch. Microbiol.,
124:1-11.

THE ROLE OF FE(III) REDUCTION

IN ANAEROBIC PROCESSES

Joanna Potekhina

Institute of Ecology
of the Volga River Basin
Togliatti, USSR

The last decade has been marked by significant developments
in ferric iron reduction studies. In view of a variety of research
(Jones et al. 1983, 1984; Lovley 1987; Lovley and Phillips   1986,
1987), ferric iron reduction appears to be an essential, and in many
instances, a most preferential pathway of the terminal metabolism of
organic matter in anaerobic environments.

Various aspects of ecology, biochemistry and physiology of ferric iron-reducing bacteria have been adequately covered by many ex-
perimenting authors (Munch and Ottow, 1982, 1983; Tugel et al. 1986;
Bell and Mills 1987); the number of strains capable of ferric iron reduction is steadily increasing.

Hydrogen and acetate have been found to be the principal substrates for (i) sulfate reducers (Christensen 1984; Jorgensen 1983;
Taylor and Parkes 1985), (ii) ferric iron-reducing bacteria (Lovley
and Phillips 1986) and (iii) methane-generic bacteria (Lovley   and
Klug 1982, 1983; Daniels et al. 1984) in microbial communities of
natural habitats.

As it is evidenced from the data available nowadays on the competitive mechanisms for common substrates between the above three
most significant bacterial groups (Lovley and Klug 1983; Reeburgh
1983), the most outcompeting are the ferric iron reducers: in the presence of excess substrates and terminal acceptors of all the   three
processes, the electron flow is directed first of all to ferric iron reduction whereas methanogenesis and sulfate reduction do not occur
at all. Lovley and Phillips (1987) reported that such a high compe-
titiveness of Fe(III)-reducing bacteria can be primarily attributed to
their greter affinity to hydrogen and acetate as well as to the substrate depletion for sulfate reducers and methanogenes under   the
predominating Fe(III) reduction process.

A profound understanding of relationships between the most important anaerobic processes occurring in ecosystems is central both
for fundamental and applied microbiology, as it is closely related to
the problem of biocorrosion of metals in aquatic environments.

A novel concept of biocorrosion control   in industrial aquatic
ecosystems by microbial management is being developed now (Potekhina 1989). This concept is based on the results of a series of fundamental studies in situ which have covered the following aspects:
(i) peculiarities in the functioning of microbial communities,
(ii) mutual impacts between microorganisms and their habitats,

(iii) interrelations and interactions between microorganisms within a community, and (iv) competitive relationships between the microorganisms in situ, along with the establishment of laws governing the dominance of certain functional groups.

Competition is considered to be a limiting factor for the suppression of activities of corrosion-related microorganisms. This can be exemplified by the inhibition of the growth and activity of the most aggressive, sulfate-reducing bacteria by diverting the electron flow from sulfate reduction to some alternative processes.

In this connection, ferric iron reduction deserves the keenest attention, as it was supported by our multiple experiments with certain strains when bacteria tested had failed to cause corrosion on some mild steels (Potekhina 1984).

According to Lovley and Phillips (1987), ferric iron reduction usually proceeds in the presence of adequate concentrations of substrates as well as with $Fe(III)$ compounds available for the bacterial reduction. While the exogenic substrates are to be introduced into an ecosystem, the source for $Fe(III)$ compounds, as it was previously shown (Potekhina 1989), normally involves the corrosion products, such as $Fe(OH)_2$, $Fe(OH)_3$, Fe_2O_3, $FePO_4$, etc. available on the metalic surfaces in aquatic environments. Our experiments have demonstrated 80-90% reduction of $Fe(III)$-containing corrosion products followed by the transfer of $Fe(II)$ compounds into a soluble form.

The results obtained allowed us to develop a technology for removing the corrosion products off the metalic surfaces of heat-exchanging equipment of cooling water systems and to put it into use at some chemical enterprises.

A very important consequence of such phenomena as loosening and exfoliation of corrosion products arising due to ferric iron reduction is the distruction of ecological pockets which are formed under the rust layer.

Our method of sulfate reduction suppression is comprised of the following procedures:
1. Introduction of a ferric iron-reducing bacterial culture and exogenic substrates (hydrogen and acetate) into a system.
2. Distruction of ecological pockets of sulfate reducers as a result of the bacterial reduction of ferric iron compounds of corrosion products.
3. Maintaining the steady-state concentrations of the substrates required for the predominance of $Fe(III)$ reduction processes in the medium.

Thus, $Fe(III)$-reducing bacteria play a dual role in anaerobic processes in industrial aquatic ecosystems. Firstly, they distruct the ecological pockets of sulfate reducers by means of the uptake of ferric iron compounds of corrosion products, and secondly, they suppress the activity of these sulfate reducers by diverting the electron flow for ferric iron reduction.

Creation of conditions for the predominance of $Fe(III)$ reduction process is a reliable prerequisite for the prevention of biocorrosion caused by sulfate-reducing bacteria. The same principle can be used for the development of other alternative biological methods of corrosion suppression which are favourably distinguished by their ecological safety as opposed to the present-day techniques involving the use of highly toxic biocides for the same purposes.

REFERENCES

Bell, P.B., Mills, A.L., and Herman, J.S., 1987, Biogeochemical conditions favouring magnetite formation during anaerobic iron reduction, Appl. Environ. Microbiol., 53:2610–2616.

Christensen, D., 1984, Determination of substrates oxidized by sulfate reduction in intact cores of marine sediments, Limnol. Oceanogr., 29:189–192.

Daniels, L., Sparling, R., Sprott, G.D., 1984, The bioenergetics of methanogenesis, Biochim. Biophys. Acta, 768:113–163.

Jones, J.G., Gardener, S., and Simon, B.M., 1983, Bacterial reduction of ferric iron in stratified eutrophic lake, J.Gen. Microbiol., 129:131–139.

Jones, J.G., Gardener, S., and Simon, B.M., 1984, Reduction of ferric iron by heterotrophic bacteria in lake sediments, J. Gen. Microbiol., 130:45–51.

Jorgensen, B.B., 1983, "Microbial Geochemistry", W.E. Krumbein, ed., Blackwell, London.

Lovley, D.R., 1987, Organic matter mineralization with the reduction of ferric iron: A review, Geomicrobiol. J., 5:375–399.

Lovley, D.R., and Phillips, E,J.P., 1986, Organic matter mineralization with the reduction of ferric iron in anaerobic sediments, Appl. Environ. Microbiol., 51:683–689.

Lovley, D.R., and Phillips, E.J.P., 1987, Competitive mechanisms for inhibition of sulfate reduction and methane production in the zone of ferric iron reduction in sediments, Appl. Environ. Microbiol., 53:2636–2644.

Lovley, D.R., Dwyer, D.F., and Klug, M.J., 1982, Kinetic analysis of competition between sulfate reducers and methanogenes for hydrogen in sediments, Appl. Environ. Microbiol., 43:1373–1379.

Lovley, D.R., and Klug, M.J., 1983, Sulfate reducers can outcompete methanogenes at freshwater sulfate concentrations, Appl. Environ. Microbiol., 45:187–192.

Munch, J.C., and Ottow, J.C.G., 1982, Einfluss von Zellkontakt und Eisen(III) Oxidform auf die Bakterielle Eisenreduktion, Z. Pflanzenernaehr. Bodenkd., 145:66–77.

Munch, J.C., and Ottow, J.C.G., 1983, Reductive transformation mechanism of ferric oxides in hydromorphic soils, Ecol. Bull. (Stockholm), 35:383–394.

Potekhina, J.S., "The role of bacteria of distinct functional groups in biocorrosion of metals", Nauka, Moscow, 1989.

Potekhina, J.S., 1984, Inhibition of corrosion in mild steels by aerobic microorganisms, J. Protec. Met., 3:469–470.

Reeburgh, W.S., 1983, Rates of biogeochemical processes in anoxic sediments, Annu. Rev. Earth Planet. Sci., 11:269–298.

Taylor, J., and Parkes, R.J., 1985, Identifying different populations of sulfate–reducing bacteria within marine sediment system, using fatty acid biomarkers, J. Gen. Microbiol., 131:631–642.

Tugel, J.B., Hines, M.E., and Jones, G.E., 1986, Microbial iron reduction by enrichment cultures isolated from estuarine sediments, Appl. Environ. Microbiol., 52:1167–1172.

Winfrey, M.R., and Ward, D.M., 1983, Substrate for sulfate reduction and methane production in intertidal sediments, Appl. Environ. Microbiol., 45:193–199.

GLYCEROL DEGRADATION BY *DESULFOVIBRIO* SP. IN PURE CULTURE AND IN COCULTURE WITH *METHANOSPIRILLUM HUNGATEI*

A.I. Qatibi , J.-L. Cayol and J.-L. Garcia

Laboratoire de Microbiologie ORSTOM , case 87
Université de Provence , 3 , place Victor Hugo
13331 Marseille Cédex 3 , France

INTRODUCTION

Among the sulfate - reducing bacteria (SRB), the genus *Desulfovibrio* is known to grow on a limited range of oxidizable substrates including hydrogen, ethanol, lactate, formate, malate, fumarate and succinate (Postgate, 1979). Recently, some *Desulfovibrio* strains have been isolated, that can utilize glycerol (Nanninga and Gottschall, 1987 ; Ollivier *et al.* , 1988). But to date few results have been reported on the utilization of glycerol by SRB. Furthermore, no data are available concerning glycerol use in a syntrophic association with a methanogen as an alternative H_2 sink.

Here, we present some results on the effects of terminal electron acceptor on glycerol dissimilation by *D. carbinolicus, D. fructosovorans* and *Desulfovibrio* sp. strain M, utilizing sulfate or *Methanospirillum hungatei* as H_2 - scavengers.

Growth yields of glycerol dissimilation by SRB were determined both in pure culture on a sulfate medium and in coculture with *M. hungatei* .

MATERIALS AND METHODS

Source of microorganisms : *D. carbinolicus* (DSM 3852) and *D. fructosovorans* (DSM 3604) were purchased from the D.S.M. in Braunschweig (F.R.G.). *Desulfovibrio* strain M was recently isolated in our laboratory ; *M. hungatei* was obtained from the collection at our laboratory.

Media and growth conditions, cell material determinations and chemical determinations were performed as described by Qatibi *et al.*, (1989).

RESULTS AND DISCUSSION

The six figures in next page show glycerol dissimilation :
- by *Desulfovibrio* sp. strain M
  * in pure culture, with sulfate (a);
  * or in coculture with *M. hungatei* without sulfate (b);
- by *Desulfovibrio carbinolicus*
  * in the presence of sulfate (c);
  * or utilizing *M . hungatei* as H_2 - scavenger (d); .

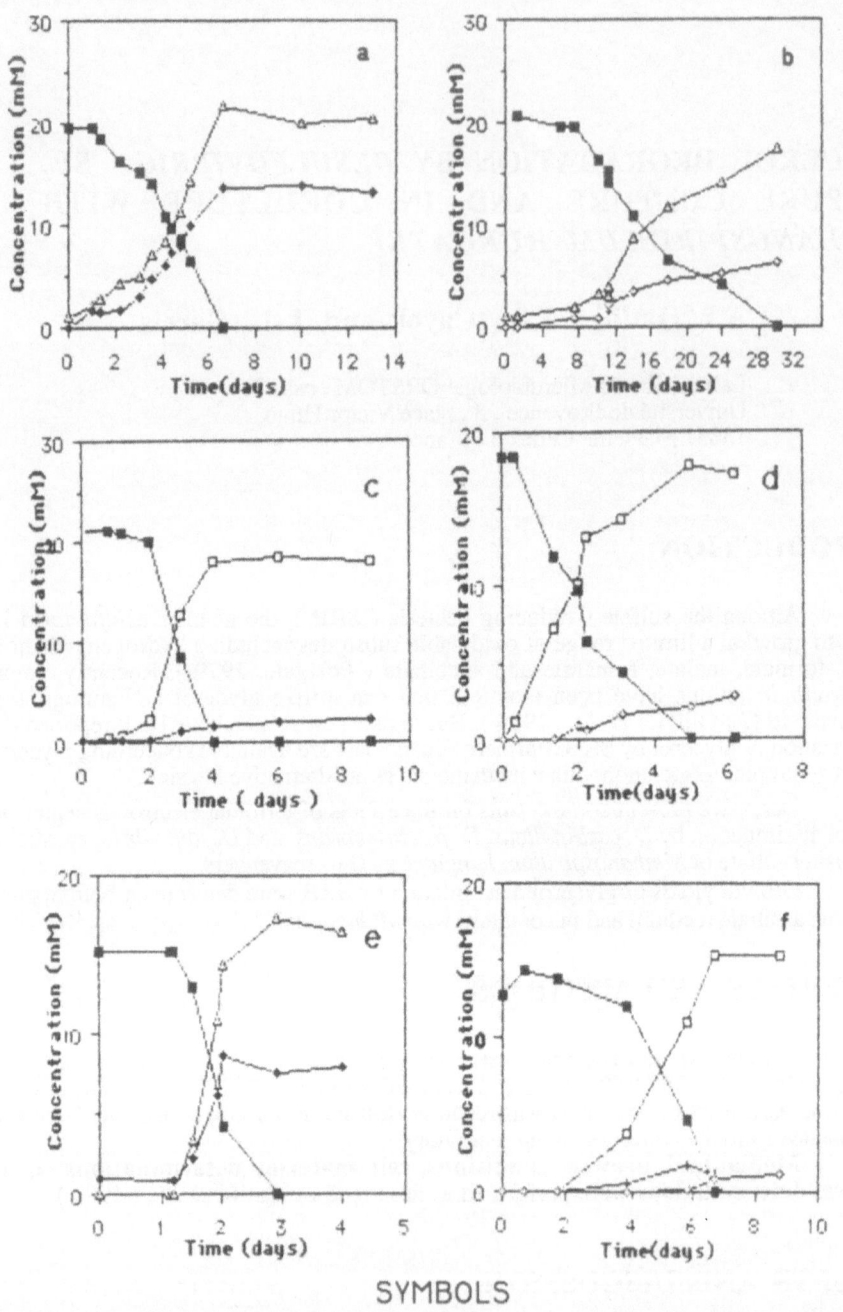

SYMBOLS

glycerol : ■ ; 3-hydroxypropionate : □ ; 1-3 propanediol : + ;
acetate : △ ; sulfide : ◆ ; methane : ◇

Table 1. Maximum specific growth and molar growth yields on glycerol, of *D. fructosovorans*, *D. carbinolicus* and *Desulfovibrio* sp. strain M

Strains	μ (h-1)	Y (g / mol)
+ sulfate		
D. fructosovorans	0.057	8.88
D. carbinolicus	0.063	8.95
Desulfovibrio sp. *strain M*	0.061	15.80
+ *Methanospirillum hungatei*		
D. fructosovorans	0.008	6.00
D. carbinolicus	0.014	6.00
Desulfovibrio sp. *strain M*	0.002	n.d.

- by *Desulfovibrio fructosovorans* in pure culture
* when sulfate was present (e);
* or in syntrophic association with *M. hungatei* (f).

In the presence of sulfate or *M. hungatei, Desulfovibrio carbinolicus* and strain M produced 3 - hydroxypropionate and acetate, and sulfide or methane respectively, from glycerol. When grown on glycerol in the absence of sulfate, *D.fructosovorans* was also able to use *M. hungatei* as an alternative hydrogen sink, but unlike the other two strains, *D.fructosovorans* oxidized glycerol into 3 - hydroxypropionate and traces of 1,3 - propanediol (the 1,3 - propanediol production is not constant) and methane, whereas in the presence of sulfate, glycerol was degraded into acetate and sulfide. This incomplete interspecies hydrogen transfer suggests a limitation of the process by the inability of *D. fructosovorans* to efficiently reduce the proton into hydrogen, rather than on the ability of *M. hungatei* to remove this hydrogen.

During additional experiments (not reported in this paper) the glycerol dissimilation by *D. carbinolicus* and *D.fructosovorans* was found to be quite complex : after several transfers on glycerol in the presence of sulfate, glycerol was oxidized into an acetate and 3 - hydroxypropionate mixture, and the ratio of the end - products seems to have depended on the medium redox - potential rather than on the available terminal electron acceptor.

Growth yields of *D. fructosovorans* and *D. carbinolicus* on glycerol in presence of sulfate or *M. hungatei* were identical (table 1), and resulted in a similar ATP gain. A higher growth yield was obtained with *Desulfovibrio* strain M on glycerol in the presence of sulfate.

These results suggest the existence, in *D. fructosovorans* and *D. carbinolicus*, of pathways and enzymes necessary for the synthesis of both acetate and 3 - hydroxypropionate.

REFERENCES

Nanninga , H.J. and Gottschall , J.C., 1986. Isolation of sulfate-reducing bacterium growing with methanol. FEMS Microbiol. Ecol. 38 : 125-130.

Ollivier , B., Cord-Ruwisch , R., Hatchikian , C.E. and Garcia J.L., 1988. Characterization of *Desulfovibrio fructosovorans* , sp. nov. Arch. Microbiol. 149 : 447-450.

Postgate , J.R., 1979.The sulfate reducing bacteria. Cambridge Univ. Press, London.

Qatibi , A.I. and Garcia , J.L. , 1989. 1,2 - and 1,3 - propanediol degradation by *Desulfovibrio alcoholovorans* , sp. nov., in pure culture or through H2 interspecies transfer.This Symposium.

1, 2 - AND 1, 3 - PROPANEDIOL DEGRADATION BY *DESULFOVIBRIO ALCOHOLOVORANS* SP. NOV. , IN PURE CULTURE OR THROUGH H$_2$ INTERSPECIES TRANSFER

A.I. Qatibi and J.-L. Garcia

Laboratoire de Microbiologie ORSTOM , case 87
Université de Provence , 3, place Victor Hugo
13331 Marseille Cédex 3 , France

INTRODUCTION

Few data have been published on utilization of reduced products such as 1,2 - and 1,3 - propanediol by sulfate-reducing bacteria; *Desulfovibrio carbinolicus* (Nanninga and Gotschall , 1987), and *Desulfovibrio fructosovorans* can oxidize 1,3 - propanediol into 3 - hydroxypropionate, but they do not convert 1,2 - propanediol. The purpose of the present study was to describe the anaerobic degradation of 1,2 - and 1,3 - propanediol by a new species of *Desulfovibrio, D. alcoholovorans* in pure culture or in syntrophic coculture with *Methanospirillum hungatei* . The growth yields and stoichiometries were determined.

MATERIALS AND METHODS

Source of microorganisms : *Desulfovibrio alcoholovorans* sp. nov., (DSM 5433) was isolated recently in our laboratory. *Methanospirillum hungatei* was purchased from the collection at our laboratory.

Media and growth conditions : the anaerobic Hungate technique (Hungate 1950), modified for the use of syringes (Macy *et al* ., 1972) was used. The growth medium was as described by Nanninga and Gottschall (1987), but it was supplemented with yeast extract at 0.01% and dithionite was omitted. Substrates were added from freshly anaerobically prepared and autoclave - sterilized solutions. *D. alcoholovorans* was tested in completely filled and sealed 100 ml - serum bottles with stoppers, in a sulfate medium. In the experiments on the coculture, sulfate was omitted ; cells were grown in 500 ml - serum bottles sealed with black rubber stoppers containing 200 ml of medium under an anaerobic atmosphere (80% N$_2$ and 20% CO$_2$). Vessels were incubated at 37° C. Inoculates used for coculture experiments were carried out from a coculture previously adapted to each substrate.

Cell material determinations : Dry weights were determined using 2 liter screw - cap bottle cultures containing one of the substrates (glycerol, 1,2 - and 1,3 - propanediol), with and without gas phase for coculture and monoculture experiments, respectively. Cells were centrifuged and washed twice with 50 mmol / l of phosphate buffer at pH 7.0. Pellets were dried to constant weight at 80°C.

Chemical determinations were carried out as described elsewhere (Ollivier *et al.*, 1988).

RESULTS AND DISCUSSION

In sulfate media, 1,2 - propanediol was degraded into acetate and propionate, sulfide and presumably CO_2 (Fig. 1). Besides acetate and propionate, no organic acid or alcohol were detected. Without sulfate, M. hungatei served as an alternative acceptor of reducing equivalents liberated by D. alcoholovorans from 1,2 - propanediol degradation (Fig. 2). But only propionate was produced as acid, with methane and presumably CO_2. At the beginning of incubation in the sulfate medium, 1,3 - propanediol led only to acetate, sulfide and presumably CO_2 (Fig. 3). As soon as about 4 mmol of 1,3 - propanediol per liter were degraded, 3 - hydroxypropionate appeared as a new end - product, and the acetate / 3- hydroxypropionate ratio was about 3.4. With M. hungatei, 1,3 - propanediol was first degraded to acetate, 3 - hydroxypropionate (acetate / 3-hydroxypropionate ratio approximately 1), methane and presumably CO_2 (Fig. 4). The accumulated 3 - hydroxypropionate was then degraded into acetate and methane.

* The stoichiometries of 1,3 - propanediol degradations by *D. alcoholovorans* in pure culture with sulfate or with *M. hungatei* were studied :

$$CH_2OHCH_2CH_2OH + SO_4^{2-} \longrightarrow CH_3COO^- + HCO_3^- + HS^- + H^+ + H_2O$$

$$CH_2OHCH_2CH_2OH \longrightarrow CH_3COO^- + H^+ + CH_4$$

1,3 - propanediol is oxidized stoichiometrically to 1 mol acetate and presumably 1 mol CO_2 during 1 mol sulfate reduction.

* 1,2 - propanediol was converted into variable amounts of propionate and acetate mixture, when sulfate was present, and the acetate / propionate ratio was not constant .

* With *M. hungatei,* the following stoichiometry can be proposed :

$$CH_2OHCHOHCH_3 + 0.25HCO_3^- \longrightarrow CH_3CH_2COO^- + 0.25CH_4 + 0.75 H^+ + 0.75 H_2O$$

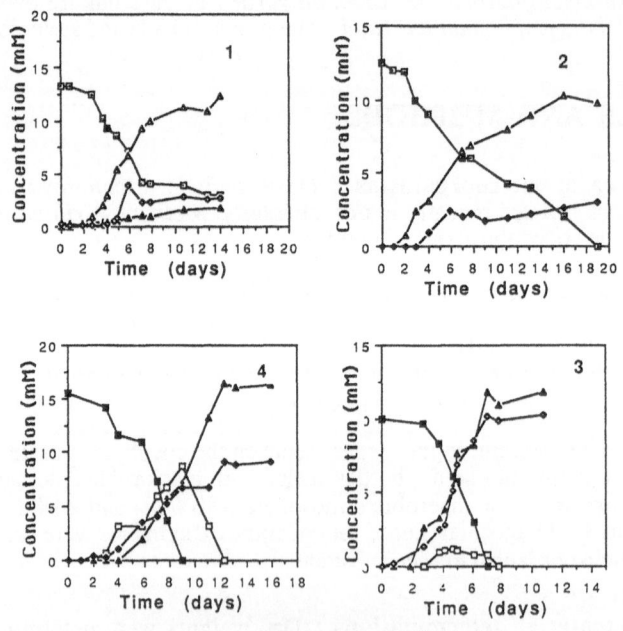

SYMBOLS

1-2 propanediol : ▣ ; 1-3 propanediol : ■ ;
propionate : △ ; 3-hydroxypropionate : ▲ ; acetate : △
SH_2 : ◇ ; CH_4 : ◇ .

Table 1. Maximum molar growth yields of *D. alcoholovorans* on glycerol, 1,2- and 1,3- propanediol

Experimental conditions		Y (g/mol)
1,3 - propanediol	+ sulfate	14.5
glycerol	+ sulfate	13.6
1,2 - propanediol	+ sulfate	10.4
1,3 - propanediol	+ *M. hungatei*	12.8
glycerol	+ *M. hungatei*	10.2
1,2 - propanediol	+ *M. hungatei*	3.0

Degradation of these reduced compounds by *D. alcoholovorans* requires a terminal electron acceptor such as sulfate or a syntrophic association with a H_2-scavenger organism. The highest growth yield (Table 1) was obtained on 1,3 - propanediol; the growth yield on 1,2 - propanediol was lower; an average growth yield was measured on glycerol. Anaerobic digestion of glycerol - containing media at low sulfate levels, often involves the accumulation of 1,3 - propanediol and propionate and a restriction of the methanization of the waste; but it was demonstrated that in the presence of high sulfate levels, sulfate - reducers may modify the fermentation patterns of glycerol and of 1,3 - propanediol (Qatibi and Bories , 1988).

The direct use of 1,2 - and 1,3 - propanediol by *D. alcoholovorans* is possible. In the absence of sulfate, *M. hungatei* served as an alternative acceptor of reducing equivalents liberated from substrate oxidation by *D. alcoholovorans* . In this case , 1,3 - propanediol was oxidized into acetate and methane; whereas, 1,2 - propanediol was converted only into propionate and methane. This incomplete interspecies H_2 - transfer indicates that the process is limited by the H_2 - consumer rather than by the H_2 - producer. 3 - hydroxypropionate was observed as an intermediate during 1,3 - propanediol degradation, and its concentration was influenced by the available terminal electron acceptor. Nevertheless, these 2 substrates seem to have been converted into H_2 and acids (acetate and propionate), and then H_2 oxidized sulfate seems to have acted as electron acceptor. This assumption fits the theory of hydrogen cycling involved in *Desulfovibrio* species, as described by Odom and Peck (1981) and confirmed by our experiments in coculture with *M. hungatei* .

REFERENCES

Hungate , R.E., 1950. The anaerobic mesophilic cellulolytic bacteria. Bacteriol. Rev. , 14 : 1 - 49.

Macy , I.M., Snellen , J.E and Hungate , R.E. , 1972. Use of syringe methods for anaerobiosis. Am . J. Clin . Nutr. , 25 : 1318 - 1323.

Nanninga , H.J. and Gottschall , J.C. , 1987. Properties of *Desulfovibrio carbinolicus* , sp. nov. and other sulfate - reducing bacteria isolated from anaerobic purification plants. Appl. Environ . Microbiol ., 53 : 802-809.

Odom , J.M and Peck , H.D. Jr ., 1984. Hydrogen cycling as a general a mechanism for energy coupling in the sulfate - reducing bacteria, *Desulfovibrio* sp. FEMS Microbiol. Lett. 12 : 47 - 50.

Ollivier, B., Cord-Ruwisch , R., Hatchikian , C.E. and Garcia , J.-L., 1988. Characterization of *Desulfovibrio fructosovorans* sp. nov. Arch . Microbiol ., 149 : 447 - 450.

Qatibi , A.I., Bories , A., 1988. Glycerol fermentation and sulfate utilization during the anaerobic digestion process. In : " Fifth Int. Symp. Anaerobic Digestion ", pp 69 - 73, A. Tilche & A. Rozzi (ed.), Monduzzi Editor (Pub.), Bologna, Italia.

LONG CHAIN FATTY ACID DEGRADATION BY A MESOPHILIC SYNTROPHIC COCULTURE ISOLATED FROM MARGIN

M. Ragot , B. Ollivier and J.-L. Garcia

Laboratoire de Microbiologie ORSTOM , case 87
Université de Provence , 3, Place Victor Hugo
13 331 Marseille Cédex 3 , France

INTRODUCTION

The fat industry, particularly olive oil processing, leads to by - products such as margins which are not easily degradable. The methanic fermentation of these wastewaters is worth investigating, since they contain 3 to 12 % of organic matter. Anaerobic degradation of long chain fatty acids (LCFA) by ß-oxidation (Weng *et al.,* 1976) occurs through syntrophic association between proton reducing and hydrogen utilizing bacteria (methanogenic or sulfate-reducing bacteria).

In the present study, enrichments on LCFA (C16) were performed with samples from by-products of Cravenco Olive Mill near Arles (France). Colonies of fatty acids degrader could only be obtained in coculture with a *Desulfovibrio* strain, after a two - month incubation.

Here we report on the taxonomical position of the isolate among the syntrophic bacteria.

MATERIALS AND METHODS

The anaerobic technique by Hungate (1969), as modified by Macy *et al.* (1972) and Miller *et al.* (1974), was used to prepare media and cultures of organisms.

Basal medium contains KH_2PO_4 (0.2 g/l), NH_4Cl (0.3 g/l), KCl (0.5 g/l), NaCl (1 g/l), $MgCl_2,6H_2O$ (0.4 g/l), $CaCl_2, 2H_2O$ (0.15 g/l), Na_2SO_4 (3 g/l), $NaHCO_3$ (2.5 g/l), trace element Widdel solution (1.5 ml/l). pH of the medium was adjusted to 7.2 with NaOH 10 N. Stock solutions of Na_2S (15%), $CaCl_2$ (200 mM) and fatty acids were autoclaved separately and added to growth medium. The final $CaCl_2$ concentration was twice the fatty acid concentration, especially with long chain fatty acids (>10) (Roy *et al.* , 1985). Dithionite and vitamins were filter sterilized.

VFA measurements were performed by gas chromatography, using an ionisation detector and a Porapak Q column .

Table 1. Substrates used as energy source by the syntroph cocultured with *Desulfovibrio* strain

Substrate added	Final H$_2$S concentration (mM)	Final Acetate concentration (mM)	Final Propionate concentration (mM)
Control	1.57	0.71	0
Acetate (C2, 20 mM)	1.57	nd	nd
Propionate (C3, 20 mM)	1.18	nd	nd
Butyrate (C4, 10 mM)	4.90	26.54	0
Valerate (C5, 10 mM)	1.77	5.00	3.04
Caproate (C6, 5 mM)	4.12	15.19	0
Heptanoate (C7, 5 mM)	5.29	11.73	3.91
Caprylate (C8, 5 mM)	5.68	24.04	0
Nonanoate (C9, 5 mM)	3.72	16.92	3.04
Decanoate (C10, 5 mM)	5.09	25.77	0
Laurate (C12, 2 mM)	3.92	15.05	0
Myristate (C14, 2 mM)	4.70	19.70	0
Palmitate (C16, 2 mM)	3.52	19.29	0
Stearate (C18, 2 mM)	2.35	7.11	0
Oleate (C18-1, 2 mM)	4.31	20.77	0
Linoleate (C18-2, 2 mM)	4.31	21.34	0
Arachidate (C20, 2 mM)	1.47	nd	nd
Yeast Extract (5g/l)	1.96	nd	nd
Biotrypcase (5g/l)	2.54	nd	nd
Casaminoacids (5g/l)	1.37	nd	nd
Fructose (5g/l)	1.17	nd	nd

nd = not determined
Results were obtained after one month of growth at 35°C .

Table 2. Differences with *Syntrophomonas sapovorans*

Strain	aminoacid requirements for growth	CaCl$_2$ requirements for fatty acids degradation	
		C < 8	C > 8
Syntrophomonas sapovorans	+	-	+
Strain CrZ	-	+	+

RESULTS

The substrates used as energy source by the syntroph cocultured with *Desulfovibrio* strain are defined in Table 1 .

Differences observed between CrZ strain and *Syntrophomonas sapovorans* , concerning amino acids and/or $CaCl_2$ requirements for growth and fatty acids degradation are expressed in Table 2.

CONCLUSION

Strain CrZ is a syntrophic bacterium which incompletely oxidizes long chain linear fatty acids up to 18 carbon atoms. As described in the case of *Syntrophomonas sapovorans* (Roy *et al.* , 1986), strain CrZ ß-oxidizes odd and even numbered fatty acids to acetate + H_2 and to acetate, propionate and H_2 respectively.

Cells are straight to curved motile rods that grow only with hydrogenotrophic bacteria (*Desulfovibrio sp.* , *Methanospirillum hungatei*). Other enrichments on LCFA under mesophilic conditions from other olive mill wastewaters led to the selection of the same type of microorganism. Strain CrZ does not use sugars or aromatic compounds. Its growth time on C16 with *M. hungatei* is 24 h. Strain CrZ grows in a defined mineral medium in the presence of vitamins.

Contrary to *S. sapovorans*, strain CrZ does not require casamino-acids for growth and the presence of $CaCl_2$ is necessary to degrade fatty acids with carbon chains shorter than 8. These minor differences do not suffice to constitute a new species. However, our results indicate that *S. sapovorans* is probably one of the dominant bacteria involved in long chain fatty acid oxidation in various ecosystems.

REFERENCES

HUNGATE , R.E. , 1969. A roll tube method for cultivation of strict anaerobes. In : " Methods in Microbiology ", vol3B : 117 - 132. J.R. Noris & D.W. Ribbons (eds), Academic Press, London.

MACY , J.M. , SNELLEN , J.E. and HUNGATE , R.E. , 1972. Use of syringe methods for anaerobiosis.Am. J. Clin. Nutr. 25 : 1318 - 1323.

MILLER , T.L and WOLIN, M.J., 1974. A serum bottle modification of the Hungate technique for cultivating obligate anaerobes. Appl. Microbiol., 27 : 985 - 987.

ROY , F., ALBAGNAC , G. and SAMAIN , E. , 1985. Influence of calcium addition on growth of highly purified syntrophic cultures degrading long chain fatty acids. Appl. Environ. Microbiol. 49 : 702 - 705.

ROY , F., SAMAIN , E., DUBOURGUIER , H.C. and ALBAGNAC , G., 1986. *Syntrophomonas sapovorans* sp. nov., a new obligatly proton-reducing anaerobe oxidizing saturated and unsaturated long chain fatty acids. Arch. Microbiol. 145 : 142-147.

WENG , C.N. and JERIS , J.S. , 1976. Biochemical mechanisms in the methane fermentation of glutamic and oleic acids . Water Res. 10 : 9 - 18.

EFFECT OF SULFIDE AND REACTOR OPERATIONAL PARAMETERS ON SULFATE REDUCING BACTERIA

Reis, M.A.M., P.C.Lemos, J. Almeida, M.T.J.Carrondo

Laboratorio de Engenharia Bioquímica, Faculdade de Ciências e Tecnologia da Universidade Nova de Lisboa 2825 Monte da Caparica, PORTUGAL

INTRODUCTION

In effluents containing large amounts of sulfate, high sulfide levels are produced in anaerobic digestion by sulfate reducing bacteria (SRB). The undesirable effect of sulfides on methanogenic reactors is widely known[1]. This effect can be minimized if the process is performed in a two phase anaerobic reactor, where produced sulfides are precipitated after the first phase (acidogenic)[2]. Therefore, this phase must be operated in order to obtain maximum sulfate removal and to produce the more favourable intermediates for methanogenic bacteria.

The understanding of the kinetic and environmental factors that influence the SRB activity on the acidogenic phase was the main objective of this study.

EXPERIMENTAL

Continuous experiments were performed in a continuous stirred tank reactor (CSTR) with automatic pH control. Batch experiments were carried out in a glass vessel reactor with continuous magnetic stirring and automatic pH control. Under H_2S stripping conditions, a gas disperser was used to sparge high purity N_2.

Distillery molasses slops were used as substrate in the continuous studies. In the batch experiments Postgate's medium C^3 was used, suplemented with trace elements and sodium lactate as organic source.

The inoculum used for continuous tests was obtained from an anaerobic digestor and for batch tests a selected culture of SRB was used.

RESULTS AND CONCLUSIONS

The CSTR was operated at different dilution rates (D) in the range of 0.035 to 0.167 h^{-1}. Volatile fatty acids (VFA) and sulfate removal were analised at steady state. Results showed that acetic acid production and sulfate removal were more influenced by changing the dilution rate than butyric and propionic acids production. Sulfate removal is associated to acetic acid production.

The pH effect, on the range of 5.4 to 6.6, on sulfate removal and VFA production was also studied at D=0.035 h^{-1}.

Maximum sulfate removal and acetic acid production were achieved at pH 6.6. At pH 5.4 butyric acid predominates and sulfate reduction was very low. The acetic acid produced in this series of experiments was also associated to sulfate removal.

To assess the contribution of SRB activity on acetic acid production, a CSTR was operated at pH 6.2 and D=0.042 h^{-1} and fed with different sulfate concentrations. Results showed that acetic acid produced was a linear function of sulphate removed, and butyric acid was mainly produced at low SRB activity. Sulfides produced are also inhibitory to SRB. Batch tests carried out with pH control showed that uptake rate decreases as sulfide production increases, and after H_2S stripping the inhibition was reversed, and sulfate uptake rate increased.

Using a general inhibition model:

$$\mu = \mu_{max} \left(1 - \frac{H_2S}{H_2S_{max}}\right)^n$$

it is possible to calculate the H_2S concentration (H_2S_{max}) which is completely inhibitory to SRB ($\mu=0$). An H_2S_{max} concentration of 543 and 562 mg/l at pH 6.2 and pH 6.6 respectively were found. A deviation of only 3% between these two values shows that it is the unionized form of sulfide that has the inhibitory effect on SRB.

From this study, it can be conclude that reactor operational parameters such as pH and dilution rate influence sulfate removal and VFA production. High pH values and low dilution rate are more favourable for this process as higher sulfate removal and higher acetic acid concentration are achieved.

Moreover, acetic acid production is always strongly associated to the sulfate removal, meaning that this acid is mainly produced by SRB. Being acetic acid the main precursor for methanogenic bacteria, there are obvious advantages of operating the acidogenic phase in presence of high SRB activities, providing that produced sulfides are removed before the methanogenic reactor.

Sulfides produced, depending on concentration and pH, are inhibitory to SRB. The sulfide inhibitory effect on SRB can be overcome by using an H_2S stripping device.

ACKNOWLEDGEMENTS

    The authors acknowledge the financial support of Direção
Geral de Qualidade do Ambiente and NATO - Science for
Stability Program. The support of Fundação Calouste
Gulbenkian, Soc. Lusitana de Destilação and Vieira & Irmãos is
also gratefully acknowledged.

REFERENCES

1. I.W.Koster, A.L.Rinzema De Vegt, G. Lettinga, Sulfide
    inhibition of the methanogenic activity of granular
    sludge at various pH levels, Water Res.20:1561 (1986).

2. M.A.M.Reis,L.M.D., Gonçalves, M.J.T., Carrondo, Sulfate
    removal in the acidogenic phase anaerobic digestion,
    Environ. Technol. Letters 9:775 (1988).

3. J.R.Postgate, The sulfate reducing bacteria, 2nd ed.,
    Cambridge university press, Cambridge (1984).

HYDROGEN TRANSFER BETWEEN NEOCALLIMASTIX FRONTALIS AND

SELENOMONAS RUMINANTIUM GROWN IN MIXED CULTURE

Anthony J. Richardson and Colin S. Stewart

Rowett Research Institute, Bucksburn
Aberdeen, AB2 9SB, United Kingdom

The anaerobic fungus *Neocallimastix frontalis* ferments cellulose to H_2, CO_2, formate, acetate, lactate and ethanol (Bauchop & Mountfort, 1981). Succinate is also a fermentation product of some anaerobic fungi (Prins & Marvin-Sikkema, personal communication, Richardson *et al.*, 1989). Theoretically, all of these products could be utilised by other microorganisms. However not all of these potential interactions are likely to occur in the gut. For example, acetate supports the growth of some methanogens in anaerobic ecosystems with slow turnover times, but in the gut acetate is not an important substrate for methanogenesis (Reviewed by Stewart *et al.*, this symposium). Some other potential interactions based on the utilisation of fungal fermentation products are summarised schematically in Fig 1, which is compiled from the known properties of the named microorganisms (Hungate, 1966; Stewart & Bryant, 1988).

Of the bacteria shown in Fig. 1, *Selenomonas ruminantium* is of particular interest because this organism is capable of utilising several of the fermentation products of fungi. *S. ruminantium* does not degrade cellulose, but ferments sugars to acetate, propionate, succinate H_2 and CO_2. Extracellular H_2 can be used by *S. ruminantium* for the formation of propionate (Henderson, 1980). Propionate is produced via the succinate pathway in *S. ruminantium*, and this bacterium can utilise succinate produced by other microorganisms in co-culture (Scheifinger & Wolin, 1973). In addition, some strains of *S. ruminantium* ferment lactate. Experiments have therefore been carried out to characterise the fermentation products of co-cultures of *Neocallimastix frontalis* and strains of *S. ruminantium*. The cultures were grown on cellulose (filter paper), contained in a nutrient medium (containing rumen fluid) similar to that of Bauchop & Mountfort (1981).

In mixed cultures containing *N. frontalis* strain RE 1 and the lactate-utilising *S. ruminantium* strain JW 13, cellulose was fermented mainly to acetate, formate, propionate and CO_2. Neither H_2 nor lactate accumulated. In comparable mixed cultures of strain RE 1 with *S. ruminantium* strain JW 2, which is unable to utilise lactate, the major products found were also acetate, formate, propionate and CO_2. Only traces of H_2 were detected and, despite the fact that strain JW 2 does not utilise lactate, the amount of lactate found in the mixed culture (up to 2 μM/ml) was only around 15 to 20% of that detected in axenic cultures of strain RE 1. Succinate was not detected in any of the mixed cultures.

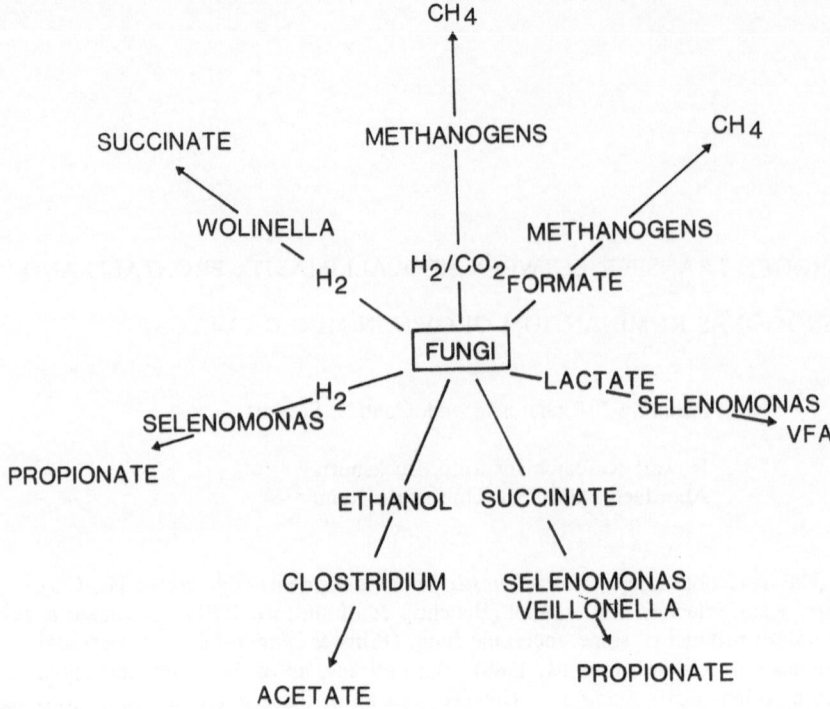

Fig. 1. Potential fate of some fermentation products of anaerobic rumen fungi.

When the time-course of degradation of filter paper was measured in the co-cultures containing strains RE 1 and JW 13, it was clear that the presence of strain JW 13 delayed cellulose digestion. Pure cultures of strain RE 1 degraded 90% of the filter paper in 6 to 7 days. In the mixed cultures with strain JW 13, similar degradation required 12 days. The effect of strain JW 2 was not measured, but the appearance of the co-cultures during incubation suggested that this strain had a similar effect.

It seems that in mixed cultures of *N. frontalis* and *S. ruminantium*, interspecies transfer of H_2 is likely to greatly exceed lactate cross-feeding, as lactate production by *N. frontalis* was largely suppressed in the presence of a hydrogen sink in the form of *S. ruminantium*. Succinate produced by the fungi is likely to be metabolised to propionate by *S. ruminantium*. There is evidence of a decrease in the rate of cellulolysis by *N. frontalis* in the presence of *S. ruminantium*, possibly as a result of competition for sugars released from the cellulose.

We thank Professor Rudolf Prins and Femke Marvin-Sikkema (Groningen) for helpful discussion of the experiments.

REFERENCES

Bauchop, T., and Mountfort, D.O., 1981, Cellulose fermentation by a rumen anaerobic fungus in both the absence and presence of rumen methanogens, Appl. Environ. Microbiol., 42: 1103-1110.

Henderson, C., 1980, The influence of extracellular hydrogen on the metabolism of *Bacteroides ruminicola, Anaerovibrio lipolytica* and *Selenomonas ruminantium*, J. Gen. Microbiol. 119: 485-491.

Hungate, R. E., 1966, "The Rumen and its Microbes,"Academic Press, N.Y.

Richardson, A. J., Calder, A. G., Stewart, C. S., and Smith, A., 1989, Simultaneous determination of volatile and non-volatile fermentation products of anaerobes by capillary gas-chromatography, Lett. Appl. Microbiol., 9: 5-8.

Scheifinger, C. C., and Wolin, M. J., 1973, Propionate formation from cellulose and soluble sugars by combined cultures of *Bacteroides succinogenes* and *Selenomonas ruminantium*. Appl. Microbiol., 26: 739-795.

Stewart, C. S., and Bryant, M. P., 1988, The rumen bacteria, in: "The Rumen Microbial Ecosystem," P. N. Hobson, ed., Elsevier Applied Science, London.

PHYLOGENETIC ANALYSIS OF METHANOGENIC BACTERIA

P. E. Rouvière, L. C. Mandelco and C. R. Woese

Microbiology Department, University of Illinois

407 S. Goodwin, Urbana, IL.61801   USA

The most recent extensive phylogenetic tree of the methanogenic bacteria presented the relationship of twenty two species and was derived from the comparison of 16S ribosomal RNA oligonucleotide catalogs (Whitman, W. B., 1985, Methanogenic bacteria, in The Bacteria, vol. 8, Archaebacteria, C. R. Woese and R. S. Wolfe ed., Academic Press, Inc., New York). This work has been expanded here. The complete sequence of the 16S ribosomal RNA was obtained for twenty nine species and a new tree was derived. The overall structure of the tree is con- served. However since the comparison of complete sequences dis- tinguishes between fast and slow evolving organisms, additional information can be obtained. The hierarchy of branching shows that the Methanococcaceae family has the deepest branching, followed by the Methanobacteriaceae, the Methanomicrobiaceae and the Methanosarcinaceae. It appears that the groups branching more recently (Methanomicrobiaceae and Methanosarcinaceae) have evolved relatively faster than the Methanococcaceae and Methanobacteriaceae This observation matches the increase in structural and metabolic diversity observed from the Methanococ- caceae to the Methanosarcinaceae. Also, extreme thermophili- city, a characteristic assumed to be ancestral and shared with all the sulfur-dependent Archaebacteria, is only found among the Methanococcaceae and the Methanobacteriaceae.

We have looked more closely to the order of the Methano- microbiales. The two families it contains, the Methanomicro- biaceae and the Methanosarcinaceae are separated as deeply as the Methanococcaceae and the Methanobacteriaceae. On this basis, we suggest that they be elevated at the level of the order, becoming respectively the Methanomicrobiales per se and the Methanosarcinales orders. This analysis also showed that the genus Methanogenium is heterogeneous since it comprises members of the Methanoplanus and the Methanomicrobium genera. We propose that the name Methanogenium be reserved for the species clustering with Methanogenium marisnigri and Methanogenium thermophilum. We also propose that the species related to Methanogenium cariaci, Methanoplanus limicola and Methanomicrobium mobile be placed in the single genus Methano- microbium.

AN ACETATE - DECOMPOSING SULPHIDOGENIC SYNTROPHIC ASSOCIATION

E. Rozanova , A. Galushko and T. Nazina

Institute of Microbiology
U.S.S.R. Academy of Sciences
Moscow, U.S.S.R.

An anaerobic sulphidogenic microbial association 1105 B was isolated from the water - flooded stratum of the Apsheron peninsular oil field (1). The bacteria reduced sulphate by decomposing acetate in the Widdel - Pfennig medium (2), containing 10 mM /l CH_3COONa and 20 mM /l Na_2SO_4. The batch culture produced up to 10 mM /l H_2S for three weeks. The bacterial cells developed only in FeS sediment. Inoculation of the association in the medium without $SO4^{2-}$ resulted in accumulation of small amounts of H_2 (0.027 - 0.041 µM H_2 / ml gas phase).

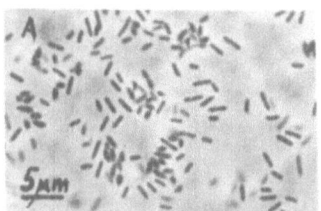
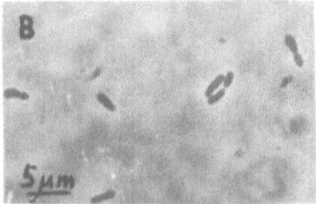
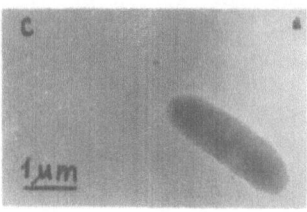

Fig. 1. Morphology of the bacterial cells of the sulphidogenic association decomposing acetate (A & B = phase - contrast micrographs ; C = transmission electron micrographs) ;

A = cells of a single colony of the association grown on a medium with acetate and SO_4^{2-} ;

B = cells of the acetogenic bacterium grown on H_2+CO_2 ;

C = cells of *Desulfotomicrobium apsheronum* gen. nov., sp. nov., strain 1105 .

The association contained no methanogenic bacteria. It included non -sporogenic rod - shaped cells of two morphological types (Fig. 1 A). The first type was represented by rods of the regular shape with the rounded ends, while the second, by shorter and thicker, sometimes swallen, rods.

The association included sulphate - reducing bacteria that were isolated as pure culture (3) and classified as *Desulfomicrobium apsheronum* gen. nov., sp. nov., strain 1105 (Fig. 1C). The microorganisms are motile non - sporogenic rods of the first morphological type. When studying the ultrastructure of the cells, we found multiple inner membranes forming lamellar packs. The bacteria utilized lactate, pyruvate, malate and ethanol by oxidizing them to acetate and CO_2. The culture grew autotrophically owing to formiate oxidation and lithoheterotrophically owing to hydrogen oxidation in sulphate - containing media. The culture did not utilize fatty acids, including acetate. The bacterial cells were found to contain cytochromes b and c. Sulphite reductase is represented by desulphorubidin. The DNA base ratio (G + C) is 52 mol.%.

Besides the sulphate - reducing microorganisms, the association contained an autotrophic acetogenic bacterium whose cells should be attributed ,to the second morphological type (Fig. 1B). The acetogenic bacterium grew on the mineral Widdel - Pfennig medium containing vitamins but without sulphate under a $H_2 + CO_2$ atmosphere (75 : 25 v %). No growth of the culture was observed in the medium containing lactate.

The evidence obtained indicates that the association made decompose acetate by the following reactions (4) :

1. $CH_3COO^- + 4 H_2O \longrightarrow 2 HCO_3^- + H^+ + 4 H_2$

$\Delta G'_0 = + 80.2 \text{ kJ}$;

2. $SO_4{}^{2-} + H^+ + 4H_2 \longrightarrow HS^- + 4 H_2O$

$\Delta G'_0 = - 151.7 \text{ kJ}$;

The total reaction :

$CH_3COO^- + SO_4{}^{2-} \longrightarrow 2 HCO_3^- + HS^-$

$\Delta G'_0 = - 71.5 \text{ kJ}$.

Thus the association studied is the second association described in litterature; the first was the thermophilic methanogenic association characterized by Zinder and Kock (5).

REFERENCES

1. E. Rozanova & T. Nazina. Acetate degradation by a binary syntrophous association including sulfate - reducing bacteria. Microbiology, 54, 497 - 499 (in Russian). (1985).
2. F. Widdel & N. Pfennig. A new anaerobic sporing acetate oxidizing sulfate reducing bacterium *Desulfotomaculum acetoxidans* (emend). Arch . Microbiol. 112, 119 - 122. (1977).
3. E. Rozanova , T. Nazina & A. Galushko. A new genus of sulfate -reducing bacteria and the description of its new species, *Desulfomicrobium apsheronum*, gen. nov., sp. nov.. Microbiology , 57, 634 - 641 (in Russian). (1988).
4. R. Thauer , K. Jungermann & K. Decker. Energy conservation in chemotrophic anaerobic bacteria. Bacteriol. Rev., 41 , 100 - 180. (1977).
5. S. Zinder & M. Koch. Non - aceticlastic methanogenesis from acetate : acetate oxidation by a thermophilic syntrophic coculture. Arch . Microbiol., 138 , 263-272. (1984).

ANAEROBIC DIGESTION OF PROTEINS, PEPTIDES AND AMINO ACIDS

J.P. Schwitzguébel and P. Péringer

Génie biologique
Ecole Polytechnique Fédérale de Lausanne
CH-1015 Lausanne, Switzerland

Waste waters from agro- and food industries often contain mixtures of proteins and carbohydrates. For example, cheese whey is composed of lactose (40-45 g/l) and proteins (6-8 g/l). Anaerobic digestion could be an interesting process for cheese factories to eliminate the increasing amount of whey: it is a convenient way of reducing the pollution load and it generates biogas as an energetic by-product. However, if lactose is a good substrate for acidogenic bacteria, the hydrolysis of proteins appears to be the limiting factor of the process. Actually, proteins present in whey are difficult to hydrolyze because of the rigid conformation imposed by their secondary structure and content of disulfide bridges.

A two-stage bioreactor was inoculated with cowpat and sewage sludge, and run continuously with unsupplemented undiluted fresh cheese whey. The retention times were 4 days in the acidogenic stage (pH 6.5, volume = 20 l) and 16 days in the methanogenic stage (pH 7.5, 80 l). The proportion of the different proteins and peptides was determined by measuring their absorbance at 215 nm, according to Waddell, after separation by chromatography on Sephadex G-75 and G-25 columns.

Ammonium, ethanol and lactate were measured enzymatically, volatile fatty acids by gas chromatography, and amino acids by HPLC on a Bondapak C18 column, after pre-column derivatization with O-phthalaldehyde.

Bacteria present in the acidogenic stage hydrolyzed 53 % of the proteins initially present: 56 % of the lactoglobulin, 45 % of the serum albumin, but only 31 % of the lactalbumin (Table 1). The proportion of small peptides (3 to 6 amino acids) increased more than 4 times. Bacteria growing in the second stage hydrolyzed totally the remaining proteins and 62 % of the peptides.

In order to improve protein hydrolysis, experiments were carried out to isolate bacteria having high protease and peptidase activities. Cowpat filtered through muslin was inoculated in vials containing only whey proteins as unique source of energy, carbon and nitrogen (8 g/l), and incubated for 96 days at 30⁰ C on a rotary shaker under anaerobic conditions. Observations under microscope revealed the development of a few species: rods, cocci and vibrios. After 48 days of cultivation, 90 % of both lactoglobulin and serum albumin, but only 46 % of the lactalbumin were hydrolyzed. Small peptides appeared (2 to 6 amino

acids), hydrophilic as well as hydrophobic. Amino acids did not
accumulate, but were either further degraded or incorporated into
bacterial proteins. As a consequence, ammonia and acetate
concentrations raised up to 50 mmol/l (Table 2). Surprisingly, very
large amounts of ethanol were also produced. Only serine is known as a
potential source of ethanol, but it is only a minor component of whey
proteins. Since pyruvate is a key metabolite in the degradation of
several amino acids, it could be converted into a more reduced compound
than acetate when hydrogen production is hindered. Iso-butyrate and
butyrate were produced in small amounts after 48 and 72 days, but the
concentration of butyrate increased markedly after 96 days. Valine
should be the main source of iso-butyrate. Iso-valerate and caproate
were also produced during the degradation of whey proteins, both
probably resulting from the degradation of leucine.

Table I

Degradation of cheese whey proteins in a two-stage plant

Fraction	Serum albumin	Lactoglobulin	Lactalbumin	Peptides
Cheese whey	1.20	3.70	0.70	0.80
1st stage effluent	0.66	1.63	0.48	3.80
2nd stage effluent	0.00	0.00	0.00	1.40

Table II

Cultivation of proteolytic bacteria on cheese whey proteins (8g/l)

Parameter	48 days	72 days	96 days
Protein (g/l)	0.56	0.36	0.31
Peptides	0.55	0.85	0.90
Amino acids (mmol/l)	0.75	0.50	0.95
Ammonium	39.4	45.8	47.9
Acetate	39.6	54.7	29.7
Ethanol	0.0	219.0	55.0
Propionate	4.4	0.0	0.0
Lactate	0.2	0.5	0.4
I-Butyrate	3.6	3.5	7.0
N-Butyrate	5.9	5.9	23.5
I-Valerate	8.5	23.9	5.4
Caproate	5.9	6.7	1.2
pH	5.6	6.1	6.0

ISOLATION OF SYNTROPHIC BACTERIA ON METABOLIC INTERMEDIATES

Alfons J.M. Stams and Caroline M. Plugge

Department of Microbiology
Agricultural University
Hesselink van Suchtelenweg 4
7603 CT Wageningen; The Netherlands

INTRODUCTION

Both obligate and facultative proton-reducing anaerobic bacteria are affected by changes in the hydrogen partial pressure. The first group of microorganisms includes mainly acetogenic bacteria which oxidize compounds like ethanol, propionate, butyrate, benzoate or phenol to acetate, (CO_2) and hydrogen (Dolfing, 1988). These reactions are only feasible at low hydrogen partial pressures and therefore carried out in syntrophy with hydrogenotrophic bacteria. The second group consists of a wide variety of fermentative bacteria able to dispose reducing equivalents as hydrogen or in the form of reduced organic compounds. The presence hydrogenotrophs leads to a shift in fermentation products.

The effect of the hydrogen partial pressure on the metabolism of proton-reducing bacteria can be attributed to the unfavourable energetics of hydrogen formation from reduced electron carriers like $FADH_2$ or NADH, with $\Delta G'$ values of +37.4 and +18.1 $kJ.mol^{-1}$ at a pH_2 of 1 atm and of +8.9 and -10.4 $kJ.mol^{-1}$ at a pH_2 of 10^{-5} atm, respectively. Knowing the pathway of substrate degradation, it can be calculated which steps in the metabolism are energetically unfavourable. E.g. in the oxidation of propionate hydrogen is generated in the oxidation of succinate to fumarate ($\Delta G^{\circ\prime}$= +86.2 kJ), malate to oxaloacetate ($\Delta G^{\circ\prime}$= +47.7 kJ) and pyruvate to acetyl-CoA ($\Delta G^{\circ\prime}$= -11.6 kJ). Pyruvate oxidation is relatively easy and may allow of syntrophic propionate oxidizers in pure culture. Succesful attempts in this direction, however, are not yet reported.

Dehydrogenation which is the initial step in the oxidation of some organic acids (lactate, malate), alcohols and amino acids (alanine, valine, leucine) is an energetic barriere for proton-reducing bacteria able to grow on these compounds, because the $\Delta G^{\circ\prime}$ of these conversions is 40 to 60 kJ per mol. As a consequence such compounds are only degraded in syntrophically, whereas the dehydrogenation products allow growth of the proton reducer in pure culture (Bryant et al., 1977; Stams and Hansen, 1985)

This study was undertaken to investigate whether syntrophic proton-reducing bacteria can be isolated directly from natural sources by the use of metabolic intermediates. Preliminary results of the isolation of pyruvate-, oxaloacetate- and α-ketoglutarate-degrading strains are presented.

MATERIALS AND METHODS

Isolation Procedure

Granular sludge from a sugar refinery (CSM, Breda, The Netherlands) was chosen as source for isolation. Such densely packed microbial biomass with high methanogenic activity forms an ideal niche for syntrophic bacteria. Granules were crushed mechanically under anaerobic conditions and serial dilutions were made in triplicate in 28-ml tubes containing 9 ml medium with 20 mM carbon source (pyruvate, oxaloacetate or α-ketoglutarate) and 20 mM bromoethanesulfonic acid (Bres) and a gasphase of 80 % N_2 and 20 % CO_2. The basal bicarbonate-buffered medium was described by Houwen et al. (1986). After 6 weeks of incubation at 30 °C in the dark, the gasphase was analyzed for hydrogen and from the highest positive tube serial dilutions were made in roll tubes with agar-medium. After another 6 weeks colonies were picked up and transferred to 120-ml vials with 50 ml liquid medium without Bres. After incubation one of the bottles in which hydrogen production was highest was chosen for further experiments.

Syntrophic Growth

Cultures obtained after isolation with ketoacids were tested to grow on the corresponding hydroxy- and amino acids, either alone or in the presence of Methanospirillum hungatei (DSM 864), added 5 % from a dense H_2-grown culture. After 6 weeks of incubation bottles were analyzed for residual substrate and products.

Analytical Methods

Fatty acids were analyzed gaschromatographically and by an LKB HPLC system equiped with an organic acid column. The latter was also used for the determination of organic acids. Amino acids were analyzed colorimetrically with a Biotronik amino acid analyzer. Ammonium was measured colorimetrically with a Technicon autoanalyzer with the indophenol-blue method (Keeney and Nelson, 1982). Hydrogen and methane were determined gaschromatographically.

RESULTS AND DISCUSSION

This study was set up to investigate whether syntrophic bacteria can be isolated directly from natural sources by the use of metabolic intermediates which are easier to degrade than the naturally occurring substrates. Isolates growing on ketoacids were obtained by one serial dilution in liquid medium and one in agar medium and screened for hydrogen as fermentation product. The obtained cultures were highly enriched but up to now not yet pure. Table 1 summarizes substrates which were utilized by the isolates in the presence and absence Methanospirillum hungatei.
The isolate obtained with pyruvate, converted pyruvate stoichiometrically to acetate, hydrogen and presumably CO_2. The isolate did not grow on lactate or alanine; but in the presence of M. hungatei growth was observed with alanine as substrate.

Table 1. Substrates utilized by isolates obtained with pyruvate, oxaloacetate and α-ketoglutarate as substrates. Growth was tested in the presence and absence of M. hungatei.

Growth substrate	Test substrate	Growth
pyruvate	pyruvate	growth
	alanine	syntrophic growth
	lactate	no growth
oxaloacetate	oxaloacetate	growth
	aspartate	no growth
	malate	no growth
	pyruvate	growth
	lactate	syntrophic growth
	alanine	no growth
α-ketoglutarate	α-ketoglutarate	growth
	glutamate	growth
	hydroxyglutarate	no growth

The oxaloacetate-degrading isolate did not grow on malate and aspartate, neither alone nor in the presence of the methanogen. As oxaloacetate is unstable and presumably rapidly decarboxylated to pyruvate, also other substrates were tested. Growth was found with pyruvate and in the presence of the methanogen also on lactate.
The isolate obtained on α-ketoglutarate grew on α-ketoglutarate and glutamate but not on hydroxyglutarate, irrespective of the presence M. hungatei. The substrate conversion and product formation by this isolate is given in Table 2.

Table 2. Substrate conversion and product formation by an α-ketoglutarate-degrading isolate.

| | substrate converted | products formed | | |
		hydrogen	acetate	propionate
α-ketoglutarate	1085[a]	404	347	801
α-ketoglutarate + M. hungatei	1085	737	276	885
glutamate	830	159	1061	125
glutamate + M. hungatei	1000	1032	1855	175

[a] Data are expressed in μmol per vials

475

Acetate, propionate, hydrogen and presumably CO2 were detected as products. Based on the mass balance it is likely that α-ketoglutarate is converted as given in equation 1 and 2.

$$\alpha\text{-ketoglutarate} \longrightarrow 2 \text{ acetate } + 1 \text{ } CO_2 \qquad (1)$$

$$\alpha\text{-ketoglutarate} \longrightarrow 1 \text{ propionate } + 2 \text{ } CO_2 + H_2 \quad (2)$$

A similar α-ketoglutarate fermentation was also found by Acidamino-bacter hydrogenoformans (Stams and Hansen, 1984). If glutamate is degraded in a similar way by the culture as α-ketoglutarate, one hydrogen is produced per two acetate formed and two hydrogen per propionate formed. This fits very nicely with the results obtained in the culture with M. hungatei, but not with those in the culture without the methanogen. A likely explanation for this discrepancy is that the culture is able to form propionate reductively.

CONCLUSION

The results show that proton-reducing bacteria with the ability to grow only syntrophically on certain substrates can be isolated on metabolic intermediates. In this study bacteria were isolated directly from natural sources, but the use of metabolic inter-mediates may even be more powerful for syntrophic enrichment cultures.

REFERENCES

Bryant, M.P., Campbell, L.L., Reddy, C.A., and Crabill, M.R. 1977. Growth of Desulfovibrio in lactate and ethanol media low in sulfate in associations with H_2-utilizing methanogenic bacteria, Appl. Environ. microbiol., 33:1162.

Dolfing, J., 1988, Acetogenesis, in: "Biology of anaerobic microor-ganisms," A.J.B. Zehnder, ed., John Wiley & Sons, New York.

Houwen, F.P., Dijkema, C., Schoenmakers, C.H.H., Stams, A.J.M., and Zehnder, A.J.B., 1987, ^{13}C-NMR study of propionate degradation by a methanogenic coculture, FEMS Microbiol. Letters, 41:269.

Keeney, D.R., and Nelson, D.W., 1982, Nitrogen inorganic forms, in: "Methods in soil analysis," A.L. Page, ed., Agronomy, Madison.

Stams, A.J.M., and Hansen, T.A., 1984, Fermentation of glutamate and other compounds by Acidaminobacter hydrogenoformans gen.nov. sp.nov., an obligate anaerobe isolated from black mud. Studies with pure culture and mixed cultures with sulfate-reducing and methanogenic bacteria, Arch. Microbiol., 137:329.

ISOLATION OF A PROPIONATE-USING, SULFATE-REDUCING BACTERIUM

*Masaharu Tasaki, Yoichi Kamagata, Kazunori Nakamura and Eiichi Mikami

*Institute of Technology Shimizu Corporation, 4-17 Etchujima 3-chome, Koto-ku, Tokyo 135, JAPAN Fermentation Research Institute, Agency of Industrial Science and Technology, Tsukuba, Ibaraki 305, JAPAN

INTRODUCTION

Propionate is one of the most important intermediates in the anaerobic digestion process because propionate accumulation may inhibit methanogenesis, reducing the efficiency of anaerobic treatment. Subsequently, propionate degrading bacteria have received increasing attention because of their important role in maintaining digester efficiency. This report describes the isolation and characterization of a propionate-using, sulfate-reducing bacterium, named strain MUD.

MATERIAL AND METHODS

Source of Organism Strain MUD was isolated from an enrichment culture inoculated with granular sludge from a mesophilic UASB reactor that was treating volatile fatty acids.
Media and Conditions for cultivation A freshwater basal medium(Widdel and Pfennig)[1] was used for all cultivations. 10 mM propionate and 10 mM Na_2SO_4 were added to the medium as the sole carbon source and electron acceptor respectively. A 20 ml tube with screw-cap was used for cultivation and all enrichment cultures were incubated at 30°C.
Isolation The pure culture was obtained after repeated application of the agar shake dilution method[1]. To check purity, the isolate was inoculated into media with either 0.1% yeast-extract and 0.1% Bacto-peptone, glucose, sucrose, lactate, or H_2 and CO_2 as substrates. After incubation, the culture were examined microscopically. All tests were carried out at 36°C.
Determinations Growth was monitored spectrophotometrically at 600nm. Substrates and fatty acids were measured by FID-GC or by HPLC with RI-detector. DNA base composition was determined by measuring deoxyribonucleosides using HPLC with UV-detector[2].

RESULTS

Enrichment and Isolation After enrichment, oval to onion-shaped bacteria like Desulfobulbus were predominant in the culture, and the 10mM of propionate was degraded within about a week. Most of colonies appearing in agar shake cultures were lens-shaped and colored dark brown. We obtained the propionate-using, sulfate-reducing strain MUD by a series of isolating operations.
Morphology Strain MUD was oval to onion-shaped, motile, single or paired, and sometimes had short chains(Fig. 1). Spores were never

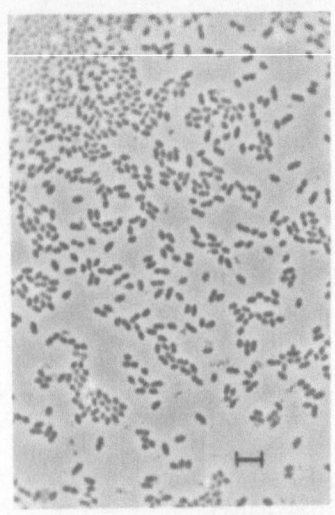

Fig. 1 Photomicrograph of strain MUD. Bar equals 5μm

evidenced by either microscopy or heat shock(10 mn at 80°C). Desulfoviridin was not detected by either the Postgate test[3] or characteristic absorption of partially purified cell extracts[4].

Growth Condition and Nutrition

Strain MUD could grow at 15°C to 44°C, and the optimum temperature was found to be 37°C. The specific growth rate(μ) was 0.05 h^{-1} under optimum conditions. Strain MUD degraded 10 mM propionate to acetate completely within about 3 days at 37°C, and it had a specific substrate consumption rate of 0.1 h^{-1}(Fig. 2). The isolate required neither organic nutrients nor growth factors like vitamins. In the presence of propionate, strain MUD was able to use sulfate or thiosulfate as electron acceptors, but growth with thiosulfate was very slow. Sulfite, sulfur, or nitrate were not utilized.

Different substrates were tested as electron donors with sulfate(Table). The isolate could utilize ethanol, propanol, butanol, pentanol, lactate and pyruvate, but could not utilize any fatty acids except for propionate. Slower growth was obtained on 1,2-propanediol or 1,3-propanediol. Lactate and pyruvate could be utilized without sulfate. In the presence of acetate as a carbon source, this strain was able to utilize H_2 or formate as an electron donor.

DISCUSSION

Four genera and six species are known as sulfate-reducing bacteria which can utilize propionate[5]. Among them, only two species are the incomplete oxidation type; one is Desulfobulbus propionicus isolated by Widdel et al.[6], and the other is Desulfobulbus elongatus isolated by Samain et al.[7]. Based on morphological characteristics, strain MUD resembled D. propionicus. The DNA base composition of strain MUD was determined by HPLC to have a content of 59 mol% guanine plus cytosine, which is the same as that of D. propionicus. However, there are some differences between strain MUD and other strains of Desulfobulbus.

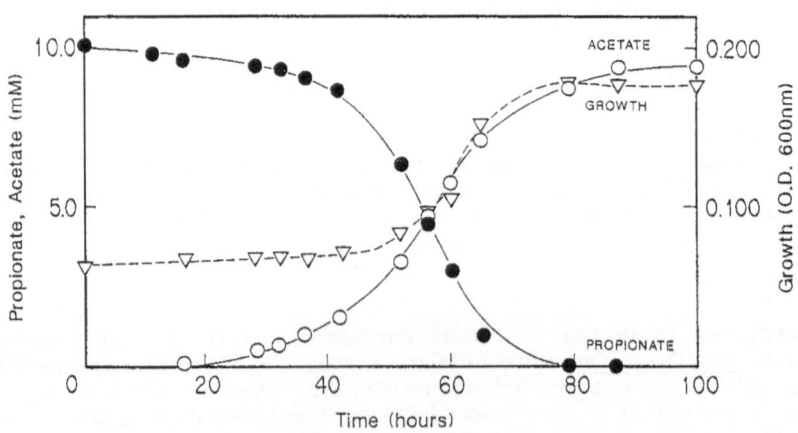

Fig. 2 Degradation of propionate by Strain MUD

Table 1 Compounds test as electron donors and sourse

E-donors and C-sources		Products	E-donors and C-sources		Products
H_2 + CO_2 + acetate	+		Ethyleneglycol	-	
Formate + acetate	+		Propanol	+	Acetate
Acetate	-		1,2-Propanediol	(+)	Acetate
Propionate	+	Acetate	1,3-Propanediol	(+)	3-HP*, Acetate
3-HP*	-		Butanol	+	Butyrate, Acetate
Acrylate	-		Pentanol	(+)	Acetate
Butyrate	-		Hexanol	-	
Crotonate	-		Pyruvate	+	Acetate
Valerate	-		Lactate	+	Acetate
Methanol	-		Pyruvate without sulfate	+	Propionate, Acetate
Ethanol	+	Acetate	Lactate without sulfate	+	Propionate, Acetate

*3-Hydroxypropionate
+, good growth: (+), slow growth: -, no visible growth
3-Chloropropionate, allylalcohol, fumarate, succinate, glucose, fructose, maltose, sorbitol and mannitol were not utilized.

Strain MUD does not require any growth factors, but D. propionicus type strains require 4-aminobenzoic acid. D. propionicus is able to use sulfate, sulfite, thiosulfate and nitrate as electron acceptors, but strain MUD can use only sulfate or thiosulfate. D. propionicus type strain can not utilize formate as an electron donor. D. propionicus strain 2pr4[6] does not require any growth factors, however its optimum temperature is 30 °C, which is lower than that of MUD. These results indicate that strain MUD is a new type of D. propionicus.

Accumulation of excess propionate inhibits methane production and reduces the efficiency of anaerobic digestion. A possible remedy to these conditions is the inoculation of propionate-degrading bacteria into the reactor. Strain MUD oxidizes propionate to acetate. This is advantageous because the acetate may be used as substrate for acetotrophic methanogens, such as Methanosarcina or Methanothrix. Other advantage to using strain MUD for this purpose include low cost and easy maintenance. Strain MUD grows in a very simple medium of tap water containing some inorganic salts and it requires no require growth factors.

ACKNOWLEDGEMENT

This research was supported by the grant of New Energy and Industrial Technology Development Organization(R&D on New Wastewater Treatment System), and performed in cooperation by Aqua Renaissance Research Association and Fermentation Research Institute.

REFERENCES

[1]F. Widdel and N. Pfennig. Studies on Dissimilatory Sulfate-Reducing Bacteria that Decompose Fatty Acids I. Arch. Microbiol. 129:395-400(1981).
[2]J. Tamaoka and K. Komagata. Determination of DNA base composition by reversed-phase high-performance liquid chromatography. FEMS Microbiol. Lett. 25:125-128(1984).
[3]J. R. Postgate. "The sulfate-reducing bacteria," Cambridge University Press(1979).
[4]Y. Seki, Y. Nagai and M. Ishimoto. Characterization of a Dissimilatory-Type Sulfate Reductase, Desulfoviridin, from Desulfovibrio africanus Benghazi. J. Biochem. 98:1535-1543(1985).
[5]F. Widdel. Microbiology and ecology of sulfate- and sulfur-reducing bacteria,in "Biology of anaerobic microorganisms," A.J.B. Zehnder, ed., A Wiley-Interscience Publication, New York (1988).
[6]F. Widdel and N. Pfennig. Studies on Dissimilatory Sulfate-Reducing Bacteria that Decompose Fatty Acids II. Arch. Microbiol. 131:360-365(1982).
[7]E. Samain, H. C. Dubourguier and G. Albagnac. Isolation and Characterization of Desulfobulbus elongatus sp. nov. from a Mesophilic Industrial Digester. System. Appl. Microbiol. 5;391-401(1984).

CHARACTERIZATION OF A SULFATE - REDUCING BACTERIUM ISOLATED FROM THE GUT OF A TROPICAL SOIL TERMITE

S.A. Traoré [1] , G. Fauque [2] , V.A. Jacq [3]
and J.-P. Belaïch [4]

1 - I.S.P. / Département de Biochimie-Microbiologie, B.P. 7021
        Université de Ouagadougou , 03 Ouagadougou , Burkina Faso
2 - A.R.B.S. , CNRS-CEA , CEN de Cadarache
        13108 , Saint-Paul-Lez-Durance , France
3 - Laboratoire de Microbiologie ORSTOM , Case 87 , Université de Provence
        3, Place Victor Hugo , 13331 , Marseille Cédex 3 , France
4 - Laboratoire de Chimie Bactérienne , CNRS
        B.P. 71 , 13277 , Marseille Cédex 9 , France

INTRODUCTION

Until now, several well known sulfate-reducing bacteria have been directely isolated from soils, water samples and other sediments. New strains were isolated in Africa from paddy soils (JACQ , 1989) and from a hypersaline lake (OLLIVIER et al. , 1989). But only three strains, described during the poster session at this Symposium (cf BRAUMAN et al. , 1989) have recently been extracted from the gut of African termites originating from two different areas : a tropical soil from a Sahelian semi arid zone in Burkina Faso (this paper) and from a soil under the Congo rain forest. The three strains differ from the other non-sporulating sulfate-reducing bacteria described elsewhere.

MATERIALS AND METHODS

* **SAMPLING** : Termites from a sub-Sahelian locality in Burkina Faso were collected and washed using 90% alcohol to eliminate any external microorganisms. After discarding the heads, the guts were extracted and used as an inoculum in a test tube containing the specific medium described elsewhere (TRAORE et al. , 1982).

* **ISOLATION AND MORPHOLOGY** : Pure culture of the sulfate-reducing bacterium X was obtained by repeated applications of the dilution technique in anaerobically Hungate test tubes as described by PFENNIG et al. (1981). The morphology was observed, using a Nikon Optiphot phase contrast microscope, on agar slides.

* **CHEMICAL ANALYSES** : The growth was monitored by measuring the optical density at 450 nm, using a Baush & Lomb Spectronic 20 spectrometer. Dry weight was obtained after a 48 hour stay in an incubator regulated at 130 °C. Biochemical properties were studied on crude extracts. The exchange activity was determined in proton-deuterium experiments performed using the direct mass-spectronic technique, as described by FAUQUE et al. (1987) and LESPINAT et al. (1986).

RESULTS

* **MORPHOLOGY** : The cells were found to be motile vibrioid rods, 0.5 μm by 2-3 μm, staining Gram-negative. They contain desulfoviridin and c-type cytochrome. No spore was observed. Optimal pH and temperature for growth are about 6.5 - 7.0 and 35 - 37 °C respectively.

* **ELECTRONS DONORS** : Among the substrates tested as possible electron donors for sulfate reduction, ethanol (5 mM), glycerol (5 mM), fumarate (10 mM), lactate (10 mM), malate (10 mM), pyruvate (10 mM) and $H_2 + CO_2$ supported the growth of the new isolate, but methanol, fructose, acetate, benzoate, butyrate, oxalate, propionate and valerate did not.

* **ELECTRONS ACCEPTORS** : As electron acceptors the strain uses sulfate, thiosulfate, sulfite, elemental sulfur and nitrate. Pyruvate and fumarate support growth in the absence of exogenous electron acceptors.

Table 1 gives the biomass production efficiency based on the electron acceptors. In lactate medium, NO_3^- was the most efficient electron acceptor. The growth yields in the lactate-nitrate medium are about 1.5 times higher than in the lactate-sulfate medium.

TABLE 1   EFFICIENCY OF BIOMASS PRODUCTION BASED ON THE ELECTRON ACCEPTORS

Electron donors	Electron acceptors	$O D_{450}$	BIOMASS PRODUCTION Dry weight of Bacteria (μg / ml)
Lactate (10 mM)	$SO_4^=$ (20 mM)	0.575 ± 0.090	115 ± 20
" " " "	$SO_3^=$ (5 mM)	0.425 ± 0.150	85 ± 20
" " " "	$S_2O_3^=$ (10 mM)	0.585 ± 0.095	117 ± 20
" " " "	NO_3- (10 mM)	0.825 ± 0.090	165 ± 20
Ethanol (5 mM)	$SO_4^=$ (20 mM)	0.465 ± 0.080	93 ± 15
Fumarate (10 mM)	none	0.630 ± 0.095	126 ± 30
" " " "	$SO_4^=$ (20 mM)	0.600 ± 0.070	120 ± 25

These data are mean values of 6 experiments, performed on 40 ml-culture incubated at 37 °C.

* **HYDROGEN METABOLISM** : **Table 2** gives the results obtained during the study of proton-deuterium exchange activity : it can be seen that this strain is very active in hydrogen metabolism. The data on the proton-deuterium exchange activity in the absence or presence of carbon monoxide as inhibitor reaveled the presence of a (Ni-Fe) catalytic center in the hydrogenase.

TABLE 2 THE ACTIVITY OF PROTON-DEUTERIUM EXCHANGE IN THE NEW ISOLATE

Specific activities , expressed in micromoles of H_2 or HD* produced / min / mg of protein

Samples	H_2	HD*	H_2 / HD* ratio
Crude extract	0.56	1.56	0.36
Soluble fraction	0.40 (a)	1.04 (a)	0.38
" "	0.35 (b)	0.85 (b)	0.41
Pellet	3.40 (a)	0.80 (a)	0.71
"	2.50 (b)	3.70 (b)	0.61

* HD = Hydrogen Deuterium

Experiments were performed using the direct mass-spectronic technique, in phosphate buffer at pH 6.0;
Data were obtained in the absence of inhibitor (a)
and in the presence of 10 mM CO as inhibitor (b).

CONCLUSIONS

As the morphological, nutritional and physiological properties of strain X suggested that this new isolate can be said to be a subspecies of *Desulfovibrio vulgaris*, we propose the name *D. vulgaris* subsp. *termitidis*. It may play an important role in the interspecific hydrogen transfer in the gut of termites.

REFERENCES

Brauman, A. , Kane , M.D. , Labat , M. and Breznak, J.A. , 1989. Hydrogen metabolism by termite gut microbes. This Symposium.

Fauque , G. , Berlier , Y.M. , Czechowsky , M.H. , Dimon , B. , Lespinat , P.A. and Le Gall , J. , 1987. A proton-deuterium exchange study of three types of *Desulfovibrio* hydrogenases. J. Ind. Microbiol. , 2 : 15 - 23.

Jacq , V. A. , 1989. Participation des bactéries sulfato-réductrices aux processus microbiens de certaines maladies physiologiques du riz inondé. Ph. D.Thesis, Univ. of Provence , Marseille.

Lespinat , P. A. , Berlier , Y.M. , Fauque , G. , Czechowsky , M.H. , Dimon , B. and Le Gall , J. , 1986. The pH dependence of proton-deuterium exchange, hydrogen production and uptake catalyzed by hydrogenases from sulfate-reducing bacteria. Biochimie , 68 : 55 - 61.

Ollivier , B. , Hatchikian , C.E. , Prensier , G. , Guézénec, J . and Garcia , J.L. , 1989. Characterization of a sulfate reducing bacterium isolated from a hypersaline african lake. This Symposium.

Pfennig, N . , Widdel , F. and Trüper , H.G. , 1981. The dissimilatory sulfate-reducing bacteria. In : " The Procaryotes " , vol.1, pp. 926 - 940 ; M.P. Starr , H. Stolp , H.G. Trüper , B.A. Balows and H.G. Schlegel (eds.) , Springler Verlag (pub.) , Berlin.

Traoré , S.A. , Hatchikian , C.E. , Le Gall , J. and Bélaïch , J.-P. , 1982. Microcalorimetric studies of the growth of sulfate reducing bacteria : comparison of the growth parameters of some *Desulfovibrio* species. J. Bact., 149 : 606 - 611.

RELATIONSHIP BETWEEN METHANOGENESIS AND SULFATE REDUCTION IN ANAEROBIC

DIGESTION OF MUNICIPAL SEWAGE SLUDGE

Katsuji Ueki and Atsuko Ueki

Faculty of Agriculture
Yamagata University
Tsuruoka 997, Japan

INTRODUCTION

Methanogenesis and sulfate reduction are terminal steps in the anaerobic degradation of organic matter. These reactions compete with each other for electron donors, i.e., H_2 and acetate, in various environments. In the presence of sulfate at available levels, sulfate reduction generally dominates over methanogenesis due to differences in kinetic and thermodynamic properties. But in anaerobic digestion of municipal sewage sludge, sulfate reduction apparently does not compete with methanogenesis (1).

In the present study, the competition for electron donors between methanogenesis and sulfate reduction in anaerobic digestion of municipal sewage sludge was investigated.

MATERIALS AND METHODS

Sewage sludge was sampled from an anaerobic digester in the Wastewater Treatment Center of Tsuruoka City in Japan. The COD_{Mn} of the sludge sample was about 3,200 ppm and the pH was about 7.4. In general, only acetate (ca. 0.8 mM) and propionate (ca. 0.3 mM), but no other volatile fatty acids(VFAs), were detected. The sludge contained sulfide at about 1.5 mM, but sulfate was below the detection limit (ca. 0.02 mM).

The sludge was incubated at 30°C under N_2 for 24 hr. Then, 10 mM Na_2SO_4, 20 mM VFA (Na-acetate or Na-propionate) and/or inhibitors (0.0005%(v/v) chloroform and/or 5 mM Na_2MoO_4) were added, and 10-ml portions of the sludge were distributed into test tubes under N_2. If necessary, the headspace gas was substituted for H_2/N_2 (40%/60%, v/v). The tubes were sealed with butyl-rubber double stoppers and incubated at 30°C on a reciprocal shaker.

The gases and VFAs were analyzed by gas chromatography (2), and sulfate was measured by high-pressure liquid chromatography using a Hitachi 655 Liquid Chromatogram with a column packed with an anion exchange resin.

RESULTS AND DISCUSSION

Effects of sulfate addition on methanogenesis and sulfate reduction in the sludge were investigated. With no sulfate added, methanogenesis proceeded actively, but sulfate reduction did not occur due to the absence of sulfate at an available level. The CH_4 produced during 8 days of incubation reached 26.7 mmol/l of sludge. In the sludge supplemented with 10 mM sulfate, methanogenesis also proceeded actively, and sulfate reduction markedly proceeded. In 8 days of incubation, 26.6 mmol of CH_4 was produced per liter of sludge and 3.8 mM of sulfate was reduced. Addition of chloroform inhibited methanogenesis by about 90%, and addition of molybdate completely blocked sulfate reduction. The inhibition of methanogenesis increased sulfate reduction, and sulfate reduction during the incubation period reached 5.8 mM. The inhibition of sulfate reduction did not significantly affect methanogenesis. These results indicate that sulfate reduction can be increased apparently without retarding methanogenesis.

Changes in concentrations of VFAs and H_2 during the incubation with or without 10 mM sulfate are shown in Fig. 1. In general, 60 to 80% of CH_4 is thought to come from acetate, and the remainder mostly from H_2/CO_2. In the present study, the inhibition of methanogenesis in the sludge without sulfate caused a temporary accumulation of H_2 and an accumulation of the VFAs, acetate, propionate, butyrate and isovalerate. But the inhibition of methanogenesis in the sludge with sulfate caused only acetate accumulation. The inhibition of sulfate reduction did not cause the accumulation of intermediary products. In the sludge with sulfate, acetate, longer-chain VFAs and H_2 accumulated, when both methanogenesis and sulfate reduction were inhibited. These results show that mostly acetate is used in methanogenesis, and that sulfate reduction utilizes H_2 and longer-chain VFAs such as propionate. And sulfate reduction competes with methanogenesis for H_2.

Thus, the sludge was incubated under N_2 or H_2/N_2 mixed gas, and the effects of exogenous H_2 on methanogenesis and sulfate reduction were examined. Exogenous H_2 promoted methanogenesis in the sludge without sulfate by more than 40% and both methanogenesis and sulfate reduction by more than 30% in the sludge supplemented with 10 mM sulfate, when compared with those under N_2. The addition of sulfate depressed methanogenesis by 10-20%. The effects of inhibitors on methanogenesis and sulfate reduction in the presence

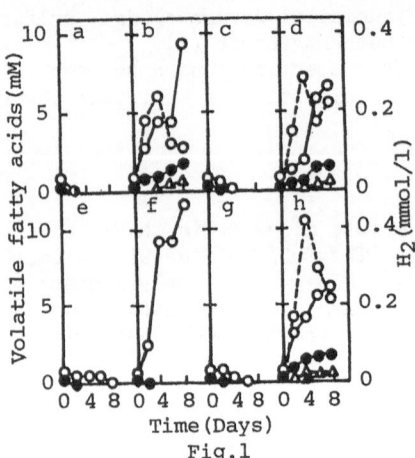

Fig.1

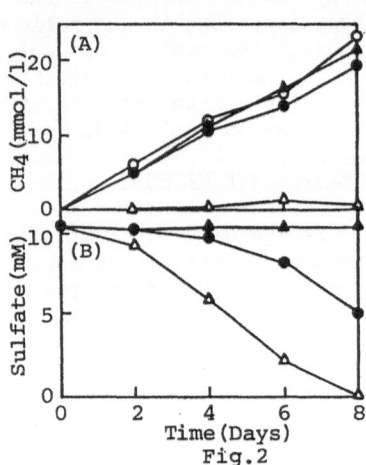

Fig.2

Fig. 1. Changes in concentrations of VFAs (solid line) and H_2 (dashed line) during anaerobic incubation of sewage sludge. The sludge supplemented with (e - f) or without 10 mM sulfate (a - d) was incubated with 0.0005% chloroform (b,f), 5 mM Na_2MoO_4 (c,g) or 0.0005% chloroform plus 5 mM Na_2MoO_4 (d,h), or without inhibitors (a,e). Acetate, (o); propionate, (●); butyrate, (△); isovalerate, (▲). The accumulated H_2 is shown as mmol/liter of sludge. The values are averages of duplicate experiments.

Fig. 2. Methanogenesis (A) and sulfate reduction (B) during anaerobic incubation of sewage sludge with H_2. The sludge supplemented with 10 mM sulfate was incubated with 0.0005% chloroform (△) or 5 mM Na_2MoO_4 (▲), or without inhibitors (●) under H_2/N_2 (40%/60%) mixed gas. The sludge was also incubated under the mixed gas without the addition of sulfate and inhibitors (o). CH_4 production is expressed as mmol/liter of the sludge. The values are averages of duplicate experiments.

of exogenous H_2 are shown in Fig. 2. The inhibition of sulfate reduction enhanced methanogenesis to the level of that without sulfate addition, and the inhibition of methanogenesis markedly enhanced sulfate reduction. The competition for H_2 accounted for the relation between methanogenesis and sulfate reduction in the presence of exogenous H_2.

The methanogenesis and sulfate reduction in the sludge supplemented with 10 mM sulfate during incubation with 20 mM acetate or 20 mM propionate are shown in Fig. 3. The addition of acetate markedly enhanced methanogenesis, but did not affect sulfate reduction, and the addition of propionate markedly enhanced both methanogenesis and sulfate reduction. In the sludge with propionate, methanogenesis proceeded at a constant rate, and the inhibition of sulfate reduction depressed methanogenesis to the level of that without the VFA addition.

Acetate was consumed without being affected by sulfate reduction, and propionate was degraded to acetate depending on sulfate reduction, as shown in Fig. 4. The inhibition of methanogenesis during incubation with propionate caused the accumulation of acetate.

These results indicate that acetate is utilized only for methanogenesis but H_2 is used for both methanogenesis and sulfate reduction. Sulfate reduction rather enhances electron flow to methanogenesis by degrading the VFAs such as propionate, and apparently does not retard methanogenesis.

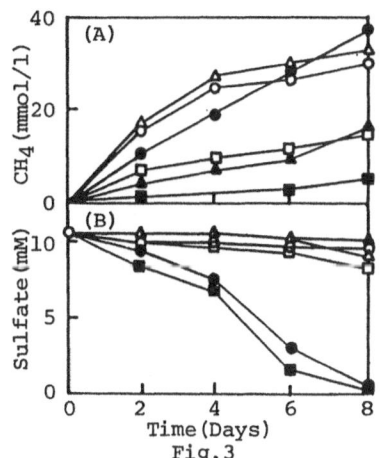 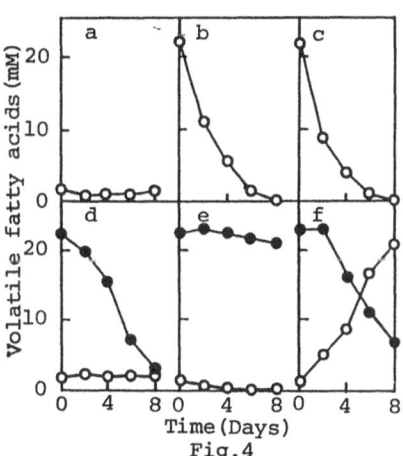

Fig.3

Fig.4

Fig. 3. Effects of acetate and propionate on methanogenesis (A) and sulfate reduction (B) in sewage sludge. The sludge supplemented with 10 mM sulfate was incubated with 20 mM Na-acetate (o), 20 mM Na-acetate plus 5 mM Na$_2$MoO$_4$ (△), 20 mM Na-propionate (●), 20 mM Na-propionate plus 5 mM Na$_2$MoO$_4$ (▲) or 20 mM Na-propionate plus 0.0005% chloroform (■), or without the addition of VFAs and inhibitors (□). The values are averages of duplicate experiments.

Fig. 4. Changes in concentrations of VFAs during anaerobic incubation of sewage sludge with acetate or propionate. The sludge supplemented with 10 mM sulfate was incubated with 20 mM Na-acetate (b), 20 mM Na-acetate plus 5 mM Na$_2$MoO$_4$ (c), 20 mM Na-propionate (d), 20 mM Na-propionate plus 5 mM Na$_2$MoO$_4$ (e) or 20 mM Na-propionate plus 0.0005% chloroform (f), or without the addition of VFAs and inhibitors (a). Only acetate (o) and propionate (●) were detected. The values are averages of duplicate experiments.

REFERENCES

1. K. Ueki, A. Ueki and Y. Simogoh, Terminal steps in the anaerobic digestion of municipal sewage sludge: Effects of inhibitors of methanogenesis and sulfate reduction, J. Gen. Appl. Microbiol. 34:425 (1988).
2. A. Ueki, K. Matsuda and C. Ohtsuki, Sulfate-reduction in the anaerobic digestion of animal waste, J. Gen. Appl. Microbiol. 32:111 (1986).

THE EFFECT OF TEMPERATURE ON BUTYRATE DEGRADATION

Peter Westermann

Department of General Microbiology
University of Copenhagen
Sølvgade 83 H 1307 Copenhagen K
Denmark

Summary

The effect of temperature on butyrate degradation was investigated in defined syntrophic cultures of Syntrophomonas wolfei co-cultured with either Methanospirillum hungatei or Desulfovibrio strain G11. The temperature response was almost linear in an Arrhenius plot from 20 - 37 ^{o}C. Below 20°C, almost no butyrate metabolism occurred. Measurements of hydrogen partial pressures in the cultures indicated that S. wolfei was more temperature sensitive than the hydrogen scavenging bacteria.
Anaerobic soil slurries enriched with 20 mM butyrate had a temperature optimum at 30°C, and metabolized butyrate at 2°C which was the lowest tested temperature.

Introduction

Although syntrophic degradation of volatile fatty acids such as butyrate is considered important in all anaerobic ecosystems, defined culture studies have, with one exception [1], been limited to mesophilic (37°C) or thermophilic (55-60°C) conditions, although temperatures in natural ecosystems rarely reach these levels. In temperate climates, temperature has been shown to be the most important variable in controlling the rate of microbial metabolism in anaerobic environments such as sediments and water-logged soils [2,3]. As anaerobic mineralization of organic matter implies the activity of a complex microbial food web, several studies have been performed to elucidate the rate limiting steps during such a degradation (e.g.[4]). These studies have, however, focused on substrate utilization under elsewise constant conditions. If the temperature response of the involved microbial groups is different, temporal variations in temperature might lead to changes in the rate limiting step.
The aim of the present study was to examine wether the butyrate degrader or two hydrogen scavenging bacteria in syntrophic coculture was the temperature sensitive step during anaerobic mineralization of butyrate.

Methods

S. Wolfei grown with either M. hungatei or Desulfovibrio G11 were
from our culture collection. The cultures were grown in mineral
medium supplemented with 2g yeast extract/l, 15-20 mM butyrate,
bicarbonate/CO_2 buffer, and cysteine and sulfide as reducing
agents. 20 mM Na_2SO_4 was added to cocultures grown with Desulfovibrio G11. Experiments were carried out in triplicate in 50 ml
serum vials (25 ml of medium in each). Growth was measured as
metabolism of butyrate by capillary gas chromatography. Hydrogen
was measured in the headspace by a mercury/mercury vapor reduction gas analyzer.
Soil slurries were prepared by anaerobic homogenisation of anaerobic waterlogged soil with interstitial water (50% vol/vol) from
a permanently waterlogged swamp. The experiments were carried out
in 500 ml serum flasks (300 ml slurry in each) in duplicate. The
rubber stoppers of the flasks were mounted with glass tubings, a
piece of butyl rubber tubing, and a hose clamp to allow sampling
of the higly viscous slurry.

Results and discussion

Fig. 1 shows Arrhenius plots of butyrate metabolism by S. wolfei
cocultured with either M. hungatei or Desulfovibrio G11, and
butyrate metabolism in soil slurries. From 20 - 27°C the slope of
the plots was almost constant for S. wolfei when grown with a
methanogen or a sulfate reducer. Below 20°C almost no metabolism
of butyrate occurred. At 20°C and below, the lag period before
the onset of butyrate metabolism was at least 7 days while no lag
period was observed at 30 and 37°C (data not shown).
At 20, 30, and 37°C the butyrate metabolic rate was higher for
S. wolfei cocultured with Desulfovibrio G11 than when cocultured
with M. hungatei, as also observed in other studies (5).
Butyrate metabolism in soil slurries amended with 20mM butyrate
was measurable at 2°C, which was the lowest temperature tested.
The optimum temperature was 30°C for soil slurries while the
optimum temperature for defined S. wolfei cocultures was at least
37°C, which was the highest temperature tested.

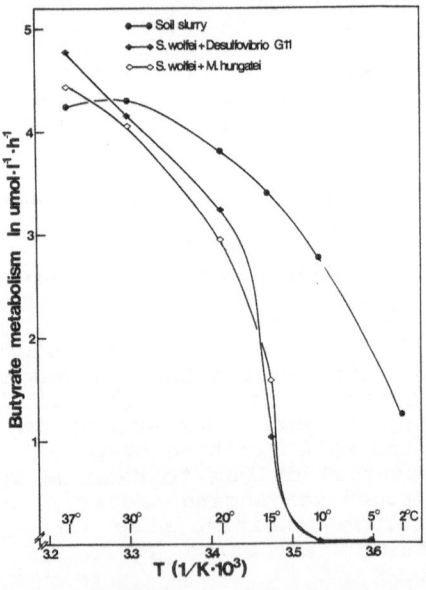

Fig.1. Arrhenius plots of butyrate metabolism in defined butyrate
degrading cocultures and in soil slurries.

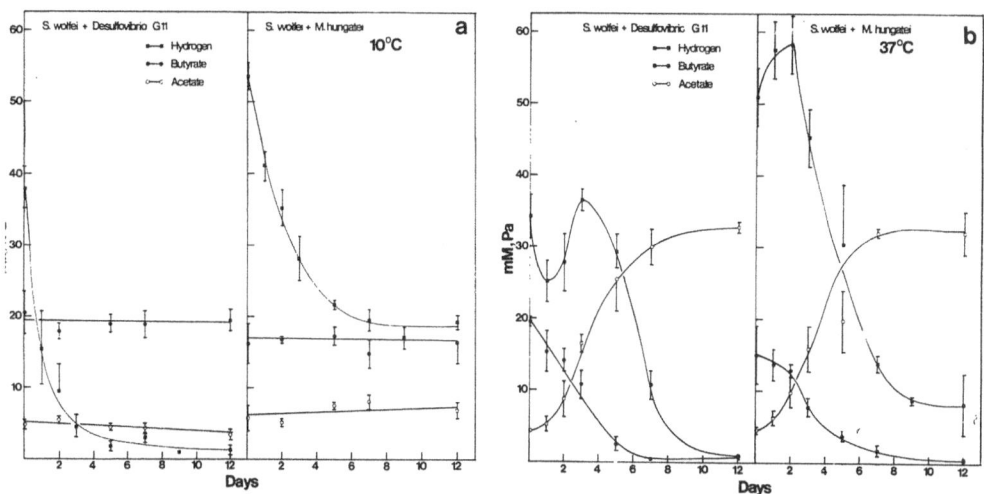

Fig.2. Butyrate metabolism in <u>S. wolfei</u> cocultures at 10°C (A) and 37°C (B).

Corresponding Q_{10} values for the comparable temperature range (20 – 30°C) were: Soil slurries, 1.65; <u>S. wolfei</u> plus <u>M. hungatei</u>, 3.04; <u>S. wolfei</u> plus <u>Desulfovibrio</u> G11, 2.52.
Fig. 2A and B show in detail butyrate metabolism by <u>S. wolfei</u> at 10 and 37°C with either <u>M. hungatei</u> or <u>Desulfovibrio</u> G11 as hydrogen scavenger. At 37°C butyrate metabolism started immediately with simultaneous accumulation of acetate. The hydrogen concentration was fairly high during active metabolism and fell off parallel to the exhaustion of butyrate. As already shown in Fig. 1 no butyrate metabolism occurred at 10 °C. The butyrate and acetate concentrations remained constant under the experiment (34 days). The hydrogen concentration was, however, reduced significantly to a stable level by both the methanogen and the sulfate reducer during the 12 days depicted on the figure. This indicates that these bacteria were active even at 10°C.
In both cultures, the hydrogen concentration after 3 days was sufficient low to allow active butyrate metabolism, even taking the increased solubility at 10°C in consideration. These results therefore indicate that <u>S. wolfei</u> and not the hydrogen scavenging sulfate reducer and methanogen is limited by low temperatures. Furthermore the results indicate that another strain or species of butyrate degrading bacteria than the tested <u>S. wolfei</u> strain must be responsible for the observed activity at at least temperatures below 20°C in the waterlogged soil.

Acknowledgement

Parts of this work was supported by The Danish Natural Science Research Council. (11-7701). Karin Vestberg is acknowledged for excellent technical assistance.

1. Boone, D.R., and L. Xun. 1987. Appl. Environ. Microbiol. 53:1589-1592
2. Abdollahi, H., and D. B. Nedwell.1979. Microb. Ecol. 5:73-79.
3. Westermann, P., and B.K. Ahring. 1987. Appl. Environ. Microbiol. 53:2554-2559.
4: Kaspar, H. F., and K. Wuhrmann 1978. Microb. Ecol. 4:241-248.
5. McInerney, M.J., M.P.Bryant, and N.Pfennig. 1979. Arch. Microbiol. 122:129-135.

THERMOPHILIC DEGRADATION OF BUTYRATE, PROPIONATE AND ACETATE IN GRANULAR SLUDGE

Margrethe Winther-Nielsen and Birgitte K. Ahring

Department of Biotechnology
The Technical University of Denmark
2800 Lyngby, Denmark

INTRODUCTION

Anaerobic degradation of butyrate and propionate is dependent on interspecies hydrogen transfer. From thermodynamic calculations it is obvious that propionate requires the lower hydrogen concentration and the degradation of propionate is, therefore, usually the rate limiting step, especially in high rate anaerobic treatment systems (2). Studies on product inhibition of butyrate by acetate and hydrogen have shown that thermophil butyrate degradation depend on both hydrogen and acetate removal (1). The concentration of propionate might also effect the degradation of butyrate. The purpose of this study was to determine specific methane production rates from butyrate, propionate, and acetate and examine the effect of propionate on the butyrate consumption rate in granular sludge from a UASB-reactor with a relatively high concentration of propionate not degraded and, therefore, found in the effluent from the reactor.

MATERIALS AND METHODS

The granular sludge was cultivated in a 5 liter thermophilic UASB-reactor at 55°C originally inoculated with thermo-philicly digested cow manure. The reactor was feed for 6 months with a medium containing butyrate, propionate, and acetate each at a concentration of 1 g COD/l. The removal efficiency exceed 90% for acetate and butyrate, while the degradation of propionate never exceeded 30% during the operation period. The reactor was operated for 2 months under steady state conditions with a hydraulic retention time of 4.5 hours before these experiments were started. Microscopic examination showed that Methanosarcina was the dominant methanogen in the granules.

Potential methanogenic activities of granular sludge cultivated in the UASB-reactor were determined by adding a slug of butyrate, propionate, acetate or a mixture of the 3 fatty acids to samples of granules in batch culture. The experiments were performed in 30 ml serum vials sealed with butyl rubber stoppers. 5 ml samples of fresh granular sludge from the reactor were distributed into vials under N_2. The vials were acclimatizated in a shaking water bath at 55°C for one hour before carbon sources were added at concentrations equal to the concentration in the influent of the UASB-reactor. Methane production was followed during the next 10 hours. The content of volatile solids was calculated from determination of the dry weight and ash content. The maximum specific methane production rate was determined from the max. slope of the graph of the accumulated methane concentration versus time.

The effect of propionate in different concentrations on the butyrate consumption rate of the granules were studied in serum vials (100 ml vials containing 25 ml of medium). The medium used in the experiments consisted of basal nutrients, trace elements, vitamins, phosphate buffer, rezasurin and sodium sulfide. Butyrate, propionate, and acetate were added as sodium salts from stock solutions. The pH in the medium was 7.2 and the gas phase was 100% N_2. 5 ml samples of granular sludge were distributed under anaerobic condition to preflushed tubes which were closed with a rubber stopper. After settling of the granules the supernatant, was removed and the granules were washed with a buffer solution. The washed granules were transferred to serum vials with medium and incubated in a shaking water bath at 55°C over night. Butyrate was added after this acclimatization period to give a concentration of approximately 6 mM. Propionate at different concentrations (10, 25, 50 and 75 mM) was added to the vials when the butyrate concumption was initiated. The exact concentration of butyrate and propionate in the vials were controlled by gas chromatography. The specific butyrate consumption rate (μ_{but}) was calculated as the slope of the semi- logarithmic plot of the butyrate concentration versus time.

Each experiment was performed at least in duplicate.

RESULTS AND CONCLUSIONS

The maximum specific methane production rates from the degradation of the different fatty acids are shown in table 1. The total consumption rate of the fatty acids in the UASB-reactor is about 12 g COD/l_r/day corresponding to 2.9 g COD/gVS/dg or a methane production rate of about 30 μmol CH_4/gVS/min. The methane production rates determined from the batch measurements are in the same order of magnitude as the reactor data.

Table 1. Maximum specific methane production rates
for various substrates of the granules

Substrate	Butyrate	Propionate	Acetate	Butyrate Propionate Acetate
Methane production rate (μmol/gVS/min)	7-10	0.45-0.49	17-18	13-26

The effect of propionate on the butyrate consumption rate of the granules is shown in table 2. Addition of propionate in high concentrations (more than 25 mM) seems to affect the butyrate consumption of the granules and to increase the variations between the data obtained.

Table 2. Effect of propionate on the specific butyrate consumption rate of the thermophilic granular sludge

Propionate conc. (mM)	μ_{but} (h^{-1})[a]
0	0.08 ± 0.02
10	0.11 ± 0.01
25	0.08 ± 0.02
50	0.04 ± 0.02
75	0.05 ± 0.05

[a] μ_{but}, Specific butyrate consumption rate. Average of 3 experiments ± std. dev.

The study showed that the degradation rate of propionate is very low in the granular sludge. However, the concentration of propionate fed to the reactor (10mM) does not seem to affect the butyrate consumption rate significantly. The growth of the propionate degrading bacteria could be inhibited by the relatively high concentration of hydrogen produced under the degradation of butyrate (2). However, in a UASB-reactor consisting of aggregates of syntrophic organisms a balance between H_2-producers and H_2-consumers would be expected to be established with time. Another explanation for the low degradation rate of propionate in the UASB-reactor could be a lower tendency to form aggregates by thermophilic propionate degrading bacteria compared with the corresponding mesophilic bacteria.

ACKNOWLEDGEMENTS

This work was supported by grants from Nordic Council of Ministers and the Danish Energy Research Program 1383/89-2.

REFERENCES

1. Ahring, B. K., and P. Westermann. 1988. Product inhibition of butyrate-degrading coculture. Appl. Environ. Microbiol. 54:2393-2397.

2. Wiegant, W.M. 1986 Thermophilic anaerobic digestion for waste and wastewater treatment. Ph.D.Thesis, Agricultural University of Wageningen.

the study, though 4 contracts (section four of appendix K...) have the same structure. However, the application of principle to [...]regulation (final case here is in[...]) [...] to manipulate. The growth in the prominent direction could be limited by the same size and concentration in [...] centres, as is also the case in [...] 1990[...]. [...] a surface defect [...] contracts and [...] a special price combined with some control regulation for the low dimension [...] of [...] conclude in that [...] state [...] [...] result in fracture or growth [...] [...] mouldings also [...] in contact continue to [...] the subsequent[...].

ACKNOWLEDGEMENT

The work is supported by [...] von Humboldt Foundation to W. [...] and the Deutsche Forschungsgemeinschaft (SFB[...]).

REFERENCES

1. Kühn, H. and L. Eiermann. 1986. Rupture Mechanical of Reinforced and their Manipulation, McGraw[...], [...].

2. Ziegert, F.K. 1986. [...] eine Untersuchung für Plaste und Kautschuk mit Experimentele, Technical University of Wandlingen.

DIFFERENTIAL EFFECTS OF SODIUM AND CARBON MONOXIDE ON THE H2-
AND GLUCOSE-DEPENDENT GROWTH OF THE THERMOPHILIC ACETOGEN
ACETOGENIUM KIVUI

Hsuichin Yang and Harold L. Drake (a)

Microbial Physiology Laboratories, Department of
Biology, The University of Mississippi
University, MS 38677 U.S.A.

ABSTRACT

Cultures of A. kivui could not be maintained at the expense of H2 in sodium-deficient medium (0.2 mM Na). Glucose
cultures did not display such a dependency on sodium. Neither
lithium nor potassium replaced the sodium requirement of H2
cultures. In the absence of growth, formate became a major
end product in sodium-deficient H2 cultures. Harmaline uncoupled acetogenesis from growth in H2 cultures, while other
metabolic inhibitors blocked H2-dependent growth and acetogenesis. Harmaline did not inhibit glucose-dependent growth
but stimulated higher acetate yields per unit biomass formed.
Carbon monoxide (CO) was inhibitory to glucose cultures but
was stimulatory to H2 cultures.

INTRODUCTION

The autotrophic mechanism by which methanogens and acetogens fix CO2 is similar and involves acetyl-CoA synthetase
(1,2,3).   Sodium plays an important role in the conservation
of energy during methanogenesis (4-8), and sodium was postulated to play an important role in the conservation of energy
during acetogenesis (8). In support of this concept, heterotrophically cultivated Clostridium thermoaceticum contains a
sodium/proton antiporter (9), metal ionophores inhibit the
growth and energy-dependent transport of nickel by acetogens
(10,11), and the H2- and CO-dependent formation of acetate by
washed cells of CO-cultivated Peptostreptococcus productus is
stimulated 2- and 3-fold, respectively, by 10 mM sodium (12).
We report here that coupling H2-dependent acetogenesis and
growth by A. kivui is strictly dependent upon sodium and may
obligately require sodium/proton antiport; conversely, growth
under heterotrophic conditions is less dependent upon sodium.

METHODS

A. kivui ATCC 33488 (13) was cultivated at 55C in defined
medium (pH 6.4; with cysteine and CO2). Sodium-enriched me-

_____

(a) Correspondence to H. L. Drake.

dium contained sodium salts; sodium-deficient medium contained potassium salts. Sodium-enriched and sodium-deficient media contained 103 and 0.2 mM Na, respectively (determined by atomic absorption spectrometry). Both 10 and 0.5 mM glucose were used. For H2 cultures, tubes were pressurized to 70 kPa over pressure with H2. Growth and cell dry weights were determined as previously described (14). Acetate and formate were analyzed by high-pressure liquid chromatography (HPLC), and CO was measured by gas chromatography (GC).

RESULTS

A. kivui could not be maintained beyond the second transfer in sodium-deficient H2 medium. Conversely, the growth of sodium-enriched and sodium-deficient glucose cultures were identical. Lithium or potassium failed to replace the sodium requirement of H2 cultures (Table 1). In sodium-deficient H2 cultures, formate became a major end product in the absence of growth. When the growth of H2 cultures was inhibited with monensin, nigericin and N,N'-dicyclohexylcarbodiimide (DCCD), formate was also produced (Table 2). In contrast, glucose cultures did not produce formate (data not shown). Harmaline, a putative inhibitor of sodium/proton antiporters, uncoupled growth from acetogenesis in H2 cultures. In contrast, harmaline did not inhibit glucose-dependent growth but yielded higher acetate yields per g biomass formed (data not shown). CO stimulated H2-dependent growth (Table 2) but was inhibitory to the growth of glucose cultures (data not shown).

DISCUSSION

The mechanism(s) by which acetogenic bacteria couple carbon and energy flow during acetogenesis is not resolved. The present study demonstrates that sodium plays a fundamental role in H2-dependent bioenergetics of A. kivui and suggests that autotrophic and heterotrophic cells may, under certain conditions, utilize dissimilar mechanisms of energy conservation. Substrate-level phosphorylation plays an important role during the heterotrophic growth of acetogens, but some form of electron transport phosphorylation may be obligatory under autotrophic conditions (2,15). That harmaline selectively inhibited the H2- rather than glucose-dependent growth of A. kivui suggests that sodium/proton antiport (9) is essential only under autotrophic conditions

TABLE 1. Effect of sodium, lithium, and potassium on growth and product formation by A. kivui cultivated at the expense of H2 in sodium-deficient medium

Salt Added	Growth (A660 nm)	Acetate (mM)	Formate (mM)
NaCl	0.07	10.3	0
LiCl	0.00	2.5	1.5
KCl	0.00	1.5	1.2
none	0.00	1.4	0.7
Na-enriched medium	0.09	11.2	0

Media were inoculated from a second transfer of a sodium-deficient H2 culture in early stationary phase. Chloride salts were injected at 24 hours postinoculation.

TABLE 2. Effect of metabolic inhibitors on H2-dependent product formation by A. kivui in sodium-enriched medium

Inhibitor (uM)	Biomass (g/l)	Acetate (mM)	Formate (mM)
none (control)	0.060	14.5	0
DCCD (700)	0.005	1.0	2.1
monensin (14)	0.004	3.3	3.0
nigericin (14)	0.000	1.5	2.5
harmaline (50)	0.005	12.5	0
CO (35 kPa)	0.080	23.0	0

with this acetogen. The synthesis of formate as an H2-dependent end product under sodium-deficient conditions supports the proposal that sodium is involved in energy conservation at some point subsequent to the formation of formate (12). However, sodium/proton antiport may not be essential to the formation of acetate per se since harmaline did not inhibit the synthesis of acetate. That sodium affects substrate consumption and product formation also implies that, under certain conditions, sodium may influence interspecies reductant transfer if sodium-dependent acetogens were syntrophic partners (2,16).

Acknowledgements. Public Health Service Grant AI21852 and Research Career Development Award AI00722 (H.L.D.) from the National Institute of Allergy and Infectious Diseases.

REFERENCES

1. Wood, H.G., Ragsdale, S.W., Pezacka, E. (1986) FEMS Microbiol. Lett. 43:345.
2. Ljungdahl, L.G. (1986) Annu. Rev. Microbiol. 40:415.
3. Fuchs, G. (1986) FEMS Microbiol. Lett. 39:181.
4. Perski, H.J., Moll, J., Thauer, R.K. (1981) Arch. Microbiol. 130:319.
5. Perski, H.J., Schönheit, P., Thauer, R.K. (1982) FEMS Lett. 143:323.
6. Müller, V., Blaut, M., Gottschalk, G. (1987) Eur. J. Biochem. 162:461.
7. Schönheit, P., Beimborn, D.B. (1985) Arch. Microbiol. 142:354.
8. Gottschalk, G. (1989) In: H.G. Schlegel and B. Bowien (ed), Autotrophic Bacteria. Science Tech Publishers, Madison.
9. Terracciano, J., Schreurs, W.J.A., Kashket, E.R. (1987) Appl. Environ. Microbiol. 53:782.
10. Lundie, L.L., Jr., Yang, H., Heinonen, J.K., Dean, S.I., Drake, H.L. (1988) J. Bacteriol. 170:5705.
11. Yang, H., Daniel, S.L., Hsu, T., Drake, H.L. (1989) Appl. Environ. Microbiol. 590:24.
12. Geerligs, G., Schönheit, P., Diekert, G. (1989) FEMS Microbiol. Lett. 57:253.
13. Leigh, J.A., Mayer, F., Wolfe, R.S. (1981) Arch. Microbiol. 129:275.
14. Savage, M.D., Drake, H.L. (1986) J. Bacteriol. 165:315.
15. Hugenholtz, J., Ljungdahl, L.G. (1989) J. Bacteriol. 171:2873.
16. Boone, D.R., Johnson, R.L., Liu, Y. (1989) Appl. Environ. Microbiol. 55:1735.

POSTERS - 2 - BIOCHEMISTRY

ONE-CARBON METABOLISM BY THE RUMEN ACETOGEN

SYNTROPHOCOCCUS SUCROMUTANS

Joël Doré[1] and Marvin P. Bryant[2]

1. INRA, CR de Theix
Laboratoire de Microbiologie
63122 CEYRAT

2. University of Illinois
Dept. of Animal Sciences
URBANA, IL 61801. USA

INTRODUCTION

The rumen acetogen *Syntrophococcus sucromutans* has an absolute requirement for an electron acceptor system to catabolize a variety of carbohydrates as electron donors [1]. It can use the O-demethylation of lignin-derived methoxybenzenoids as electron acceptor system, with the corresponding hydroxybenzenoid, acetate and CO_2 as products. Formate or a methanogen in coculture can serve as electron acceptor system. Acetate and CO_2 are the only products, with methane for the coculture.

We have studied the catabolism of the one-carbon (C1) units formate, bicarbonate, the methoxyl group of vanillate and carboxyl group of pyruvate by washed cells of *S. sucromutans* and investigated enzymatic activities of the acetyl-CoA pathway including all these activities involving tetrahydrofolate (THF) as a C1 carrier, formate- and CO dehydrogenase, and hydrogenase.

MATERIAL AND METHODS

Organism and culture techniques. *S. sucromutans* strain S195 was from the laboratory collection. The bacterium was grown using anaerobic techniques, and the medium of Krumholz and Bryant[2] was modified as formerly described[3].

Enzymatic methods. Cell-free extracts were prepared anaerobically by treatment through a French pressure cell (52,400 KPa) and centrifugation (35,000 g, 1 h, 4°C) under N2. Aliquots were kept at -20°C in butyl-rubber stoppered vials and used only once. Formate dehydrogenase (FDH) and enzymes of the THF pathway were assayed as described by O'Brien and Ljungdahl[4]. Hydrogenase was assayed according to Drake[5]. CO-dehydrogenase (CODH) was assayed in a similar manner with a 100% CO headspace in place of H2. The formation of acetate from ^{14}C-methyl THF was assayed according to Ghambeer et al.[6]. The formation of acetate from O-[methyl- ^{14}C]vanillate was tested under similar conditions.

Washed cell experiments. The cell pellet of a 3 l batch culture in late exponential growth phase was suspended in 20 ml 100 mM potassium phosphate buffer (pH 6.8) and 5 mM cysteine after two washes in the same buffer (200 ml each). Ten ml assay mixtures in 25 ml Balch tubes contained pyruvate, formate

Microbiology and Biochemistry of Strict Anaerobes Involved in Interspecies Transfer
Edited by J.-P. Bélaich *et al.*
Plenum Press, New York, 1990

and bicarbonate (5 mmole each). Labeled formate or bicarbonate were 2 and 0.5 µCi/mmol, respectively. Five ml assays in 12 ml serum vials were performed using O-methyl labeled vanillate or carboxyl labeled pyruvate as labeled substrate. Vanillate assays were run with or without bicarbonate/CO_2 and/or pyruvate. Assays were started by addition of the cell suspension (1 ml per 10 ml) and stopped by acidification with 3N $HCLO_4$ (1 ml per 10 ml) after 0, 1 and 3 h. Soluble fermentation products were separated by isoionic exchange chromatography[7]. Acetate degradation was according to Abraham and Hassid[8]. All counts were performed in Beckman LS 5801 scintillation counter and corrected for quench and background. Organic acids were assayed using HPLC.

RESULTS AND DISCUSSION

Cell extracts were observed to effectively synthesize acetate from methyl-THF and pyruvate. The incorporation of the methyl label from methyl-THF into acetate using extracts of vanillin- grown cells was furthermore dependent on the presence of pyruvate, CoASH and ferrous ions (Table 1). With O-methyl labeled vanillate, only poor convertion occured even in the presence of pyruvate.

Table 1. Synthesis of acetate from methyl-THF or vanillate by crude extracts of S. sucromutans.

| Growth conditions | dpm Acetate [% convertion] | |
	from methyl-THF	from vanillate
Cellobiose-formate	6,997 [44]	3,256 [1.65]
Cellobiose-vanillin	4,336 [27]	1,273 [0.65]

Assays were incubated 10 min at 37°C in the dark. Complete reaction (µmoles): pyruvate (30), dithiothreitol (10), ferrous ammonium sulfate (5), CoA-SH (3.3), THF (0.5) in assays with vanillate only, potassium phosphate buffer pH 7.0 (50) in 1 ml volume. Assays contained 8.70 and 8.17 mg proteine for the formate and vanillin grown cells, respectively. One µmole labeled methyl-THF (25,323 dpm) or vanillate (217,000) were added to start the reaction. The data shown were corrected for blank without cell extract.

Cell extracts were also shown to contain all of the enzyme activities of the Wood pathway for metabolism of C1 compounds on THF carriers (Table 2). The specific activities measured were in the range reported in former studies of heterotrophically grown acetogens[9]. The CODH activity measured gave further strong evidence in favor of the contribution of the Wood pathway for acetate synthesis. Uptake hydrogenase activity was measured but we were unable to detect FDH. This latter enzymatic activity has been measured in all other acetogens using the Wood pathway[9]. Its absence in exctracts of S. sucromutans is consistent with the use of formate as electron acceptor.

Table 2. Specific activities of enzymes of the THF pathway in cell extracts of S. sucromutans after growth with cellobiose and a C1 electron acceptor[a]

| Enzymatic activity assayed | C1 electron acceptor for growth | |
	Formate	Vanillin
Carbon monoxide dehydrogenase[b]	2.91	2.31
Formyl-THF synthetase	2.30	15.45
Methenyl-THF cyclohydrolase	0.47	0.36
Methylene-THF dehydrogenase[c]	8.13	6.30
Methylene-THF reductase[d]	0.29	0.06

a Specific activity : µmoles of substrate converted or product formed per min per mg of protein.
b Activity given as µmole MV reduced per min per mg protein.
c NAD but not NADP was the electron acceptor.
d FAD was used as electron acceptor and [14C-methyl]THF was used as the substrate.

In assays with labeled bicarbonate, an exchange of label with pyruvate was observed. The exchange with formate was negligible and the acetate formed was predominantly labeled in the carboxyl group. Similar results were obtained with carboxyl-labeled pyruvate as electron donor (Fig. 1). Labeled formate or vanillate, as electron acceptors, resulted in a predominant to complete labeling of the methyl group of acetate. No labeled CO_2 was observed during metabolism of the C1 of vanillate. This synthesis of position labeled acetate is consistent with the use of formate or methoxyl groups as electron accepting moieties.

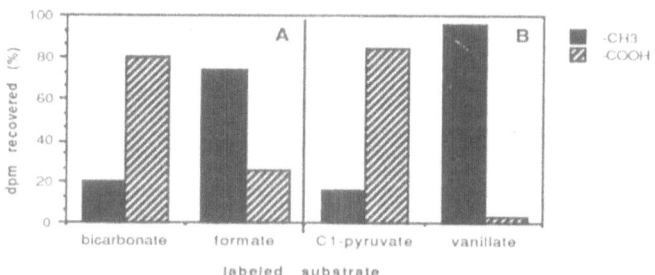

Fig. 1. Distribution of label in the carbons af acetate produced by washed cells of S. sucromutans in assays using pyruvate and formate (**A**) or vanillate (**B**).

Clearly *S. sucromutans*, which is unable to use H_2 as an electron sink product of carbohydrates or pyruvate breakdown in pure culture would not benefit a pathway converting the O-methyl group of methoxyaromatics to CO_2. This is a major difference with other acetogens such as strain TH001[10] and *A. woodii*[11] that can use the methoxyl group of vanillate as sole energy source. The poor convertion of vanillate by cell extracts while washed whole cells metabolize it efficiently suggest a close association between the methyl-transfering and acetate synthesizing complexes, with a likely involvement of the membrane. Finally, *S. sucromutans* appears enzymatically equally well adapted to use formate or methoxyaromatics as electron acceptor, and observations should be made to assess the natural electron acceptor used by the bacterium.

REFERENCES

1. L.R. Krumholz and M.P. Bryant. 1986a. *Syntrophococcus sucromutans* sp. nov. gen. nov. uses carbohydrates as electron donors and formate, methoxybenzenoids or *Methanobrevibacter* as electron acceptor system. Arch. Microbiol. 143:313-318.
2. L.R. Krumholz and M.P. Bryant. 1986b. *Eubacterium oxidoreducens* sp. nov. requiring H_2 or formate to degrade gallate, pyrogallol, phloroglucinol and quercetin. Arch. Microbiol. 144:8-14.
3. J. Dore and M.P. Bryant. 1989. Lipid growth requirement and the influence of lipid supplement on the fatty acid and aldehyde composition of *Syntrophococcus sucromutans*. Applied Environ. Microbiol. 55:927-933.
4. W.E. O'Brien and L.G. Ljungdahl. 1972. Fermentation of fructose and synthesis of acetate from carbon dioxide by *Clostridium formicoaceticum*. J. Bacteriol. 109:626-632.
5. H.L. Drake. 1982. Occurence of nickel in carbon monoxide dehydrogenase from *Clostridium pasteurianum* and *Clostridium thermoaceticum*. J. Bacteriol. 149:561-566.
6. R.K. Ghambeer, H.G. Wood, M. Schulman, and L. Ljungdahl. 1971. Total synthesis of acetate from CO_2. III. inhibition by alkyl-halides of the synthesis from CO_2, methyltetrahydrofolate, and methyl-B12 by *Clostridium thermoaceticum*. Arch. Biochem. Biophys. 143:471-484.

7. R.K. Thauer, E. Rupprecht, and K. Jungermann. 1970. Separation of 14C-formate from CO2 fixation metabolites by isoionic-exchange chromatography. Anal. Biochem. 38:461-468.

8. S. Abraham and W.Z. Hassid. 1957. The synthesis and degradation of isotopically labeled carbohydrate intermediates. Methods Enzymol. 4:489-560.

9. G. Fuchs. 1986. CO2 fixation in acetogenic bacteria: variations on a theme. FEMS Microbiol. Reviews. 39:181-213.

10. A.C. Frazer and L.Y. Young. 1986. Anaerobic C1 metabolism of the O-methyl-^{14}C-labeled substituent of vanillate. Appl. Environ. Microbiol 51:84-87.

11. R. Bache and N. Pfennig. 1981. Selective isolation of *Acetobacterium woodii* on methoxylated aromatic acids and determination of growth yields. Arch. Microbiol. 130:255-261.

STRUCTURAL AND FUNCTIONAL PROPERTIES OF A FERREDOXIN ISOLATED FROM

Methanococcus thermolithotrophicus

M.L. Fardeau[1], M. Bruschi[1], R. Cammack[2], J.P. Belaich[1]
M. Frey[3], and E.C. Hatchikian[1]

1. LCB/CNRS, BP71,13277Marseille Cdex9,F. 2.Dpt. Biochem
King's college, London W8 7AH, UK. 3. LCCMB/CNRS, Fac de
Médecine, Secteur Nord, 13326 Marseille Cedex 15, F

Ferredoxins have been isolated from organisms belonging to each major
group of archaebacteria, including the extreme halophiles, the
thermoacidophiles, and the methanogens. So far, two species of
methanogenic bacteria were known to contain ferredoxin: Methanosarcina
(3,7) and more recently Methanobacterium thermoautotrophicum ΔH (8).
In this paper, we report the structural and functional properties of the
first ferredoxin isolated from a chemolithotrophic methanogen only
growing on CO_2 and H_2 or formate, Methanococcus thermolithotrophicus
(4).

M. thermolithotrophicus ferredoxin has a molecular weight of 7262 Da,
calculated from the amino-acid composition. The amino-acid sequence of
the protein was determined (figure 3). It contains 60 residues and is
characterized by the presence of five lysines -an uncommon feature in
ferredoxins- and like the ferredoxin of Methanosarcina thermophila , by
the lack of histidine and arginine residues. The characteristic pattern
of the eight cysteine residues per molecule suggested the presence of
two 4(Fe-4S) clusters. This hypothesis was confirmed by ESR spectroscopy
analysis, which showed a typical signal of a ferredoxin where there is a
spin-spin interaction between 2 (4Fe-4S) clusters. The oxidized protein
exhibited a minor ESR signal corresponding to an oxidized (3Fe-4S)
cluster, (figure 1). Methanococcus ferredoxin shows a thermal stability
similar to that of highly thermostable ferredoxins isolated from
thermophilic bacteria.

The biological activity of the ferredoxin was investigated in different
low-redox-potential reactions, including pyruvate dehydrogenase,
hydrogenase and CO dehydrogenase activities(4). Though the extract of M.
thermolithotrophicus exhibited these different enzymatic activities,
ferredoxin has been found to be only involved in electron transport from
CO dehydrogenase complex. The reduction of pure ferredoxin by CO
dehydrogenase activity present in the extract free of ferredoxin and F420
was monitored at 385 nm. A low activity leading to a partial reduction of
ferredoxin was detected. This reduction was further substantiated by
using metronidazole, an artificial electron acceptor which is chemically
reduced by ferredoxin. The rate of metronidazole reduction depended on
the ferredoxin concentration (figure 2) as the activity increased
sixfold when the concentration of ferredoxin was increased from 0.15 to 2
nmol.

Microbiology and Biochemistry of Strict Anaerobes Involved in Interspecies Transfer
Edited by J.-P. Bélaich et al.
Plenum Press, New York, 1990

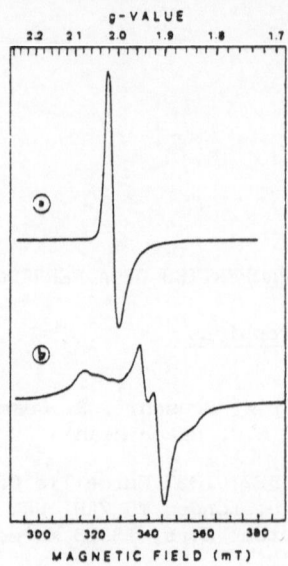

Figure 1   ESR spectra of M. thermolithotrophicus ferredoxin, 0.18 mM in
0.5 M Tris-HCl (pH 7.6). (a) Ferredoxin as prepared. (b) Ferredoxin
reduced with 2 mM dithionite at 20°C for 30 min.

This suggests that the ferredoxin may function as an electron donor,
while the ferredoxin of acetate-grown Methanosarcina thermophila, which
also requires electron transport from the CO dehydrogenase complex,
functions as an electron acceptor.

 Since the determination of the first three-dimensional structure of a
ferredoxin (from Peptococcus aerogenes) by Adman et al (1), several
structures of ferredoxins have been solved through X-ray crystallographic
methods (2,5,9,10). The four atomic models which have been established
so far show remarkable similarities in the chain folding and in the
chelation of the iron-sulfur clusters in spite of substantial differences
in their amino-acid sequences and in the content and the number of their
iron sulfur clusters ((4Fe-4S) or (3Fe-4S),).
          The three-dimensional structure of the similar ferredoxin from
M. thermolithotrophicus (FdMt)(figure 3) has been modelled with   the
X-ray structure of the ferredoxin from Peptococcus aerogenes (FdPa, 1)
as the initial template. Computer graphics (program TOM) and energy
minimization techniques were used.

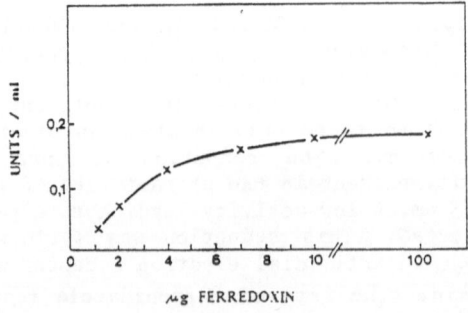

Figure 2   Requirement of ferredoxin for the reduction of metronidazole
by CO dehydrogenase activity of the extract free of carriers.

508

In the resulting model the two (4Fe-4S) clusters are chelated to the protein through four cysteic residues respectively. But unlike the clusters of FdPa, their hydrophobic environments differ from each other.

FdPa   .AYVINDSCIA..CGACKPECPVNCIQQ..GSIYAIDADSCIDCGSCASVCPVGAPNPED

FdMt   SVTIDYDKCKGPECAECVNACPMEVFEIQGDKVVVAKEDDCTFCMVCVDVCPTDAITVKE

Figure 3. Alignement of the amino-acid sequence of Peptococcus aerogenes (FdPa) and Methanococcus thermolithotrophicus (FdMt) ferredoxins, (a dot refers to a gap to make all alignments most probable)

One remarkable feature of the present model is also the patch of three of the five lysines which lie on one side of the molecule, which shows thus a dipolar character. Both these modifications should correspond obviously to important differences between the redox-potentials and the electron transfer properties of FdMt and FdPa.

It is now generally admitted that thermal stability in proteins results from many small changes over the polypeptide chain (see ref.6). Our atomic model suggests that the major stabilizing factors in FdMt -by comparison with FdPa- could be the hydrophobic core of the molecule and an hydrogen bonding network which stabilizes the common N and C termini regions.

References

1 Adman, E.T., Sieker, L.C. and Jensen L.H..1973. The Structure of a Bacterial Ferredoxin. J. Biol. Chem. 248: 3897-3996.

2 Fukuyama, K.,Nagahara, Y., Tsukihara, T. and Katsube Y.. 1988. Tertiary structure of Bacillus thermoproteolyticus (4Fe-4S) ferredoxin.   J. Mol. Biol. 199: 183-193.

3 Hatchikian, E.C., Bruschi M., Forget N., and Scandellari M. . 1982. Electron transport components from methanogenic bacteria: the ferredoxin from Methanosarcina barkeri (strain Fusaro). Biochem. Biophys. Res. Commun. 109 : 1316-1323.

4 Hatchikian E.C., Fardeau M.L., Bruschi M., Belaich J.P., Chapman A., and Cammack R.. 1989. Isolation, characterization, and biological activity of the Methanococcus thermolithotrophicus ferredoxin. J. Bacteriol. 171: 2384-2390.

5 Kissinger, C.R., Adman, E.T., Sieker, L.C. Jensen L.H. and LeGall J. 1989. The crystal structure of the three-iron ferredoxin II from Desulfovibrio gigas. FEBS Letters. 244: 447-450.

6 Menéndez-Arias, L. and Argos P.. 1989. Engineering protein thermal stability. J. Mol. Biol. 206: 397-406

7 Moura J., Moura J.J.G., Huynh B.H., and Santos H. 1982. Ferredoxin from Methanosarcina barkeri: evidence for the presence of a three-iron center. Eur. J. Biochem. 126: 95-98.

8 Reeve J.N., Beckler G.S., Cram D.S., Hamilton P.T., Brown J.W., Krzycki J.A., Kolodziej A.F., Alex L. Orme-Johnson and Walsh C.T. 1989. A hydrogenase linked gene in Methanobacterium thermoautotrophicum H encodes a poly-ferredoxin. P.N.A.S., in press.

9 Stout, G.H. Turley, S., Sieker L.C. and Jensen L.H.. 1988. Structure of ferredoxin I from Azotobacter vinelandii. Proc. Natl. Acad. Sci. USA. 85: 1020-1022.

10 Stout C.D.. 1989. Refinement of the 7 Fe Ferredoxin from Azotobacter vinelandii at 1.9 A° resolution.J. Mol. Biol. 205: 545-555.

11 Terlesky K.C., and J.G. Ferry. 1988. Ferredoxin requirement for electron transport from the carbon monoxide dehydrogenase complex to a membrane-bound hydrogenase in acetate-grown Methanosarcina thermophila J. Biol. Chem. 263: 4075-4079.

DESULFOVIBRIO GIGAS HYDROGENASE: CRYSTALLOGRAPHIC STUDIES

Michel Frey [1], Christian Cambillau [1], Vincent Nivière [2] and Claude Hatchikian [2]

[1] LCCMB-CNRS Faculté de Médecine- Nord Bd. Pierre Dramard 13326 MarseilleCDX15

[2] LCB-CNRS 31 Chemin Joseph Aiguier Marseille CDX 9

Hydrogenases, which are widely distributed in bacterial and algal species catalyse the reversible oxidation of molecular hydrogen as indicated by the equation :

$$H_2 \rightleftharpoons 2\,H^+ + 2\,e^-$$

They play an important role in energy-linked electron transfer of many organisms that utilize hydrogen as a source of energy (Adams, Mortenson and Chen, 1981; Odom and Peck, 1984; Fauque et al. , 1988).

Two types of hydrogenases have been characterized in different species of sulfate-reducing bacteria belonging to the genus Desulfovibrio: A type containing one nickel atom and several iron- sulfur clusters - including in some cases a selenium atom- and a type containing iron- sulfur clusters only (for review see Hatchikian et al., this meeting).

The hydrogenase from D. gigas (M. W. 88 kD) consists of two subunits of 62 kD and 26 kD respectively. It contains one nickel atom, one (3Fe -4S) cluster and two (4 Fe- 4S) clusters per molecule (Hatchikian et al., 1978; Cammack et al., 1982; LeGall et al. 1982; Teixeira et al., 1983). The enzyme, which is oxygen stable is capable of undergoing reversible oxidation in the presence of reducing agents. Three different states of the enzyme have been identified according to the catalytic activity (Fernandez et al., 1985, 1986): two inactive forms termed "unready" and "ready" which are converted, more or less easily, under reducing conditions to the active form capable of reacting catalytically with hydrogen. The activation of the enzyme involves reduction of the nickel atom and possibly of the iron-sulfur centres, followed by an alteration of the coordination state of the nickel atom (Fernandez et al., 1986).

The genes encoding the two subunits of the D. gigas hydrogenase have been cloned and sequenced and the amino- acid sequence determined from the nucleotide sequence (Voordouw et al. 1989). The crystallization and the determination through X- ray crystallographic methods of the three- dimensional structure have been undertaken to get structural data on the detailed architecture of the molecule : localization of the four redox centres within - or between - the two subunits and folding of the two polypeptidic chains. This should also pro-

Microbiology and Biochemistry of Strict Anaerobes Involved in Interspecies Transfer
Edited by J.-P. Bélaich *et al.*
Plenum Press, New York, 1990

511

vide a structural basis for a better understanding of the electronic exchanges between the hydrogenase and the tetrahaemic cytochrome c3 (M.W. 13000) which is its obligate physiological partner and the three-dimensional structure of which is known (Haser et al., 1979).

Single crystals of D. gigas hydrogenase have been produced at pH 6.5 (Nivière et al. , 1987) with either polyethylene- glycol (PEG, m.w. 6000; form A) or ammonium sulfate (form B) as precipitants . The form A crystals are orthorhombic : a= 125 Å , b = 200 Å, c= 136 Å; Space group -S.G.- C222 or C222₁ with probably two hydrogenase molecules per asymmetric units. The form B crystals are monoclinic : a= 257 Å, b= 185 Å, c= 148 Å and ß= 101°, S.G. C2 with probably eight hydrogenase molecules per asymmetric unit. Partial X-ray diffraction patterns have been collected at the L.U.R.E in Orsay and at Daresbury in Great- Britain with synchrotron radiation. They extend to 6 Å resolution for the form A crystals and to 3 Å - 4.5 Å for the form B crystals.

To check the activity of the crystallized hydrogenase we washed , dried and dissolved a single crystal in 100 mM-Tris.HCl buffer at pH 8.5. The kinetic of hydrogen- uptake activity of the solution (Fernandez et al., 1985) is comparable to the one of the freshly isolated protein. This solution could also be activated after a five hours incubation in the presence of 1mM reduced methyl viologen or 1 atmosphere hydrogen .

On the other hand, we incubated under hydrogen , overnight, a crystal in its mother liquor maintained in a sealed vial. The crystal kept its shape and color. We dissolved it, again, in 100 mM-Tris.HCl buffer at pH 8.5. The measured activity of this solution was clearly characteristic of the active form, thus indicating that the crystallized enzyme is fully activated after a prolonged incubation in reducing conditions, like the freshly purified enzyme. All these experiments were performed under anaerobic conditions (with argon or gas-tight syringes).

It is now established that hydrogenases can be crystallized (Higuchi et al., 1987 ; Nivière et al. 1987) and that, therefore, a three- dimensional structure could be solved in the near future, all the more as the corresponding chemical sequences have been determined. Moreover it has been shown that the crystallized D. gigas hydrogenase can be reactivated like the freshly purified enzyme, indicating that the enzyme has been crystallized in the native form.

REFERENCES

Adams , M.W.W. , Mortenson , L.E. and Chen , J.-S. , 1981 . B.B.A. , 594 , 105 - 176 .

Cammack , R. , Patil , D. , Aguirre , R. and Hatchikian , E.C. , 1982 . FEBS Letters , 142 , 289 - 292 .

Fauque , G. , Peck , H.D. , Moura , J.J.G. , Huynh , B.H. , Berlier , Y. , DerVartanian , D.V. , Teixeira , M. , Przybyla , A.E. , Lespinat, P.A. , Moura , I. and Le Gall , J. ,1988 . FEMS Microbiology Reviews , 54 , 299 - 344 .

Fernandez , V.M. , Hatchikian , E.C. and Cammack , R. , 1985 . Biochim. Biophys. Acta , 832 , 69 - 79 .

Fernandez , V.M. , Hatchikian , E.C. , Patil , D.S. and Cammack , R. , 1986 . Biochim. Biophys. Acta , 883 , 145 - 154 .

Haser , R. , Pierrot , M. , Frey , M. , Payan , F. , Astier , J.P. , Bruschi , M. and Le Gall, J. , 1979 . <u>Nature</u> , 282 , 806 - 810 .

Hatchikian , E.C. , Bruschi , M. and Le Gall , J. , 1978 . <u>Biochem. Biophys. Res. Commun.</u> , 82 , 451 - 461 .

Higuchi , Y. , Yasuoka , N. , Kakudo , M. , Katsube , Y. , Yagi , T. and Inokuchi , H. , 1987 . <u>J. Biol. Chem.</u> , 262 , 2823 - 2825 .

Le Gall , J. , Ljungdahl , P.O. , Moura , I. , Peck , H.D. , Xavier , A.V. , Moura , J.J. , Teixeira , M. and DerVartanian , D.V., 1982 . <u>Biochim. Biophys. Res. Commun.</u> 106 , 610 - 616 .

Nivière ,V. , Hatchikian , E.C. , Cambillau , C. and Frey , M . , 1987 . <u>J. Mol. Biol.</u> 195, 969 - 971 .

Odom , J. M. and Peck , J.D. , 1984 . <u>Ann. Rev. Microbiol.</u> 38 , 551 - 592 .

Teixera , M. , Moura , J.J. , Fauque , G. , Czechowski , M. , Berlier , Y. , Lespinat , P.A. , Le Gall , J. , Xavier , A.V. and Moura , I., 1986 . <u>Biochimie</u> , 68 , 75 - 84 .

Voordouw , G. , Menon , N.K. , Choi , B.S. , Peck , H.D. and Przybyla , A.E. , 1989 . <u>J. Bact</u> . 171 (under press) .

Kinetic studies of electron transfer between hydrogenase and cytochrome

c_3 by electrochemistry

J. Haladjian, P. Bianco, F. Guerlesquin, M. Bruschi, V.

Nivière and C. Hatchikian

Laboratoire de Chimie Bactérien-
ne, B.P. 71 - 13277 Marseille Cedex 9 - France

Three soluble molecular forms of hydrogenase have been isolated from various Desulfovibrio species (1). Well characterized hydrogenases representative of each of these forms are : the periplasmic hydrogenase from D. vulgaris (2) which contains exclusively non-heme iron (Fe hydrogenase) ; the periplasmic nickel non-heme iron selenium hydrogenase ((Ni Fe Se) hydrogenase) from D. gigas (3, 4) ; and the nickel iron selenium hydrogenase (Ni Fe Se) hydrogenase) from D. desulfuricans (5). These hydrogenases differ in their metal centre composition, mechanistic properties (6), sensitivity to inhibitors (7), amino acid sequences (1) and immunological properties (8). All these hydrogenases have in common the capability to reduce directly the tetrahemic low potential cytochrome c_3 under hydrogen atmosphere (9, 10).

The two nickel-type hydrogenases exhibit a molecular mass of approx. 90 kD and consist of two distinct subunits of 26 kD and 62 kD (5, 11). The (NiFe) hydrogenase typified by the periplasmic enzyme from D. gigas, contains 1 nickel atom, 11 iron atoms and 12 acid-labile sulfur atoms organized in a (3Fe-4S) centre and two (4Fe-4S) clusters (12-14). The (Ni Fe Se) hydrogenase (D. desulfuricans Norway) lacks the (3Fe-4S) cluster but contains one atom of selenium in addition of the nickel centre and two (4Fe-4S) clusters (5).

Nickel-containing hydrogenases from Desulfovibrio species lose most of their H_2-uptake activity under oxidizing conditions ; this loss of activity is a reversible process. D. desulfuricans Norway soluble hydrogenase is rapidly activated in the presence of strong reductants (5, 15) whereas D. gigas enzyme requires a long-time incubation under reducing conditions in order to express its full activity (15, 16).

Electrochemical techniques offer a convenient way for measuring fast rates of electron transfer between proteins and are an alternative to stopped-flow methods. The electrochemical approach has been used to study the interactions between two physiological partners from two Desulfovibrio electron transfer chains, hydrogenase and cytochrome c_3 from D. desulfuricans Norway and from D. gigas.

The electron transfer chain can be simulated using the following electrochemical model.

$$\frac{1}{2} \text{H}_2 \quad \Bigg) \Bigg(\begin{array}{c} \text{Hase}_{ox} \uparrow \\ \\ \downarrow \text{Hase}_{red} \end{array} \Bigg) \Bigg(\begin{array}{c} \uparrow \text{cyt } c_{3,\,red} \\ \\ \text{cyt } c_{3,\,ox} \searrow \end{array} \Bigg) \Bigg[\text{ electrode}$$

$$\text{H}^+$$

In the presence of substrate (here hydrogen), cytochrome c_3 enzymatically reduced by hydrogenase is electrochemically re-oxidized and an enhanced anodic current is detected. (Fig. 1A). (17,18). The dependence of kinetic current on time is shown in Fig. 1B. The establishment of the catalytic current is virtually instantaneous for D. desulfuricans Norway system but an evolution upon the addition of D. gigas hydrogenase is observed over a relatively-long time.

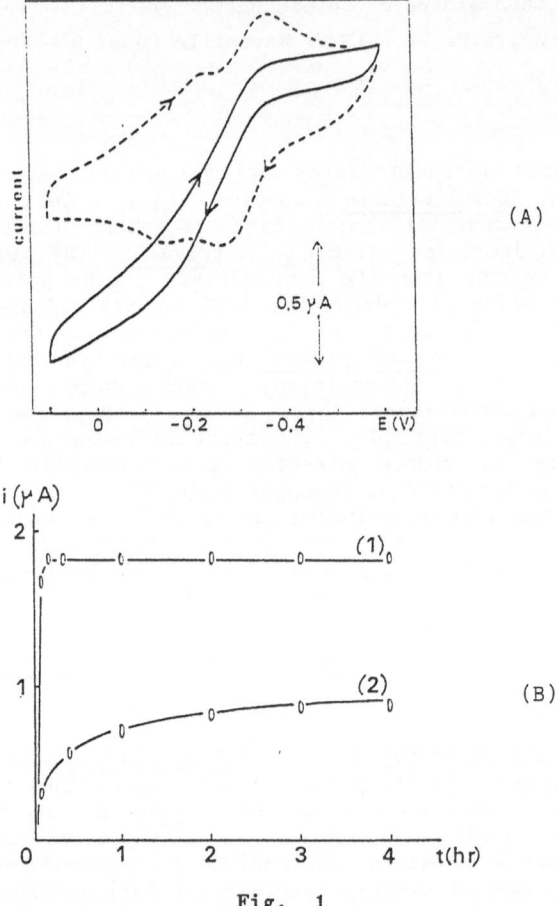

Fig. 1

(A) Cyclic voltammograms at the PG electrode of 42 μM D. gigas cytochrome c_3 (...) without hydrogenase, (—--) with 15 nM D. gigas hydrogenase after 3h. Scan rate : 5 mVs^{-1}. (B) Dependence of the kinetic current on time (1) 42 μM D. desulfuricans Norway cytochrome c_3 + 17 nM hydrogenase (2) 47 μM D. gigas cytochrome c_3 + 15 nM hydrogenase.

The second-order rate constant of the electron transfer between both physiological partners can be determinated. Results are as follows :

	Hase/cyt c_3 from <u>D.d.</u>Norway	Hase/cyt c_3 from <u>D.gigas</u>
k	3×10^7 M^{-1} s^{-1}	6.5×10^7 M^{-1} s^{-1}

It is assumed that both partners interact specifically via a mechanism which involves the formation of a complex, followed by a rapid intramolecular exchange.

Moreover it must be underlined that the high reactivity and the sensitivity of the technique used permit to detect hydrogenase concentrations as low as 10 nM (1 µg cm^{-3}).

REFERENCES

1. Prickril, B.C., He, S-H., Li, C., Menon, N., Choi, E-S., Przybyla, A.E., Der Vartanian, D.V., Peck, H.D. Jr., Fauque, G., Le Gall, J., Teixeira, M. Moura, I., Moura, J.J.G., Patil, D. and Huynh, B.H. (1987). Biochim. Biophys. Res. Commun. 149, 369-377.
2. Van der Westen, H., Mayhew, S.G. and Veeger, C. (1978). FEBS Lett. 86, 122-126.
3. Cammack, R., Patil, D., Aguirre, R. and Hatchikian, E.C. (1982). FEBS Lett. 142, 289-292.
4. Le Gall, J., Ljungdahl, P.O., Moura, I., Peck, H.D. Jr. Xavier, A.V. Moura, J.J.C., Teixeira, M., Huynh, B.H., and Der Vartanian, D.V. (1982). Biochem. Biophy. Res. Commun. 106, 610-616.
5. Rieder, R., Cammack, R. and Hall, D.O. (1984). Eur. J. Biochem. 145, 637-643.
6. Lespinat, P.A., Berlier, Y., Fauque, G., Czechowski, M., Dimon, B. and Le Gall, J. (1986). Biochimie, 68, 55-61.
7. Berlier, Y., Fauque, G., Le Gall, J., Choi, E.S., Peck, H.D., Jr. and Lespinat, P.A. (1987). Biochem. Biophys. Res. Commun. 146, 147-153.
8. Aketagawa, K.J., Kobayashi, K. and Ishimoto, M. (1983). J. Biochem. 93, 755-762.
9. Yagi, T., Honya, M. and Tamiya (1968). Biochim. Biophys. Acta 153 : 699-705.
10. Bell, G.R., Lee, J.P., Peck, H.D. Jr. and Le Gall, J. (1978). Biochem. Biophys. Res. Commun. 82, 451-461.
11. Hatchikian, E.C., Bruschi, M. and Le Gall, J. (1978). Biochem. Biophys. Res. Commun. 82, 451-461.
12. Teixeira, M., Moura, I., Xavier, A.V., Huynh, B.H., Der Vartanian, D.V., Peck, H.D. Jr., Le Gall, J. and Moura, J.J.G. (1985). J. Biol. Chem. 260, 8942-8950.
13. Fernandez, V.M., Hatchikian, E.C., Patil, D.S. and Cammack, R. (1986). Biochim. Biophys. Acta 883, 145-153.
14. Cammack, R., Patil, D.S., Hatchikian, E.C. and Fernandez, V.M. (1987). Biochim. Biophys. Acta 912, 98-109.
15. Cammack, R., Fernandez, V. and Schneider, K. (1986). Biochimie 68, 85-91.
16. Fernandez, V.M., Hatchikian, E.C. and Cammack, R. (1985). Biochim. Biophys. Acta 832, 69-79.
17. Haladjian, J., Bianco, P., Guerlesquin, F. and Bruschi, M. (1987). Biochem. Biophys. Res. Comm., 147, 1289-1294.
18. Nivière, V., Hatchikian, E.C., Bianco, P. and Haladjian, J. (1988). Biochim. Biophys. Acta. 935, 34-40.

HYDROGENASE IN *DESULFOBACTER*

Torleiv Lien and Terje Torsvik

University of Bergen, Norway
Department of Microbiology and Plant Physiology
Jahnebakken 5, N-5007 Bergen, Norway

Abstract

Hydrogenase activity has been detected in all the *Desulfobacter* species tested. The results indicate that *Desulfobacter* species contain only one hydrogenase with low specific activity.

Introduction

Our knowledge of hydrogenases in sulfate reducing bacteria is mainly derived from studies with the genus *Desulfovibrio* from which three types of hydrogenases have been isolated. The present work gives results from the assay of hydrogenase activity in the following *Desulfobacter* species: *D. postgatei, D. hydrogenophilus, D. curvatus* and strain B 54 (isolated from an oilfield in the North Sea).

Results and discussion

The bacteria were grown at 30^0 C in a mineral medium with acetate as the carbon and energy source. Hydrogen consumption was assayed by following the reduction of benzyl viologen in anaerobic cuvetts with hydrogen in the gas phase. Significant specific hydrogen-consumption activity was found in whole cells of all the *Desulfobacter* species tested. Although *D. curvatus* is unable to grow on hydrogen, it had the highest activity in whole cells, 350 nmole H_2 x min^{-1} x mg^{-1} protein, and higher than found for the hydrogenutilizer *D. hydrogenophilus*.
Hydrogen-evolution activity, from reduced methyl viologen, with whole cells of *D. posgatei* and strain B 54, was lower than the hydrogen-consumption activity.
For *D. curvatus* and *D. postgatei* the cytoplasma fraction had a higher specific hydrogen-consumption activity than the membrane fraction. The opposite was the case for *D. hydrogenophilus*.

Native polyacrylamide electrophorecis, followed by staining of the hydrogenase activity, revealed only one band with activity and with the same migration for the membrane and cytoplasma fractions.

Antibodies against the three different hydrogenases from *Desulfovibrio* were challenged with extracts from *D. postgatei* and strain B 54 in an immublotting (Western blots). A crossreaction was observed with the heaviest component (62 kD) of the nickel-iron

Microbiology and Biochemistry of Strict Anaerobes Involved in Interspecies Transfer
Edited by J.-P. Bélaich *et al.*
Plenum Press, New York, 1990

519

hydrogenases from *Desulfovibrio gigas* and *D. vulgaris* indicating that the *Desulfobacter* hydrogenase may be a nickel-iron hydrogenase.

A part of this work has been carried out in the laboratory of prof. Peck, Department of Biochemistry, University of Gerorgia, USA.

ATP-SYNTHESIS COUPLED TO THE TERMINAL STEP OF METHANOGENESIS

S. Peinemann and G. Gottschalk

Institut für Mikrobiologie
Grisebachstrasse 8
D-3400 Göttingen, FRG

Introduction

The universal methanogenic reaction from all substates is the reductive demethylation of methyl-CoM by the methyl-coenzyme M methylreductase system according to (Gunsalus and Wolfe, 1980):

$$CH_3\text{-}S\text{-}CoM + H_2 \longrightarrow CH_4 + HS\text{-}CoM$$

This highly exergonic reaction has always been considered as the step coupled with ATP synthesis. Investigations with whole cells of *Methanosarcina barkeri* led to the conclusion that methanogenesis from methanol plus H_2 gives rise to an electrochemical proton gradient, which is subsequently used to synthezise ATP (Blaut and Gottschalk, 1984).
To study energy conservation reactions on the subcellular level we have taken advantage of the methanogenic strain Göl from which everted vesicles could easily be obtained (Mayer et al., 1987). These crude vesicle preparations converted methyl-CoM and H_2 at a significant rate and independent of the addition of ATP (Deppenmeier et al., 1988). Here we report on substrate-dependent ATP synthesis by these preparations .

Results

1) ATP synthesis coupled to methane formation from H_2 plus methyl-CoM by vesicles

Vesicle preparations of strain Göl formed methane from H_2 + methyl-CoM at a rate of 33.8 nmol min^{-1} mg $protein^{-1}$ and ATP was concomitantly synthesized. Without substrate neither ATP nor CH_4 was produced.
Addition of 2-bromoethanesulfonate, an inhibitor of methyl-CoM reduction to methane, reduced the rate of CH_4 formation to 0.2% and abolished ATP synthesis completely (Table 1) indicating the dependency of ATP formation on methanogenesis.
If the proton motive force was the driving force for ATP formation via an ATP synthase it should be possible to uncouple methane formation from ATP synthesis by a protonophore. As evident from Table 1 addition of the uncoupler SF 6847 decreased ATP synthesis to 15% of the control. If SF 6847 was added in combination with the K^+-ionophor valinomycin ATP synthesis was completely abolished. However, in both cases methane formation was not affected.
The participation of an ATP synthase in methyl-CoM reduction-dependent ATP synthesis was tested by adding the ATPase inhibitor DCCD. Since DCCD did not only interfere with ATP synthesis but also with methanogenesis (Table 1) another ATPase inhibitor was used.

Diethylstilbestrol (DES) blocked ATP formation but did not affect methane production significantly (Table 1).

Table 1. Effect of various inhibitors on ATP synthesis and CH_4 formation from methyl-CoM and H_2 by vesicle preparations.

Additons	ATP (nmol min^{-1} mg protein^{-1})	CH_4 (nmol min^{-1} mg protein^{-1})
	0.34	33.8
2-bromoethanesulfonate (10 mM)	0.00	0.1
SF 6847 (5 nmol/mg prot.)	0.05	34.1
Valinomycin (1.25 nmol/mg prot.)	0.20	34.2
SF 6847 (5 nmol/mg prot.) + Valinomycin (1.25 nmol/mg prot.)	0.00	34.5
DCCD (100 nmol/mg protein)	0.00	1.2
Diethylstilbestrol (60 nmol/mg protein)	0.00	28.0
Na-vanadate (60 nmol/ mg protein)	0.33	32.5
Sulfobetaine (0.32 %)	0.00	0.2
Sulfobetaine + 10 mM Titan(III)citrate	0.00	22.8

Each value is an average of three determinations.

Addition of the detergent sulfobetain decreased ATP and CH_4 formation by 100% (Table 1). The methyl reductase activity could be restored to 68% if titanium(III)citrate was added as electron donor but this methanogenesis was not accompanied by ATP synthesis. This indicates that an intact cytoplasmic membrane has a function in methanogenic electron transfer. To confirm that the observed substrate-dependent increase in ATP concentration was due to a net synthesis of ATP from ADP and Pi, the incorporation of added [32P]phosphate was monitored. [32P]phosphate was incorporated at a rate five times higher if methyl-CoM was present. Addition of SF 6847/valinomycin or DES reduced the rate to the level observed without substrate.

2) ATP synthesis driven by an artificial transmembrane gradient of protons

To further substantiate the role of a proton gradient in these vesicle preparations artificial pH gradients were imposed across the vesicular membranes. This was achieved by diluting vesicles loaded with 0.5 M NH_4Cl in a buffer containing 0.5 M choline chloride (Nakamura et al., 1986). A ΔpH of 3.0 units was able to drive ATP synthesis at a rate of 3.9 nmol/min mg protein and this ATP formation was inhibited if DES, the combination SF 6847/valinomycin or the detergent sulfobetaine were added.
To monitor the transmembrane pH gradient the fluorescent dye acridine orange was used. The formation of a ΔpH could be visualized from fluorescence quenching of this weak base. Addition of SF 6847/valinomycin prevented quenching of fluorescence. On the other hand DES did not affect the magnitude of acridine orange fluorescence .

Conclusions

1. ATP synthesis by vesicle preparations of strain Gö1 depends on methanogenesis from methyl-CoM and H_2

2. The increase in the ATP concentration is due to a net formation of ATP from ADP and P_i.

3. Substrate-dependent ATP synthesis requires an uncoupler-sensitive transmembrane proton gradient and an intact ATP synthase. This gradient is formed in the process of methanogenesis from methyl-CoM and H_2.

4. Artificially created transmembrane H^+-gradients are capable of driving ATP synthesis in vesicles, demonstrating the presence of an H^+-ATPase.

References

Blaut, M. and Gottschalk, G., 1984, Coupling of ATP synthesis and methane formation from methanol and molecular hydrogen in *Methanosarcina barkeri*. Eur. J. Biochem. 141, 217-222

Deppenmeier, U., Jussofie, A., Blaut, M. and Gottschalk, G., 1988, A methyl-CoM methylreductase system from methanogenic strain Gö1 not requiring ATP for activity FEBS Lett. 241, 60-64

Gunsalus, R.P. and Wolfe, R.S., 1980, Methyl coenzyme M reductase from *Methanobacterium thermoautotrophicum*. J. Biol. Chem. 255, 1891-1895

Mayer, F., Jussofie, A., Salzmann, M., Lübben, M., Rhode,M. and Gottschalk, G., 1987, Immunoelectron microscopic demonstration of ATPase on the cytoplasmic membrane of the methanogenic bacterium strain Gö1. J. Bacteriol. 169, 2307-2309

Nakamura, T., Hsu, C. and Rosen, B.P, 1986, Cation/proton antiport systems in *Escherichia coli*. J. Biol. Chem. 261, 678-683

POSTERS - 3 - GENETICS

EXPRESSION OF AN EUBACTERIAL PUROMYCIN RESISTANCE GENE IN THE ARCHAEBACTERIUM *METHANOCOCCUS VOLTAE*

Odile Possot, Petra Gernhardt[†], Maryline Foglino[†], Albrecht Klein[†] and Lionel Sibold

Unité de Physiologie Cellulaire, Département des Biotechnologies, Institut Pasteur, 25 rue du Dr. Roux, 75724 Paris Cedex 15, France
[†]Molekulargenetik, Fachbereich Biologie, Philipps-Universität, D-3550 Marburg/Lahn (FRG)

INTRODUCTION

Physiological and biochemical studies of methanogenic archaebacteria at the molecular level are in constant progress[1]. However, further advances are hampered by the lack of genetic tools in these microorganisms. Recently, low frequency transformation has been reported in *Methanococcus voltae*[2] and *Methanobacterium thermoautotrophicum Marburg* sp.[3]. An important need is the development of cloning and expression vectors for gene tranfer. A long term goal is the construction of methanogenic strains with properties useful for applied purposes[4]. A major obstacle for the construction of an useful vector was the lack of suitable selective markers since methanogens are resistant to most of the antibiotics that are used for selection of recombinant plasmids in eubacteria. However, we have recently shown that the transformable strain *M.voltae* is sensitive to puromycin and fusidic acid, two antibiotics for which eubacterial resistant genes are known[5].

We report here that the puromycin resistant gene from *Streptomyces alboniger* encoding the puromycin acetyltransferase (*pac* gene) can be expressed in *M.voltae*. This gene is a suitable selectable marker for genetic engineering of methanogenic archaebacteria.

RESULTS AND DISCUSSION

Puromycin inhibits at the level of protein synthesis the growth of *M. voltae* with a MIC of 2 μg/ml[5]. The puromycin is converted to acetyl- puromycin by the product of the *S. alboniger pac* gene[6]. We have shown that acetyl-puromycin is also inactive *in vivo* and *in vitro* against *M. voltae*[5].

These results prompted us to check if the eubacterial *pac* gene might confer the resistance to puromycin to *M. voltae*. For this purpose, we constructed a plasmid in which the *pac* gene is flanked by the strong promoter and terminator of the gene encoding the component C of the *M. voltae* methyl CoM reductase[7,8] (Fig.1). This expression unit was inserted in *M. voltae* genes previously cloned and sequenced (*his*A, *trp*B,A and *gln*A)[9-11] to introduce a region of homology with the chromosome and promote recombination *in vivo*. The construction with *his*A, is reported in Fig.1.

Plasmids Mip 1 and 2 were used to transform *M.voltae* according to the method II described by Bertani and Baresi[2] and drug resistant clones on solid complex media containing 10 μg/ml of puromycin were selected. Seven clones were analyzed: 1A, 1B, 6, 7 and 8 deriving from Mip 1; 2C and 2D deriving from Mip 2. All of them were His[+] except

Microbiology and Biochemistry of Strict Anaerobes Involved in Interspecies Transfer
Edited by J.-P. Bélaich *et al.*
Plenum Press, New York, 1990

527

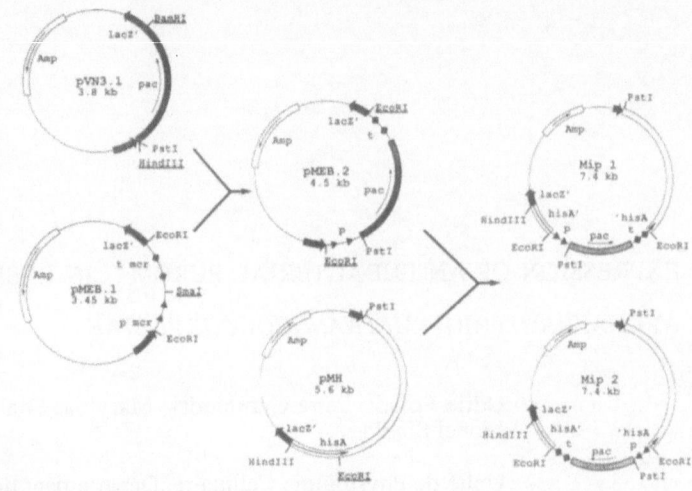

Fig.1. Construction of the "expression unit" pMEB.2 and its cloning into the *M. voltae hisA* gene, yielding Mip 1 and Mip 2.

pVN3.1 was obtained from Dr. A. Jimenez.

Abbreviations: p mcr (▶) and t mcr (◆) promoter and terminator regions of the methyl - CoM - reductase transcription unit of *M. voltae*. Underlined restriction sites were used for the constructions.

Fig.2 . Characterization of the *M. voltae* puromycin resistant clones 1A, 1B, 6, 7, 8, 2C and 2D isolated after transformation by Mip 1 and Mip 2. (The restriction patterns of transformants 6 and 8 were similar to 1A and those of 7 was similar to 1B, not shown).

I. Southern blot analysis of the integration sites of Mip. A, a *pac* probe (1.1 kbp *Pst*I-*Bam*HI electroeluted fragment from pVN3.1).
B, a *hisA* probe (2.3 kbp *Eco*RI-*Pst*I electroeluted fragment from pMH). The 2.7 kbp bands are due to a pUC vector contamination of the *hisA* probe. In line 2D, the 2.9 kbp signal must be a double band.
C, a pUC probe (2.7 kbp pUC18 vector hydrolysed by *Hind*III). Total *M. voltae* transformants and wild type (wt) DNAs were restricted by both *Hind*III and *Pst*I.

II. the supposed integration patterns of the vectors in the *M.voltae* chromosomes.

Sizes of linear fragments are in kb.

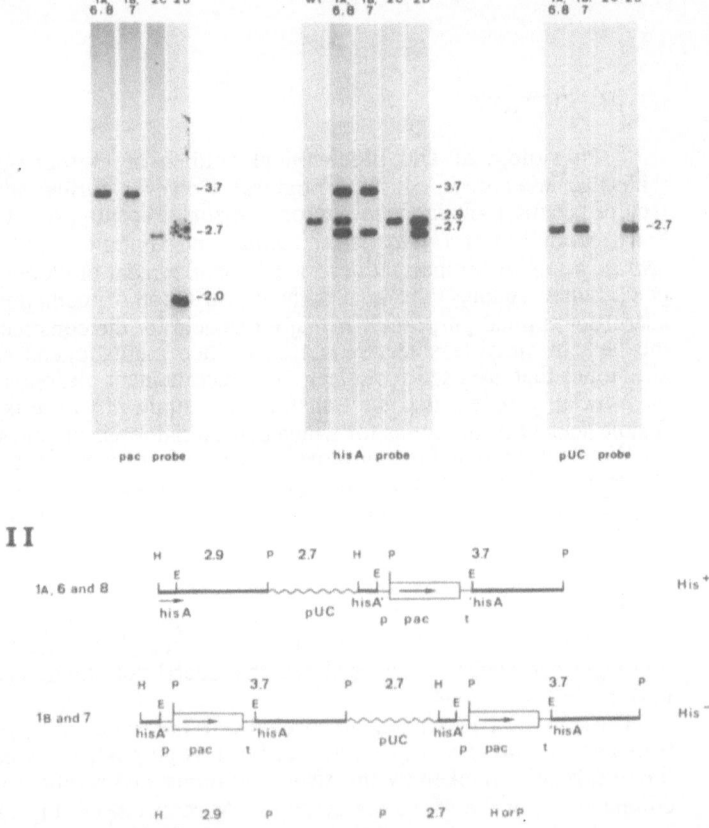

Southern hybridization of total DNA isolated from these clones was performed using *pac, hisA,* and pUC probes. From results reported in Fig. 2 and additional restriction analyses not shown, the puromycin resistant clones appear to belong to three categories:

i) Clones 1A, 6, 8 and 2D result probably from the integration of Mip 1 into the chromosomal *hisA* region by a single recombination event. One may speculate that the recombination event occurred downstream from the expression unit, such that the reconstituted *hisA* gene remains under the control of its own promoter. Indeed, the restriction pattern of the *hisA* region of these clones is modified as compared to the wild type. The DNA fragments hybridizing with the *pac* probe hybridize also with the *hisA* probe and a complete pUC sequence is detected.

ii) In clones 1B and 7, all DNA fragments are similar to those observed with clones 1A, 6 and 8, except the 2.9 kb *Hind*III-*Pst*I fragment which is not detected. This might suggest an event of gene conversion.

iii) In clone 2C, the DNA fragment hybridizing with the *hisA* probe has a normal size and no hybridization is observed with the pUC probe. It is therefore likely that recombination occurred outside of the *hisA* region, but no explanation can account yet for the excision of the pUC vector.

Stability of the *pac* marker was checked by growing the recombinant strains for 20 generations without selection pressure.

CONCLUSION

Our work shows that the *pac* gene from the eubacterium *S. alboniger* can be expressed and maintained in the methanogenic archaebacterium *M.voltae,* conferring puromycin resistance up to 10 μg/ml. This is the first demonstration of the expression in an archaebacterium of a gene isolated from an organism which belongs to one of the two other primary kingdoms.

The demonstration that the *pac* gene from *S. alboniger* is a suitable selectable marker for *M. voltae* opens the way for the construction of cloning vectors in this archaebacterium. This marker might be used for genetic engineering of other methanogenic strains since all methanogens are sensitive to puromycin[12].

References

1. J.W. Brown, C.J. Daniels and J.N. Reeve, Gene structure, organization and expression in archaebacteria, *CRC, Crit. Rev. Microbiol.* 16: 287 (1989).
2. G. Bertani and L. Baresi, Genetic transformation in the methanogen *Methanococcus voltae* PS, *J. Bacteriol.* 7: 398 (1987).
3. U.E. Worrell, D.P. Nagle, D. Mc Carthy and A. Eisenbraun, A genetic transformation system in the archaebacterium *Methanobacterium thermoautotrophicum* strain Marburg, *J. Bacteriol.* 170: 653 (1988).
4. J. Konisky, Methanogens for biotechnology: application of genetics and molecular biology, *Trends Biotechnol.* 7: 88 (1989).
5. O. Possot, P. Genhardt, A. Klein and L. Sibold, Analysis of drug resistance in the archaebacterium *Methanococcus voltae* with respect to potential use in genetic engineering, *Appl. Environ. microbiol.* 54: 734 (1988).
6. R. A. Lacalle, D. Pulido, J. Vara, M. Zalacain and A. Jimenez, Molecular analysis of the *pac* gene encoding a puromycin *N*-acetyltransferase from *Streptomyces alboniger*, *Gene* 79: 375 (1989).
7. R. Allmansberger, S. Knaub and A. Klein, Conserved elements in the transcription initiation regions preceding highly expressed structural genes of methanogenic archaebacteria, *Nucl. Acid. Res.* 16: 7419 (1988).
8. B. Müller, R. Allmansberger and A. Klein, Termination of a transcription unit comprising highly expressed genes in the archaebacterium *Methanococcus voltae, Nucl. Acid. Res.* 13: 6439 (1985).
9. D. Cue, G.S. Beckler, J.N. Reeve and J. Konisky, Structure and sequence divergence of two archaebacterial genes, *Proc. Natl. Acad. Sci. USA,* 82: 4207 (1985).
10. L. Sibold and M. Henriquet, Cloning of the *trp* genes from the archaebacterium *Methanococcus voltae* : nucleotide sequence of the *trp*BA genes, *Mol. Gen. Genet.* 214, 439 (1988).
11. O. Possot, L. Sibold and J.P. Aubert, Nucleotide sequence and expression of the glutamine synthetase structural gene, *gln*A of the archaebacterium *Methanococcus voltae, Res. in Microbiol.* (in press).
12. J.L. Oliver, J.L. Sanz, R. Amils and A. Marin, Inferring the phylogeny of archaebacteria : the use of ribosomal sensitivity to protein synthesis inhibitors, *J. Mol. Evol.,* 24: 281 (1987).

CLONING AND SEQUENCING THE LOCUS ENCODING FOR THE LARGE AND SMALL SUBUNIT GENES OF THE PERIPLASMIC [NiFeS] HYDROGENASE FROM DESULFOVIBRIO FRUCTOSOVORANS

ROUSSET M., DERMOUN Z., HATCHIKIAN C.E. and BELAICH J.P.

Laboratoire de Chimie Bacterienne, 31 chemin J. Aiguier, B.P. 71 F-13277 Marseille Cedex9 FRANCE

Hydrogenases play a crucial role in the mineralization of organic matter by microbial communities. They make electron circulation possible between hydrogen donors and hydrogen oxidizing bacteria. Hydrogenases from sulfate reducing bacteria have been extensively studied. They participate in the interspecies hydrogen transfer, which is an intercellular metabolic activity, as well as in the intracellular hydrogen cycling involved in energy metabolism. We studied Desulfovibrio fructosovorans hydrogenase because of its ability to function as well as hydrogen producing or hydrogen consuming bacteria in syntrophic associations.

E. coli DH5α was used as recipient strain. The chromosomal DNA from D. fructosovorans was partially digested with first EcoRI and secondly by HindIII. Fragments ranging from 4 to 8 kb recovered from agarose gels by electroelution were ligated in pUC18 and used to transform E. coli. The Amp^r and βGal$^-$ transformants were selected.
In the first cloning campaign, about 3,000 recombinants clones were tested by immunological screening using purified antibodies directed against purified hydrogenase. One positive clone harboring a 3.5 kb EcoRI/EcoRI insert was identified and called pEH15. A Western immunoblot analysis showed that the cloned gene was coding only for the large subunit.
pEH15 recombinant plasmid was digested at SphI and XbaI polylinker sites. Overlaps were generated by exonuclease III/ exonuclease VII digests. Then, deleted double stranded recombinant plasmids were denatured by NaOH. The sequencing reaction was routinely performed with Sequenase. The pEH15 plasmid insert sequence confirmed that the whole large subunit was present, preceded however by the 300 last bp of the small subunit.

A second cloning campaign using 7 kb HindIII
DNA fragments that strongly hybridized with the pEH15 insert
led to the selection of the pHH7 recombinant plasmid
harboring a 7 kb DNA insert coding for both subunits. The
small subunit sequence was then completed by walking on the
chromosome using 17 mer oligonucleotides as primers.

The 0.94 kb gene of the small subunit preceds the 1.69
kb gene of the large subunit. The two genes are separated by
66 nucleotides and a ribosome binding site (GGAGG) is
centered at -9 bp with respect to the translational
initiator ATG codon of the two subunits. No transcriptional
initiating sequences were found to exist in the 66 bp
regions upstream from the large subunit. Nevertheless, a 12
bp inverted repeated sequence with a ΔG of -21.6 kCal/mol
which was present between the large subunit and the third
ORF, might constitute either a terminator or an
intercystronic regulatory element. A possible promoter
region, resembling a weak E. coli promoter, is present 180-
200 bp upstream from the translational start of the small
subunit. The -35 sequence seems to be very similar to E.
coli consensus sequence TTGACA, whereas, the pribnow box is
rather unlike the consensus TATAAT. The purine A176 might be
the start point of DNA transcription. All these results
strongly suggest that the two genes may constitute an
operon.
After a 66 bp no coding region downstream from the large
subunit TAA stop codon, a Shine & Dalgarno sequence is
centered at -7 bp from the ATG of a third open reading
frame.
The small subunit sequence showed a 50 amino acid leader
peptide which was sized by comparison with the protein N-
terminal sequence. The calculated molecular masses were
28.411 kDa in the case of the mature small subunit and
61.640 kDa in that of the large subunit. The hydrogenases of
D. fructosovorans and D. gigas have been compared. 64.7% and
62.7% homology was found to exist in tne case of the small
and large subunits, respectively.

```
CGTGCTGCATACGATTTCTTTATTCATATCCTTTACTTCCCCGGCCTCATCACCCCGGTA
ACACCCTTGAATCACGTGAAAAATTTGCCAAACCCCATTGACGTTCAAACAAACTCGTGA
TTAACTGCAAAAGGGAAATCGAACCCGGCGACGCAGGAACGCACGCGTCCGTCACATAGA
GGCGGCGGCGGCCGGTTCGCCTGCGACGGCCGCGTGCCGAACGGGTCAACGGATGCTACG
TGGCCTGGGGATCATGGGACGCCCCGGACCATGCCTCTCGTAAAACAAAGGAGGACGTTT
ATGAACTTTTCCGTGGGTCTTGGCAGGATGAATGCGGAAAAACGGCTTGTGCAAAACGGC
 M  N  F  S  V  G  L  G  R  M  N  A  E  K  R  L  V  Q  N  G
GTCTCCCGCCGCGACTTCATGAAATTTTGCGCCACCGTGGCCGCGGCCATGGGCATGGGC
 V  S  R  R  D  F  M  K  F  Ⓒ  A  T  V  A  A  A  M  G  M  G
CCGGCGTTCGCGCCCAAGGTCGCCGAAGCATTGACGGCCAAACACCGTCCGTCGGTGGTC
 P  A  F  A  P  K  V  A  E  A  L  T  A  K  H  R  P  S  V  V
TGGCTGCACAACGCCGAGTGCACCGGCTGCACCGAAGCGGCGATCCGGACGATCAAACCT
 W  L  H  N  A  E  Ⓒ  T  G  Ⓒ  T  E  A  A  I  R  T  I  K  P
TATATAGACGCGCTCATTCTCGACACCATCTCCCTGGATTACCAGGAGACCATCATGGCC
 Y  I  D  A  L  I  L  D  T  I  S  L  D  Y  Q  E  T  I  M  A
GCGGCCGGCGAGACGTCCGAGGCGGCCCTGCACCAGGCCCTCGAAGGCAAGGACGGCTAC
 A  A  G  E  T  S  E  A  A  L  H  Q  A  L  E  G  K  D  G  Y
TACCTCGTGGTCGAGGGCGGCCTGCCCCACCATCGACGGCGGCCAGTGGGGCATGGTTGCC
 Y  L  V  V  E  G  G  L  P  T  I  D  G  G  Q  W  G  M  V  A
GGCCATCCCATGATCGAGACCACCAAGAAGGCCGCGGCCAAGGCCAAGGGCATCATCTGC
 G  H  P  M  I  E  T  T  K  K  A  A  A  K  A  K  G  I  I  Ⓒ
```

532

```
ATCCGGCACCTGCCTCACGGCGGCGTCCAGAAGGCCAAACCCAATCCCAGCCAGGCCAAG
 I  R  H  L  P  H  G  G  V  Q  K  A  K  P  N  P  S  Q  A  K
GGCGTGTCCGAAGCCCTCGGCGTCAAGACCATCAACATCCCCGGCTGCCCGCCCAACCCC
 G  V  S  E  A  L  G  V  K  T  I  N  I  P  G ©  P  P  N  P
ATCAACTTCGTGGGCGCCGTGGTCCATGTCCTGACCAAGGGCATCCCGGATCTCGACGAG
 I  N  F  V  G  A  V  V  H  V  L  T  K  G  I  P  D  L  D  E
AACGGCCGTCCGAAGCTCTTCTACGGCGAGCTGGTCCACGACAACTGTCCGCGCCTGCCC
 N  G  R  P  K  L  F  Y  G  E  L  V  H  D  N ©  P  R  L  P
CACTTCGAGGCCTCCGAATTCGCGCCCTCCTTCGATTCCGAAGAGGCCAAGAAAGGCTTC
 H  F  E  A  S  E  F  A  P  S  F  D  S  E  E  A  K  K  G  F
TGCCTCTACGAACTCGGCTGCAAGGGCCCCGTTACCTACAACAACTGCCCCAAGGTGCTG
©  L  Y  E  L  G ©  K  G  P  V  T  Y  N  N ©  P  K  V  L
TTCAACCAGGTCAACTGGCCCGTCCAGGCCGGCCACCCCTGCCTCGGCTGCAGCGAGCCG
 F  N  Q  V  N  W  P  V  Q  A  G  H  P ©  L  G ©  S  E  P
GACTTCTGGGACACCATGACGCCGTTCTACGAGCAGGGCTAACCCCCTCCTTTGTAGCGG
 D  F  W  D  T  M  T  P  F  Y  E  Q  G  ⎯⎯⎯⎯⎯⎯
CCAGTACCCGGACACCTTCAACAGCACCGAACGTCTCGTGACGGAGGAAGCATATGGCTG
                                                          M  A
AGAGCAAACCCACGCCACAATCCACCTTCACCGGCCCCATCGTGGTCGACCCCATTACCC
 E  S  K  P  T  P  Q  S  T  F  T  G  P  I  V  V  D  P  I  T
GGATCGAAGGTCACTTGCGGATCATGGTCGAGGTGGAAAACGGTAAGGTCAAGGACGCCT
 R  I  E  G  H  L  R  I  M  V  E  V  E  N  G  K  V  K  D  A
GGAGCTCCTCGCAGCTCTTCCGCGGCCTGGAAATCATCCTCAAAGGCCGCGATCCCCGCG
 W  S  S  S  Q  L  F  R  G  L  E  I  I  L  K  G  R  D  P  R
ACGCCCAGCACTTCACCCAGCGCGCCTGCGGCGTGTGCACGTACGTCCACGCCCTGGCCT
 D  A  Q  H  F  T  Q  R  A ©  G  V ©  T  Y  V  H  A  L  A
CCAGCCGCTGCGTCGATGACGCCGTCAAGGTCAGCATCCCGGCCAACGCCCGCATGATGC
 S  S  R ©  V  D  D  A  V  K  V  S  I  P  A  N  A  R  M  M
GCAACCTGGTCATGGCCTCCCAGTATCTCCATGACCACCTCGTCCACTTCTATCACCTGC
 R  N  L  V  M  A  S  Q  Y  L  H  D  H  L  V  H  F  Y  H  L
ACGCCCTCGACTGGGTCGACGTGACCGCCGCCCTCAAGGCCGATCCCAACAAGGCCGCCA
 H  A  L  D  W  V  D  V  T  A  A  L  K  A  D  P  N  K  A  A
AACTGGCAGCCTCCATCGACACGGCTCGGGACCGGCAACTCGGAAAAGGCCCTCAAGGCAG
 K  L  A  A  S  I  D  T  A  R  T  G  N  S  E  K  A  L  K  A
TCCAGGACAAGCTGAAAGCCTTCGTCGAGTCCGGACAGCTCGGCATCTTCACCAACGCCT
 V  Q  D  K  L  K  A  F  V  E  S  G  Q  L  G  I  F  T  N  A
ACTTCCTCGGCGGCCACAAAGCCTACTACCTGCCGCCCGAGGTCAACCTCATCGCCACCG
 Y  F  L  G  G  H  K  A  Y  Y  L  P  P  E  V  N  L  I  A  T
CCCACTACCTGGAAGCCCTGCACATGCAGGTCAAGGCGGCCAGCGCCATGGCCATCCTCG
 A  H  Y  L  E  A  L  H  M  Q  V  K  A  A  S  A  M  A  I  L
GCGGCAAGAACCCCCACACCCAGTTCACCGTCGTGGGCGGCTGCTCCAACTACCAGGGCC
 G  G  K  N  P  H  T  Q  F  T  V  V  G  G ©  S  N  Y  Q  G
TGACCAAGGACCCGCTGGCCAACTACCTGGCCCTGAGCAAGGAAGTCTGCCAGTTCGTCA
 L  T  K  D  P  L  A  N  Y  L  A  L  S  K  E  V ©  Q  F  V
ACGAGTGCTACATCCCTGACCTGCTGGCCGTGGCCGGCTTCTACAAGGACTGGGGCGGCA
 N  E ©  Y  I  P  D  L  L  A  V  A  G  F  Y  K  D  W  G  G
TCGGCGGCCACCAGCAACTACCTGGCCTTCGGCGAGTTCGCCACCGACGACAGCTCCCCCG
 I  G  G  T  S  N  Y  L  A  F  G  E  F  A  T  D  D  S  S  P
AGAAACACCTGGCCACCTCGCAGTTCCCGTCCGGCGTCATCACCGGCCGCGACCTCGGCA
 E  K  H  L  A  T  S  Q  F  P  S  G  V  I  T  G  R  D  L  G
AGGTGGATAACGTGGACCTCGGCGCCATCTACGAAGACGTCAAGTACTCCTGGTACGCC
 K  V  D  N  V  D  L  G  A  I  Y  E  D  V  K  Y  S  W  Y  A
CCGGCGGCGACGGCAAGCACCCCTACGACTGCGTCACCGATCCCAAGTACACCAAGCTCG
 P  G  G  D  G  K  H  P  Y  D ©  V  T  D  P  K  Y  T  K  L
ACGACAAGGACCACTACTCCTGGATGAAGGCCCCCCGCTACAAGGGCAAGGCCATGGAAG
 D  D  K  D  H  Y  S  W  M  K  A  P  R  Y  K  G  K  A  M  E
TCGGTCCCTTGGCCCGCACCTTCATCGCCTACGCCAAGGGGCAGCCCGACTTCAAAAAGG
 V  G  P  L  A  R  T  F  I  A  Y  A  K  G  Q  P  D  F  K  K
TCGTGGACATGGTCCTCGGCAAACTCTCCGTCCCGGCCACGGCCCTGCATTCGACCCTCG
```

```
V  V  D  M  V  L  G  K  L  S  V  P  A  T  A  L  H  S  T  L
GACGCACCGCCCGCGGCATCGAGACCGCCATCGTCTGCGCCAACATGGAGAAGTGGA
G  R  T  A  A  R  G  I  E  T  A  I  V  Ⓒ  A  N  M  E  K  W
TCAAGGAAATGGCCGACAGCGGCGCCAAGGACAACACCCTGTGCGCCAAGTGGGAGATGC
I  K  E  M  A  D  S  G  A  K  D  N  T  L  Ⓒ  A  K  W  E  M
CCGAGGAGTCCAAGGGCGTCGGCCTGGCCGATGCTCCCCGCGGCTCCCTGTCCCACTGGA
P  E  E  S  K  G  V  G  L  A  D  A  P  R  G  S  L  S  H  W
TCCGCATCAAGGGCAAGAAGATCGACAACTTCCAGCTGGTTGTCCCCTCGACCTGGAACC
I  R  I  K  G  K  K  I  D  N  F  Q  L  V  V  P  S  T  W  N
TCGGTCCCCGGGGGCCCCAGGGCGACAAGAGCCCGGTGGAAGAGGCCCTTATCGGCACGC
L  G  P  R  G  P  Q  G  D  K  S  P  V  E  E  A  L  I  G  T
CCATCGCCGATCCCAAACGCCCGGTCGAAATCCTGCGCACGGTCCACGCCTTCGACCCCT
P  I  A  D  P  K  R  P  V  E  I  L  R  T  V  H  A  F  D  P
GCATCGCCTGCGGCGTGCACGTCATCGAGCCCGAGACCAACGAGATCCTCAAGTTCAAGG
Ⓒ  I  A  Ⓒ  G  V  H  V  I  E  P  E  T  N  E  L  L  K  F  K
TTTGCTAAGAGCGGCAACCCTCTAATACGAAGGACCCGCTTCGGGCCGGGTTCTTCGTTT
V  Ⓒ
CTCAAGGAGGTCGCATGTCCGATACCCCGCCCAAAATCCTCATCCTCGGCCTCGGCAACA
TCCTCTACACCGACGAGGGCGTCGGCGTGCGGGCCGTGGAGCGCCTTCTCGAAACCCACG
```

AUTHOR INDEX

E

Ellis J.E.   posters (1. 7 & 1. 8)

F

Fajardo C.   poster (1. 10)
Fardeau M.L.   poster (2. 2)
Fauque G.   poster (1. 36)
Fernandez V.M.   L5
Foglino M.   poster (3.1)
Fonty G.   poster (1. 4)
Ford T.E.   poster (1. 12)
Frey M.   posters (2. 2 & 2. 3)
Frey G.   G1
Fukuzaki S.   poster (1. 19)

G

Galuschko A.   poster (1. 32)
Garcia J.L.   B4 + posters (1. 23 , 1. 26 , 1. 27 & 1. 28)
Gernhardt P.   poster (3. 1)
Gibson C.R.   M9
Gottschalk G.   L1 + poster (2. 6)
Gouet Ph.   poster (1. 4)
Granderath K.   poster (1. 2)
Greksák M.   poster (1. 9)
Gros E.   M11
Grotenhuis J.T.C.   poster (1. 24)
Guerlesquin F.   B2 + posters (2. 4)
Guézénec J.   poster (1. 23)
Guiot S.R.   M7
Guillaume J.B.   G2
Guyot J.P.   poster (1. 10)

H

Haladjian J.   poster (2. 4)
Halboth S.   L3
Hansen A.   M12
Hansen T.A.   M8
Harper S.R.   poster (1. 11)
Hausner W.   G1
Haser R.   B2
Hatchikian E.C.   L5 + posters (1. 23, 2. 2 , 2. 3 , 2. 4 & 3. 2)
Heitz P.   B4
Henry E.   poster (1. 12)
Hillman K.   posters (1. 7 , 1. 8 & 1. 13)
Hormann K.   poster (1. 2)
Houwen F.P.   B5

I

Imbert M.   B6

J

Jacq V.A.   poster (1. 36)
Joncquiert J.C.   G2
Jud G.   poster (1. 14)

K

Kamagata Y.   M10 + poster (1. 35)
Kane M.D.   poster (1. 6)
Kaspar H.F.   M1
Klein A.   L3 + poster (3. 1)
Kothe , E.   L3
Kozánková J.   poster (1. 9)
Kremer D.R.   M8
Krishnan S.   poster (1. 15)
Kuhner C.H.   B3

L

Labat M.	poster (1. 6)
Laine B.	B6
Lalitha K.	poster (1. 15)
Larsen S.	poster (1. 16)
Le Bloas P.	M11
Leisinger T.	L2
Leluan G.	poster (1.18)
Lemos P.C.	poster (1. 29)
Lien T.	posters (1. 3 & 2. 5)
Lloyd D.	M6 + posters (1. 7 , 1. 8 & 1. 13)
Lindley N.D.	M11
Loubière P.	M11

M

Macfarlane G.T.	M9
Mackie R.I.	poster (1. 17)
Mac Leod F.A.	M7
Mah R.A.	L8 + poster (1. 1)
Mandelco L.C.	poster (1. 31)
Mandrand M.A.	G4
Matsuda K.	M1O
Meile L.	L2
Meyer M.	poster (1. 2)
Mikami E.	M10 + poster (1. 35)
Mitchell R.	poster (1. 12)
Mountfort D.O.	M1

N

Nagai S.	poster (1. 19)
Nakamura K.	M1O + poster (1. 35)
Nazina T.	poster (1. 32)
Ney U.	M4
Niel P.	poster (1. 18)
Nishio N.	poster (1. 19)
Nivière V.	posters (2. 3 & 2. 4)
Noyola A.	poster (1. 10)
Nozhevnikova A.N.	posters (1. 20 , 1. 21 & 1. 22)

O

Ogata M.	B1
Ollivier B.	posters (1. 23 & 1. 28)
Onderka V.	poster (1. 9)

P

Pauss A.	M7
Péringer P.	poster (1. 33)
Peinemann S.	poster (2. 6)
Pheulpin P.	G2
Plokker J.	B5
Plugge C.M.	posters (1. 24 & 1. 34)
Pohland F.G.	poster (1. 11)
Possot O.	poster (3. 1)
Potekhina J.	poster (1. 25)
Prensier G.	poster (1. 23)

Q

Qatibi A.I.	posters (1. 26 & 1. 27)

R

Ragot M.	poster (1. 28)
Reis M.A.M.	poster (1. 29)
Richardson A.J.	M2 + poster (1. 30)
Richaud P.	G3
Rimbault A.	poster (1.18)